Letts study aids

A-Level Modern History
(1815 to present day)

Course Companion

David Weigall MA

Senior Lecturer in History, Cambridgeshire College of Arts and Technology

Michael J Murphy MA

Head of Department of History, Cambridgeshire College of Arts and Technology

Charles Letts Books Ltd
London, Edinburgh & New York

First published 1982
by Charles Letts Books Ltd
Diary House, Borough Road, London SE1 1DW

Design: Ben Sands
Illustrations: Illustra Design
Editor: Michael Croza-Ross

ISBN 0 85097 423 2

Printed and bound by
Charles Letts (Scotland) Ltd

Preface

This book is designed to meet the needs of students covering the popular late modern period of history for A-Level and Scottish Higher examinations. It is intended to be used as an aid throughout the course, as a book for constant reference and as something which will be particularly helpful for revision.

It is essential to have a clear idea of the full scope of the demands of A-Level study in this subject – both of the knowledge required and of the specific skills expected of the good candidate. The purpose of the following units is to provide the student both with the necessary information and with practice in learning to analyse, select and present what he or she knows in order to achieve the best examination results. We hope this book will also give the student the confidence to follow his or her interests and studies in history further, beyond the A-Level examinations.

Though this book is primarily a companion guide to the European and British History outline syllabuses, it will also be useful to those who are studying for prescribed and special subject papers, as indicated on pages 3 and 4. Readers should note that units 14 and 15 deal with developments after, as well as before 1951 – unit 15 extending to 1973.

In preparing this book we have received indispensable help and advice from A-Level History examiners and specialist historians in schools, colleges and universities. This means that we have been able to benefit from the practical experience of teachers and examiners and the insights of professional historians.

Our particular thanks are due to Mr. Harry Browne and to Dr. Boyd Hilton for their invaluable contributions and advice throughout, to Mr. Ritchie Greig for his advice and help over the Scottish syllabuses, and, for their special contributions to particular topics, to the following: Professor Norman McCord, Mr. Stephen Beller, Mr. Christopher Catherwood, Mr Alan Griffiths, Mr. Euan O'Halpin, Mr. Martin Moriarty, Mr. John Pollard, Dr. David Stevenson and Dr. Don Watts.

We are grateful, too, to the staffs of all the Examination Boards for their prompt helpfulness and advice.

We are, further, very indebted to all the staff of Charles Letts and Co. Ltd who have helped with this publication, and to their advisers, for their encouragement, expertise and forbearance.

Use of past questions

We are grateful to the following examination boards for allowing us to use past A-Level and Scottish Higher questions in this publication: the University of London University Entrance and School Examinations Council, the Scottish Examination Board, the Welsh Joint Education Committee and the Southern Universities' Joint Board. With the exception of units 18 and 24, the document questions are taken from London Board past papers, as are the following essay questions: unit 1, question 1; unit 2, question 2; unit 10, question 2; unit 11, question 3; unit 12, question 2; unit 14 question 1; unit 15, question 3; unit 21 question 1; unit 24, question 1 and unit 29, question 2. The following questions have been taken from Scottish Higher (traditional) papers: unit 7, question 2, unit 9, question 1; unit 13, question 1. We have used the following from Welsh Joint Education Committee papers: unit 3, question 3; unit 10, question 3; unit 23, question 3; unit 26, question 3; unit 28, question 2; unit 29, question 2. The questions from the Southern Universities' Joint Board are as follows: unit 7, question 3; unit 14, question 2; unit 18, question 2; unit 22, question 3. The authors, not the examination boards concerned, are responsible for the question practice exercises.

Acknowledgement is due to the Longman Group Ltd for permission to use the diagram on page 212.

David Weigall

Michael J. Murphy

Contents

How to use this book

The purpose of this book is to help you to study the subject with maximum effectiveness throughout your course of A-Level study. The 30 unit topics have been selected after a complete and systematic analysis of syllabuses and past examination papers for the outline papers from all the Examination Boards. They do not pretend to be an introduction to all aspects of the European and British history outline syllabuses for the period since 1815. The units, however, do deliberately concentrate on topics which are essential to a proper understanding of the nineteenth and twentieth centuries and which, year after year, are recognized as such in the questions set by the examiners.

The period from the end of the Napoleonic Wars is an extremely popular and widely studied one. It is also a very complex one, for which there is inevitably a far greater abundance of historical material than for any previous period. The units in this book are intended to guide you to greater comprehension and to give you help and practice in acquiring those historical skills which are discussed in the next section.

The years after 1815 were years of unprecedented and dramatic transformations, rapidity of social and economic development and of major political change. In many respects this period forms a natural unit of historical development for study. When one considers such topics as the growth of Nationalism, parliamentary democracy and Communism, it can also be seen to offer the basic background of knowledge for an understanding of present-day developments, institutions and ideas.

Each unit contains the following sections:

(1) a brief *introduction*;
(2) a comprehensive list of *study objectives*;
(3) a *commentary*;
(4) a *chronology*;
(5) *question practice*, both essay and document; and
(6) a *reading list*, both for textbook and other sources.

Where appropriate, maps and diagrams are also included.

HOW TO USE THE UNITS

The study objectives

The **Study objectives** section in a unit is intended to provide you with a comprehensive synopsis for that unit. It indicates what you should be looking out for in your reading and the considerations which your examiners, at the end of your course, will require you to bear in mind. It encourages you to approach your reading analytically, as explained in the following section. It also challenges you to relate the various aspects one to another.

Before you embark on your reading for a given topic you should read through the relevant list of study objectives very carefully to assess the full scope of the topic. When you have completed your reading and note-taking, check through the objectives and ask yourself if you have mastered enough knowledge and acquired enough understanding of what you have read to be able to give an account of each of them. When you have mastered all these points you should be confident that you will be prepared to answer questions on that topic at A Level, from whatever angle they may come.

The commentaries

The commentaries should be read as a supplement to your textbook and other reading. It is important to remember that the commentaries are *not* intended as an alternative to other reading but as a guide to what is of significance to your needs for A-Level History. They are condensed and written with an emphasis on analysis of the topics, covering such considerations as the causes, effects and wider historical significance of events and developments. They include definitions of key concepts and, organised under clear headings, the commentaries are intended to help you with the process of historical selection, comprehension, comparison and interpretation. Remember, though, that the important thing is for you to master the work. These assist you in that process but should not be regarded as doing your work for you!

The chronologies

The chronologies include a list of events, enactments, etc. arranged by topic. Those in small capitals are of special significance. You may use these to ensure, with the help of your textbook and other sources of information, that you make notes on these key facts. In note-taking you should record clearly (a) the key details, *e.g.* the major clauses of the Treaty of Versailles, and (b) the wider significance for an understanding of that topic, *e.g.* the influence of the Treaty of Versailles on public opinion, particularly in Germany.

As you work through your reading, you will find it very helpful to refer to the chronologies in order to clarify the order of events. Remember that an unclear grasp of historical sequence is a sure recipe for disastrous confusions! You will also, of course, find these chronologies an excellent aid to memory in your pre-examination revision.

The question practice sections

These sections offer two types of question essay and document:

The essay questions

Each of the essay questions includes three parts:

1 *Understanding the question* – an analysis of what the question is really getting at. This draws attention to the wording of the question and to any confusions that may arise from it; it points out any implications you should be aware of and also gives you general advice on how to answer the question.

2 *Knowledge required* – a summary of those areas of the topic which you should have mastered in order to be able to answer the question proficiently.

3 *Suggested essay plan* – an essay plan showing you how to construct a comprehensive and relevant answer to the question. The emphasis is on the use of evidence to present a historical argument – on the use of what you have read and learned to present a well-reasoned response to the question *as set*.

You should note that these are *not* full model answers. In A-Level History, examiners are looking for knowledgeable convincing and well-thought-out answers. But in history essay writing at this level there is no such thing as an *exclusively right answer*, in contrast, for instance, with the solution of mathematical problems. The suggested essay plans are presented, therefore, as exercises for you to consider, discuss and work through – and emphatically *not* as the only right answer. They should be studied when you have done your reading and note-taking and thought out the topic. Again, they are not intended as substitutes for your own efforts.

You may choose to draft your own essay plan after reading the *Understanding the question* and *Knowledge required* sections and then compare it with the points and construction of the *Suggested essay plan* section. Alternatively, you may wish to work through the plan, referring to your own notes and reading. You should remember that the construction, presentation and ordering of historical material are most important. Practice in constructing and appropriately supporting a historical argument is an essential skill. Very often A-Level History candidates fail to do justice to what they know because of weak presentation and incompleteness in their essays.

You are encouraged to select further questions on each topic from past papers and to draw up your own essay plans. You will find that there is no more effective way of building up your confidence in your ability to use what you know to answer specific questions and to produce essays that are wholly relevant.

The document questions

These may be used either with your textbooks to hand as an exercise to find out more about the topic, or, after you have finished your reading for the topic, to test your grasp of it. You will note the weighting of marks for these questions. They give you some indication of the amount of space you should be prepared to allocate to each. There is no hard-and-fast rule about this but – while the straightforward factual questions on the document, for which two marks are awarded, probably only require several lines – clearly you may well need to spend more than half a page on the demanding analytical question which offers eight marks.

The reading lists

The reading list at the end of each unit provides relevant references to the most commonly used textbooks for nineteenth- and twentieth-century European and British History outlines, as well as to titles recommended for *further* reading on the topic concerned. Wide reading is essential for the good history student – you will find that in-depth study of the units obliges you to make use of sources additional to your textbooks.

Maps and diagrams

These are included where necessary and should be studied very carefully when you are following the units. Modern European History, in particular, involves you in coming to terms with very complicated territorial issues such as the resettlement of Europe after the First World War. If you do not take good care to discover the exact location of events and territorial changes you will find a great deal of history, at best, unclear and, at worst, incomprehensible.

The tables on pp. 3 and 4 are based on the 1983 examination syllabuses for A Level and Scottish Higher. Some Examination Boards offer more than one syllabus and, where this is so, to avoid confusion, the code is entered in the second column below.

Please note that syllabuses and examination rubrics may be changed from year to year. You should ensure that you are aware of all the requirements – for instance, whether there will be a compulsory documents question or any restriction on the questions you will be allowed to answer in the outlines papers. If you are in any doubt you should consult your teacher or contact the Examination Board concerned.

Analysis of examination syllabuses
European History *Special subject and prescribed topic papers*

Examination Board	Syllabus	Paper	Unit numbers in this book (inclusive)
AEB	630	11	7,10
AEB	630	13	11–13
Cambridge		20	6
Cambridge		21	13
JMB			
London	B 267	6	4,6
London	B 267	8	9–13
Oxford		14	9,13
O and C		7	12
SUJB			
WJEC	C	A2 (d)	1,2
WJEC	C	A2 (e)	7,8
WJEC	C	A2 (f)	9–15
SEB	Alternative	5	6
SEB	Alternative	7	13
NIGCE			

Analysis of examination syllabuses
European History *Outline papers*

Examination Board	Syllabus	Paper	Unit numbers in this book (inclusive)
AEB	630	04	1–15
AEB	Alternative 673	04	1–15
Cambridge		13	1–15
Cambridge		14	1–15
	A	1	1–15
JMB		Alternative J	1–8
JMB		Alternative K	7–13
JMB		Alternative L	8–15
London	A 266	13	1–15
London	C 268	2	1–15
Oxford		7	1–7
Oxford		8	8–15
O and C		9	1–15
SUJB		2	1–15
WJEC	A	A2	1–13
WJEC	B	A2	1–13
SEB	Traditional	2	1–15
SEB	Alternative	Period B	1–5
SEB	Alternative	Period C	2–9
SEB	Alternative	Period D	9–15
NIGCE		2	1–15

Analysis of examination syllabuses
British History Special subject and prescribed topic papers

Examination Board	Syllabus	Paper	Unit numbers in this book (inclusive)
AEB	630	07	29
	630	10	17–19
Cambridge		18	17,18
		21	29
JMB			
London	A 266	6	17–25
	A 266	7	23,25–28
Oxford		13	23
O and C		5	23,25
		6	23
SUJB			
WJEC	B	A1 (d)	16–19
	B	A1 (e)	21–25
	B	A1 (f)	26–30
SEB	Alternative	4	16–19
	Alternative	6	23,25
	Alternative	8	30
NIGCE			

Analysis of examination syllabuses
British History Outline papers

Examination Board	Syllabus	Paper	Unit numbers in this book (inclusive)
AEB	630	02	16–30
	Alternative 673	02	16–30
Cambridge		3	16,17
		4	16–30
		5	16–30
JMB	A	2:U6	16–25
		U7	21–30
		D4	16–21
		D5	25–29
	B	I	17,18
		II	26–28,30
London	B 267	13	16–30
	C 268	1	16–30
Oxford		7	16–29
		8	25–30
O and C		8	16–30
SUJB		1	16–30
WJEC	A	A1	16–29
	C	A1	16–29
SEB	Traditional	1	16–30
	Alternative	Period B	16–21
		Period C	20–25
		Period D	26–30
NIGCE		1	16–30

EXAMINATION BOARDS

GCE Boards *(England and Wales)*

AEB
Associated Examining Board
Wellington House, Aldershot, Hampshire GU11 1BQ

Cambridge
University of Cambridge Local Examinations Syndicate
Syndicate Buildings, 17 Harvey Road, Cambridge CB1 2EU

JMB
Joint Matriculation Board
Manchester M15 6EU

London
University Entrance and Schools Examinations Council
University of London, 66–72 Gower Street, London WC1E 6EE

Oxford
Oxford Local Examinations
Delegacy of Local Examinations, Ewert Place, Summertown, Oxford OX2 7BX

O and C
Oxford and Cambridge Schools Examination Board
10 Trumpington Street, Cambridge; and Elsfield Way, Oxford

SUJB
Southern Universities' Joint Board for School Examinations
Cotham Road, Bristol BS6 6DD

WJEC
Welsh Joint Education Committee
245 Western Avenue, Cardiff CF5 2YX

Scotland and Northern Ireland

SEB
The Scottish Examination Board
Ironmills Road, Dalkeith, Midlothian EH22 1BR

NIGCE
Northern Ireland General Certificate of Education Examinations Board
Examinations Office, Beechill House, Beechill Road, Belfast BT8 4RS

Study hints for A-Level History

Before you start your A-Level History course it is essential for you to be absolutely clear about the sort of skills that will be required of you. What are the qualities which, at the end, your examiners will be looking out for? What is being tested? How do the demands of A Level differ from those of any history you may have done before?

It is obvious that anything you write, either in course essay work or for your exams, should display a sound factual grasp, a clear sense of the order of events and their relationship to one another – that, in short, you should be able to *describe* and to *narrate* accurately. But it is a great mistake to imagine that these qualities on their own will satisfy A-Level History requirements. While a sound factual and chronological grasp is essential, it is only the beginning.

Historical analysis

As the reports drawn up year after year by the A-Level History examiners show, the most frequent failing among candidates is an unwillingness, or an inability, to *apply* what they know and to *analyse* it in the context and within the bounds of a specific question. Again and again, it is not so much a lack of knowledge but an apparent absence of thought about that knowledge, and of practice in using it, which lose the marks. These skills can only be mastered with practice and experience, not least in sorting out what is relevant in your reading. Answers to A-Level History questions come from the process of selection and analysis.

You have only to consider the style and demands of these questions: they involve you in discussing, comparing, judging and justifying. A grasp of the narrative of historical events and the ability to prescribe are presupposed in the good A-Level candidate. At O-Level and CSE the basic skills of memorising, understanding historical sequence and presenting it in an orderly fashion may be adequate accomplishments for examination success. This is not to say that the good O-Level or CSE candidate may not attempt more than this, either in course work or in exams. The A-Level student, however, *must* attempt more and must always resist the temptation to drift into a *simply narrative* response to a question where *analysis* is required.

This is not to disparage narrative, nor to say that narrative and analysis cannot be very satisfactorily combined. They can; and in history – which is the study of individuals, institutions and developments in time – narrative sequence is an essential component. However, A-Level History is not simply a matter of offering historical details but of using what you know in interpretation and explanation – not just of knowing but of giving the clearest indications that you understand the implications both of what you know and of the further questions to which that knowledge may give rise.

Examination Board comments

The University of London GCE Board emphasized this point recently: examiners, they stated, were 'not looking for the mere regurgitation of received truths... The highest marks are reserved for answers which clearly provide a cogent and coherent response to the question as set out on the examination paper, and which are supported by substantial illustrative material intelligently selected'. They warn that 'other work may display accurate and relevant knowledge but fails to use this knowledge as an effective contribution to an answer to the question set. Such work will be considered for Grade E equivalent but seldom more. On the other hand, any answer which shows real endeavour to focus relevantly on the question set, and achieves some success in this, will always be treated sympathetically'. The Joint Matriculation Board is even more specific on this, saying that 45% of the marks awarded for an A-Level History answer will be given for knowledge and 55% for the ability to assess evidence and to select and organise material. A-Level History papers, it insists, demand more than narrative answers and a wholly narrative answer would only secure a maximum of 10 out of 25.

Historical assessment

A-Level History requires the assessment of problems in their context. The demands of an A-Level essay, in terms both of depth and of sophistication, mark a considerable advance on what you were prepared for in O- or CSE-Level History. While a typical O-Level question may ask you to describe, for instance, how the Great Reform Act came about, an A-Level examiner will want you to discuss its significance and implications, *e.g.* for subsequent reforms in the nineteenth century. At O Level you were expected to describe how the unification of Germany came about; at A Level you must be prepared, for example, to estimate the various contributory causes and to weigh their significance – to consider the extent to which it was the result of long-term economic developments, and the part played in it by French diplomatic calculations.

In addition to analysis and a capacity for well-supported and soundly-constructed historical argument, you should be able to show an appropriate sense of historical period and a clear understanding of the framework of ideas and attitudes of the age about which you are writing. The good historian is able to intuit and to convey (1) a feeling of the way that people's minds reacted in a particular period, and (2) an understanding of their situation.

He is able, in short, to display historical imagination.

The less able student all too often shows himself confined in his judgement by contemporary perspectives. This shows particularly in those essays where students deliver harsh (and often unsubstantiated) verdicts on past figures, institutions or developments. To criticise the past out of hand for not living by what you consider to be the values of the present demonstrates a failure of historical imagination. It will be more important for you to show the examiner that you understand how the unreformed pre-1832 British parliamentary system worked than to tell him how 'undemocratic' you consider it was! Similarly, an intelligent appraisal of Metternich's policies after 1815 is not really helped by denouncing him, from the vantage of hindsight, for not grasping what would happen in Europe 50 years after the Congress of Vienna.

Definition of concepts

A-Level history, particularly with all the '-isms' of the late modern period, requires you to be able to understand and handle concepts. Questions often demand a mastery of technical historical vocabulary, and the instructions of examiners or essay-setters to 'discuss', 'consider' and 'examine' make essential a high level of discrimination in the meaning of words and in their use in historical writing. An extremely common failing among history students is the misuse of staple words in the history textbook vocabulary – words such as *absolutist, totalitarian, radical, reactionary* and *protectionist*. An imprecise grasp of such concepts as abound in the study of modern history leads to desperately confused and inadequate answers in course or examination work. You have only to consider the following A-Level questions to appreciate that any proper answers to them are inconceivable without a clear definition in one's mind of what the concepts mean and imply in their context:

'*To what extent were liberal ideas in Europe in the first half of the nineteenth century an expression of Romanticism rather than of the Enlightenment?*'

'*Explain and illustrate what you understand by the term 'Radical' in the period 1815–1906.*'

'*A demi-semi-constitutional monarchy.*' *To what extent does this phrase appropriately summarize the achievement of the Dumas in the reign of Nicholas II?*'

One should add that it is not just the misunderstanding and misuse of concepts which lead to poor and waffly answers and essays. Another frequent weakness is the attempt to explain historical events by the sweeping use of broad concepts, such as *Nationalism* and *technological change*, without any specific examples given or without sufficient discrimination between different sorts of nationalism or kinds of technological change. It is essential for you to master sufficient detail to be able to illustrate and clarify such general concepts.

Historical debate and interpretation

The topics which you cover in your history course, and on which you will be questioned in the examination, will frequently have produced very considerable controversy and a wide variety of interpretations among historians. There are few entirely simple problems in modern history and you should be aware of the major debates, at least in outline. At A Level, one of the purposes of studying history is to introduce you to historical interpretation and to particular approaches to the study. The skill of constructing and sustaining a historical argument depends, in the first instance, on analysis and the sifting of available evidence. You will rapidly gain confidence in coming to terms with history through practice in evaluating the arguments for and against particular interpretations. You will also discover that it is ill-advised to opt unhesitatingly for one interpretation against all others unless you can sustain this view with a compelling display of evidence supporting it. At all events, one should avoid giving the impression of one-sidedness.

The wider context of 'Factors'

Though the analytical approach to history will lead you to identify separate factors – as in the causes of a war or revolution – you should be prepared to discover the wider context, *i.e.* the interrelatedness and linkage between them, as well as identifying their individuality. Though they help to clarify and define, such categories as *economic, political, social, constitutional* and *ideological* overlap and, in explaining any event or development, are often found to be interrelated.

To illustrate this point: a good essay on the causes of the revolutions of 1848 will identify political demands and social and economic pressures but it will not treat them as totally separate and compartmentalized factors. It will not fail to show

how they interact – how, for instance, the economic crises are reflected in the political programmes of that year.

A rote learning approach to your studies tends, quite simply, to make for pedestrian and unimaginative answers when you come to write a course essay or take your examination. The uninspired reeling-off of factors will suggest that you have not thought either sufficiently deeply or thoroughly about what you have been studying.

The importance of relevance

The most frequent loser of marks at A Level is the sort of answer which is either plainly irrelevant to the question, or only of marginal relevance. Many students who may well have written comprehensive essays on a topic during their course are disconcerted to find that the questions in the examination cover aspects for which they were not prepared or are posed from an unexpected angle. All examiners are only too well aware of the candidates who have prepared a number of major topics and virtually memorised answers on them – quite frequently with stock quotations. In such papers a question on Peel becomes *the* answer on Peel. This fundamental failing goes back to the question of historical analysis.

You should remember that, in preparing course work and in examinations, your historical material should always be shaped to provide a relevant answer, supported by apposite allusion and factual illustration. The narrative and descriptive parts of your essays should be at the service of your analysis of the question. In good historical writing the writer is seen to be in control of the evidence in presenting a historical argument. It is very obvious to the examiner who those candidates are who have sorted out the relevant considerations, who have analysed what they have been reading, have compared it with other reading and who have developed the skills of identifying the significant points and of relating them.

The questioning approach

The good student develops, from his reading and thought about the subject, a questioning habit of mind. He goes on from any question to ask further questions. He requires the clarification of any ambiguous term. To give an example: when faced with a question on the causes or results of any particular event or development he will take care to distinguish between the immediate and the longer-term causes or results. If he is asked to explain the outbreak of the First World War, he has an awareness of the far-reaching considerations – of all those things which produced the international climate in which the Powers were quite likely to go to war – as well as the immediate occasion, being the course of events following the assassination of Archduke Francis Ferdinand in 1914. The good answer illuminates clearly the relations between the longer-term and the immediate causes.

Conclusion

A sound piece of work for A-Level History will show an intelligent understanding of the past, a clear analysis based on a sound grasp of fact and a capacity for lucid expression. The criteria by which you will be judged in your examinations will be: the relevance, accuracy and quantity of factual knowledge; its effectiveness or presentation and the ability to communicate knowledge in a clear and orderly fashion with maximum relevance. You should be able to show that you can present and justify a historical argument, that you are capable of exercising historical judgement and are aware of period and context. These are a variety of skills. They can only be acquired over the course of time with wide reading and much practice.

Revision and the examination

REVISION

Effective revision for A-Level History is not simply passive absorption, but a very active process. It is an opportunity for you to reassess what you have written in your notes and essay assignments, not just an exercise in re-reading.

You should remember, as explained in the previous section, that you are being examined not simply on your memory but on your capacities for thought, analysis and judgement. After revising each topic you should be able to feel confident that you will be capable of answering questions on it from whatever perspective they may come.

In order to encourage this confidence and flexibility of approach you will find it helpful to practice drafting skeleton essay plans for past questions. You should list the main points you would make in the order in which you would present them (remember that a clear structure is most important in essay writing), and jot down such key information as you would use to support them.

In drafting either full or outline answers in your revision, remember that your material should always provide a fully specific and relevant answer. Bear in mind what was said earlier, that any narrative and descriptive parts of your essays should be used to serve your analysis of the question. At the same time, though you should in no circumstances allow the weight of factual information to obscure the argument of your answer, you must never assume that the examiner will be aware that you know the facts if you have not stated them, or that he will find your assertions and generalisations self-evidently correct.

Further, you should allow yourself practice in fully answering questions under 'mock' examination conditions, timing yourself for this exercise. You will find that these exercises act as a constant guide to you in your revision. They will make you actively reassess what you have done over the course in the light of the sort of questions which will be asked and of the considerations which will be uppermost in the examiners' minds.

A planned programme

Revision should also be clearly and systematically planned. If you are taking the Summer A-Level examinations you should have covered the great majority of the syllabus topics by the end of the Spring term (if you are taking the Winter examinations, by the beginning of the Autumn term). You are best advised to draw up a schedule for revision, starting, in the case of the Summer examinations, at the beginning of April.

Try to give an hour or more each day specifically to your history revision – you will obviously have more time available during the Easter vacation. If you have taken a 'mock' examination, give particular attention to those topics where you have found conspicuous gaps in your knowledge and understanding.

A clearly thought-out programme of revision is one which you can complete without panic or a hasty scramble, and which you can combine with additional reading for the subject. Leave a period of, say, a week to a fortnight at the end for comprehensive re-revision, during which you can refine key points from your notes and clarify any final difficulties. A thorough schedule along these lines will give you the confidence which is such an important element in achieving success.

THE EXAMINATION

Some of the following points may appear very obvious to you. They are, nevertheless, frequently disregarded by candidates and the neglect of them is again and again noted in examiners' reports as leading to poor or indifferent answers and to failure in the examination.

You should try and ensure that you observe the following rules:

1   Read the rubric for each paper and section of paper very carefully and answer the full number of questions required of you. Pay particular attention to subdivisions, *i.e.* where you are asked to answer questions from more than one section.

2   Be sure that you have an absolutely clear idea of what any question you choose to answer means and answer it *as set*. Pay particular attention to the precise requirements of the examiners, *i.e.* are you being asked to 'explain', 'comment upon', 'illustrate', etc.? If the wording of a question is genuinely ambiguous or confusing be prepared to point this out in your answer.

3   Choose the questions you are confident you can answer rather than those, however interesting, which you *think* you might be able to answer.

4   Attempt as far as possible to apportion your time equally between the questions, unless there are indications in the rubric or marks (as in some document questions) to the contrary. Avoid particularly the over-long first answer which forces you to scramble to finish the last question.

5   Answer all parts of the question. For instance the following: *'Show why and how a 'cold war' developed after 1945. What were its consequences?'* Note *why, how* and *consequences*. The examiner will be looking for answers which deal with all three.

6   This is equally true of questions which require you to 'compare' or 'contrast'. Candidates often fall down badly by simply concentrating on one side of the question and making only a passing reference to the other(s), *e.g.*:
'Compare the circumstances in which the Irish Home Rule Bills were introduced in 1886, in 1893 and in 1912.'
'Compare the foreign policy of Bismarck between 1870 and 1890 with that of his successors in the Imperial Chancellery up to 1914.'
Any treatment of these questions which concentrated almost exclusively on one of the Home Rule Bills, or simply on Bismarck's foreign policy – however detailed and thorough – would not just be partial answers. They would not really be answers at all, since the purpose of the question is to elicit a comparison.

7   Remember to use what you know to provide an answer which is strictly relevant to the question. It is useful to draft an essay outline and to check its relevance to the question, as set, before you answer it.

8   Avoid essay conclusions which are *simply* repetitions of what you have already written.

9   Provide a clear structure and sequence in your answer. The art of presenting a historical argument is extremely important.

10   Always write clearly, avoiding vagueness either of expression or in factual allusions, *e.g.*: *'Austria lost several battles'* – which *battles, when and why?*

11   In answering document questions read the extract(s), not just the questions, very carefully. If asked to explain them, do so in your own words. Make sure the examiner sees that you understand what the document means and its relevance. Look out particularly for differences of emphasis or tone if there is more than one extract on a topic.

12   Give yourself time at the end to re-read what you have written, to correct any errors, or to make any necessary additions.

Part I
Europe, 1815–1951
Introduction

Europe during the nineteenth century underwent a process of political, economic and social change quite unprecedented in its rapidity. It was the first period in human history in which the pace of transformation was so rapid that society was vastly altered within one lifespan – a period in which human beings living at the same time were capable of acting as though they belonged to completely different eras.

This is not to say that these changes were unheralded. There had been remarkable discoveries in the 'scientific revolution' of the seventeenth century. During the period of the Enlightenment thinkers had argued that there could be major progress and radical improvement for the human condition, and their ideas and criticisms had helped to sap confidence in the Ancien Régime. But still, in the eighteenth century – in spite of notable advances – much of everyday life on the Continent was closer in its routines to the Middle Ages than to the twentieth century.

The essential agents of transformation and modernization were the growth of industrialization and the influence of the French Revolution. Britain was the first industrial nation. In the third quarter of the nineteenth century she was followed, among the major Powers, by France and Germany and – by the early twentieth century – by Russia. The respective scale and success of industrial development and the growth of population in the various countries affected the balance of power in Europe and were major elements in the imperialist expansion of the Powers.

At the end of the eighteenth century France was Europe's leading country, its greatest military Power and the focus of its cultural life. By 1871, the year in which the new German Empire was proclaimed after France's defeat at the hands of Prussia, France had already lost that pre-eminence she had enjoyed in European affairs since the seventeenth century. During the first half of the twentieth century it was to be German aggrandisement, based on a rapidly increasing population, great economic advance and military efficiency which threatened the European balance and led to two wars.

The underlying developments during the nineteenth and twentieth centuries, which form the background to Units 1–15 of this book, were not only very rapid by comparison with previous centuries; they were also complex and closely interrelated. There was a very great increase in population in Europe – of the order of about 40% during the first half of the nineteenth century. This – combined with industrialization, urbanization, the increase of communications and the much greater mobility of labour, the spread of education and the growing influence of public opinion – compelled change in the old social structures and led to the demand for social reforms. These were accompanied by a growing challenge to the political order and hierarchy and a demand for more representative government.

Though the economic structure of European life was still – overall – predominantly agricultural during much of the nineteenth century, the very great increase in population provided industry with both a labour force and a growing market. Industrial advance was stimulated by the revolution in transportation with, in particular, the notable growth of railway construction in the 1840s and 1850s (something which also, among other things, helped to transform military planning and strategy). Land, which had formally been the basis of social organisation in all European states, became, therefore, rela-tively less important as industrialization progressed and finance capital came to play an ever-increasing role. At the same time, as the sum total of human wealth dramatically increased and the social consequences of industrialization became fully apparent, social reformers and radical politicians called more and more for the redistribution of wealth. Materialistic achievement was accompanied by growing criticism of the inequalities of wealth, by the evolution of socialistic ideas and by demands for the organization of state welfare.

In 1815 there were six great Powers in Europe – the multinational empires, the Austrian, Russian and Turkish (all were to collapse during the First World War); the constitutional monarchies, Britain and France; and the dynastic state of Prussia. There was also the rest of Germany and the Italian peninsula, both divided into separate states. The unification of these territories were to be central events in the mid-nineteenth century.

The ideas of the French Revolution and the experience of Napoleonic rule left an indelible impression on European society. While the ideas of the French revolutionaries threatened the old social order with its hereditary hierarchy of status, they also posed a challenge to the European order of states.

The Revolution had proclaimed that the state belonged to the people and that the people correspondingly owed loyalty to the state. They were identified in the concept of the nation-state. As the century progressed the national idea became the dominant force in international relations in Europe. More and more peoples came to be influenced by the idea that each nation has its own character and that each has the right to run its own affairs.

The nationalist idea presented a particular threat to those established monarchies where the unifying principle was loyalty to the dynasty, as with the Habsburg Empire. In the first half of the nineteenth century Nationalism tended to go hand in hand with Liberalism in the revolts against the conservative order which followed the Vienna Settlement, major upheavals occurring in 1820, 1830 and, above all, in 1848.

The impact of industrialization and the great population increase combined, then, with new political and social ideas and with cultural advance to transform Europe in the nineteenth century. Together with these developments went the general movement towards the consolidation of state power and extension of state controls over society, with the growth of professional bureaucracies. This was a process whose fuller implications were only to become apparent in the 20th century.

At the same time as the 19th century was witnessing the ascendancy of state power, it also saw the apogee of European influence across the globe and a burst of new imperial acquisition during its later years. The decades after 1900, however, witnessed a progressive eclipse of the European role in the world. Through the rise of other major Powers, the revolt of colonial dependencies against imperial rule, and the depredations of two world wars, the Western European nations, individually, could no longer aspire to their nineteenth century status on the international scene.

1 The Congress of Vienna and the Concert of Powers

1.1 INTRODUCTION

After the defeat of Napoleon the primary task of the Allies was to secure a lasting peace. The Vienna Settlement of 1815 inaugurated a period of political and social conservatism in Europe – a period in which the maintenance of international stability was closely linked with resistance to revolution within states. Diplomacy in this period – personified in the career of Metternich – worked to preserve the status quo.

The territorial settlement was not fundamentally shaken until the 1860s, when Germany and Italy were united, and there was no European-wide war until 1914. The Holy Alliance and the Quadruple Alliance were other significant agreements. The latter, among other things, provided for regular meetings between the Powers. At the first of these – at Aix-la-Chapelle in 1818 – France was readmitted to the Concert of Powers.

Great Britain progressively distanced herself from the continental states, however, over the question of the right of intervention in the internal affairs of other countries. This rift became fully apparent at the subsequent congresses of Troppau, Laibach and Verona. See Unit 20 and the map on page 140.

The continuing underlying rivalries between the major Powers were clearly displayed over the Greek revolt. The post-Napoleonic Concert of Powers did not in practice outlive Tsar Alexander I, who died in December, 1825 – though there were later attempts to revive it in times of international crisis.

1.2 STUDY OBJECTIVES

1   The events from the Treaty of Chaumont to the Treaty of Vienna; the nations represented at Vienna and the statesmen representing them; the divisions and alliances among these nations; the influence of Metternich, Castlereagh, Talleyrand and Gentz, especially.

2   The terms of the Vienna Settlement as it affected France, the Netherlands, Germany, Italy, Poland and Norway; the movement of peoples and national boundaries; the respect for legitimacy; distrust of liberal and nationalist movements; fear of renewed French aggression (especially after the 'Hundred Days') and concern to preserve a balance of power between the states.

3   The allegations that the Congress of Vienna, under Metternich's influence, tried to 'put the clock back'; alternative view that Napoleon had tried to put it forward too quickly; other alternative view that Metternich's real fault lay rather in trying to *keep* the clock back *after* 1815.

4   The success of the Congress of Vienna in helping to ensure European peace for 40 years: no Power was utterly humiliated, so there were no extreme desires to upset the settlement; the general attitude of war-weariness after Napoleon.

5   The formation of the Holy Alliance; Tsar Alexander's political ideas and ambitions; Metternich adopted the Alliance as a means to check Russia's south-eastern expansion; Castlereagh's attitude to the Alliance.

6   The formation of the Quadruple Alliance as the basis for the Concert of Europe; Castlereagh's role in this.

7   The career and political philosophy of Metternich; the interaction between his domestic and foreign policies; his desire to turn the Concert of Europe into a device to 'police' the world and to suppress revolutionary movements.

8   The Congress of Aix-la-Chapelle and the admission of France to the Concert of Europe.

9   The impact of revolutionary movements in Europe 1819–20 (the assassinations of Kotzebue and de Berry, Spanish and Neapolitan revolts, the rising of the Semenovsky regiment) and the Congresses of Troppau and Laibach.

10   The Congress of Verona sanctions French and Austrian intervention in Spain and Italy respectively, but keeps Russia out of Greece.

11   Canning's political philosophy, attitudes to Greece, Spain and the South American colonies; reactions to the Monroe Doctrine; declaration that the Concert of Powers is dead ('Each nation for itself . . .').

12   Concert of Europe threatened also by the conflict of interest of Russia and Austria; in 1812 Russia had reached the Danube, thus facing Austria across the Dobrudja, which Austria needed for commercial reasons. As Gentz pointed out, the end of Turkey would spell the end of Austria, so Russia's southern ambitions dismayed Metternich.

13   The revolutions of 1830; the Treaty of Münchengrätz resurrecting the Holy Alliance of Austria, Russia and Prussia – formally signifying the end of the post-Napoleonic Concert of Europe.

1.3 COMMENTARY

The Congress of Vienna

The statesmen who met at Vienna in 1814–15 to create the post-Napoleonic Europe were determined to ensure that no power would be able to dominate the Continent as Napoleon had done. They wanted to secure a lasting peace.

At the same time they were concerned with the maintenance of order within states and the prevention of revolution.

In the minds of all the statesmen of this period the threat of war and the fear of domestic political upheavals went together. In the shadow of the French Revolution and its Napoleonic aftermath, this is one of the reasons why peace was preserved for so long after 1815. There was no major conflict between the powers until the Crimean War 1854–56.

Never before had so many rulers and principal ministers met to work out a comprehensive peace settlement. The key figures were the Austrian Foreign Minister Metternich, the British Foreign Secretary, Castlereagh, Tsar Alexander I and the French representative Talleyrand. It was Metternich who became most identified with the postwar order, regarded by some as the saviour of Europe from war and revolution and by others as the repressive upholder of the rights of monarchs and dynasties against the rights of peoples. This was not least because he survived as a statesman until the revolutions of 1848 and because the multinational Habsburg Empire which he served had most to lose through the revolutionary changes which he resisted.

Territorial changes

In 1815 there was no possibility of simply reconstructing the European map of 1789. A number of changes were irreversible. In Germany, for instance, 300 states had been reduced to only 39. The major territorial changes of the Congress were made with a view to preventing any future French expansion. The treaty of Chaumont of March 1814 had already indicated that Britain, Russia, Austria and Prussia were prepared to form a permanent league to contain France.

In order to achieve this the Kingdom of the United Netherlands was established in the north; the Kingdom of Savoy was enlarged to the south and the control of northern Italy was given to the Habsburgs in the form of the newly-created Kingdom of Lombardy – Venetia. Swiss neutrality was guaranteed by the powers.

One of the most significant results of the settlement in the longer term was the strengthening of *Prussia*. Apart from her other acquisitions, she was given the Rhineland. This placed her in a position where she would have to take the lead in any future war against France. Her territories were now so distributed that by consolidating them she would bring about the economic unification of the greater part of Germany.

In the period after 1815, however, *Austria* remained the most powerful German state. She not only regained territory she had lost to Bavaria but received the Presidency of the German Confederation or Bund.

It is important to note that Metternich and Castlereagh, while anxious to defend Europe against any revival of French power, were also keen to safeguard Central Europe against any expansion of Russian power.

The major achievement on the Continent after Vienna was that both France and Russia appeared sufficiently contained to allow a balance of power to emerge.

The problems of Poland and Saxony

From the start there were serious rivalries between the victorious powers. By the winter of 1814 disagreement over Poland and Saxony had become so acute that it seemed quite possible that relations between Russia and Prussia on the one hand and Austria and Britain on the other would break down completely.

The large Polish state which Alexander wanted would have meant a considerable expansion of Russian power westwards. If Prussia had been allowed, as she wanted, to take Saxony, her position would have been greatly strengthened at the expense of Austria. Talleyrand intervened on this issue and in January 1815 Britain, France and Austria signed a secret agreement to resist the demands of Russia and Prussia by force of arms if necessary. In the event a compromise was arrived at over both territories and a Kingdom of Poland (Congress Poland) was set up under Alexander I with a constitutional regime.

The Holy Alliance

There was widespread pessimism about whether peace could be preserved. In addition to the territorial and other agreements arrived at in Vienna, a number of other proposals were put forward. The most important of these were the Holy Alliance of Alexander I and the Quadruple Alliance suggested by Castlereagh.

The Holy Alliance was conservative in a double sense:

1   it set itself against changes of the European frontiers as laid down in 1815; and

2   it was opposed to political changes *within* states.

In this Alliance the Russian, Austrian and Prussian leaders promised to treat one another in accordance with 'the sublime truths which the Holy Religion of our Saviour teaches' and to watch over their respective peoples 'as fathers of families'. All the European rulers were invited to subscribe to it and all did except the Sultan, the Pope and the Prince Regent. The last explained that such a personal agreement between monarchs was incompatible with constitutional government.

Absolute and constitutional government

Though it was privately dismissed by Metternich as a 'loud-sounding nothing', the Holy Alliance underlined a growing division between the absolute and constitutional monarchies. That is, Russia, Austria and Prussia on the one hand and Britain and France on the other. Liberals and constitutionalists who believed in representative government argued that it was an agreement between rulers, not peoples. In their view its alleged defence of 'decent Christian order' would be used as an excuse for repression within states.

The Quadruple Alliance

By this the victorious powers pledged themselves to prevent any attempt by France to overthrow the peace settlement. Castlereagh regarded this as *the* essential arrangement on which all others

must rest. It made provision for the periodic meetings of the powers. Other statesmen joined him in hoping that the settlement would signal the creation of a permanent system of consultation – a Concert of Europe.

The Concert of Europe

What was attempted here in the immediate postwar period was a system of collective security among governments: as opposed to free competition among states in pursuit of their individual interests, they would have to hold some common interest to be more important than individual state ambition. Underlying this was the spectre of revolution. Conflict between states would lead to revolution within states. The wider object of the peacemakers was, in the words of Gentz, Metternich's secretary, to contain the 'restlessness of the masses and disorders of our times.'

The congresses

Four congresses were convened under Article VI of the Quadruple Alliance, but there were differences right from the start. The leaders of the conservative, absolutist monarchies argued that peace between states and social and political order within states could not be separated. They insisted on the right of intervention in the affairs of states. Britain – and this was clearly spelled out in Castlereagh's State Paper of May 1820 – held that there should be no general right of interference in the affairs of states.

The Congress of Aix-la-Chapelle

This marked the end of the postwar treatment of France as a defeated enemy. Though the Quadruple Alliance was renewed as a precaution against France, another alliance, the Quintuple, brought her in.

The Congress agreed, in return for final settlement of the indemnity, that the army of occupation should leave France. At the same time, a marked divergence opened up between Britain and Russia: the Tsar called for an international army to uphold the 'sacred principles of order'. Castlereagh replied that the British government would resist all efforts 'to provide the transparent soul of the Holy Alliance with a body'.

The Congresses of Troppau and Laibach

These underlined the growing difference of view between Britain and the continental Powers. Both concerned the issue of intervention. Austria was authorized to suppress the Naples uprising. The powers of the Holy Alliance affirmed their determination to intervene wherever a legitimate regime was in danger of being overthrown. Britain opposed the principle of intervention.

The Congress of Verona

In October 1822, following the revolt of the Spanish insurgents against King Ferdinand VII, the five Great Powers came together for what proved to be the last of the era's congresses which involved them all.

George Canning had taken the place of Castlereagh. In opposing the French intervention against the Spanish rebels, he completed the movement away from the Concert of Powers which Castlereagh had begun. Great Britain's constitutional and political system and her dominant interest in developing her empire and commerce set her apart not only from Russia but from the other allies as well. Her independence was confirmed by her cooperation with the North American Republic to prevent continental intervention in the New World, on the principles expressed in the Monroe Doctrine.

Subsequent congresses and the Concert of Powers

The idea of a Concert of Powers survived the immediate post-

Napoleonic period. There were subsequent congresses, such as those at Paris (1856) and Berlin (1878). A clear distinction should be drawn, though, between the subsequent international conferences of heads of states and the congress system.

The key question about the post-Napoleonic period was whether a *genuinely international system of peacekeeping* could be made to work. The attempt failed primarily because of the disagreement over intervention. The three major areas of disagreement were:

1   The response of the Powers to revolts in Spain and Italy.

2   The Greek War of Independence against Turkey. (If the Russians intervened to aid the Greek rebels the whole balance of power in Europe might be destroyed). (See Unit 3, 'The Eastern Question to 1856'.)

3   The extent to which European powers should impose their will on colonial territories. President Monroe stated that the American continents were not to be considered for future colonization by any European Power and he opposed intervention in already established settlements.

Metternich's 'System'

The revolutions of 1830 further widened the rift between the Powers and the division of Europe into liberal constitutionalist and conservative blocs.

This put an end to Metternich's hope of building a conservative Concert of Powers. His views had been formed by the cosmopolitan attitudes of the eighteenth century and the horror of revolution. His personal influence was particularly apparent in the German Confederation. He came to be seen as the arch-upholder of an oppressive order by the liberals and nationalists in the years before 1848. These emphasized the freedom of nations, the legal and political rights of individuals and constitutionalism.

A balanced assessment of this period requires that such criticisms should be studied in the light of the very real achievement of the Vienna Settlement.

1.4 CHRONOLOGY

1814	**March**	THE TREATY OF CHAUMONT; arranged by LORD CASTLEREAGH, British Foreign Secretary, to prevent any break-up of the Fourth Coalition of Great Britain, Russia, Austria and Prussia against the Emperor Napoleon. Guarding against a separate peace with the French ruler, the alliance was to continue for 20 years. Allied entry into Paris.
	April	Napoleon abdicated unconditionally; the Allied Powers granted him the island of Elba as a sovereign principality.
	May	THE FIRST TREATY OF PARIS: France to retain her frontiers of 1792.
1815	**June**	THE CONGRESS OF VIENNA: the major decisions of peacemaking. The Congress was interrupted by the return of Napoleon from Elba and the 'Hundred Days', brought to an end by the battle of Waterloo (June 18th) and Napoleon's second abdication (June 22nd). THE ACT OF THE CONGRESS OF VIENNA (June 8th).
	September	THE HOLY ALLIANCE: drawn up by TSAR ALEXANDER I. The Russian, Austrian and Prussian monarchs promised to treat one another in accordance with the 'sublime truths which the Holy Religion of Our Saviour teaches' and to watch over their respective peoples 'as fathers of families'. Subscribed to by all European rulers except the Prince Regent, the Pope and the Sultan of Turkey.
	November	THE SECOND PEACE OF PARIS: France restricted to the boundaries of 1790 and made to pay an indemnity for the war. THE QUADRUPLE ALLIANCE: between Great Britain, Russia, Austria and Prussia. Each of the signatories promised to supply 60,000 men should France attempt to violate the Treaty of Paris. This was concluded for 20 years for the preservation of the territorial settlement. The principle of government by international conference was agreed: the Powers to meet periodically to discuss common interests and problems.
1818		THE CONGRESS OF AIX-LA-CHAPELLE: arranged the withdrawal of allied troops from France and finally settled the French indemnity payments. France was admitted to the newly-constituted QUINTUPLE ALLIANCE.
1819	**July**	THE CARLSBAD DECREES: following the murder of August von Kotzebue, previously a tsarist agent, by a patriotic student. These decrees were proposed by Metternich and sanctioned by the Germanic Confederation. They bound sovereigns to repress Liberalism. Control of universities; censorship and the establishment of an inquisition into secret societies.
1820–21		THE CONGRESSES OF TROPPAU AND LAIBACH: called on the insistence of Tsar Alexander, alarmed at the outbreak of revolts in Naples, Portugal, Piedmont and Spain. Metternich persuaded the three Eastern Powers, Prussia, Russia and Austria to accept the TROPPAU PROTOCOL. This was directed against revolts and uprisings which might disturb the European Peace. It stated that any state which had succumbed to revolution had ceased to be a member of the Holy Alliance and that the powers could intervene to restore order. Castlereagh refused to accept the policy of interference in the affairs of other states. This dissent was clearly expressed in the BRITISH STATE PAPER OF MAY 5TH 1820.
1821	**February**	OUTBREAK OF THE GREEK REVOLT (see topic unit 'The Eastern Question' to 1856)
1822	**October**	THE CONGRESS OF VERONA: convened to deal with the situations in Spain and Greece. Lord Castlereagh's successor, GEORGE CANNING, denounced the 'European Areopagus'. Though

		unable to prevent French military intervention in Spain, he refused to cooperate with the other powers. Collapse of the Congress System.
1823	December	THE MONROE DOCTRINE: the background to this to be found in the threat of the Powers of the Holy Alliance to restore the Spanish American colonies to Spain and the aggressive attitude of Russia on the north-west coast of America. It stated that 'the American continents by the free and independent condition which they have assumed and maintained are henceforth not to be considered as subjects of future colonization by any European powers.'
1825	August	Portugal recognized the independence of Brazil.
	December	Death of Tsar Alexander I.
1827	July	THE TREATY OF LONDON: between England, Russia and France to secure the independence of Greece.
1830	February	THE LONDON CONFERENCE: Greece declared independent under the joint protectorate of England, Russia and France.
1830–31		REVOLUTIONS: Parisian revolt of July 1830 against the policies of the Bourbon Charles X and his minister Polignac. Orleanist King Louis-Philippe succeeded to power. Rulers forced to abdicate and constitutions introduced in the German states of Brunswick, Hesse-Cassel and Saxony. Revolt in Brussels (August to September 1830) assisting the INDEPENDENCE OF BELGIUM. In Italy, risings in Modena, Parma and the Papal States. The Russian garrison driven out of Warsaw in November 1830. Revolutions suppressed during 1831. November 1831 Britain and France agree to the separation of Belgium and Holland.
1832	November	British and French capture Antwerp to force Holland to recognize the independence of Belgium.

1.5   QUESTION PRACTICE

1   The Quadruple Alliance of 1815 agreed to 'reunions devoted to the great common interests' of its members. How far did the events of the next 10 years show that these members had very few 'great common interests'?

Understanding the question

This is a question on the workings of the Concert of Europe circa 1816–26; you must develop a narrative of events and analyse the differences of interest and ideas among the Great Powers.

Knowledge required

Diplomacy 1816–26. Details of the various congresses. Attitudes to the Greek revolt and to international intervention in the internal concerns of small states.

Suggested essay plan

1 Assess the situation in 1815; the common fear of Napoleon and of possible French movement for revenge; financial and psychological war weariness; this leading to a belief in institutionalized diplomacy.

2 The formation of the Quadruple Alliance; seen by Castlereagh as a counter to Alexander I's mystical Holy Alliance of Austria, Russia and Prussia, and as a means of containing Russian expansionism.

3 Metternich and Castlereagh were the most active members of the Alliance, but differed fundamentally over its purpose; Castlereagh saw it as an instrument to guarantee the territorial arrangements made in 1815; Metternich believed it constituted a licence to intervene in the domestic affairs of nations, and to suppress revolutionary movements.

4 Aix-la-Chapelle; concord maintained; agreement to admit France into the European concert.

5 1819–20 revolutionary movements: the Burschenschaft, the murders of Kotzebue and de Berry; Neapolitan revolution; military revolution in Spain and grant of a liberal constitution. All this turned Alexander I from liberalism to blackest reaction ('Today I deplore all that I said and did in 1815–18'); Alexander became Metternich's pawn; the Carlsbad Decrees.

6 British protests against the new direction of the Quadruple Alliance; Castlereagh's 1820 State Paper; the Congress of Troppau; alliance splits over Protocol; Congress of Laibach; Austria suppresses Italian revolutions; Canning – his policies were similar to Castlereagh's but more genuinely isolationist.

7 The Greek revolt: Britain and Austria combined to thwart Alexander; the Congress of Verona; sanctions interventions by France in Spain and by Austria in Italy, and the Congress of Verona allowed Canning to encourage Latin American colonies – but kept Russia out of Turkey.

8 Canning refused to attend 1824 Congress on South America – which marked the virtual end of Congress rule. In 1825 Prussia, Austria, France and Russia met but the meeting broke up in anger. In Canning's words: 'The age of the Areopagitica is gone by; each nation for itself, and God for us all'.

9 Conclusion: the proposition contained in the question is true; there were few *permanent* common interests. The Alliance operated so long as Britain, Austria, Prussia and Russia were afraid, first of Napoleon, then of a French war of revenge. By 1818 that fear had subsided. A common fear of Russia enabled Britain, Austria and France to combine at times after that and a common fear of revolution led to Austro-Russian cooperation from time to time. By 1826 the Alliance was virtually dead.

2   'I ruled Europe sometimes, but I never governed Austria.' How accurate is this assessment of Metternich's influence?

Understanding the question

The point is to compare Metternich's considerable success in manipulating international diplomacy, at least until the middle twenties, with his failure to secure necessary political and constitutional change within the Habsburg Empire; you need to explain what his ideals for the Empire were, and what he wanted to do.

Knowledge required

Metternich's personality and policies. Austria's domestic power structure and politics. The Congress of Vienna, Concert of Europe and general European diplomacy 1815–48.

Suggested essay plan

1 Metternich was given to cryptic and melodramatic utterances

(like 'I came into the world either too soon or too late'), so one has to be sceptical; nevertheless there is considerable truth in the proposition under discussion.

2 Metternich's career and political philosophy: a paternalist philosophy – he believed absolute power was essential for protecting the *people* against middle-class industrialists and constitutionalists; he realised that his views went against 'the spirit of the age'; supported police-state, censorship, repression, but not a too unkindly one.

3 The power structure in Austria and Metternich's position: he was believed to be all-powerful, and identified with the regime, so that his fall in 1848 was itself sufficient to satisfy many revolutionaries; but in fact he was much less powerful domestically than Finance Minister Kolovrat; power was centred in Coreisenregiment (Archduke Lewis, Kolovrat, Metternich), with Metternich the least powerful.

4 His lack of influence was shown by the suppression of his schemes for reform. For example, he wanted devolution of government to solve the nationalist problems; a Reichsrat, Council of State, and provincial diets; and privileges for German merchants in Budapest; all such schemes were thwarted by his rivals.

5 He tried to prevent the rise of capitalism, especially in agriculture; he clung to serfdom, but could not prevent the development of the money economy.

6 Metternich's boast about his European role also had some truth; as Austria's Foreign Minister he undoubtedly took the lead at Vienna; he succeeded for a while in turning the Concert of Europe into a policing institution against subversives (e.g. the Carlsbad Decrees) and also into a means for checking Russian expansion.

7 But all this was due to his able diplomacy, not to Austrian power; the balancing act was bound to fail – *e.g.* he encouraged France to intervene in Spain in order to persuade Alexander that Spain posed a serious revolutionary threat to world order, and succeeded in persuading Alexander not to add to the danger by assisting Greece; *but* France's invasion of Spain induced Canning to quit the Concert of Europe, which effectively undermined Metternich's influence.

8 So while Metternich exercised great influence, he never *ruled*. Moreover, his foreign policy reacted unfavourably on his domestic political ambitions. Austria's widespread European commitments meant that she could not be isolated. So western ideas of Nationalism and Liberalism infiltrated and doomed the regime.

3   How much attention did the Congress of Vienna pay to the principle of nationality?

Understanding the question

Having discussed what you understand by the principle of nationality, you should look in turn at the various problems with which the Congress had to deal, deciding how far they were respectful of Nationalism in each case. Distinguish between the attitudes of the different countries and statesmen. (See: Liberalism, Nationalism and the Revolutions of 1848.)

Knowledge required

Nationalist movements in the early nineteenth century. The ideas of Nationalism. The negotiations at Vienna and the Vienna settlement.

Suggested essay plan

1 Define what is meant by nationality in the early nineteenth century: Germany and linguistic-cultural nationalism; ideas of the power of the state, still mainly confined to the universities in 1815, but already given practical reality by the Prussian War of Liberation against Napoleon.

2 Similar liberation movement in Spain, where Nationalism was more liberal than in Germany; Napoleonic rule also awakens Nationalism in Italy; stirrings of nationality elsewhere.

3 The circumstances of 1814–15 treaty-making before and after the 'Hundred Days'; the countries represented at Vienna and statesmen representing them; secret diplomacy and behind-the-scenes alliances; the responsibility for the decisions contained in the Treaty.

4 Incorporation of Belgium in Holland: fear that France might soon overrun an independent Belgium; not really against Nationalism because few Belgians were *then* very concerned; it was to be Dutch misgovernment of Belgium (no religious toleration, religious equality, etc.) which caused the national revolution in 1830.

5 The German Confederation was at least a step towards unification compared with the Holy Roman Empire; no prospects of unification in 1815, whatever the Congress had done, because Germany was still very divided.

6 Italy: only Piedmont and the Papal States were *not* handed over to foreign rule; this was a clear frustration of nationalist sentiment, which Napoleon's influence there had stimulated.

7 The transfer of Norway from Denmark to Sweden was a good example of the Congress's arbitrary treatment of various peoples in its rearrangement of the map of Europe; but Norway had no national tradition, and anyway she was to retain her own government and army; her alliance with Sweden was to work until 1905.

8 Poland was 'liberated', but with the foreseeable result that it fell under Russian influence. Here in effect, if not in intention, was the Congress's greatest frustration of nationality.

9 Conclusion: the Congress of Vienna was moved more by pragmatism than by principle; so, though it did often flout the ideals of nationality (as of Liberalism), this was hardly a conscious decision. The main principle followed was legitimacy (e.g. the Bourbon restoration, Austrian rule in Italy), but far more important was a pragmatic concern for the balance of power.

4   Document question (The Congress of Vienna): study extracts I and II below and then answer the subsequent questions (a) to (g):

Extract I

'Among these four powers, Austria found herself in the most awkward position. She could no longer regard the Emperor Alexander…as other than a declared enemy, and Prussia, carried along by her own greed and her own ambition, as the inseparable ally of this enemy. She was nervous of too close a 'rapprochement' with France…for fear of offending public opinion by aligning herself openly with a power who had recently been the common enemy of Europe…. Another factor also restrained Austria. Though she agreed completely with France in her views on the affairs of Poland and of Germany, she held different views on the affairs of Italy…. There remained therefore only England as Austria's sole support; but England wanted peace, peace above all, peace…at any price and on nearly any conditions.'

(Metternich's *Memoirs concerning the Congress of Vienna*)

Extract II

'…There were three particular interests which were vital to British security and which must at all costs be safeguarded. The first was British maritime rights or a solution in our favour of the freedom of the seas. The second was the creation in the Low Countries of a unitary State, closely allied to Great Britain, and capable of forming a barrier against any further French aggression. And the Third was the exclusion so far as was possible, of French influence from the Iberian Peninsula....'

(Harold Nicolson, *The Congress of Vienna*)

Maximum marks

(a) What ambitions had the Emperor Alexander revealed at Vienna which led Metternich to regard him as an 'enemy'? (3)

(b) Regarding which territories did Prussia, at the Congress, show 'greed and ambition'? How far was this ambition satisfied? (4)

(c) Explain for what reasons and with what results Austria and France agreed completely...on the affairs of Germany. (4)

(d) Why did the French and Austrians have 'different views on the affairs of Italy'? (4)

(e) To which 'British maritime rights' in particular had neutral nations objected during the Revolutionary and Napoleonic Wars? (3)

(f) How far does an examination of the second and third British 'interests' mentioned in Extract II support the view expressed by Metternich in Extract I that 'England wanted peace, peace above all'? (5)

(g) Discuss whether, despite Metternich's complaints about Austria's 'awkward position', the Congress of Vienna can be regarded as having produced a settlement of Europe which was satisfactory to Austria. (8)

1.6 READING LIST

Standard textbook reading

David Thomson, *Europe Since Napoleon* (Penguin, 1977), chapters 1–6.
Anthony Wood, *Europe 1815–1945* (Longmans, 1975), chapters 4–8.

Further suggested reading

R. Albrecht-Carrié, *The Concert of Europe* (Harper and Row, 1968). •

F.B. Artz, *Reaction and Revolution 1814–32* (Harper and Row, 1963).

C.W. Crawley, *The New Cambridge Modern History 1793–1830*, volume IX (Cambridge University Press, 1974), chapter 25.

Henry A. Kissinger, *A World Restored: Metternich, Castlereagh and the Problems of Peace 1815–22* (Gollancz, 1973).

J. McManners, *Lectures on European History 1789–1914* (Blackwell, 1974).

Mack Walker (ed.), *Metternich's Europe* (Harper and Row, 1968).

G. de Bertier de Sauvigny, *Metternich and his Times* (Darton, Longman and Todd, 1962).

L.C.B. Seaman, *From Vienna to Versailles* (Methuen, 1972).

2 Liberalism, Nationalism and the revolutions of 1848

2.1 INTRODUCTION

In 1848 revolution spread across the European continent from Sicily to Denmark and from France to Hungary. The background to these upheavals was one of severe economic crisis, of disastrous harvests, commercial failure and very high unemployment. This was combined, though, with considerable political unrest.

The dominant political ideas of the revolutionaries were Liberalism and Nationalism. These ideas had been closely identified with each other in the first half of the nineteenth century in the agitations against the political order upheld by the Vienna Settlement, and epitomised in the Carlsbad Decrees.

'Metternich's Europe', which had been shaken in 1830, now faced major revolutions in France, Italy, Germany and Central Europe, and in the last three Nationalism played a very major part. There were, however, very marked divisions between the moderate liberal and the radical revolutionaries – between those who wanted constitutional changes and those who envisaged a new order of society.

In spite of the failure of the revolutions to achieve such objectives as the unification of Germany, they had considerable achievements to their credit – for instance, the abolition of serfdom in the Habsburg Empire. The year 1848, which has been described as the 'seedplot of modern history', also had a very considerable influence on subsequent social and political developments.

2.2 STUDY OBJECTIVES

1 The idea of Liberalism; its political and economic aspects, both based on individualism; usually a middle-class doctrine, useful for attacking traditional vested interests. The basic belief in progress through free institutions, toleration and natural rights (the freedoms of speech, movement, assembly, press and worship); also the freedom of trade. Often associated with ideas of democracy, but many liberals were distrustful of full universal suffrage.

2 Nationalism; its debt to the French Revolution; Napoleon stimulated it in Italy and provoked it (against himself) in Prussia and Spain; especially strong in areas of Habsburg rule; often associated with, but sometimes incompatible with, Liberalism (as in the German liberals' attitude to the Polish question).

3 The intellectual strain of Nationalism, stressing linguistic tradition and culture; poets, philologists, historians; examples of Greece and Serbia.

4 Businessman's nationalism, seeking efficiency through centralisation. Customs unions and economic consolidation as preludes to Nationalism; Nationalism and the middle class.

5 The economic and social causes of the 1848 revolutions.

6 The role of Liberalism in the 1848 revolutions, especially in Germany; the Frankfurt Parliament; Mazzini and democratic-republican Liberalism in Italy; Thiers and the Liberal opposition in France; the Viennese Liberals; the attack on the Carlsbad Decrees; demands for citizen militias to replace standing armies.

7 The role of Nationalism in 1848; Poles, Italians, Magyars, Czechs, Danes; many German liberals were also expansionist nationalists with Grossdeutsch ambitions; the Panslav Congress; the conflict of nationalisms. Nationalism was not fulfilled in 1848 but the year was extremely influential in its development and it gained martyrs.

8 Liberalism and Nationalism were especially strong among intellectuals; 1848 as 'the revolution of the intellectuals'; overproduction of lawyers, teachers, journalists without sufficient employment for them.

9 Other factors causing revolution; the rise of Anarchism and Socialism among new proletariats or 'classes dangereuses'; the power of Paris to influence events in the rest of Europe, e.g. the

Journal des Debats read by students as far afield as those in Moldavia and Wallachia; improved transport speeded the movement of ideas.

10 The suppression of revolutions and the reasons for failure; Liberalism and Nationalism are only able to unite disparate classes while common enemies, like Metternich, stand; once successful, revolutionaries fall apart.

2.3 COMMENTARY

Liberalism

Nineteenth-century Liberalism had its origin in the French and American revolutions.

It was associated with human rights and liberties: the right to property, freedom of speech, of the press, worship and political assembly. It involved the demand for free institutions in society and equality before the law.

During the French Revolution the principle of the sovereignty of the people had been put forward. The underlying notion here was that the only legitimate foundation for a government were the interests and consent of the people it governed.

The term 'Liberal' only became a widespread political description after the Spanish liberal revolt of 1820. It was an expression of the resistance to autocratic and absolutist (non-representative) governments and rulers and of the movement in favour of constitutions and free institutions. It became the battle-cry for the opponents of the post-Napoleonic order and Metternich's 'System'.

Another important aspect of this period was the growth of *economic liberalism*: the demand that industry, trade and agriculture be freed from government regulation and supervision. This was a principle that could be traced in the immediate past to the French economic philosophers of the eighteenth century and to Adam Smith, author of 'The Wealth of Nations'.

Liberalism developed as an international movement of ideas, transcending the boundaries of the European states. At the same time, during the period 1815–48 the liberal ideal and ideas of *national* independence and identity were closely linked.

Nationalism

Nineteenth-century nationalism was particularly encouraged by the influence of the French Revolution. It was the revolutionaries in 1792 who had first clearly formulated the principle of the nation's 'natural frontiers' within which the 'sovereignty of the people' should be exercised.

The two most striking examples of Nationalism during the century were the creation of a united Italy and a united Germany. Diplomacy, war, economic developments and revolution were all to assist in this process. But there was a strong underlying movement of ideas: the emergence of the notion of the *Nation-State*.

This was that a people speaking the same language, of the same race and with the same customs should correspond to an independent state in an identifiable territorial unit.

A clearly defined geographical area, as with the Italian peninsula, was only one criterion.

The strongest element in the early nineteenth century was *linguistic* nationalism – the common indentity of a people speaking the same language.

A common religion was another factor, but it was by no means to be found in all nationalist movements – e.g. Germany was divided between Catholic and Protestant confessions.

The term 'Nationalism' came into current usage in the 1830s. The most influential nationalist of this period was the Italian Giuseppe Mazzini, founder of 'Young Italy' and 'Young Europe'.

His ideas united Nationalism and Liberalism. He eloquently supported the aims of all nationalist groups in Europe. He believed that the popular desire for national independence and freedom was the motive force of nineteenth-century history.

He was also convinced that free nations would not go to war with one another and looked forward to the evolution of a united Europe.

At the same time there was another strand of Nationalism, strongly exemplified, for instance, in later German policies, which was anti-liberal in essence. It tended to glorify the individual nation's traditions and demands at the expense of neighbouring peoples. This was to lead to the demand for the suppression of the nationhood of other peoples.

In the period up to 1848 Nationalism was closely allied with Liberalism. The enemies of individual rights and of the freedom of nations were seen to be the same: the conservative order of the Vienna Settlement, the Holy Alliance, autocratic monarchs, the separate governments of the German and Italian states.

The revolutions of 1848

The background and causes

Though there were common political aims among the revolutionaries in 1848, there were also specific demands. The first outbreak, for instance, which was in Palermo, was essentially a Sicilian move for independence from Naples.

The fall of King Louis-Philippe in France, which precipitated the spread of revolutions throughout Germany and Central Europe, had as its immediate cause the demand for the extension of the suffrage.

There were moderate demands, such as for liberal constitutions, and far more radical ones, such as the Communist manifesto of Marx and Engels.

The economic situation These political demands would not have attracted widespread support among the masses but for the grave economic situation and social hardship at this time.

Since 1845 Europe – everywhere from Poland to Ireland – had suffered a series of poor, and sometimes catastrophically bad, harvests. High food prices resulted and these in turn reduced the demand for manufactured goods, leading to high unemployment in the towns and cities.

In the urban centres there was often marked overcrowding. The rapid growth of towns and cities is one of the great social characteristics of this period.

The revolutions of 1848 were essentially *urban* affairs. To some extent they began as large-scale riots. There had already in the immediately previous period been a number of food riots – for instance in Berlin in 1847.

The support for revolution Many of the most militant supporters of the revolutions came from among the skilled workers, displaced in increasing numbers by the growth of power-driven machinery. There was widespread machine-breaking.

In 1848 the large-scale modern factory employing hundreds and thousands, rather than tens, of workers was very much the exception and factory workers were not a large proportion of the urban population.

The social distress of these years is emphasized by the marked increase of emigration from Europe.

The various crises were linked and this explains the extent of upheaval and discontent: in 1847 the agricultural crisis was followed by a financial crisis which checked investment and made credit hard to come by. This was accompanied by a major industrial depression.

The peasantry Over large areas of Europe the decayed feudal order in the countryside had not been abolished.

In the Habsburg Empire the situation was acute. There the peasants were determined in 1848 to sweep away the surviving feudal burdens and jurisdictions. The restriction of landed property to the nobility, the limitation of personal rights, the maintenance of tithes and forced labour services encouraged grave discontents among the peasantry of Central Europe. Their emancipation by the Habsburgs in 1848 was to be one of the great achievements of that year.

The course of the revolutions

Though the first outbreak of revolution in 1848 was at Palermo in Sicily, it was the campaign of political banquets in Paris in favour of the extension of the suffrage and against the government of M. Guizot which precipitated both the events in France and the revolts which spread throughout Germany and Central Europe.

By mid-March 1848 Metternich was forced to flee from Vienna; Frederick William IV, faced with civilian commotion, had withdrawn his soldiers to barracks in Berlin. In Italy the independent or semi-independent states, the Papal States, Tuscany, Piedmont and Naples were forced to grant liberal constitutions and there were successful risings against Austrian rule in Venice and Milan.

The revolutionary demands The new governments which came into existence in the year of revolutions were above all interested in political reform, constitutional liberties, parliamentary assemblies and bills of rights.

More particularly in Italy and Germany and the Habsburg lands the revolutionaries sought national unity as well. So in Germany, for instance, the revolutions aimed at political reforms within the states and also produced a parliament at Frankfurt committed to national unification. In Italy the risings within states coincided with a Piedmontese war of liberation against Austria.

The Habsburg Empire The events in the Habsburg Empire are particularly complex. There were four dominant 'master' nationalities within it. These were the Germans, Hungarians (Magyars), Poles and Italians. Their upper and middle classes also covered the territories of the subject races, the Czechs, Slovaks, Yugoslavs, Rumanians and Ruthenes.

The four 'master' races demanded a united Germany, Italy, an independent Hungary and a reunited Poland, including between them all the territories of the subject races.

There was a clash between the various nationalities *and* a historic divide between the subject and master races. Nationalism threatened the Habsburg dynasty with complete disruption.

In essence, during the 1848–9 upheaval the subject races supported the Habsburgs against the master races.

The divisions and failure of the revolutions

In their attacks on the established order there was no real harmony between the objectives of the urban populations and the peasantry.

In the cities there was also a conflict of purpose between the hard-pressed artisans, who wanted government assistance, and the middle classes, who wanted an end to economic restrictions.

Moderate revolutionary governments moved quickly to quell more radical uprisings.

Different programmes were put forward for national unification in Italy and Germany. It was disputed whether or not a united Germany could incorporate Austria. The nationalities came into conflict.

Further, in many parts of Europe, as in Italy, the revolts were more inspired by local loyalties than national patriotism. In the last resort, the Romans were fighting for Rome and the Venetians for Venice. Sicily rejected Naples and Prussia refused to merge into Germany.

The two ideas, Liberalism and Nationalism, which had seemed inseparably linked in the pre-1848 period, were now effectively in conflict.

By and large in Germany and the Habsburg Empire, liberals – when faced with a challenge to their national group – jettisoned their liberal principles. In Vienna the German revolutionaries bitterly rejected Czech demands for autonomy. In the Frankfurt Assembly, many Germans came to recognize that their nationalist objective might well have to be achieved by illiberal means and with the support, or at least acceptance of the least liberal institution in German life – the Prussian army.

There was a fundamental division between the moderate advocates of limited political reforms and the radical supporters of major social upheaval and transformation. In Italy, for instance, some nationalists came to fear democratic republicanism more than they feared the domination of Austria.

Objectively, radical republicanism had no chance in the France of 1848 and certainly none elsewhere in a continent which was still so very largely agrarian and in which military power lay in the hands of conservative monarchs.

The revolutions failed to gain lasting mass support, not least because the economic situation improved.

By the spring of 1849 the revolutionary impetus had largely exhausted itself. The most important exception was Hungary. And there, Russia, afraid of the influence of Magyar success on Poland, intervened to put down the revolution.

The achievements of 1848

The Europe that emerged from 1848–9 was very considerably changed. In many respects 1848 did for Europe what 1789 had done for France.

The abolition of serfdom and of all checks on personal freedom enabled worker and peasant to move about at will. One of the consequences was greatly increased emigration.

Some of the most obvious constitutional gains were in countries such as the Netherlands which had not undergone revolutions. The impact of 1848 was nowhere forgotten.

In France universal suffrage and the principle of popular sovereignty were established; in Italy encouragement was given to the process of the Risorgimento; in Germany the national movement drew the conclusion that unification would be achieved by other than liberal parliamentary means.

The social influence of these events was also very significant. The 'June Days', for instance, confronted people in a very dramatic way with the major problems of an urbanised society.

While it encouraged the liberalization of society, it also signified the future greater political consciousness and activity of the masses.

2.4 CHRONOLOGY

France

1848	**27 January**	De Tocqueville warned the Chamber of Deputies of impending social and political revolution.
	22 February	THE OUTBREAK OF REVOLUTION IN PARIS.
	23	Dismissal of M. Guizot.
	24	Flight of King Louis-Philippe; LAMARTINE secured the approval of a provisional government.

Chronology cont.

	25 February	DECLARATION OF THE RIGHT TO WORK.
	26	Declaration of the setting up of NATIONAL WORKSHOPS (Ateliers Nationaux).
	28	Blanqui founded the Central Republican Committee on release from prison; the National Workshops and the LUXEMBOURG COMMISSION set up.
	2 March	INTRODUCTION OF UNIVERSAL MANHOOD SUFFRAGE; reduction of working day in Paris to ten hours and in the provinces to twelve.
	4	LAMARTINE'S MANIFESTO TO EUROPE.
	23 April	Elections held for National Assembly; defeat for the radical revolutionaries.
	9 May	Institution of the Commission of Executive Power.
	24	DECISION TO CLOSE THE NATIONAL WORKSHOPS.
	23 June	THE 'JUNE DAYS' in Paris began.
	24	GENERAL CAVAIGNAC opened offensive against the insurrection.
	26	Defeat of the Paris revolt.
	4 November	Adoption of a new constitution by the National Assembly.
	10 December	LOUIS NAPOLEON ELECTED PRESIDENT OF THE REPUBLIC. (See Unit 5: Napoleon III and the Second Empire).
1851	**15 March**	THE FALLOUX LAW.
	31 May	New electoral law reducing electorate from 9.5 million to 6 million.
	2 December	COUP D'ETAT OF LOUIS NAPOLEON.

Italy

1848	**2 January**	'Cigar riots' in Milan.
	12	Revolt in Sicily and declaration of independent government.
	10 February	Constitution proclaimed in Naples by Ferdinand II.
	14	Pius IX created a Commission of Reform.
	17	Constitution announced in Florence.
	4 March	CHARLES ALBERT proclaimed constitution in Piedmont.
	14	Piux IX granted constitution in Rome.
	17	Daniele Manin lead revolution in Venice.
	18	'THE FIVE DAYS'–uprising in Milan; Radetzky evacuated city.
	20	Revolt in Parma.
	22	Republic proclaimed in Venice.
	24	Sardinia declared war on Austria.
	8 April	Piedmontese troops defeated Austrians at Goito.
	13	Sicily declared independence from Naples.
	25	Papacy joined war against Austria.
	29	Pius IX withdrew support from the nationalist movement and condemned the Italian war.
	15 May	Collapse of the Naples revolt.
	29	Battle of Curtatone; Austrians defeated Tuscany.
	3 July	Venice supported Charles Albert.
	25	BATTLE OF CUSTOZZA. Major victory for Radetzky. Sardinian troops driven from Milan and the remainder of Lombardy.
	9 August	Armistice of Vigevano between Austria and Sardinia.
	11	Sardinian troops expelled from Venice.
	15 November	Assassination of Count Rossi, Prime Minister of the Papal States.
	24	Pius IX escaped to Gaeta.
1849	**8 February**	Proclamation of the Tuscan Republic.
	9	MAZZINI PROCLAIMED REPUBLIC IN ROME.
	12 March	Sardinia ended truce with Austria.
	23	BATTLE OF NOVARA; major Austrian victory. Abdication of Charles Albert; accession of VICTOR EMMANUEL II.
	25 April	French troops landed in Papal States.
	15 May	Neapolitan troops occupied Palermo.

4 July	French troops entered Rome; Piux IX restored.
6 August	PEACE OF MILAN concluded Austro-Sardinian conflict.
28	Venice finally surrendered to Austrians.

Germany

1848 18 March	UPRISING IN BERLIN; FREDERICK WILLIAM IV GRANTED CONSTITUTION.
31	The VORPARLAMENT met at Frankfurt.
2 May	Prussia invaded Denmark over the SCHLESWIG-HOLSTEIN QUESTION.
18	GERMAN NATIONAL ASSEMBLY MET AT FRANKFURT.
22	Berlin meeting of Prussian National Assembly.
14 June	Frankfurt: the First Democratic Congress.
29	Archduke John of Austria elected REICHSVERWESER at Frankfurt.
16 August	Berlin: the JUNKERPARLAMENT.
23	Berlin: The Congress of Workers' Associations.
26	TREATY OF MALMÖ between Denmark and Prussia.
16 September	Frankfurt Parliament ratified the Malmö armistice.
18	Republican riot and state of siege in Frankfurt.
5 December	Dissolution of Prussian National Assembly.
1849 23 January	PRUSSIA ADVOCATED UNION OF GERMANY WITHOUT AUSTRIA.
15 February	Frankfurt: formation of the GROSSDEUTSCHLAND group.
17	Frankfurt: formation of the KLEINDEUTSCHLAND group.
27 March	GERMAN NATIONAL ASSEMBLY OFFERED THE TITLE 'EMPEROR OF THE GERMANS' TO AN UNWILLING FREDERICK WILLIAM IV.
5 April	The Vienna government recalled the Austrian deputies from Frankfurt.
28	FREDERICK WILLIAM REFUSED THE IMPERIAL CROWN.
10 May	Heinrich von Gagern resigned.
14	Frederick William annulled the mandates of the Prussian deputies at Frankfurt.
30	THE PARLIAMENT LEFT FRANKFURT; in Berlin the edict of the three estates was promulgated.
6 June	The National Assembly moved to Stuttgart.
18	Troops dissolved the Stuttgart Assembly.
23 July	Baden rebels surrendered to Prussia.
1850 31 January	A LIBERAL CONSTITUTION GRANTED IN PRUSSIA.
20 March	German Parliament summoned to Erfurt by Frederick William.
29 April	Erfurt Parliament opened.
2 July	Peace of Berlin between Prussia and Denmark.
28 November	CONVENTION OF OLMÜTZ; Prussia subordinated to Austria and recognized the Frankfurt Diet

The Habsburg Empire

1848 12 March	OUTBREAK OF REVOLUTION IN VIENNA.
13	Resignation of Metternich.
15	The Hungarian Diet accepted the reforms of March 1847.
8 April	BOHEMIAN CHARTER.
11	SANCTION OF THE HUNGARIAN CONSTITUTIONAL LAWS; ABOLITION OF THE SEIGNORIAL REGIME.
25	Constitution, including responsible government, granted to Austria.
15 May	Second rising in Vienna.
17	The Emperor Ferdinand fled from Vienna to Innsbruck.
2 June	THE PAN-SLAV CONGRESS met in Prague.
10	Vienna confirmed the integrity of the Kingdom of Hungary.
12	Prague riot and the end of the Czech movement.
17	Czech rising suppressed by the Austrians.
28	DISSOLUTION OF THE PRAGUE CONGRESS.
22 July	Vienna: the opening of the Reichstag Constituent Assembly.
12 August	The Emperor Ferdinand returned to Vienna.

Chronology cont.

	23 August	Vienna: the workers' insurrection.
	7 September	ABOLITION OF SERFDOM IN AUSTRIA.
	24	KOSSUTH proclaimed President of the Committee for the National Defence of Hungary.
	6 October	Third rising in Vienna; assassination of Latour.
	31	GOVERNMENT IN FULL CONTROL AGAIN; VICTORY OF WINDISCHGRÄTZ.
	9 November	Vienna: execution of Robert Blum.
	21	SCHWARZENBERG appointed Prime Minister and Minister of Foreign Affairs.
	27	AUSTRIAN DECLARATION ON THE 'STATE UNITY OF THE EMPIRE'.
	2 December	Abdication of Ferdinand; accession of Franz Joseph.
1849	**5 January**	Windischgrätz occupied Pest.
	26	Defeat of the Hungarians of Kapolna.
	4 March	PROCLAMATION OF AUSTRIAN CONSTITUTION.
	7	Dissolution of the Reichstag at Kremsier and the concession of a centralized constitution to the Empire.
	9	Austria proposed the formation of a Central European Federation.
	14	Austria confirmed the abolition of seignorial dues.
	14 April	Hungarian Diet proclaimed independence, with Kossuth as leader.
	13 August	BATTLE OF VILAGÒS: Hungarians defeated by the Austrians who were aided by the Russians.
1851	**31 December**	Austrian constitution abolished.

2.5 QUESTION PRACTICE

1 Explain why there were so many revolutions in 1848.

Understanding the question

This is not a question on the causes of *all* the revolutions of 1848. It asks rather which causes were *common* to most or all of those revolutions, and what factors there were which encouraged the spread of revolution.

Knowledge required

The causes of the revolutions of 1848 and the general history of Europe in the previous 30 years.

Suggested essay plan

1 Make the point that *Nationalism*, which received a great boost in the events of 1848 (gaining martyrs), has coloured subsequent views of the events of that year. Because of the strength of nationalism *since* then, it is easy to forget how integrated (i.e. supra-national) upper-class Europe was in the first half of the century. Ukrainian nobles had more in common, say, with Spanish grandees than either had with their own peasants.

2 The 'old order' therefore had a unity. In attacking it, revolutionaries in different countries often thought of themselves as attacking the same thing: the 'system' symbolized by Metternich.

3 Moreover, revolutionary ideologies developed an international flavour; *Communist Manifesto* for workers of the world to unite; Liberalism, Socialism and Anarchism all had an international flavour. Paris was especially influential in this.

4 The example of Paris was all the greater because of the growth of communications; telegraphs and railways (though the latter *also* helped to move troops and so to repress revolutions).

5 Economic factors common throughout Europe; beginnings of industrial revolution break up old monopolistic methods of economic regulation (guilds etc.); only in Britain and Belgium, however, (where revolutionary discontent was limited) was the economy sufficiently advanced to guarantee employment in mass-producer industries.

6 There was an economic crisis throughout Europe during 1845–6, with deflation causing severe unemployment in 1847–8 especially among the artisan handicraft classes (e.g. the handweavers of Silesia and Britain).

7 There was agricultural failure, especially in the Balkans, Habsburg lands, Poland and Ireland.

8 So there was no need to blame, as contemporaries liked to do, international conspiracies. Apart from the common problems facing different countries, the nature of the pre-1848 regime throughout Europe was such that the revolution was bound to be contagious.

2 'France was bored': is this a sufficient explanation of the downfall of Louis-Philippe?

Understanding the question

This is a fairly straightforward question on the causes of the French revolution of 1848. It pushes you in the direction of explanation along the lines of mass psychology, but this should not prevent discussion of all other factors behind the revolution.

Knowledge required

Louis-Philippe's policies and political system generally. Opposition movements. The condition of French society in 1848. The influence of the French historical tradition and the intellectual life of Paris.

Suggested essay plan

1 The meaning of Lamartine's criticism and of the allegation that 'France was bored'; the growth of European anarchist movements centred on Paris; the influence of the French revolutionary tradition.

2 Louis-Philippe personally could not measure up to the glamour of French history; the regime with its pragmatism and expediency, after the extremism of the legitimist Charles X, seemed unequal to the pageant of French traditions.

3 Louis-Philippe tried to tap the Napoleonic legend by the expedition to Algeria and by bringing Napoleon's bones home to *Les Invalides*. But it remained obvious that the regime's only real ideology was the materialism of 'enrichissez-vous', of great

increase of personal fortunes, of economic competition and of private enterprise; note Alexis de Tocqueville's view that the government was a joint-stock organization for the profit of its members.

4 Boredom stemmed from restrictions on political participation: there were brilliant debates in the Chambre, but only a narrow franchise; only ¼ million had the vote in a population of 32½ million; as 92% of the voters were landed, so urban political participation was even lower; the lesser bourgeoisie were also disfranchised, including schoolteachers, who also helped to influence youth against the regime.

5 There was boredom also with a humiliating foreign policy; Algiers did not compensate for Palmerston's public contempt.

6 Other explanations of revolution: Republican opposition (especially by Thiers) was frustrated by Guizot's success in consolidating the Centre-Right government after 1840; Thiers prepared to stir up revolutionary feeling against the regime, simply because he was unable to gain office within the system.

7 There were economic difficulties; the deflationary crisis of 1845–6, social problems and growth of working-class consciousness; there was an absence of social legislation, except the 1841 Labour Law.

8 Corruption: there was a narrow oligarchy in control of the regime.

9 The immediate causes of revolution were: the role of the press, the Banquet campaign and the row over the fortifications of Paris.

10 Conclusions: a regime which has no fixed body of support depends on public opinion ultimately; but a regime of compromises finds it hard to satisfy opinion an a country like France prone to extremism. Boredom did not cause the revolution, but it helps to explain why few bothered to rally round Louis-Philippe when he was in trouble.

3   To what extent should the revolutions of 1848 be explained in social rather than political terms?

Understanding the question

This is almost a straight question on the causes of the 1848 revolutions. Go through these causes one by one, grouping them under the headings 'political' and 'social and economic'. It is probably best to deal with the revolutions as a group rather than country by country, but as you discuss each cause do not forget to point out the differences between countries.

Knowledge required

The causes of revolution in France, Spain, the Habsburg Empire, Italy, Poland, and perhaps Ireland. The social and economic conditions in the 1840s. The political effects of liberal and nationalist movements.

Suggested essay plan

1 A brief survey of the revolutionary outbreaks.

2 Distinguish between the instigators and leaders of revolution, and the masses whose support alone makes revolution possible. The leaders were mainly motivated by political aims, their inarticulate followers motivated more by social pressures.

3 Liberalism as a political explanation: the middle classes aspired to political power after the British 1832 example; frustration with the narrowness of the franchise under Louis-Philippe; concern for economic opportunities, the promotion of railways; the desire for basic human freedoms (speech, conscience etc.), especially strong in the German states; the ideas of Mazzini.

4 Nationalism as a political explanation; various national movements in Europe. Distinguish the cultural, linguistic, romantic nationalism from middle-class economic and bureaucratic nationalism.

5 The alienation of the intellectuals: the expansion of education and of professional training (as in law and medicine), without a corresponding expansion of employment opportunities for professionals, provides another *political* explanation.

6 Also 1848 saw the birth of political anarchism and the cult of violence. This led the middle classes to desert the revolution.

7 As Marx realised, 1848 was also a social phenomenon, e.g. 'The June Days'.

8 But the main support for revolution comes from those classes beginning to be left behind by industrial development, and unable to compete with factory operatives; Louis Blanc and 'the right to work'; handloom weavers of Silesia; the Artisan Congress demanded the restoration of guild controls over production; a clear clash between these social interests and those of the political revolutionaries of Frankfurt who desired free trade.

9 The peasants as a social factor; the state of serfdom in Europe; the failure of the potato crop in the 1840s causes starvation and suffering, (e.g. in Ireland and Galicia).

10 The economic depression throughout Europe in 1847; the recovery in 1848 undermined the revolution; this suggests that while *politics* provides a focus, the social condition of the people mainly explains what happened.

4   Document question (Prussia and the 1848 revolutions): study the extract below and then answer the subsequent questions (a) to (g):

'To my Beloved Berliners... By my patent of convocation this day, you have received the pledge of the faithful sentiments of your King towards you and towards the whole of the German nation. The shout of joy which greeted me from unnumbered faithful hearts still resounded in my ears, when a crowd of peacebreakers mingled with the loyal throng, making seditious and bold demands, and augmenting in numbers as the well-disposed withdrew.

'As their impetuous intrusion extended to the very portals of the Palace... and insults were offered to my valiant and faithful soldiers, the courtyard was cleared by the cavalry at walking pace and with their weapons sheathed; and two guns of the infantry went off of themselves, without, thanks be to God!, causing any injury. A band of wicked men, chiefly consisting of foreigners... have converted this circumstance into a palpable untruth, and have filled the excited minds of my faithful and beloved Berliners with thoughts of vengeance for supposed bloodshed; and thus have they become the fearful authors of bloodshed themselves. My troops, your brothers and fellow-countrymen, did not make use of their weapons until forced to do so by several shots fired at them from the Königsstrasse. The victorious advance of the troops was the necessary consequence.

'It is now yours, inhabitants of my beloved native city, to avert a fearful evil. Acknowledge your fatal error; your King, your trusting friend, enjoins you, by all that is most sacred, to acknowledge your fatal error. Return to peace; remove the barricades which are still standing: and send to me men filled with the genuine ancient spirit of Berlin, speaking words which are seemly to your King; and I pledge you my royal truth that all the streets and squares shall be instantaneously cleared of the troops, and the military garrisons shall be confined solely to the most important buildings – to the Castle, the Arsenal and a few others and even here only for a brief space of time. Listen to the paternal voice of your King, ye inhabitants of my true and beautiful

Berlin; and forget the past, as I shall forget it, for the sake of that great future which, under the peace-giving blessing of God, is dawning upon Prussia, and through Prussia upon all Germany…'

Written during the night of 18/19 March 1848.

(Frederick William IV addresses the Berliners)

Maximum marks

(a) What events of the previous day in Berlin led to the writing of this letter? (3)

(b) What were 'the faithful sentiments' which the King pledged 'towards you' (i.e. the Prussians) 'and towards the whole of the German nation'? (4)

(c) Why were such pledges insufficient to reassure the 'crowd of peace-breakers' who made such 'seditious and bold demands'? (3)

(d) How do you explain the King's conciliatory tone in the second paragraph of this letter, and his insistence that his cavalry had acted 'at walking pace and with their weapons sheathed'? (4)

(e) What significance may be attached to the King's appearance, shortly after the date of this letter wearing the black, red and gold sash of the former Holy Roman Empire? (3)

(f) In which German city at this time did the hopes of the liberal nationalists chiefly centre? Discuss the suggestion that the efforts in that city to promote the cause of unification in 1848–9 were too narrowly based. (6)

(g) In what ways did the events of the next three years run contrary to the royal prediction that a 'great future' was 'dawning upon Prussia, and through Prussia upon all Germany', and how do you account for these events? (8)

2.6 READING LIST

Standard textbook reading

J. A. S. Grenville, *Europe Reshaped 1848–78* (Fontana, 1976), Part 1.

David Thomson, *Europe Since Napoleon* (Penguin, 1977), chapters 9–11.

Anthony Wood, *Europe 1815–1945* (Longmans, 1975), chapters 9–13.

Further suggested reading

M. S. Anderson, *The Ascendancy of Europe* (Longmans, 1972).

J. P. T. Bury (ed.), *The New Cambridge Modern History 1830–70*, volume X (Cambridge University Press, 1964), chapters 9 and 15.

E. J. Hobsbawm, *The Age of Revolution, Europe 1789–1848* (Sphere, 1977).

H. Kohn, *Nationalism: Its Meaning and History* (Anvil, 1965).

W. L. Langer, *Political and Social Upheaval 1832–52* (Harper and Row, 1972).

L. B. Namier, *1848 – The Revolution of the Intellectuals* (Oxford, 1962).

Roger Price (ed.), *1848 in France* (Thames and Hudson, 1975).

L. C. B. Seaman, *From Vienna to Versailles* (Methuen, 1972).

P. N. Stearns, *The Revolutions of 1848* (Weidenfeld and Nicolson, 1974).

J. L. Talmon, *Romanticism and Revolt* (Thames and Hudson, 1967).

3 The Eastern Question to 1856

3.1 INTRODUCTION

The Eastern Question was the dominant diplomatic issue among the major European Powers in the nineteenth century. It had political strategic, economic, religious and cultural implications. The Ottoman Empire, whose growing weakness had long encouraged Russian expansion southwards, continued to decay. The Serbian rising of 1804 had posed the threat of Balkan nationalism, but it was the Greek struggle for independence which created the major confrontation. It underlined the conflict of interest between the Powers in this area, particularly the British, French and Austrian apprehensions over Russian aims.

The establishment of Greek independence was followed by the Mehemet Ali crises. The crucial strategic, balance of power – question of the Straits was dealt with in the Treaty of Unkiar-Skelessi and the Straits Convention. Subsequently, Napoleon III's interest in the holy places and Tsar Nicholas I's claim to the right of protection over the Orthodox minorities precipitated the Crimean War between Russia and the Western Powers in support of Turkey. This was the first conflict among the major European Powers for 40 years. The war, and the Treaty of Paris which concluded it, had far-reaching effects on European diplomatic relations.

3.2 STUDY OBJECTIVES

1 The origins of the Eastern Question in the late eighteenth century and the basic elements of the problem: Christians' rights and Balkan unrest; the growing weakness of the Sultan's authority; Russian involvement; the attitude of Britain, France and Austria; British and Austrian agreement to shore up 'the sick man of Europe'.

2 A clear grasp of the geography of the Near East with an appreciation of the various nationalities in the Ottoman Empire (Moldavia, Wallachia, Serbia, Bulgaria, Greece, Egypt, etc.).

3 An analysis of the delicate balance of power in Europe after 1815 and of the ways in which the Eastern Question threatened to disrupt it.

4 The causes of the Greek Revolt (1820–30) and the involvement of Egypt, Russia, Britain, Austria and France.

5 The reorientation of Russian policy under Nicholas I between 1829 and 1840; his cooperation with Metternich to sustain the Ottoman Empire and secure a peaceful extension of influence there on the basis of the Treaty of Unkiar-Skelessi (July 1833).

6 Mehemet Ali and the Egyptian revolt against the Sultan (1831–40); Palmerston's intervention and French protestations; British encouragement of Syrian rebellion against Egypt; the decline of Egyptian power and the terms of the Straits Convention of July 1841.

7 The causes of the Crimean War; disputes concerning the Holy Places and the protection of the Christians; the role of diplomats on the spot such as Lord Stratford de Redcliffe, public opinion in Britain and France and the xenophobic hatred of Russia; the reasons for Tsar Nicholas's stand on the question of the Sultan's Christian subjects; his occupation of the Principalities in July 1853; circumstances leading to the declaration of war.

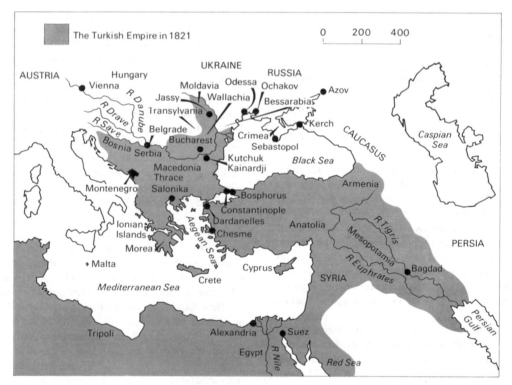

Fig. 1 The Eastern Question to 1856. The geography of the Near East showing the extent of the Ottoman Empire in 1821

8 The course of the war including the long siege and eventual fall of Sebastopol; the tactics and technology employed; the condition of the troops.

9 The outcome of the war; the terms of the Treaty of Paris.

10 The longer-term consequences and developments: failure of Turkey to reform; persistence of unrest and rebel movements within the empire; Russia's repudiation of the Black Sea clauses in 1871. The impact of the war on the domestic politics of the participants. The results for the international standing of Piedmont and Austria's position among the Powers.

3.3 COMMENTARY

Background to the Eastern Question

More than any other single issue in international affairs during the nineteenth century, the Eastern Question was a continual preoccupation of the Powers.

In essence it resulted from the progressive decline of the Turkish Empire. As the control of the Sultan of Constantinople over his territories, subject peoples and local governors weakened, the Empire became increasingly an object of anxious speculation and rivalry among the Powers.

The balance of power

During the later eighteenth century the spread of Russian influence in the area had aroused growing apprehension among the Western Powers (Britain and France and Austria).

Particularly at issue were the fate of Constantinople and the Straits and the future of the Balkan territories. What happened here would affect the balance of power in Central Europe and the Mediterranean. It could have even wider consequences: the growth of Russian influence in the Near East was regarded by the British as posing a distinct threat to their imperial position in India.

Constant factors

Constant factors in the Eastern Question were: (1) the repeated failure of the Turkish Empire to be effectively reformed; (2) the existence of a very large Christian Greek Orthodox minority under Mohammedan rule (approximately two out of five of the Sultan's subjects were Christian); and (3) increasing restlessness of the Balkan peoples and a tendency to fight among themselves the more they liberated themselves from the common enemy, Turkey.

The Greek Question

In April 1821 the Orthodox Archbishop Germanos openly rejected Turkish rule over the Greeks. Popular unrest erupted into a long war of independence.

Turkish control of South-Eastern Europe had been under increasingly sharp attack since the turn of the nineteenth century. There were various reasons for this: (1) the growing awareness of individual nationhood which was to develop into strong nationalist movements; (2) the influence of the example of the French Revolution; and (3) a surge of foreign sympathy for the Greek cause, both for the Christian Orthodox and because of the spread of Philhellenism, the veneration for the Greek contribution to civilisation and culture.

Philiké Hetairia

Greeks were being made aware of their own ancient literature and traditions and the demand for independence was combined with a strong cultural revival. In 1814 the 'Philiké Hetairia' was founded to press for the liberation of Greece. This movement gained influence in many cities and attempted to enlist the support of Tsar Alexander I whose foreign minister, Capodistrias, was a Greek.

Philhellenic societies all over Europe collected arms and supplies to support Greeks, who were pictured in the West as the defenders of civilisation. Their ambition was to reconstruct the Byzantine Empire with Constantinople as its capital.

The intervention of the Powers

The rebels were soon weakened by factional differences among themselves and the growing resistance of the Turks. In 1825 the Sultan appealed to the Pasha of Egypt, Mehemet Ali, for help, promising him Syria and Crete as a reward. By 1827 it appeared that the Greek struggle was lost.

The European Powers intervened at this point to prevent the outbreak of a wider struggle and to safeguard their own interests. These were Russia, Great Britain, France and Austria and they had conflicting objectives.

Russia For obvious strategic and commercial reasons the dominant Russian aim was the freedom of her ships to pass the Straits and access to the Mediterranean. Some advisers to the Tsar called for the outright seizure of Constantinople.

Other significant considerations in the Russian view were that the major part of the Balkan peoples belonged to the Eastern Orthodox Church while many of them were also, as the Russians were, of Slav race. In the Treaty of Kutchuk-Kainardji of 1774 the Russian monarchy had sought to secure certain privileges for itself as protector of the Christian Orthodox in the Turkish Empire.

Great Britain The Napoleonic period had alerted Great Britain to the major strategic significance for her of the Eastern Mediterranean. As a maritime and imperial power she had secured Malta and the protectorate of the Ionian islands at the Treaty of Vienna. Control of this area by another Power could easily threaten her imperial possessions, particularly India. She was therefore opposed both to Russian and French ambitions in the Near East. The exclusion of these Powers from a controlling position in the area became a fixed aim of British policy in the nineteenth century.

France The French interest in the Levant was long-standing. The French monarchy had traditionally enjoyed a position as protector of the interests of the Catholic Christians in the Near East. France already had an established interest in the area when Napoleon led his expedition to Egypt. This interest was well illustrated during the Mehemet Ali Crisis, in which she pursued an independent policy, and in the subsequent French colonization of the Mediterranean area.

Austria Austria was opposed to any extension of Russian influence in the Balkans. The close kinship between the Russian people and the Balkan Slavs, later to be emphasized by the Panslav movement, was already posing a long-term threat to the multi-national Habsburg Empire. At the same time, Austria had a strong interest in the free navigation of the Danube and the Principalities, Moldavia and Wallachia.

The Russo-Turkish War

In July 1827 Russia, Great Britain and France signed a treaty to enforce an armistice and to make Turkey agree to the establishment of an autonomous Greece. The allies destroyed the Egyptian fleet at Navarino and in April 1828 the Russians declared war against Turkey. In September 1829 Turkey was forced to concede the independence of Greece and greater autonomy for the Principalities in the Treaty of Adrianople. With the accession to the Greek throne of Otto of Bavaria, Greece became the first independent Christian Balkan kingdom in the nineteenth century.

The Greek revolt and the status quo

Throughout the Greek crisis one should bear in mind not simply the course of events and conflict of interest among the Powers but also the way in which it was related in the minds of statesmen to the question of order within states.

International opinion came to support the national and constitutional demands of the Greek rebels. Metternich, on the other hand, regarded the revolt as a possible threat to the status quo generally. And much as Tsar Alexander might sympathize with the Christian Greeks, he recoiled from the implications of supporting revolution against the principles of the Holy Alliance.

The Crimean War

The Crimean War originated in 1853 with a Russian attempt to protect the rights and privileges of the Orthodox communities in Turkey. It brought to a head the underlying conflict of interests. Russia occupied Moldavia and Wallachia, which were already protectorates and after the Russian Black Sea fleet destroyed a Turkish fleet at Sinope, Britain and France came to Turkey's help. The entry of the British and French fleets into the Black Sea amounted to their claiming predominance in the area.

The attitudes of the Powers

Great Britain saw herself as acting to protect the Straits and Constantinople from Russian control. Public opinion in Britain was incensed at Russian actions and credited Russia, among other things, with a desire to destroy Great Britain's influence in the Near East. Liberals and Radicals in particular saw the war more as an issue of principle with Russia than a defence of the Turkish Empire. For them it assumed the character of an ideological struggle, a battle between the ideas of autocratic government and constitutional authority based on popular representation.

Napoleon III's claims on the Sultan over the Holy Places were interpreted by Tsar Nicholas I as an attempt to replace Russian influence by French.

In the eyes of Russian nationalists, the Western Powers came to the help of the 'infidel' Turk out of a deep-seated distrust of Russia and her desire to 'liberate' the Christian Balkan peoples.

Tsar Nicholas feared the revival of Bonapartist sentiment in France and that Napoleon's actions would spread revolution throughout Europe and destroy the 1815 settlement.

Russia demanded that the Turks respect not only the religious rights of the Orthodox but that the Sultan accept the Tsar as protecting *all other rights* that he might feel properly belonged to the Christian Orthodox. This was a claim unjustified by the terms of the Treaty of Kutchuk-Kainardji.

The consequences

The main battleground was the Crimean Peninsula. The war revealed the general backwardness of Russia and the military unpreparedness of the Western Powers. The British forces in particular suffered more from the effects of disease than from battle.

Nicholas had mistakenly counted on Austrian support. The Austrians mobilized part of their army to protect the mouth of the Danube but stayed neutral. After the fall of the fortress of Sebastopol in September 1855, peace negotiations started in Paris.

The significance of the peace terms

Turkey Turkey was admitted as a member of the Concert of Europe and the Powers undertook to respect Turkish independence. The Sultan promised to improve the conditions of his subjects without discrimination of race or creed.

Russia Not until the end of the First World War was any state forced to submit to such an obvious and humiliating limitation of its freedom of action as Russia in the settlement of 1856.

Russia lost the right of sending her Black Sea Fleet into the Mediterranean. She also lost control of the mouth of the Danube. The Straits were closed to all foreign warships when Turkey was at peace and the Principalities were no longer a Russian Protectorate. They were to have 'an independent and national' administration. The foundation was laid at the Paris Congress for the future independent state of *Rumania* (recognized officially in 1878).

The significance of the clauses neutralizing the Black Sea was that Russia was neither allowed to construct a navy nor support its operations there. On the other hand, if Russia in any way infringed the terms of the Treaty the British and French navies could enter the Black Sea at will. The nullification of these clauses was Russia's primary ambition over the next 15 years.

Wider considerations

The Crimean War put an end to the Holy Alliance. Russia gave up her general mission of defending the Vienna Settlement and preserving Europe from revolution.

She now concentrated on her great expansion in Central Asia and took an increasing interest in the problems of the Balkan Slavs – an interest which was to be of the utmost significance for the later evolution of the Eastern Question.

Alexander II, Tsar from 1855–81, was prompted by the revelation of Russia's weakness in the war to give priority to reforms within Russia, most notably the emancipation of the serfs.

Austria found her position relatively weakened. Without Russian support she became increasingly isolated as a Power. Indirectly, the Crimean War helped to make possible the emergence of the united nation-states of Italy and Germany, both achieved at Austria's expense.

3.4 CHRONOLOGY

1774	**July**	THE TREATY OF KUTCHUK KAINARDJI: Russia received territory in the Crimea and obtained the right of free navigation for trading vessels in Turkish waters. Turkey recognised the right of Russia to make representations on behalf of Christian subjects in the Ottoman Empire. This was used as justification for later Russian involvement in Ottoman affairs.
1783		Russia incorporated the Crimea. Sebastopol became a Russian naval base.
1812	**May**	THE TREATY OF BUCHAREST: following a six-year Russo-Turkish war. Turkey surrendered the province of Bessarabia to Russia. Limited autonomy was granted to Serbia.
1814		PHILIKÉ HETAIRIA: founded at Odessa with the blessing of the Russian government. Led by Alexander Hypsilanti, a member of a Greek family from Moldavia and an officer in the Russian army.
1815		THE VIENNA SETTLEMENT: British protectorate over the Ionian islands and the possession of Malta.
1820		Britain assumed defensive responsibilities in Persia and the Persian Gulf.
1821	**February – March**	OUTBREAK OF THE GREEK REVOLT in the Danubian Principalities and the Morea.
1825	**February**	EGYPTIAN INTERVENTION IN GREECE: Ibrahim, son of PASHA MEHEMET ALI, landed in the Morea and subdued the revolt in the peninsula.
1826	**April**	THE ST. PETERSBURG PROTOCOL: signed by Britain and Russia. The two Powers agreed to mediate between Greece and Turkey on the basis of autonomy for Greece under Turkish suzerainty.
1827	**July**	THE TREATY OF LONDON: Britain, France and Russia agreed that if the Turks refused an armistice the three Powers would threaten to support the Greeks.
	October	THE BATTLE OF NAVARINO: destruction of the Egyptian fleet by British, French and Russian squadrons.
1828	**April**	OUTBREAK OF THE RUSSO-TURKISH WAR.
1829	**March**	THE LONDON PROTOCOL: Greek autonomy guaranteed by the Powers.
	September	THE TREATY OF ADRIANOPLE between Russia and Turkey: the Russians abandoned their conquests in Europe; Russia to occupy the Danubian Principalities pending the payment of an indemnity; the Turks accepted the London Protocol.
1830	**February**	New three-power protocol provided for Greek independence.
1831	**November**	MEHEMET ALI DECLARED WAR ON THE SULTAN, having demanded Syria as a reward for his help against the Greeks.
1832	**March**	OTTO OF BAVARIA appointed King of Greece.
	July	Turkey finally accepted Greek independence.
	December	The battle of Konieh. Ibrahim completely routed the main Turkish army.
1833	**February**	The Sultan requested Russian protection against Mehemet Ali – the Muraviev mission.
	May	Russian intervention forced the Egyptians to make peace.
	July	THE TREATY OF UNKIAR SKELESSI: this was between Russia and Turkey for a term of eight years. Each party was to come to the other's aid in event of attack. A secret clause excused the Turks from fulfilling this obligation provided they closed the Dardanelles to foreign warships. This represented the height of Russian influence at Constantinople. Britain and France interpreted the treaty to mean that the Bosphorus was to remain open to Russian warships and that Turkey was to be at the mercy of Russia.
	September	THE MÜNCHENGRÄTZ AGREEMENT: the Tsar agreed with Austria to maintain the Ottoman Empire; if partition became inevitable, action would only be taken after consultation between the two powers.

Chronology cont.

1839	**April**	The Sultan declared war on Mehemet Ali with the invasion of Syria.
	June	Egyptian victory at Nizib.
1840	**July**	THE TREATY OF LONDON: Britain, Russia, Austria and Prussia agreed to force a settlement on the Egyptians. Mehemet Ali rejected the demand to give up Crete, northern Syria, Mecca and Medina, hoping for French support for his ambitions.
	November	Mehemet Ali makes peace.
1841	**June**	THE STRAITS CONVENTION: signed by the five great Powers. The Straits were to be closed to all foreign warships in time of peace.
1844	**June**	TSAR NICHOLAS I visited Britain. He proposed to Lord Aberdeen that Britain and Russia should consult together in the event of the collapse of the Ottoman Empire.
	September	THE NESSELRODE MEMORANDUM: The Russian foreign minister summed up the supposed Anglo-Russian agreement.
1848	**September**	Russian troops suppressed revolt in Moldavia and Wallachia.
1849	**May**	Russian troops defeated the Hungarian revolution.
	October	Crisis over Hungarian Kossuth's presence in Turkey and Russian and Austrian demands that the revolutionaries be extradited. British and French naval demonstration at Besika Bay.
1850–52		DISPUTE OVER THE HOLY PLACES: the Sultan yielded to pressure from Napoleon III to grant privileges to Roman Catholic monks. Quarrel with the Greek Orthodox supported by the Tsar.
1853	**January**	TSAR NICHOLAS'S CONVERSATIONS WITH LORD SEYMOUR, the British ambassador. The Tsar predicted the collapse of the Ottoman Empire and made a bid for British agreement. He stated that while Russia would not take Constantinople, he would not allow any other power to occupy it.
	February – May	THE MENSHIKOV MISSION TO CONSTANTINOPLE: Prince Menshikov attempted to gain Turkish concessions over the Holy Places and to secure a treaty recognising a Russian protectorate over Orthodox churches in Constantinople. ARRIVAL OF LORD STRATFORD DE REDCLIFFE (British Ambassador to Turkey), who advised the Turks to reject the wider demands of Menshikov. Russians decided to occupy the Principalities.
	June	British and French fleets went to Besika Bay.
	July	RUSSIAN TROOPS INVADED MOLDAVIA AND WALLACHIA. THE VIENNA NOTE.
	October	The Turks declared war on Russia.
	November	THE BATTLE OF SINOPE between the Russian and Turkish fleets. Outburst of anti-Russian indignation in Britain.
1854	**January**	BRITISH AND FRENCH FLEETS ENTERED THE BLACK SEA to protect the Turkish coasts and transports. Russians rejected request to evacuate Moldavia and Wallachia.
	March	ALLIANCE OF BRITAIN AND FRANCE WITH TURKEY AND DECLARATION OF WAR ON RUSSIA.
	August	THE VIENNA FOUR POINTS: the suggested conditions of peace were to be (1) a collective guarantee of the position of the Principalities and Serbia; (2) free passage of the mouth of the Danube; (3) a revision of the Straits Convention in the interest of the balance of power; (4) Russian abandonment of claims to a protectorate over the Sultan's subjects. Russians rejected these demands. Russians evacuated Moldavia and Wallachia.
	September	LANDINGS IN THE CRIMEA BY THE BRITISH AND FRENCH. The battle of Alma and the siege of Sebastopol.
	October	The battle of Balaclava.
	November	The battle of Inkerman.
1855	**March**	Accession of TSAR ALEXANDER II.
	September	The Russians evacuated Sebastopol.
	November	Austrian alliance with Britain and France. Austria mobilized all her forces and issued an ultimatum to Russia, but did not engage in hostilities.
1856	**February – March**	PEACE AGREED AT CONGRESS OF PARIS. Joint declaration by the British, French and Austrian governments that they regarded any infringement of Turkish independence and integrity as a 'casus belli'. The Turkish reform edict, the HATTI-HUMAYUN: the most significant Turkish reform edict of the nineteenth century, offering Christian subjects security of life, honour and property. Civil offices thrown open to all subjects of the Sultan.

3.5 QUESTION PRACTICE

1   Explain why the Eastern Question loomed so large in European affairs between 1815 and 1841.

Understanding the question

Consider not only the instability of the Near East and the problems faced by the Sultan, but the reasons why the Great Powers became involved. The question is as much about the balance of power as about developments within the Turkish Empire.

Knowledge required

The various national movements against the Sultan and the internal weaknesses of the Empire. Economic, religious and territorial reasons for Great Power involvement in Turkish affairs. General diplomatic history 1815–41. The course of Greek and Egyptian revolts. The details of European attitudes and intervention.

Suggested essay plan

1   The state of Europe after the Napoleonic War; emphasize war-weariness, both economic and psychological; consequent attempts to maintain peace at all costs by congress and cooperation; the Eastern Question as the main threat to peace after 1815, and as finally shattering it in 1854.

2   The main reasons for this were to be found in the internal political weakness of the Ottoman Empire, and also in the rise of Balkan nationalisms; discuss the latter, beginning with the Serbian revolt of 1804–12 and append a sketch map.

3   The interest of the Great Powers in the fate of Turkey; Russian and French involvement as rival protectors of the Sultan's Christian subjects, and Russia's territorial ambitions in the Balkans; Britain's commercial interests in the Near East and her concern to maintain the freedom of the Straits and the route to India; France's conquest and settlement of Algeria (1830) and North African ambitions generally, as a means of offsetting her loss of power inside Europe; Austria's own historic role as protector of Europe against the Turks and her anxiety to prevent an extension of Russian power in the Balkans.

4   A brief narrative of the Greek national revolt 1820–30; attitudes of Alexander and Nicholas, and also of Metternich, Canning and Polignac.

5   A brief account of the Egyptian revolt under Mehemet Ali and Ibrahim Pasha; attitudes and involvement of the Powers, leading to the Straits Convention and settlement of 1841.

6   Show that, whereas the Greek episode had ultimately encouraged cooperation among the Powers, the Egyptian affair led to a division of Europe into opposing camps. The problem was that Russia and Britain, the two main victors against Napoleon, the one with her great territorial empire, the other with her immense commercial preponderance, were bound to dispute over the spoils of the crumbling Turkish empire; realistic, Nicholas recognized this by offering Greece to Britain in 1844, if only he could have Constantinople.

Meanwhile, France supported Britain, but was anxious to maintain her own say in the matter; just as Austria supported Russia, but more to check her influence over the Porte than to encourage it. The outcome was the Straits Convention and 12 years of uneasy and unusual quietness on the Eastern Question.

2   Do you consider that the policy of Tsar Nicholas I towards Turkey led inevitably to the Crimean War?

Understanding the question

This is a straight question on the causes of the Crimean War,

slightly disguised. You will need to sort out the short-term and immediate factors leading to the outbreak of war from the longer-term factors which made a conflict inevitable sooner or later. Emphasis in both cases should be on Nicholas's role, but you should also set the answer in the context of the responsibility of other participant countries, especially France and Britain.

Knowledge required

The background to the Eastern Question. Its development 1815–53. Russian attitudes and policy under Nicholas I. The attitude of the Great Powers. Diplomatic manoeuvres preceding the war.

Suggested essay plan

1   The internal situation in Turkey since the late eighteenth century; Balkan nationalism and the plight of Christian peoples.

2   The development of Russian policy since Catherine the Great; Moscow's status as the 'Third Rome' and protector of Christians in Constantinople; Russia's territorial ambitions towards the feeble Turkish Empire.

3   The accession of Nicholas I; his and Nesselrode's Near-Eastern policy; cooperation with Britain and France against the Turkish navy at Navarino; Unkiar-Skelessi and the reorientation of policy to secure the peaceful penetration and exploitation of Turkey; Mehemet Ali and the Egyptian crisis; the strong hostility of Britain (especially Palmerston) to Russian policy, and her fear that it would lead to separate French and Russian 'spheres of influence' in the Near East; the 1841 Straits Convention established the Eastern Question as a European 'problem'.

4   The immediate build-up to the war; Nicholas's extreme demands in the dispute with France over the Holy Places; Russia's occupation of Moldavia and Wallachia; a brief account of diplomatic manoeuvres leading to the Turkish declaration of war and the engagement of the the Powers.

5   The view that Napoleon III was more to blame than Nicholas; Napoleon used the dispute over the Holy Places as an excuse for the exercise of 'la gloire' and to attack the Vienna Settlement of 1815; against this, Napoleon was dependent on Catholic support – i.e. on French Catholic opinion that France should protect the Latins in the Holy Places.

6   The view that Britain was mainly to blame for the Crimean War; economic interests' desire to penetrate Turkish markets; the roles of Stratford de Redcliffe and Palmerston; hostility of British public opinion to Russia.

7   Conclusion: it is sometimes said that the Crimean War was an 'accident'; on the other hand, Russia and Turkey had fought about every 20 years for the previous two centuries; the war in 1854 was different because the fear of Russian aggrandizement led Britain and France to back Turkey. In western eyes, Russia, the greatest Slav nation, represented all that was barbaric; war became inevitable when even Francis Joseph of Austria, whom Nicholas had rescued in 1849 when he put down the Hungarian revolution, refused to support Russia.

3   Examine whether the Crimean War had greater consequences for Western Europe than for Eastern Europe.

Understanding the question

'Consequences' here infer not simply the direct and immediate, but also the indirect and long-term results. You should be ready to comment on any significant influence of the war on domestic developments in the countries concerned, as well as on the more obvious international consequences. Eastern Europe may be defined as (1) the Russian sphere of influence and (2) the Balkans.

Knowledge required

The results of the Crimean War: the Paris Congress, especially its territorial and strategic implications; the effects on the Ottoman Empire; its long-term implications for relations between the Powers and developments within them.

Suggested essay plan

1   The Congress of Paris: the conclusion of peace terms and the imposition of a European solution on the Turks; emphasize that the participants in the Congress had aims of their own: Britain wished to weaken Russia strategically in every way; France was anxious to treat Russia leniently; Austria wished to strengthen her hold on the Principalities.

2   The terms affecting Russia and the Balkans; the clause regarded as most significant by the victorious Powers was the neutralization of the Black Sea; but Britain and France could enter the Black Sea at will should Russia attack Turkey or otherwise infringe the terms of the treaty; there was a vulnerability, therefore, of the South Russian coastline.

3   Navigation of the Danube was freed from Russian influence; Serbian rights were guaranteed; the emergence of an independent Rumania (formally recognized by the Powers at the Congress of Berlin). Stress the growth of Balkan nationalism; by denying both Russia and Austria an extension of influence in the Balkans, conditions came about which permitted the creation of independent states from the declining Ottoman Empire.

4   The collapse of the Ottoman Empire, which Nicholas I had predicted, was postponed; some genuine efforts at internal reform were made; the Sultan announced the *Hatti-humayun*, proclaiming religious equality throughout the Empire, economic reforms and abolition of bribery and tax-farming; but progress towards reform was too slow and incomplete.

5   Wider consequences of the Crimean War: Russia became a revisionist Power; the war ended the period during which the conservative solidarity of governments and the maintenance of the 'status quo' were emphasized; the final blow to the Holy Alliance. Russia now concentrated on expansion in Central Asia and on a policy of general support for the Slavs; the reforms under Alexander II were a direct result of the exposure of Russian weakness by the war.

6   The diplomatic impact on Western Europe of the war: Austria found herself isolated; the war alienated Russia; it was not the revolutions of 1848 but the Crimean War which finally destroyed the balance of the Metternich era; only after the Crimean War did those political upheavals become possible which the Austrian minister worked so hard to postpone; the subsequent successes of Cavour and Bismarck were helped by the weakening of the Habsburg Empire after the Crimean War.

7   Conclusion: the consequences of the war were of major significance in both Eastern and Western Europe, though, on balance, the impact on developments in the West was probably the greater – more particularly in indirectly encouraging the emergence of a united Italy and a united Germany; the results for Russia and the Balkans should also be given full weight.

4   Document question (The Great Powers and the Greek Problem): study the extract below and then answer the subsequent questions (a) **to** (g):

Extract

'In the Name of the Holy and undivided Trinity.
His Majesty the King of the United Kingdom of Great Britain and Ireland, His Majesty the King of France and Navarre, and His Majesty the Emperor of all the Russias, penetrated with the necessity of putting an end to the sanguinary contest which, by delivering up the Greek provinces and the isles of the Archipelago to all the disorders of anarchy, produces daily fresh impediments to the commerce of the European States, and gives occasion to piracies, which not only expose the subjects of the High Contracting Parties to considerable losses, but besides render necessary burthensome measures of protection and repression; (the Kings of the United Kingdom and France) having besides received, on the part of the Greeks, a pressing request to interpose their mediation with the Ottoman Porte, and being as well as His Majesty the Emperor of all the Russias, animated by the desire of stopping the effusion of blood, and of arresting the evils of all kinds which might arise from the continuance of such a state of things, have resolved to unite their efforts and to regulate the operation thereof by a formal treaty, with the view of re-establishing peace between the contending parties by means of an arrangement which is called for as much by humanity as by the interest of the repose of Europe....

'1. The Contracting Powers will offer to the Ottoman Porte their mediation with a view to bringing about a reconciliation between it and the Greeks. This offer of mediation shall be made to this Power immediately after the ratification of the Treaty, by means of a collective declaration signed by the Plenipotentiaries of the allied Courts at Constantinople: and there shall be made, at the same time, to the two contending parties, a demand of an immediate armistice between them, as a preliminary condition indispensable to the opening of any negotiation.

'2. The arrangement to be proposed to the Ottoman Porte shall rest on the following bases: the Greeks shall hold of the Sultan as their suzerain; and in consequence of his suzerainty they shall pay to the Ottoman Empire an annual tribute, the amount of which shall be fixed, once for all by a common agreement. They shall be governed by the authorities whom they shall themselves choose and nominate, but in the nomination of whom the Porte shall have a determinate voice.'

(The Treaty of London for the Pacification of Greece, 6 July, 1827)

Maximum marks

(a) What caused the Greeks to send to the British and French 'a pressing request to interpose their mediation' at this time?   (3)

(b) What, in the area affected by the Greek revolts, were the special interests of *each* of 'His Majesty the King of the United Kingdom of Great Britain and Ireland, His Majesty the King of France and Navarre, and His Majesty the Emperor of all the Russias'?   (6)

(c) How had Russian policy been affected by the accession of Nicholas I as 'the Emperor of all the Russias'?   (4)

(d) Why was it that Britain, Russia and France 'resolved to unite their efforts'?   (5)

(e) What were the dangers for the future in the London decision that the Greeks should 'hold of the Sultan as their suzerain'?   (2)

(f) How, and with what results, did 'the Contracting Powers' try to enforce 'an immediate armistice'?   (4)

(g) How successfully was the Greek problem settled by 1832, and what did Russia and Britain respectively contribute to the settlement?   (7)

3.6 READING LIST

Standard textbook reading

J.A.S. Grenville, *Europe Reshaped 1848–78* (Fontana, 1976), chapters 10 and 11.

David Thomson, *Europe Since Napoleon* (Penguin, 1977), chapter 12.

Anthony Wood, *Europe 1815–1945* (Longmans, 1975), chapter 15.

Further suggested reading

M.S. Anderson, *The Eastern Question 1774–1923* (Macmillan, 1966).

J.P.T. Bury (ed.), *The New Cambridge Modern History 1830–1870*, volume IX, (Cambridge University Press, 1964), chapter 10.

D. Clayton, *Britain and the Eastern Question from Missolonghi to Gallipoli* (University of London, 1971).

C.W. Crawley (ed.), *The New Cambridge Modern History 1793–1830*, volume IX, (Cambridge University Press, 1974), chapter 19.

L.C.B. Seaman, *From Vienna to Versailles* (Methuen, 1972).

A.J.P. Taylor, *The Struggle for Mastery in Europe* (Oxford, 1971).

4 The Risorgimento and the unification of Italy

4.1 INTRODUCTION

The Vienna Settlement replaced the Napoleonic rule of Italy by establishing the Habsburg Empire as the dominant power in occupation of Lombardy and Venetia. During the years after 1815, Italian nationalist sentiments found expression in conspiratorial societies, such as the Carbonari, in a general cultural reawakening, and, in the 1830s and 1840s, in specific political programmes for the unification of the peninsula. The best-known romantic revolutionary and nationalist of this period in Europe was Mazzini, founder of 'Young Italy'.

After the failure of the 1848 revolutions and the discrediting, in nationalist eyes, of Pope Pius IX there was growing recognition that Italy could not, unaided, liberate herself. In the event, the unification was brought about under the leadership of Piedmont and her Prime Minister, Count Cavour.

In 1859, following his secret meeting with Cavour at Plombières, Napoleon III assisted Piedmont in defeating Austria and gaining Lombardy from the Habsburgs. This was followed by the incorporation of the central Italian principalities. In 1860 Garibaldi and his volunteer troops landed in Sicily and then, crossing to the mainland, proceeded to capture most of the Kingdom of Naples. Cavour forestalled his march on Rome and Garibaldi surrendered his conquests to King Victor Emmanuel.

The new Kingdom of Italy was proclaimed in March, 1861. Venetia was gained as a result of the Austro-Prussian War and Rome, which had been garrisoned by the French, followed after the Prussian defeat of France. The Pope, deprived of his temporal power, refused to recognise the new Italian state. The Piedmontese administrative system was extended to the rest of Italy. The new kingdom was weakened by deep divisions and deeply disappointed many nationalists, like Mazzini.

4.2 STUDY OBJECTIVES

1 The impact of the French Revolution and Napoleonic rule on Italian nationalist feelings and on the government and administration of the Italian states.

2 Italy in 1815: the effect of the Congress of Vienna; the restoration of Austrian, Bourbon and Papal rule; the nature of this government.

3 The Risorgimento proper, or the romantic phase of Italian nationalism; the Carbonari, the literati and the 1820–21 revolutions; the republican and democratic nationalism of Mazzini and 'Young Italy'; the risings of the 1830s.

4 Two rulers briefly took up the cause of Italian nationalism: Pius IX and Charles Albert of Piedmont; the influence of the neo-Guelph writers (Gioberti, Farini, Balbo); the Piedmontese war against Austria and the 'constitution' of 1848.

5 The 1848 revolutions in the peninsula and their suppression by Radetzky; an analysis of support for nationalism in Naples, Tuscany and Milan.

6 Piedmont and the businessman's phase of Italian nationalism; the abandonment of the democratic revolutionary approach of Mazzini; the desire for an Italian Zollverein, or customs union with a common tariff and communications system, uniform weights and measures; frustrations with the economic restrictions of Austrian rule; the National Society.

7 The restoration of the credibility of the Piedmontese monarchy after the defeat of the 1848 revolutions by D'Azeglio; his replacement by Cavour; Cavour's financial and administrative reforms, economic expansion and the development of the bourgeois liberal state; his anticlerical measures (the Siccardi Laws); his admiration for Britain and France; his relations with Victor Emmanuel, who forced him to engage in the Crimean War.

8 Cavour's announcement of Piedmont's claims at the Treaty of Paris; his decision to secure unification through diplomacy; the attitude of the Powers to Italy; the secret diplomacy with Napoleon (Plombières) and provocation of the war with Austria; the French motives for supporting Piedmont; the reasons for Napoleon's disengagement from the war and Villafranca.

9 The plebiscites of 1859, completing the first stage of Italian nationalism.

10 The political and social background of the Two Sicilies; Mazzini, the revolution and Garibaldi's adventure; his republicanism and distaste for Piedmont; the attitude of Britain and France; Cavour's skill in exploiting the situation; the intervention of the Piedmontese army; the incorporation of the Two Sicilies in the Kingdom of Italy.

11 The diplomatic and military manoeuvres leading to the incorporation of Venice and Rome; the attitude of the Powers, especially France to these events.

12 The nature of the new Italian state; its bureaucratic and centralized structure, which alienated many patriots; the domination of North over South; Piedmont's free trade competition ruins the South's precarious economy; the estrangement between the Papacy and the new Italian state.

4.3 COMMENTARY

The 'Risorgimento'

The term 'risorgimento' in the sense of a movement of national rebirth or resurgence – was first used in the eighteenth century.

It is used to describe the general movement of growing national consciousness and revival from that period, but more

Fig. 2 The stages of Italian unification 1859–70 including the gains of 1919

particularly from 1815 onwards. Very frequently it is employed more comprehensively to include the actual process of territorial unification through war and diplomacy between 1859 and 1870.

Very many Italians during the nineteenth century neither comprehended nor sympathized with the idea of a united Italy, and many after the unification opposed it.

During the first part of the century there seemed to be almost insuperable barriers to unity which included: (1) the division into seven separate states; (2) strong regional loyalties; (3) economic backwardness and divisions; (4) mass illiteracy; (5) the institution of the Papacy, a primarily international institution, whose territories cut the peninsula in two; (6) the domination of Italy either directly or indirectly following the Vienna Settlement. (Metternich described Italy as a 'geographical expression'.)

Napoleon I and Italy

The revolutionary conquests of Napoleon did much to destroy the old regime in Italy. His imperial rule strengthened the idea of national unification in two ways: (1) it aroused the intense opposition of Italians to foreign domination – this expressed itself after 1815 against the Austrians; and (2) the idea of the nation-state propagated by the French Revolution and the introduction of administrative modernization and reforms under the Emperor encouraged the breakdown of the traditional Italian system of separate states. It accustomed the educated classes in particular to a greater degree of unity and efficiency.

The Restoration

After 1815, when Italy came under Austrian domination,

patriots and reformers became increasingly insistent on the need for independence and at least some degree of unification. This was the period of the secret societies, such as the Carbonari, or Charcoal Burners, and of revolts in Naples, Sicily and Piedmont. But among many of the opponents of the restored governments the unification of Italy was in no sense a priority. They were more concerned with reforms and constitutionalism.

It was only in the 1830s and 1840s that clear and positive alternatives were put forward for the future of the peninsula.

The alternative programmes for unification

The basic choices which emerged were these: (1) was Italy to be a fully unified and centralized state or was it to be a federation? (2) Was it to be a monarchy or a republic?

Three major alternatives were put forward:

(1) The achievement of a single unified republic through popular insurrection. This democratic idea of unification was put forward by the most influential prophet of the Risorgimento and the leading exponent of early nineteenth-century nationalism, Giuseppe Mazzini, founder of 'Young Italy'.
(2) A confederation of states under the leadership of the Pope. This solution was proposed by Vincenzo Gioberti, who wrote *Of the Moral and Civil Primacy of the Italians* (1843). In this he argued that the Papacy and Roman Catholicism were the glories of Italy and must lead the national revival.
(3) The unification of Italy under the King of Piedmont.

All of these ideas found supporters during the revolutions of 1848, but following the rejection by Pope Pius IX of Gioberti's role for him, the second alternative was very largely discredited. In the event it was the Piedmontese solution which was to succeed.

Piedmont and Cavour

After the 1848 revolutions, Piedmont was alone among the Italian states in preserving her constitution.

Her government remained free of Austrian domination and during the 1850s she underwent important reforms and modernization.

Cavour, with the reputation of being a moderate liberal and supporter of parliamentary institutions, established himself as the leading Piedmontese politician. By 1852 he had become Prime Minister. As a believer in free trade and constitutionalism he achieved much during the 1850s. In this period Piedmontese trade trebled and she became the leading, as well as the most liberal, state in Italy.

Italian independence

Though he had little experience of foreign affairs when he came to power, Cavour hoped for the achievement of Italian independence.

As a skilled practical politician he was very conscious of the reasons for the failure of the 1848 revolutions, when it had optimistically been assumed by many of the revolutionaries that an Italy that was divided, and in many areas backward, could liberate itself.

He recognized that it was only when one or more of the Powers developed a serious interest in helping Italy would it become a possibility – and he looked to France for that support.

The Peace Congress of Paris

Cavour had contributed Piedmontese troops to the allied side against Russia in the Crimean War. This war, in turn, had helped to make possible the unification of Italy by transforming the relations between the Powers.

The Vienna Settlement had been shaken: Russia was no longer committed, as in the earlier period, to upholding the established order in Europe. Austria had been weakened and isolated. It became possible for Napoleon III, who considered support for Italian freedom to be part of the hereditary policy of his dynasty and who had been involved in his youth in the revolt of 1831 in the Romagna, to ally with Piedmont against Austria.

The Plombières Agreement

In 1858 an Italian nationalist revolutionary, Felice Orsini, made an attempt to assassinate Napoleon III as a dramatic gesture in favour of Italian liberation. Orsini regarded Napoleon as a traitor to the Italian cause, with which he had earlier been associated, and the destroyer of the Roman Republic as President of the French Second Republic. Napoleon allowed Orsini to be executed but authorized the publication of Orsini's appeal to him to free Italy.

Six months later Napoleon met Cavour at Plombières and laid down the basis for the future alliance and reorganization of the peninsula. The essential terms of this were: (1) that, with French help, Piedmont would annex Italy above the Apennines (Lombardy and Venetia); (2) Savoy and Nice were to be given to France; (3) the territories of Central Italy, except Rome and the surrounding area, would constitute a separate kingdom; (4) Southern Italy would conserve its unity and frontiers, though the ruling dynasty would be changed.

At this stage it seems clear that neither Cavour nor Napoleon anticipated the unification of *all* Italy, nor even desired it.

The war with Austria

Austria allowed herself to be provoked into a declaration of war by demanding that Piedmont disarm. The conflict lasted from April to July 1859, with the French and Piedmontese capturing most of Lombardy following the battles of Magenta and Solferino. Simultaneously, revolts broke out in the states of Parma, Tuscany, Modena and the Romagna. The rulers of the Duchies left their states, and Piedmontese authorities moved in to run provisional governments. This went beyond the arrangements agreed at Plombières.

Villafranca

In July 1859 Napoleon made a truce with the Austrian Emperor, Francis Joseph, at Villafranca.

Two major reasons for his halting the war after Solferino were his fear of Prussian mobilization against France and his distrust of Cavour's activities in the Duchies, especially Tuscany, which according to Plombières was not to be part of the enlarged Piedmont.

Villafranca arranged that only Lombardy should be ceded to Piedmont. Venetia was not included and the future of central Italy was left uncertain. The agreement of Plombières, therefore, was only partly fulfilled and Cavour, who was excluded from the negotiations, resigned.

The annexation of central Italy

Cavour's supporters and the nationalist group, the 'National Society', worked very actively in central Italy, campaigning for annexation to Piedmont. Increasingly, ideas of reform and Liberalism, characteristic of the period up to 1848, were giving way in risorgimentist circles to aggressive Nationalism and militarism.

In January 1860 Cavour returned to power and negotiated this annexation. Plebiscites were held and pronounced in favour of unification. In March 1860 the Romagna, Tuscany, Modena and Parma became incorporated in the North Italian Kingdom. Savoy and Nice were handed over to France, though the agreement at Plombières had specified that this would be in return for liberating Italy to the Adriatic (i.e. Lombardy *and* Venetia).

Garibaldi and the 'Thousand'

Up to this point the political initiative was in the hands of Cavour.

In the next phase of unification he found himself challenged by Mazzini and the democratic programme of complete unification.

Mazzini had formed the idea of an expedition to the South which would work its way up to Rome and Venetia. In April 1860 Garibaldi's volunteer army of the 'Thousand' left Genoa for Sicily, conquering first the island and then Naples. He appeared determined to march on Rome, where a French garrison had been stationed since 1849.

Both Cavour and Napoleon were extremely opposed to anything like a Mazzinian republican movement gaining control in Italy.

Secondly, if Garibaldi were allowed to march on Rome this would almost certainly involve French intervention to protect the Pope.

Cavour was also aware of the problems, social and economic, that the annexation of the southern territories would bring to the government of a united Italy.

In September 1860 Cavour forestalled Garibaldi by advancing along the Adriatic coast and taking the rest of the Papal State except the area immediately around Rome. Piedmontese troops then linked up with Garibaldi, who surrendered his conquests to the Piedmontese King, Victor Emmanuel. Plebiscites ratified the annexations and the Kingdom of Italy (excluding Rome and Venetia) was proclaimed in March 1861.

Some underlying considerations

Cavour employed the force of Italian patriotism which Mazzini had done so much to encourage. He linked it with the traditional desire of Piedmont to expand in northern Italy. By enlisting Napoleon's help he was able to exploit the balance of power and create a Kingdom of Italy.

The essential condition for this was that Austria, defeated in 1859, was unwilling to resume the war.

Napoleon recognized that the other Powers would not allow France to extend her influence on a large scale in Italy.

In the actual form which the unification took, the preparatory work of the 'National Society' was of considerable significance. Its supporters argued that it was essential for the national movement to rally round Piedmont and her monarchy. They set themselves against Mazzini's democratic and republican ideas.

Their propaganda for Italian unity laid particular emphasis on the economic benefits which would come from it.

In Sicily Garibaldi's achievement was largely due to the revolt of the peasants. When he reached the island it was already out of the control of the government. To start with he allied himself with the peasantry, abolished the hated milling tax and promised land redistribution. Later he suppressed peasant uprisings. In the event it was not surprising that the land-owning class came to the conclusion that the re-establishment of order in the countryside would be better entrusted to the Piedmontese monarchy and its army.

The constitution of the new Italy

The government of the new Kingdom of Italy after 1861 was based on a very restricted franchise (approximately 2% of the population). It was supported by an élite.

The work of the Risorgimento came to be seen not, as Mazzini had hoped, as an achievement of 'the people of Italy, but of an upper- and upper-middle-class minority.

The Piedmontese constitution (Il Statuto) was extended to the rest of Italy together with the legal system and bureaucracy. A rigidly centralized system of administration was adopted. This and the process of integration were costly and imposed a heavy burden of taxation which was much resented.

In the South the relationship of the classes remained semifeudal. The social and economic predicament of the South was to loom very large in the history of the new Italy. The failure of politicians to understand – or, if they did understand, to act effectively towards meeting the problem of the South, helps to account for its continuing backwardness and for the serious civil war which broke out in the 1860s.

The completion of unification and the Papacy

The completion of unification came with (1) the gain of Venetia in 1866 as a result of the Austro-Prussian War, and (2) the acquisition of Rome in 1870 during the Franco-Prussian War, when the French garrison withdrew.

The second involved the destruction of the temporal power of the Pope. In the 'Syllabus of Errors' the Pope had already declared an out-and-out opposition to such liberal and secular ideas as those on which the new Italian state was based.

By the Law of Guarantees the Italian Government unilaterally settled the Pope's affairs. He was to remain in possession of the Vatican and to draw a pension from the Italian State. He ignored the Law, refused to take his allowance and presented himself to the world as the 'prisoner of the Vatican'. This division remained a grave weakness for the Italian State until the Concordat between the Vatican and Fascist Italy in 1929.

4.4 CHRONOLOGY

1851–61		COUNT CAMILLO BENSO DI CAVOUR: entered the government at Piedmont under D'Azeglio as Minister of Agriculture and Commerce.
1852	**November**	CAVOUR PRIME MINISTER OF PIEDMONT. His government was a coalition of the Right-Centre and Left-Centre; the 'Connubio'. Programme of modernization: finances reorganized; tariffs revised; new commercial treaties negotiated; rapidly accelerated railway construction; modernization of the Piedmontese army.
1855–6		Piedmont entered the Crimean War on the Franco-British side. Cavour attended the Congress of Paris and pressed the grievances of Italy.
1856		FOUNDATION OF THE NATIONAL SOCIETY: this argued for the unification of Italy under Piedmont. It had the encouragement of Cavour. Though its policy conflicted with the ideas of Mazzinian republicanism, it also gained backing from numbers of Mazzinians.
1858	**January**	THE ATTEMPTED ASSASSINATION OF NAPOLEON III AND THE EMPRESS EUGENIE BY FELICE ORSINI. Orsini appealed to Napoleon from his death cell to liberate Italy.
	July	THE SECRET MEETING OF NAPOLEON AND CAVOUR AT PLOMBIÈRES; Napoleon agreed to join

Piedmont in a war on Austria provided it could be provoked in a manner to justify it in the eyes of French and European opinion. Austrians were to be expelled from northern Italy and Lombardy and Venetia were to be incorporated by Piedmont. France was to be rewarded with Savoy and Nice. An independent kingdom of central Italy was to be set up.

	December	The formal treaty of alliance signed by France and Piedmont.
1859	**March**	Piedmontese reserves called up.
	April	Mobilization of Austrian army.
		AUSTRIAN ULTIMATUM TO PIEDMONT: directed her to demobilize in three days. This supplied Cavour with the provocation which he and Napoleon needed. The ultimatium was rejected and the war began.
	May	Revolutions in Parma, Modena and Tuscany, encouraged by the National Society.
	June	THE BATTLE OF MAGENTA.
		Insurrections in Bologna, Ferrara, Ravenna and the Papal Legations.
		BATTLE OF SOLFERINO.
	July	PEACE OF VILLAFRANCA: Lombardy was to be granted to France, with the exception of Mantua and Peschiera. Napoleon was to cede this to Piedmont. Venetia was to remain under Austrian control. Italian princes were to be restored. Resignation of Cavour.
	August–September	Representative assemblies in Parma, Modena, Tuscany and the Romagna declared for unification with Piedmont.
	November	Treaty of Zürich finalized the agreement of Villafranca.
1860	**January**	CAVOUR, RETURNED TO POWER, NEGOTIATED THE ANNEXATION BY PIEDMONT OF THE CENTRAL STATES IN RETURN FOR THE CESSION OF NICE AND SAVOY.
	March	PLEBISCITES IN PARMA, MODENA, TUSCANY AND THE ROMAGNA VOTED FOR ANNEXATION TO PIEDMONT.
		Treaty of Turin: Piedmont ceded Savoy and Nice to France after a plebiscite.
	April	Unsuccessful rising in Sicily against the Bourbons.
	May	GIUSEPPE GARIBALDI AND HIS 'THOUSAND' SAILED FROM GENOA TO SICILY.
		Garibaldi defeated the Neapolitans at Catalfimi, captured Palermo and set up a provisional government.
	August	Garibaldi crossed to the mainland.
	September	GARIBALDI TOOK NAPLES, PLANNING TO MARCH ON ROME AND TO PROCEED TO THE CONQUEST OF VENETIA.
		Cavour, fearing French intervention on behalf of the Pope, intervened. The Papal forces defeated at Castelfidardo.
		Piedmontese forces advanced into Neopolitan territory and joined forces with Garibaldi.
	October	Naples and Sicily voted by plebiscite for union with Piedmont.
1861	**February**	The siege of Gaeta.
	March	THE KINGDOM OF ITALY PROCLAIMED BY THE FIRST ITALIAN PARLIAMENT WITH VICTOR EMMANUEL AS KING. The government was based on the Piedmontese constitution of 1848, the 'Statuto'.
	June	Death of Cavour.
1862	**August**	The battle of Aspromonte: Garibaldi and his volunteers were defeated by government troops.
1864	**September**	THE SEPTEMBER CONVENTION: Napoleon finally agreed to evacuate Rome within two years in return for an Italian promise to move the capital of Italy from Turin to Florence. Napoleon regarded this as a renunciation of Rome.
1866	**May**	Alliance of Italy and Prussia encouraged by Napoleon.
	June	ITALY DECLARED WAR ON AUSTRIA. ITALY DEFEATED AT THE SECOND BATTLE OF CUSTOZZA.
	July	VENETIA CEDED TO ITALY.
1867	**November**	THE BATTLE OF MENTANA. Garibaldi defeated by Papal troops supported by the French. The Roman question continued to be an open dispute in relations between France and the new Italy and to prevent an alliance.
1870	**August**	WITHDRAWAL OF FRENCH TROOPS FROM ROME IN THE FRANCO-PRUSSIAN WAR.
	October	Rome was annexed to Italy after a plebiscite and became the capital.
1871	**May**	THE LAW OF GUARANTEES: defined the relations between the Italian government and the Papacy. The Pope was granted full religious liberty and the Vatican was given the rights of an independent territory. He was offered an annual income from the Italian treasury. This law was not accepted by the Pope (Pius IX) who asserted that he was 'the prisoner of the Vatican'. The relations between the Papacy and the new State of Italy were not regularized until the Lateran Concordat of 1929.

4.5 QUESTION PRACTICE

1   Consider the view that Italy was unified by improvisation rather than by calculation.

Understanding the question

First you need to consider how far the great calculator, Cavour, really planned his moves ahead and how far he merely seized at opportunities as they presented themselves. Then you must assess the relative contributions of Cavour and the improvisor *par excellence*, Garibaldi. You may also weigh up other factors, such as the long-term social, economic and geopolitical forces making unification inevitable, and largely independent of political or military intervention.

Knowledge required

The details of the unification process from 1852 to 1870. The careers of Cavour and Garibaldi. The economic and diplomatic background to the unification.

Suggested essay plan

1   *Brief* survey of the course of the unification 1852–70; domestic development of Piedmont – secret diplomacy – war – plebiscites – Sicilian revolution and Garibaldi's Redshirts – intervention of Piedmont and its annexation – incorporation of Venice and Rome.

2   Evidence of forethought and planning on Cavour's part; his realization that unification was impossible without European diplomatic and even military assistance; sees how to develop the state on West European, liberal, free-trade lines.

3   It was once thought that Cavour had calculatingly engaged Piedmont in the Crimean War in order to put his Italian ambitions at the forefront of European attention; now we know that Victor Emmanuel pushed a reluctant Cavour into the war in order to divert attention in Piedmont away from domestic problems. So Cavour's actions at the Congress of Paris were more opportunist than calculated.

4   Diplomatic calculation was shown at Plombières and in engineering the Austrian attack on Piedmont. Note, however, that the policy was nearly disastrous for Italian unity; the outcome of the war was far from satisfactory. Only the decisions of Parma, Modena and Tuscany to join Piedmont (thanks to Ricasoli and Farini, not Cavour) made Cavour's diplomacy *seem* successful.

5   Garibaldi, his background and attitudes; has more fixity of purpose and is more single-minded than Cavour: 'Unity in one fell swoop – to get Rome and all will follow'. But Garibaldi's methods were those of an adventurer; his Redshirts created a situation for Cavour to exploit, and enabled Cavour to incorporate Naples within Italy; Garibaldi's march on Rome (1860); Cavour improvised on Garibaldi's improvisations. (Note also that Cavour was largely motivated by hatred of Garibaldi, and by a desire to thwart him, as much as by a desire to affix Naples to Italy; he played on the spectre of republicanism to prevent Garibaldi's solution.)

6   Did Cavour plan unification? Even if he did, it would never have begun without Napoleon III and without Garibaldi it would never have been completed.

7   Besides, there are long-term economic and geopolitical factors to be considered. The need for economic consolidation and administrative centralization pushed both Germany and Italy unwittingly towards unification; the decline of Austria also created a political vacuum in both areas. Calculation and improvisation therefore affected the timing of unification more than the fact of it.

2   Compare the nationalism of Mazzini with that of Cavour.

Understanding the question

This is a straightforward question whose meaning is clear. It is important to avoid treating Mazzini and Cavour in isolation, or lapsing into a narrative about their respective roles in the story of unification. It will help here if you measure both men against some yardstick of what nineteenth-century Nationalism was about.

Knowledge required

The careers and philosophies of Mazzini and Cavour. A working knowledge of Nationalism as a force in Italy and Europe generally.

Suggested essay plan

1   Define what is meant by Nationalism in the nineteenth century; explain the impact of the French Revolution and Napoleon and relate Nationalism to the romantic movement in literature.

2   Discuss the connection of Nationalism with Liberalism in both its political (democratic) and economic (free trade) aspects; examine the widespread identification of Liberalism and Nationalism as twin forces of revolution against legitimist and authoritarian governments. Often, as in pre–1848 Germany, Diets were the only forums for the expression of nationalist feelings; this made Nationalism seem to be liberal.

3   Consider the *il*liberal and anti-individualist aspects of nineteenth-century nationalism; the idea of the 'General Will', the state or 'Volk'. Nationalism also often encouraged centralisation, bureaucratic conformity and the suppression of individual liberties.

4   Relate Mazzini and Cavour to the above discussion: the former a democratic nationalist, the latter making use of democracy (plebiscites) but more of a businessman's and bureaucrat's nationalist; the former preferring *political* to economic liberalism; the latter encouraging economic liberalism in his bid to build Piedmont into an efficient, western industrial power.

5   Amplify this view of Mazzini with details of his life and activities: his contempt for the Carbonari; his linguistic and republican forms of Nationalism; his belief in individual liberty; he formed *Young Italy* and then *Young Europe*, with their doctrines of 'sacred nationality' and equality.

6   Do likewise for Cavour, his aims and policies, both domestic and foreign: his concern for economic efficiency and expansion, and his admiration of the West; his success in 'achieving' unification and his influence on the form of the new Italian state.

7   Explain why Cavour's type of nationalism 'won'; Mazzini's refusal to compromise; his disgust with the monarchical middle-class state established in Italy.

3   In what ways did the internal unification of Italy remain incomplete in 1870?

Understanding the question

The word 'internal' may appear perplexing, for Italy was subjected to a high degree of administrative centralization after unification. It is best to interpret the word in the widest sense, emphasizing the distinction between the *real* and *legal* Italies (as explained below). Put another way, the question asks: what divisions still existed (and were felt to exist) in Italy in the year in which the territorial unification was completed?

Knowledge required

The background: the Risorgimento and the process of

unification. The constitution and nature of the new Kingdom of Italy. The internal opposition to the new state. The political, economic, social and religious divisions in Italy in 1870.

Suggested essay plan

1   Quote Massimo D'Azeglio, in conversation with King Victor Emmanuel: 'Sir we have made Italy, now we must make Italians'. A very large proportion of the population had little or no comprehension of the reality of the new Italian state called *Italy*. Loyalty was still overwhelmingly to the locality and region and there were wide regional cultural and economic differences. Even the élite who had been most influential in the Risorgimento had often been less inspired by Italian national consciousness than by a hatred of petty oppression and of Austrian domination.

2   The process of unification had been achieved by a series of annexations to Piedmont of the various Italian states. Cavour had pushed ahead in the end to forestall Garibaldi. The new state developed from the start as an expansion of the old Piedmont rather than as the totally new political creation Mazzini and Garibaldi had hoped for. With the unification, the Piedmontese constitution and administrative system were extended over the rest of the peninsula.

3   Cavour had given Neapolitan and Sicilian leaders to believe that there would be a good degree of regional autonomy. In fact rigid centralization on the French model was instituted.

4   *However,* the new Italy, though legally and administratively one, did *not* embrace either the political support of the masses or their involvement. The solution of unification by Piedmont was imposed from above. Italy was not built by 'the people' on new foundations.

5   The parliament was elected on an extremely narrow franchise, the property qualification enfranchising only 600,000 in a population of 27 millions. The people of the Mezzogiorno (the south) were particularly under-represented; in some parts the vote was only enjoyed by a few notabilities. There was, from the start, a great divide between the rulers and the ruled.

6   Economic divisions: there was a very marked gap between the commercially and industrially developing north of Italy and the backward south. In one crucial respect the governments of the new Italy *widened* this division. They refused to discriminate in favour of the south and instead, with very damaging effects, exposed the struggling commerce of the south to free-trade competition from the rest of Italy.

7   The peasants still constituted the majority of the population but, with the exception of Sicily, they had played only a little role in the unification. They remained alienated in the new state. The sale of church lands which followed the unification benefited the 'latifundists' (large-scale landowners), who also frequently in these years misappropriated common lands. The land question generated bitter class hatred; there was widespread brigandage, particularly in the south, and revolts against heavy government taxation – *e.g.* the 'macinato' (a tax payable on the grinding of corn). There was mass resentment of the new state, whose principal personifications seemed to be the Army recruiter and the taxman.

8   The question of Church and State: with the loss of temporal power, Pope Pius IX declared out-and-out hostility towards the new state. The 'Roman Question' bedevilled Church–state relations until the Lateran pact under Mussolini in 1929. The Church encouraged the enemies of the new state (*e.g.* the Bourbons) and called on Catholics to boycott political life – *i.e.* neither to vote nor offer themselves as candidates. This was accompanied by papal denunciations of the liberal spirit in modern society, as manifested by the state.

9   The newly-united Italy, therefore, contained very deep divisions – some of which were aggravated by the unification process under Piedmont. The refusal of successive governments to decentralise emphasised both this and their particular fear of social instability and revolt in the south.

4   Document question (Italy 1815–49): study extracts I, II and III below and then answer questions (a) to (f) which follow:

Extract I

'Austria ... cuts Italy in half and is its actual mistress. ... By the re-establishment of the entire temporal domain of the Pope, two and a half millions of Italians have been plunged afresh into a state of absolute nullity, and the King of Naples, relegated to the end of the peninsula, has no longer any means of contributing to the defence of Italy, while on the other hand Austria threatens the King of Piedmont on his flank ...

'Examples ranging over several centuries prove that there will always be blood to shed until Italy is left alone with all foreigners alike excluded. Neither France nor Austria will ever consent to yield to the other – as long as this rivalry subsists Europe can hope for no real peace. The only means of extinguishing it would seem to be the establishment in the north of Italy of a state strong enough to defend the Alps and bar the gates of Italy to all foreigners.

'... Each part of northern Italy is at exactly the same stage of civilisation; there is a general consensus of opinion and a community of interests; in fact in many ways the inhabitants resemble one another far more than they do those of Tuscany, Rome or Naples. Northern Italy would have a population of seven or eight millions. A state of that size could not give rise to jealousy. Situated between two great Powers ... it would hardly be able to maintain its independence without the help of Russia.' *(Memorandum by the Piedmontese ambassador to the Tsar Alexander, March 1818)*

Extract II

'(These experiences) all confirmed me in the conviction that *carbonarism* was in fact dead ... and that it would be better ... to seek to found a new edifice upon a new basis.... I conceived the plan of the association of *Young Italy....* I was led to prefix unity and the Republic, as the aim of the association.... I saw regenerate Italy becoming at one bound the missionary of a religion of progress and fraternity, far grander and vaster than that she gave to humanity in the past.... Why should not a new Rome, the Rome of the Italian people, arise?'

(From *The Life and Writings of Mazzini*, London, 1864)

Extract III

'Italy contains within herself, above all through religion, all the conditions required for her ... *risorgimento* ... she has no need of revolutions within and still less of foreign invasions or foreign exemplars....

'The real principle of Italian unity ... the Papacy, is supremely ours.... It is concrete, living, real – not an abstraction or a chimera, but an institution, an oracle and a person....

'The benefits Italy would gain from a political confederation under the moderating authority of the Pontiff are beyond enumeration....

'(It) would increase the strength of various princes without damaging their independence ... place Italy again in the first rank of the Powers ... provide opportunities to resume expeditions and the establishment of colonies in various parts of the globe ... eliminate, or at least reduce, the difference in weights, measures, currencies, customs duties and systems of ceremonial and civil administration which ... divide the various provinces.

(V. Gioberti, *On the Moral and Civil Primacy of the Italians*, 1843)

Maximum
marks

(a) Explain briefly what the writer in Extract I meant by *each* of the following statements: 'Austria cuts Italy in half and is its actual mistress'; 'two and a half millions of Italians have been plunged afresh into a state of absolute nullity'; 'Austria threatens the King of Piedmont on his flank'. (6)

(b) Show what you understand by *each* of the following terms: 'carbonarism'; 'Young Italy'; 'Risorgimento'. (3)

(c) Which of the arguments put forward in Extract I for a unified kingdom of northern Italy were likely to appeal to the Tsar? (3)

(d) How does the plan proposed in Extract II differ from the plan envisaged in Extract I in (i) content, and (ii) in its tone? (4)

(e) Quote sections of Extract III which suggest that Gioberti's arguments were designed to appeal to *each* of the following: nationalists; the middle classes; the monarchists. (6)

(f) Explain how the events of 1848–49 in Italy revealed weakness in *each* of the plans outlined in the three extracts. (9)

4.6 READING LIST

Standard textbook reading

J. A. S. Grenville, *Europe Reshaped 1848–78* (Fontana, 1976), chapter 12.

David Thomson, *Europe since Napoleon* (Penguin, 1977), chapter 14.

Anthony Wood, *Europe 1815–1945* (Longmans, 1975), chapter 12.

Further suggested reading

J. P. T. Bury (ed), *The New Cambridge Modern History 1830–70*, volume X (Cambridge University Press, 1964), chapter 21.

Edgar Holt, *The Risorgimento: the Making of Italy 1815–70* (Macmillan, 1970).

J. McManners, *Lectures on European History 1789–1914* (Blackwell, 1974).

D. Mack Smith, *The Making of Italy 1796–1870* (Harper and Row, 1968).

G. Procacci, *A History of the Italian People* (Penguin, 1973).

Jasper Ridley, *Garibaldi* (Constable, 1974).

L. C. B. Seaman, *From Vienna to Versailles* (Methuen, 1972).

C. Seton-Watson, *Italy from Liberalism to Fascism* (Methuen, 1967), Prologue.

5 Napoleon III and the Second Empire

5.1 INTRODUCTION

Louis Napoleon, nephew of the great Emperor, who had previously made two failed attempts against the July Monarchy, was elected president of the French Republic in 1848. He benefited from the growing appeal of the 'Napoleonic legend', and appeared to offer order and stability after a period of frightening upheaval. He worked hard to establish his popularity with the mass of the French people, not least the Catholic middle classes.

Re-election of the President was prohibited by the constitution and, on 2 December, 1851, he mounted a coup d'état, extending his presidential authority, and then, a year later, took the title of Emperor. He appealed directly to the people over the heads of the politicians by means of the plebiscite and received their overwhelming support.

His ambition was to reconcile order with progress and he greatly strengthened the executive presidential power at the expense of the legislature. He imposed authoritarian controls, such as censorship, but after 1859 there was progressive liberalisation of the regime, culminating in the ministry of Emile Ollivier – though this did not mean full parliamentary rule on the British model. His rule spanned a period of considerable economic expansion and rising prosperity.

In his foreign policy Napoleon set out to undo the Vienna Settlement. He appealed to French nationalism both by seeking additional territory and by espousing the idea of a Europe reorganised on national principles. Success against Russia in the Crimean War was followed by his intervention in Italy in 1859 and the expulsion of Austria from Lombardy. His relations with the new Italian state, however, were complicated by the Roman Question. The regime suffered humiliation over Mexico.

By 1865 Napoleon was in poor health – he gravely miscalculated over the Austro-Prussian War and was brought to defeat, following the Hohenzollern candidature to the Spanish throne, in the Franco-Prussian War of 1870.

5.2 STUDY OBJECTIVES

1 The rise of Louis Napoleon; his personality, sense of historical mission; his early attempted coups against the Orleanist regime of Louis-Philippe. His writings: *Napoleonic Ideas* and *The Extinction of Pauperism*.

2 The reasons for his election as President of the French Republic in December 1848 on the platform of stability and law and order.

3 The coup of 2 December and the establishment of the Second Empire.

4 The constitution of the Empire during the 1850s. Napoleon's ability to work through administrators and to keep the new liberal ministers (Morny, Persigny) relatively powerless. The organization of bureaucratic rule through prefects, maires, etc.; the limitation of the Assembly's powers; the shrewd exploitation of plebiscites and universal suffrage; the control of elections through the use of official candidates; the control of public opinion by press controls, restriction on meetings and the superintendence of education.

5 Expansionist economic policy: deficit financing, public works (canals and railways), free trade, large-scale mobilisation of funds (through the Banque de France, Crédits Mobilier and Foncier) of funds amassed under the Orleanists; a state-directed industrial revolution, fulfilling Napoleon's dreams of a technological future (influence of Saint-Simon).

6 Paternalist social policy: benevolent attempts at the re-distribution of wealth and the protection of the working classes; permission for workers to strike and form cooperatives.

7 The court and social life; the Empress and the 'gilded beauties of the Second Empire'; Haussmann and the reconstruction of Paris.

8 Foreign policy; Napoleon's overriding desire to disrupt the humiliating Vienna Settlement of 1815; his role in the unification of Italy; his domestic and foreign policy motives in his dealings with Cavour; the annexation of Savoy and Nice; the Mexican adventure and the problems posed for France by the rise of Prussia in the 1860s.

9 Analysis of the reasons for the support of the Second Empire by: peasants, army, clergy, nouveaux riche, haute bourgeoisie (industrialists, merchants, stockholders).

10 The opposition to the regime, as reflected in the 1863 and 1869 elections: the hard core of opposition from the Republicans of Paris, Lyons, Marseilles, led by Thiers and Simon; also the Bourbons and Orleanist legitimist opposition and the contempt of figures like Victor Hugo; the disappointment after Napoleon's withdrawal from the Piedmontese/Austrian War; army resentment at foreign policy reverses; the dislike by protectionists of Cobden/Chevalier Treaty.

11 The liberal reforms of the constitution after 1867 in response to growing unpopularity, leading to the proclamation of the Liberal Empire under Ollivier in 1870.

12 Foreign policy disasters: the Hohenzollern candidature and Bismarck's success in involving Napoleon in war; the course of the Franco-Prussian War; the overthrow of the Empire; the siege of Paris and the Commune; surviving remnants of Bonapartism as an occasional force in Third Republic politics.

5.3 COMMENTARY

Prince Louis Napoleon

Prince Louis Napoleon, nephew of the great emperor, was heir to the Napoleonic tradition. As an exile under the law of 1816 which banned the House of Bonaparte from France, he had made two unsuccessful attempts to seize power as pretender at Strasbourg in 1836 and at Boulogne in 1840. Sentenced to life imprisonment in 1840, he had escaped from the fortress of Ham and taken refuge in England where he remained until the outbreak of the 1848 revolution. He had written two books which were significant for his subsequent political career: *Napoleonic Ideas* and *The Extinction of Pauperism*. 'The Napoleonic idea,' he wrote, 'is not an idea of war, but a social, industrial, commercial idea – an idea of humanity'. *The Extinction of Pauperism* served to associate Bonapartism in the public mind with the need for social reform.

The President of the Second Republic

In 1848 Louis Napoleon could rely on the support of the Napoleonic legend, which had established the emperor as the champion of progress and the defender of religion and society. His campaign for the Presidency laid the basis for his subsequent power and was cleverly conceived. He posed as a man above class or party who would put an end to revolutionary agitation, maintain order, restore prosperity and revive French glory. In *The Extinction of Pauperism* he had argued that in future it would only be possible to govern with the support of the masses.

He did not openly recommend a restoration of the empire, but declared his approval of a republic based on universal suffrage. In the majority of the Départements he received 80% of the votes.

Various factors explain his success:

1 the fame of his family name among a largely illiterate and politically unaware electorate;

2 the general desire for stability and order after the upheavals of 1848;

3 the unpopularity of the republic;

4 the mutual dislike of royalists and republicans;

5 his well-advertised interest in the 'social question'.

All these helped to secure his victory over General Cavaignac, the 'butcher' of the 'June Days', by a massive majority. He was only elected for a term of four years, however. He soon made it clear that he had no intention of being simply the figurehead President which many of the members of the Assembly wanted him to be.

The coup d'état

On December 2nd 1851, the anniversary of the coronation of Napoleon I and the battle of Austerlitz, he took the law into his own hands. With a carefully planned coup he seized power, changed the constitution and gave the Presidency overwhelming authority and a ten-year term of office.

He presented himself as the people's champion by reintroducing the universal suffrage which the assembly had refused to restore the previous summer. He had taken the solemn oath to uphold the Republic. Now he appealed to 'the solemn judgement of the only sovereign whom I recognize in France – the people'.

This was confirmed by a plebiscite in which 91% of those voting expressed approval of his actions. By an administrative order special tribunals were instituted in January 1852, the 'mixed commissions' to take action against opponents of the new order. About 10,000 were transported to Algeria.

Napoleon's political appeal through the device of the plebiscite or referendum was directly to the people over the heads of the politicians. A country which had never really reconciled itself to the Second Republic, welcomed the rule of a man whose name signified something to all men and who promised the restoration of order, the return of national confidence and the recovery of business.

The restoration of the Empire

In November 1852 another plebiscite approved the end of the Second Republic and the proclamation of the Second Empire. The promise of revived national glory, empire without war, and of stability outweighed the attraction of those political liberties which were suppressed in the new order.

Significantly, he had by this time won the strong support of the Roman Catholic vote, both through his defeat of the Roman Republic of Mazzini in 1849 and through the Falloux Law which had consolidated Church power over education. He had succeeded where Louis-Philippe had failed in winning over the clericals from their allegiance to the House of Bourbon.

His aim was to reconcile order and progress, by offering a rule above the factions. In the first period until 1859 he was to maintain an authoritarian rule, establish close relations between the government and the Church and give vigorous aid to business enterprise.

The constitution

The constitution of the Second Empire was based on that of 1800. The powers of the lower house, or Corps Législatif, were greatly reduced. The Senate acted as the guardian of the constitution and its members were, in practice, Napoleon's nominees.

The lower house could only meet for three months in the year and its membership numbered only a third of the old assembly. Their debates were not allowed to be reported in the press and ministers were not permitted to be questioned by deputies. It could discuss but not amend bills initiated by the emperor.

Everything was done to strengthen the *executive* power, the Emperor and his ministers. Only the Council of State, dominated by Napoleon, was allowed to introduce laws. Napoleon appointed to all offices, was Commander-in-Chief of the forces and had the right of declaring war and making treaties. He described his political system as a pyramid which former rulers had tried to balance upon its apex and which he now set on its base – that is the consent of the masses.

Centralized administration and control

The system of the Second Empire relied on the powers of centralized administration. Under the guidance of the Minister of the Interior, the prefects in the Départements, (or administrative provinces of France), exercised greater powers than ever before, though many of them were inherited from the July Monarchy.

Strict censorship of the press throughout France was effectively established by the requirement that no newspaper in Paris or in the provinces could be published without government authority. Any newspaper could be suspended after three warnings had been given to it by the prefects or Minister of the Interior.

At the same time, political clubs were suppressed. There was interference in education. The National Guard was disbanded except in Paris. Official candidates were nominated for elections and not a single republican deputy was elected until 1857.

Economic developments

The Second Empire was a period of considerable financial, commercial and industrial development. In 1852 Napoleon gave a charter to a joint-stock investment bank, the Crédit Mobilier, which channelled capital into numerous enterprises, particularly railways and mining.

There was a rapid development of land and sea transport associated with a boom in the metallurgical and mining industries. Foreign investment rose sixfold. Though it did not promote anything like a revolution in agriculture, the period saw the steady relative decline of the rural population in favour of the towns. The Crédit Foncier engaged in large investments in urban real estate.

To the great majority of Frenchmen Napoleon appears to have offered what they wished for, a guarantee of stability and order and a marked increase in prosperity. He was aware of the political importance of providing for the material welfare of the masses. He sought through public works another means to help the peasant and the urban worker; he was anxious to create conditions of full employment.

One of the most marked features was the rapid development of communications. In 1848 only 1800 kilometres of railway line existed. By the end of the Second Empire a network of 17,500 kilometres had been laid. A larger, more unified market stimulated production. This was also a period of major building construction – most notably the transformation of central Paris by Baron Haussmann.

Foreign policy

Before the Empire had been in existence for two years it was involved in a major war with Russia. Napoleon's involvement in the Crimean War resulted partly from his desire to rally Catholic support at home through championing Catholic interests in the Near East and also from his keenness for an alliance with England. He was particularly anxious to avoid what he regarded as his uncle's grave error of enmity with Great Britain.

By the Quadruple Alliance of 1815 the Allies had agreed to look upon the return to France of Napoleon or any of his relatives as tantamount to a French declaration of war on the rest of Europe. He went out of his way to disclaim any aggressive purpose with his statement 'the Empire means peace'. He was nevertheless clearly intent on taking any opportunity to challenge the status quo of the Vienna Settlement and was sympathetic to nationalist ideas. Just as he wanted to restore harmony within France, he was anxious to assert her diplomatic pre-eminence

in Europe. The Peace Congress of Paris in 1856, which concluded the Crimean War, seemed to confirm France's revived position on the Continent.

France and Italy

As a young exile in Rome in 1831 Napoleon had supported the Italian revolution against the Pope. After Orsini's attempt to assassinate him, he reached a secret agreement with Cavour, Prime Minister of Piedmont, at Plombières; this led to the war against Austria in 1859 (see Unit 4, *The Risorgimento and Unification of Italy*).

The Napoleonic tradition pointed to Italy as a French sphere of interest. By destroying Austrian predominance in Italy, Napoleon would be able to reverse the Vienna Settlement while at the same time being seen to 'do something for Italy' and give support to the principle of nationality.

He did not wish to create a wholly united Italy but sought – while helping to reorganize it – to keep it still divided and looking to France for support. He did not want to stir up popular nationalism and was deeply opposed to the republicanism of Mazzini.

Napoleon III's Italian policy also had a domestic political significance. Throughout his reign he attempted to reconcile two conflicting traditions of French history – that of the Catholic monarchy of Louis XIV and that of the revolution of 1789. In the Italian question this led to an impossible paradox. On the one hand he was spurring on the process of unification by his support for Piedmont. On the other, he was giving the Pope military aid against the emerging Italian kingdom.

The 'Liberal Empire'

In November 1860 Napoleon announced that he would 'give the great bodies of the state a more direct part in the formation of the general policy of our government'. The second half of his reign saw striking constitutional reforms at home which have led to the period from 1860 being described as the 'Liberal Empire'. Already in 1859 he had issued a general amnesty to his political opponents. This was followed by more moderate laws and an increase in the powers of the lower chamber for instance the right to scrutinize the budget. He also attempted to woo the workers, allowing them limited rights of association and the right to strike. In 1868 they were allowed to form unions.

In 1860 the French government signed the Cobden-Chevalier treaty encouraging freedom of commerce with Great Britain; and in subsequent years concluded agreements along similar lines with Belgium, the Netherlands, the Zollverein, Sweden, Switzerland, Italy and Spain.

Emile Ollivier

The elections of 1857 had brought into the Corps Législatif five deputies, including Emile Ollivier and Jules Favre, who were to become a nucleus of republican opposition. The elections of 1863 and 1869 saw successive increases in the numbers of republicans returned to the lower house, while Thiers, who returned to active political life in 1863, undertook to organize a conservative opposition, known as the 'Third Party'.

The election of 1869 showed strong support for liberal constitutional reform. Those who were out-and-out opponents of the Empire, the 'irreconcilables', remained a small minority. But a large number of deputies were returned who wished to combine support for the Empire with constitutional reform. They wanted a ministry which was dependent on the consent of the legislature.

In January 1870 Emile Ollivier formed a new ministry. The emperor now appointed a cabinet which enjoyed the support of the chamber. The 'Liberal Empire' fell far short, however, of the idea that the executive should be solely responsible to the majority in the chamber; it did not mean a full parliamentary regime on the British model.

The revised constitution was essentially a compromise between autocratic rule and parliamentary sovereignty, and Napoleon retained considerable powers. He could, for instance, propose a revision of the constitution, which would require a plebiscite but not the sanction of parliament. He reserved sufficient power to be able to restore his personal rule.

This new constitution was overwhelmingly approved by the French people in May 1870. Napoleon's electoral support was only 10% less than that secured in 1851.

The collapse of the Second Empire

While Napoleon greatly extended French influence overseas, for instance in West Africa, China and Indochina, the apparent foreign policy successes of the first 10 years of his rule gave way in the second decade to grave errors. His record was badly damaged by the following: (1) the Prussian success in consolidating their power in North Germany after Sadowa; (2) his failure to secure 'compensation' (*e.g.* Luxemburg) after the Austro-Prussian War; (3) the collapse of the Mexican adventure; and (4) the failure of his diplomacy to strengthen the French position before the Franco-Prussian War.

The Mexican adventure was a particularly spectacular disaster. By giving help to conservatives who were eager to overthrow the Mexican regime France would be seen as serving the Catholic Church by saving it from the menace of republican anti-clerical policies. France could also present herself as the guardian of the Catholic and Latin peoples of the New World against the Anglo-Saxon and Protestant influence of the US.

At all events, what started out as an attempt to build a liberal catholic empire had ended by 1867 with the withdrawal of French troops and the execution of Maximilian by the Mexican revolutionaries.

After 1867 there was no real possibility of friendly Franco–Prussian cooperation, but Napoleon did not wholeheartedly pursue any alternative policy. The continued presence of French troops in Rome protecting the Pope prevented the creation of an alliance with Italy.

The Franco–Prussian War was the final act of the Second Empire (see Unit 6. *The Unification of Germany* for its origins). The war started on 1 August 1870, and on 3 September news reached Paris of the surrender at Sedan and the capture of the emperor. On receipt of this news, the republican opposition in the Corps Législatif proclaimed the overthrow of the Empire and the establishment of a Republican Government of National Defence.

5.4 CHRONOLOGY

Prince Charles Louis Napoleon Bonaparte (1808–73): nephew of Napoleon I; made two failed attempts as Bonapartist pretender against the July Monarchy of Louis-Philippe at Strasbourg (1836) and Boulogne (1840).

1848	**December**	ELECTED PRESIDENT OF THE SECOND REPUBLIC with the support of over 5 million votes. Swore 'to remain faithful to the democratic Republic and to defend the Constitution.'
1849	**April–June**	French intervention against the Roman Republic.
1850	**March**	THE FALLOUX LAW, considerably extended the influence of the Roman Catholic Church in French education.
1851	**2 December**	The COUP D'ETAT: dissolution of the Assembly: restoration of universal suffrage and the convening of a plebiscite for revision of the constitution. Staged on the anniversary of Napoleon I's coronation and the battle of Austerlitz. THE PLEBISCITE: the President secured 7.5 million against 640,000 votes.
1852	**January**	THE NEW CONSTITUTION: the Chief of State declared 'responsible to the nation' but given 'free and unfettered authority'. A Council of State, Senate and Legislative Assembly set up.
	February	Repressive measures introduced; control of the press through a system of warnings to editors. Establishment of joint-stock banks, issuing long-term credit: *Crédit Foncier* and *Crédit Mobilier*. Programme of public works. (From 1854 the rebuilding of Paris by Baron Haussman.)
	2 December	THE RE-ESTABLISHMENT OF THE EMPIRE: Napoleon took the title of Napoleon III in accordance with the imperial tradition which recognized the Duke of Reichstadt as Napoleon II although he had never been crowned. The powers of the Emperor were extended by a 'Senatus-consultum': he was given authority to conclude treaties of commerce; the budget of every ministry was voted by the Legislative Assembly, but the subdivision of sums granted was to be settled by imperial decree.
1853	**January**	Napoleon married EUGÉNIE DE MONTIJO.
1854	**March**	French declaration of war on Russia (see Unit 3, *The Eastern Question to 1856*).
1855	**May–November**	The Paris International Exhibition
1856	**February–April**	THE PEACE CONGRESS AT PARIS: end of the Crimean War.
1858	**January**	Attempt of Felice Orsini to assassinate Napoleon and the Empress (see Unit 4, *The Risorgimento and the Unification of Italy*).
	July	COMPACT OF PLOMBIÈRES.
1859	**May**	GENERAL AMNESTY: those who had been exiled in 1851 allowed to return to France. War of France and Piedmont against Austria.
	July	PEACE OF VILLAFRANCA.
1860	**January**	THE COBDEN–CHEVALIER TREATY WITH GREAT BRITAIN: France revoked the general prohibition on the import of British goods and reduced the duties upon imported coal, iron, machinery and raw textiles, while Britain reduced the duties upon French wines and spirits.

Chronology cont.

	March	French annexation of Savoy and Nice.
	November	Powers of the legislature extended: the decrees empowered the Senate and Legislative Assembly to move and discuss freely a reply to the address from the throne; parliamentary debates to be fully reported.
1861	**November**	The financial powers of the legislature extended. Napoleon renounced the right to borrow money while the legislature was not in session and agreed that the budget should be voted by sections.
1861–7		THE MEXICAN EXPEDITION: owing to the refusal of the revolutionary Juarez government to meet its debts, France, Britain and Spain decided to force fulfilment of these obligations. The British and Spanish withdrew when they recognized Napoleon's plans to establish a Catholic Latin Empire in Mexico while the US was involved in the Civil War.
		French troops captured Mexico City (June 1863) and proclaimed the Habsburg Archduke Maximilian Emperor. By 1866 the US was demanding the withdrawal of the French, and Napoleon – because of his European involvements – was forced to desert Maximilian, who was captured and executed by the Mexicans (June 1867).
1864		THE SEPTEMBER CONVENTION: Reversing his attitude on the Roman question, Napoleon agreed to withdraw his troops from Rome within two years in return for a promise from the Italian government not to attack Papal territory. This move outraged French Catholics. The Pope issued the encyclical QUANTA CURA AND THE SYLLABUS OF ERRORS.
1866	**March**	Emergence of the THIRD PARTY: this wished to support the Empire but favoured the development of political liberties and the establishment of a Ministry responsible to the legislature, which was to control the general policy of the government.
	July	Prussian defeat of Austrians at SADOWA (see Unit 6, *The Unification of Germany*). The failure of Napoleon to secure 'compensation': was widely regarded by French public opinion as a national humiliation.
1867	**January**	The right of interpellation granted (the right to question the Government on policy and its actions).
	March	The Senate was given the right to examine projected laws in detail.
1868	**May**	LIBERAL PRESS LAWS: the abolition of the power of government to warn, suspend or suppress newspapers.
	June	Limited right of public meeting: public meetings allowed to be held subject to police supervision.
1869	**May**	Parliamentary elections resulted in an Assembly including 30 republicans.
	June	The Third Party demanded the creation of a responsible Ministry.
	July	THE NEW LIBERAL REGIME: by decree of the Senate the Legislative Assembly was given the right to propose laws, criticize and vote the budget and to choose its own officers. The Senate became a deliberating body with public sessions and had the right to discuss laws voted by the Assembly and send them back for consideration.
	December	EMILE OLLIVIER was entrusted with the formation of a Cabinet representative of the majority of the Legislative Assembly.
1870	**April**	The Senate was made an upper house, sharing legislative power with the Assembly; constituent authority was taken from the Senate and given to the people.
	May	PLEBISCITE ON THE LIBERAL REFORMS: by a majority of 7.3 million to 1.5 million, gave approval of the liberal reforms since 1860 and ratification of the 'Senatus-consultum' of April 1870. A sweeping victory for Napoleon.
	July	French declaration of war on Prussia (see Unit 6, *The Unification of Germany*).
	September	Napoleon and the Army capitulated at SEDAN.
		FRENCH THIRD REPUBLIC PROCLAIMED

5.5 QUESTION PRACTICE

1   'A showman and a sham dictator.' Examine this view of Napoleon III.

Understanding the question

The proposition is that Napoleon's policies were based mainly on *fantasies* (such as his longing to emulate his great uncle) and *images* (such as his lavish court life and the transformation of Paris by Haussmann). You are entitled to disagree, and to argue, if you wish, that he was a hard-headed and ruthless realist intent on consolidating his own power. A balanced answer would survey the evidence for each viewpoint.

Knowledge required

Napoleon's personality. His domestic and foreign policies. The strength of the revolutionary and counter-revolutionary traditions. The power of slogans and images – which made the French so susceptible to a 'politician of appearances' like Napoleon.

Suggested essay plan

1   Napoleon's early life and personality; Bonapartist fantasies; plots and coups against the Orleanist regime.

2   Summary of his foreign policy, emphasizing his fondness for empty gestures and secret diplomacy; encouraged Cavour but was not prepared to carry through his support for Italian unity;

the Mexican adventure; Bismarck revealed Napoleon's lack of realism and caution in the secret diplomacy of the later 1860s.

3   Summary of his domestic policies, showing his emphasis on the regime's image: Eugénie and the sumptuous court life; the transformation of Paris.

4   On the other hand, Napoleon showed keen awareness of shifts of public opinion, and a pragmatic sense of what he could get away with; he shrewdly exploited plebiscites and for a time was successful in controlling elections; his diplomatic gestures seemed empty, but were made with a careful eye to conciliating sections of opinion at home. He could be ruthless in persecuting subversives and shrewd in handling ministers.

5   Which view do you think best fits the evidence? Perhaps his decision to abandon the Piedmont/Austria War after Solferino illustrates the problem. Do you think he did it because, being humane, he could not stand the sight of blood on the battlefield? Or did he pull out because Prussia was mobilizing, and because Piedmont's success could threaten the Papacy and alienate Catholic opinion at home?

6   Many people called him a mountebank and adventurer; Bismarck said he was 'a sphinx without a riddle, a great un-fathomed incapacity'. Note how much the French had responded since 1789 to gestures and images and how Louis-Philippe had been lacking in successful 'showmanship'. Note, too, that Napoleon's rule lasted longer than average in nineteenth-century France and only ended because of external disasters.

2   Explain why the Second Republic was so soon replaced by a Second Empire.

Understanding the question

In order to explain *why* the Republic was short-lived, you need to consider both Louis Napoleon's personal political ambitions and the instability of the constitution established in December 1848. You may conclude that the word 'replaced' is slightly misleading, and that the Empire was a logical and likely outcome of the previous constitutional compromise.

Knowledge required

The causes, occasion and course of the February revolution. Louis Napoleon's life, ideas aims; the circumstances of his election as President. The various political parties and ideas in France and why, at the time, his campaign divided Frenchmen least. The domestic policies of the Republic (especially on education). Foreign policy (with particular reference to Rome). Opposition to Louis Napoleon's Presidency. The circumstances of his coup d'état and his confirmation as Emperor.

Suggested essay plan

1   Explain the origins of the Second Republic, starting with the causes of the February Revolution; explain divisions in the new provisional government between middle-class republicans (Lamartine) and full-blooded socialists (Louis Blanc); show how the National Assembly's attack on Blanc's national workshops scheme led directly to bloody insurrection and its suppression during 'the June Days'. Note how this created great fear among nearly all Frenchmen except the proletarians of Paris, Lyons and Marseilles.

2   The National Assembly decided to create a Republic with a president to be elected (for four years without the possibility of re-election) by universal suffrage. Analyse the candidates. Intro-duce Louis Napoleon – his ancestry, character, ideas (*e.g.* 'idées napoleoniennes'), and previous attempts at a coup d'état, imprisonment and forced exile. Point out that he was almost

totally unknown, except among a few score henchmen, on his arrival in France in 1848.

3   Explain his enormous success in the presidential elections; his perception that the vast majority of Frenchmen desired stability after the turmoil of the 'June Days'; fear of Socialism among the middle class and peasants; provincial France's desire for revenge on Paris which had single-handedly made the February revolution; the glamour of Louis Napoleon's name and associations contrasted with the drab boredom of Louis-Philippe; even conservative monarchists supported him owing to division in their ranks between Orleanists and Bourbons; even workers preferred him to Lamartine and Cavaignac, their June oppressors.

4   Show how Louis Napoleon was able to convert this good-will into support for – or at least tolerance of – the Empire; his own ambitions and desire to imitate his uncle.

5   Show how Louis Napoleon, by carefully fostering con-servative support through the expedition to Rome and the Loi Falloux (education law), *almost* succeeded in achieving his ambition by constitutional means; the important role of the clergy in moving public opinion towards him.

6   The coup of 2 December and the controlled use of force and repression; the plebiscite of 1852 confirmed Napoleon as emperor; his skilful management and reasons for lack of opposition.

7   Conclusion: short-lived nature of the Republic is no sur-prise; that its President should have survived it is more surprising and due mainly to his political skill and sympathy with the aspirations of ordinary Frenchmen. The basic problem was constitutional: Napoleon never solved the problem of the relationship between the executive and the legislature; moreover the constitution was so inflexible that there was no means to alter it – except by force.

3   Do you agree that Napoleon III's domestic policy was 'authoritarian but paternal'?

Understanding the question

This is largely a straight question on domestic policy; the diffi-culty lies in the *definitions* contained in the question. By definition, all paternal governments are authoritarian; it *could* be held that all authoritarian government must be paternal. The fact that the quotation is 'authoritarian *but* paternal' implies that the word 'paternal' is used here, not merely to mean 'fatherlike' (which would include authoritarianism) but 'kindly', 'caring', 'anxious to please one's subjects'. In other words, here is a moral judge-ment for you to assess – 'Napoleon was dictatorial but had his people's welfare at heart'.

Knowledge required

All aspects of Napoleon's domestic policies (political, consti-tutional, social, economic) as well as an ability to discuss the concepts 'paternal' and 'authoritarian' – with perhaps some comparative references, perhaps to Napoleon I or the Enlightened Despots.

Suggested essay plan

1   Analyse the meaning of the two terms with *brief* references to eighteenth-century enlightened despotism and Napoleon I.

2   Discuss Louis Napoleon's character and early political ideas, his strong impulse to govern dictatorially but also to do what most people wanted and to be popular. Consider the view that he manufactured public opinion and then followed it; you may prefer the view that, though egocentric, he was also soft-

hearted and genuinely sought to please and to find out what people did want.

3   Analyse the authoritarian nature of the regime, sometimes called a 'benign police state': the limitation of the powers of the Legislative Assembly; the use of prefects, mayors, etc. to control elections and influence public opinion; the cunning exploitation of plebiscites, and increased clerical control of education; press censorship and restrictions on thought – the latter mainly unsuccessful.

4   Note also the relaxation of the constitution in the 1860s, leading to the Liberal Empire of 1870, based on parliamentary democracy. Admittedly this was partly defensive, a panic response to growing opposition; but it was partly genuine – Napoleon's conscience never quite recovered from the events of 2 December, 1851, and its aftermath.

5   The paternalism of the regime was little more than senti-ment. There was some protection for striking workers, some small redistribution of wealth by fiscal methods; Napoleon had some Saint-Simonian ideas, but had more ideas of a techno-logical future than socialist ones.

6   The main direction of his economic policy was away from paternalism and protection and towards free trade and the release of capital. His regime encouraged the release of private enterprise through commercial trade treaties and a new banking system. Public enterprise was considerable but essentially to complement private economic activity. Competition, not pater-nalism, was the main aim.

7   Possible conclusions: persistent opposition suggests that his regime, thought corrupt and repressive, was neither success-fully authoritarian nor paternal.

4   Document question (Louis Napoleon): Study extracts I and II below and then answer questions (a) **to** (g) **which follow:**

Extract I

'I do not think that any of us doubted that he (Louis Napoleon) would try to become Emperor . . . . We thought that once we had made use of the popularity of his name to overthrow the revolutionary usurpers of February 1848 and restabilize the bases of a disturbed society, we should retain the influence to prevent him . . . .

'It is probable that it would have been a very simple matter if the President Prince had been such as his exploits at Boulogne and Strasburg had led us to suppose, and such as he had been judged . . . on the benches of the Constituent Assembly, an adventurer who was both mad and incapable, with the confidence of a visionary in his imperial star, but lacking experience, know-ledge or reliable resources of character and intelligence . . . .

'Once he had reached the end of his mandate and attempted to prolong it illegally, he would not have found any support in the sane and sensible section of the population . . . . (But) far from being weakened during his three years presidency, on the contrary . . . there were now added to the blind votes of the crowd the support of all the commercial and industrial interests . . . .

'We realized that an army of four hundred thousand men, and all the resources which the administration in France provides for the man in power, joined to immense popularity, made of this candidate for Empire a very redoubtable adversary for those who remained faithful to liberal and constitutional principles.'
(From the *Mémoires du Duc de Broglie*, published in collected form, 1938)

Extract II

'In the year 1851, France passed through a kind of minor trade crisis . . . . The French bourgeoisie attributed this stagnation to purely political causes, to the struggle between parliament and the executive power, to the precariousness of a merely provisional form of state . . . . But this influence of the political conditions was only local and inconsiderable.

'The apparent crisis of 1851 was nothing else but the halt which over-production and over-speculation invariably makes in describing the industrial cycle before they . . . rush feverishly through the final phase of this cycle and arrive once more at their starting point, the general trade crisis.

'Now picture to yourself the French bourgeois, how in the throes of this business panic his trade-sick brain is tortured . . . by rumours of coups d'etat and the restoration of universal suffrage, by the struggle between parliament and the executive power, by the Fronde war between Orleanists and Legitimists . . . . You will comprehend why the bourgeois madly snorts at his parliament-ary republic: "Rather an end with terror than terror without end!" Bonaparte understood this cry.'

(Karl Marx, 1852)

Maximum marks

(a) Explain the reference to 'the revolutionary usurpers of February 1848'.   (3)
(b) What were Louis Napoleon's 'exploits at Boulogne and Strasburg' and what had been their effect on his reputation?   (4)
(c) Explain what is meant by (i) 'the confidence of a visionary in his imperial star', and (ii) 'once he had reached the end of his mandate'.   (4)
(d) In what ways did the fears expressed in the last para-graph of Extract I come true, both in 1851 and in later years?   (6)
(e) What do you understand by 'Orleanists and Legiti-mists'?   (2)
(f) Describe briefly why, according to the writer of Extract II, 'the French bourgeoisie' supported Louis Napoleon and explain what mistake the writer thought that they had made.   (5)
(g) How far does an examination of the subsequent history of France (1852–70) suggest that 'the commercial and industrial interests' had been wise to support Louis Napoleon?   (7)

5.6 READING LIST

Standard textbook reading

J. A. S. Grenville, *Europe Reshaped 1848–78* (Fontana, 1976), chapters 9, 16, 17.

David Thomson, *Europe Since Napoleon* (Penguin, 1977), chapter 13.

Anthony Wood, *Europe 1815–1945* (Longmans, 1975), chapter 15.

Further suggested reading

J. P. T. Bury, *France 1815–1940* (Methuen, 1969).

Alfred Cobban, *History of Modern France*, volume 2, 1799–1871 (Penguin, 1970).

A. Horne, *The Fall of Paris: the Siege and the Commune* (Macmillan, 1965).

M. Howard, *The Franco-Prussian War* (Hart-Davis, 1961).

J. McManners, *Lectures on European History 1789–1914* (Black-well, 1974).

L. C. B. Seaman, *From Vienna to Versailles* (Methuen, 1972).

J. M. Thomson, *Louis Napoleon and the Second Empire* (Black-well, 1954).

6 The unification of Germany

6.1 INTRODUCTION

The 1815 settlement created a Germanic Confederation in which Austrian influence was dominant. In the subsequent period of Romantic Nationalism there was growing agitation against the disunity of Germany, particularly among the student clubs, the Burschenschaften, in the universities, The largest step in the direction of German unification in these early years, however, was the creation of the Prussian-dominated customs union, the Zollverein, from which Austria was excluded.

During the revolutions of 1848 conflicting programmes for a united Germany were advanced in the Frankfurt Assembly and were rejected by the Austrian and Prussian monarchs. In the event, the actual process of unification owed more to diplomatic rivalry and war than to nationalist sentiment.

A major constitutional crisis between William I of Prussia and the Prussian House of Representatives over Army reform brought Otto von Bismarck to power as Minister-President in 1862. He was determined to uphold Prussian interests and the existing conservative monarchical order and he proceeded to finance an expanded Army without parliamentary authorisation.

With a well-equipped Army and a rapidly developing industrial economy to support it, Prussia went to war against Denmark over the old issue of the Danish possession of Schleswig-Holstein, territories with a partly German population. Joint occupation of the Duchies with Austria enabled Bismarck to provoke the Seven Weeks' War against Austria. This, which led to the Austrian defeat at the battle of Sadowa, brought all of the German states north of the river Main into the North German Confederation – a new entity under Prussian leadership.

Deterioration of relations with Napoleon III and, in 1870, the issue of the Hohenzollern candidature to the Spanish throne precipitated the Franco–Prussian War. This completed the unification of Germany, which now incorporated the South German states, Alsace and most of Lorraine. The new German Empire was declared in the Palace of Versailles in January, 1871.

6.2 STUDY OBJECTIVES

1   As background: the rearrangement of Germany into 39 states in 1815; the increased influence of Prussia and Austria; the constitution of the new Germanic Confederation – the Bund.

2   Nationalist movements in Germany, 1815–48, and the social classes and groups supporting them; the intellectual and literary traditions of Nationalism in Prussia; the influence of the 'War of Liberation'.

3   Economic expansion and integration, 1815–48; the Zollverein; the ideas of Friedrich List.

4   The 1848 revolutions (see Unit 2); the Frankfurt Parliament and the rival Artisan Congress; pressing respectively for free trade and guild controls; the Kleindeutsch versus the Grossdeutsch ideas of unity; Frederick William IV's refusal of the Crown of a united Germany; the reasons for the failure of the revolutions.

5   The 1850s: political disappointments for Prussia and economic disappointments for Austria; Prussia's 'humiliation' at Olmütz; the badly-handled mobilization of the Prussian Army

during the Italian War of Independence; on the other hand, Austria failed to break into the Zollverein and suffered badly from the Vienna financial crash of 1856–7.

6   Otto von Bismarck: his Junker background, personality and political ideas; his reactions to Prussia's 'humiliations' while he was in the Bundestag at Frankfurt; his devotion to the established Prussian political and social order; the debate as to whether he planned the unification of Germany or merely seized opportunities, later claiming that he planned it.

7   Bismarck's appointment as Prime Minister-President of Prussia in 1862; Roon, Manteuffel and the Army budget crisis; the liberal wish for a Landwehr instead of the aristocratically-controlled regular Army; Bismarck's solution, by-passing the Chamber and collecting taxes directly; the constitutional significance of the crisis.

8   The first stage of unification: the Polish and Schleswig-Holstein crises.

9   The second stage of unification: the Austro-Prussian War; the militarization of Prussia, Moltke and the General Staff seeking to exert Army control over the politicians; improvements in weaponry and communications; the international significance of the battle of Sadowa; the extent and constitution of the North German Confederation; the domestic implications, *e.g.* the Indemnity Bill.

10   The final stage of unification: relations with France 1867–70; the Hohenzollern candidature crisis; the Franco–Prussian War and its results.

11   The constitution of the new German Empire, 1871; the influence of Prussia within it.

12   The industrial development of Germany in the 1850s and 1860s: the extent to which Germany was united by 'coal and iron' rather than by 'blood and iron'.

6.3 COMMENTARY

The impetus to unification

The transformation of Germany from the 39 separate states of the Confederation (Bund), as it was in 1815, into the German Empire of 1871 was a process which comprised several different elements. It is important to analyse them and see how they relate to one another.

1   First, on one level, it was a resolution of the long-standing rivalry between the dynasties of Prussia and Austria over the German states that lay between them.

(*a*)   **Austria** with its extensive Slav, Italian and Hungarian dependencies, was primarily a non-German Power. From the Congress of Vienna, however, it had maintained a controlling influence in German affairs in the Bund. This dominance was re-established after the 1848 revolutions, at Olmütz (1850).

(*b*)   **Prussia**, on the other hand, after the acquisition of the Rhineland in 1815, gradually consolidated her position as a German Power, particularly in the economic sphere. Between 1866 and 1870, through three wars, she dramatically extended her authority over the rest of Germany and excluded Austria.

2   The process of unification also reflected the strong force of popular *German nationalism* which had been so evident during the revolutions of 1848. This emphasized the unity of Germany and allegiance to the nation as a whole rather than its particular states. It was both a political and a cultural force. Towards the end of the century this was to find an even wider expression in German imperialism. While, to begin with, it appeared that Prussia had annexed Germany, in the end Germany and German nationalism were to absorb Prussia.

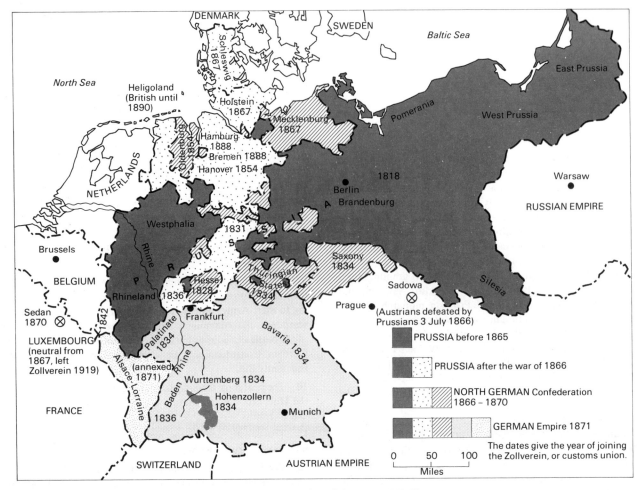

Fig. 3 The Unification of Germany 1818–71

3   There was the powerful economic impetus of rapid industrial growth, of improved communications, particularly railways, and of the commercial integration of the various parts of the country. This was particularly encouraged by the *Zollverein*, or customs union, of 1834 which continuously strengthened the Prussian position throughout this period.

'Grossdeutsch' and 'Kleindeutsch'

As the events of 1848–9 very clearly showed, German nationalists were divided among themselves over both the extent and the constitution of a united Germany.

The 'Grossdeutsch' or 'greater German' solution looked to Austria for leadership. To incorporate Austria in a German national state was impossible, however, so long as the Habsburgs insisted on retaining – as they did – their non-German territories and peoples.

Increasingly, therefore, German patriots turned to Prussia for the achievement of unification: the 'Kleindeutsch' or 'lesser German' solution which excluded Austria.

While the growth of nationalist feeling was contributing to a sense of common German identity, it was also undermining the multinational Habsburg Empire. The force of Nationalism, which was the dominant political development of the nineteenth century, was later – in the wake of the First World War – to bring about the Empire's collapse.

The example of the Italian War of Independence did a good deal to reawaken national ambitions in Germany and, in 1859, a national society was formed which campaigned for a parliamentary regime in a united Germany under Prussian leadership.

Otto von Bismarck

The determination of the Prussian monarchy to assert itself in Germany had a long history. Bismarck, who played a unique role in the process of unification, saw himself as upholding this tradition of strengthening Prussia and conserving its institutions.

He had become involved in politics as a representative of the Junkers (landed gentry) in Berlin in 1847 and was noted for his extremely conservative views. He showed himself to be passionately opposed to German liberalism and the Frankfurt Parliament of 1848. In 1851 he was chosen to serve as Prussia's ambassador to the revived German Diet. In 1857 he was sent as ambassador to St. Petersburg. In 1862 he was transferred to Paris and then invited back to Berlin in September of that year to push through the violently disputed army reforms in the face of Liberal opposition.

The army reforms and the constitutional crisis

In 1860 the Prussian Landtag was presented with a Bill for army reforms. Though the population of Prussia had doubled since 1815 the number of recruits had not risen.

The Bill proposed to increase annual recruitment by more than a half, to raise the period of active service to three years (rather than two) and to integrate the Landwehr, the semi-civilian reserve force, with the regular army.

The Liberal majority passed the budget for increased expenditure for a year but refused to accept the principle of military reform. They insisted that the two-year term of service should be retained and rejected the forcing of the militia out of the field army.

This precipitated a major constitutional crisis and the Prussian King, William I, threatened abdication. In his view the army must remain a strong and reliable instrument in the hand of the monarch. It must strengthen the traditional position of the Crown, not the people. A new political party, the Progressive Party, was founded in 1861 to fight what was a constitutional battle of Liberal parliamentarians against the established order of Junkerdom and entrenched institutions.

Bismarck assumed ministerial power to resolve this conflict in the King's favour. For four years (1862–6) he conducted the government of Prussia without a constitutionally sanctioned budget. He waged two wars (against Denmark and Austria) without any grant of money from the representative assembly. He imposed press controls, dismissed numerous officials and intervened in elections to force through the royal policy.

Prussian leadership in Germany

It is important to consider Bismarck's contribution to unification of Germany in context. His role, particularly as diplomatist, was extremely important – indeed, crucial. But there are other factors – besides this and the modernization of the Prussian army – which would have favoured Prussian dominance whoever had been Chief Minister at Berlin:

1 The Zollverein had already made most of Germany an economic unit and the industrialization of North Germany was already well under way by 1862. The starting-point for this was laid in the financial reforms of Maassen in 1818. In order to unite the scattered provinces of Prussia he created a new tariff system which abolished all internal customs and established free trade throughout Prussia.

2 The German population was increasing with great rapidity, notably more, for instance, than that of France.

3 A good railway network had been constructed.

4 The acquisition of the Rhineland in 1815.

5 An extremely efficient civil service at the disposal of the Prussian monarch.

6 The relative weakening of the Habsburg Empire due to economic factors, the unresolved problems of the nationalities and of internal rivalries, and Prussia's growing diplomatic isolation.

The Schleswig-Holstein Question

As Prussian ambassador to the revived Germanic Diet at Frankfurt, Bismarck had first-hand experience of the Austro-German rivalry. The first step to Prussia's renewed rise to power, as he saw it, would be to make her master north of the river Main.

In a memorandum to William I, Bismarck explained that the more Prussia was bound by German federal arrangements the less her real power would be because Austria could always organize the other members of the Bund against her.

In January, 1864, he induced the Austrians to sign an alliance with Prussia against Denmark on the grounds that the new King of Denmark, Christian IX, had infringed the London agreement of 1852 over the duchies of Schleswig and Holstein.

This complex issue of competing national claims was then manipulated by Bismarck to bring about war with Austria after the Convention of Gastein. The dispute with Austria was both over Schleswig-Holstein and over rival proposals for a constitution for Germany to replace or reform the constitution of the Confederation. Beneath this there was the perennial rivalry of the two major German Powers.

The diplomatic background

The war of Austria and Prussia against Denmark convinced Bismarck that Britain and Russia were unlikely to intervene in an Austro-Prussian conflict. The diplomatic scene favoured Prussia in three main ways:

1 Russian relations with Austria had been cool since the neutrality of the Habsburgs in the Crimean War; while Bismarck had moved quickly to offer Russia support at the time of the Polish uprising in 1863.

2 In October, 1865, he had met Napoleon III at Biarritz and prepared the way for French neutrality in the event of a struggle with Austria. (By February, 1866, Bismarck was declaring that such a conflict had now become only a matter of time.)

3 In April 1866 a secret alliance was concluded between Prussia and Italy. Article 2 provided that, if Prussia's plans for reforming the German Confederation failed and Prussia took up arms, Italy would follow Prussia in declaring war on Austria. As a reward Venetia was to be ceded to Italy.

The outbreak of war with Austria

Bismarck's weapon against the Austrian-dominated German Confederation, which was an alliance of sovereigns, was to threaten to place German's future in the hands of a German parliament elected on the basis of universal suffrage.

His clear aim, as the Austrians saw it, was to exclude them from any say in German affairs. Prussia, by force of her size and her economic and military power, would then be able to dominate the 'lesser Germany'.

They broke off negotiations over Schleswig-Holstein and proposed to the German Diet that all the German States, except Prussia, should be mobilized.

Sadowa

The 'Seven Weeks' War' was arguably the central event in nineteenth-century German history. The battle of Sadowa finally resolved the long Austro-Prussian rivalry and assured the victory of the 'Kleindeutschland' solution.

Austria was excluded from the future development of German affairs and North Germany was consolidated under Prussian leadership.

At the same time, the triumph of the Prussian army – through superior planning, efficiency and mobilization – destroyed the possibility that Prussia might become a parliamentary state on the British model: it was also a domestic victory of established institutions over the Liberal opposition.

The desire for a strong Germany under Prussian leadership in the event proved stronger than the wish for a liberal Germany. The Indemnity Bill emphasized this shift of opinion. The majority of the Liberals became government supporters and for more than a decade the National Liberal and Free Conservative parties gave Bismarck the support he needed to consolidate the foundations of the German Empire.

The Treaty of Prague

Bismarck imposed a moderate peace. The Habsburgs were not required to sacrifice any territory.

Prussia annexed Hanover, Schleswig-Holstein, part of Hesse-Darmstadt and the City of Frankfurt; the independence of the remaining North German states was lost except for a measure of local autonomy in the North German Confederation.

In this, central authority was exercised by the *Bundesrat*, the Federal Council, in which Prussia could always secure a majority.

The King of Prussia assumed the title of President of the Bund and the function of Commander-in-Chief. He appointed, and could dismiss, the Federal Chancellor. Officially he controlled foreign policy and could declare war and make peace.

The constitution also provided for the election of a representative assembly, the *Reichstag*, on the basis of universal manhood suffrage. But this was only allowed a limited role. It did not, for instance, have control over the military budget, which comprised the major part of the Confederation's taxation.

Bismarck and Napoleon III

Before the Treaty of Prague the French Emperor had won the assurance that France would not face a united Germany that included the territories to the south of the Main.

Behind Napoleon's back, however, Bismarck concluded secret alliances with the Southern States. In return for Prussia not insisting on large indemnities from them, it was agreed that – if Prussia found herself in a war endangering her territory – the Southern States would place their armies under the command of the King of Prussia.

There has been much historical debate over Bismarck's precise intentions after 1867. Did he, for example, see a war with France as necessary and inevitable?

In fact, the North German Confederation already included two-thirds of the whole of Germany. It was obvious from the start that the remaining states would find it very difficult to enjoy the 'independent sovereign existence' as laid down in the Treaty of Prague.

In this period Bismarck proposed elections throughout all the German states for a gathering of a 'Zollparlament', or customs parliament, with a view to this assembly extending its jurisdiction over non-commercial questions.

At the same time he refused to 'compensate' Napoleon for French neutrality during the Austro-Prussian War. He rejected the 'hotel-keeper's bill', as he called it. Among other things he stirred up German national feeling against the idea of ceding Luxemburg to France.

The Hohenzollern candidature

This diplomatic crisis which led to the Franco-Prussian War originated in September, 1868, when a revolution in Spain drove the reigning Queen out of the country. The idea of approaching a Hohenzollern prince for the Spanish throne came from the leading minister of the revolutionary government.

Bismarck's assertion that he had nothing to do with the affair until the full crisis broke in July 1870 is false. In March 1870 he had already sent a memorandum to William I arguing the case for the candidature and he subsequently sent agents to Spain.

The French government argued that a German prince on the Spanish throne would mean the 'encirclement' of France. The French played into Bismarck's hands by demanding that King William not simply repudiate the candidature but also offer an assurance that he would not authorize a renewal of it.

William's refusal to meet the second of these demands, which was emphasized by the Ems telegram, left Napoleon and his ministers with only two ways out of the situation which they had created – either to climb down or to fight.

By the time of the declaration of war Bismarck had contrived to make it appear that all the blame for the war was French. This allowed him to bring into force the military alliances between Prussia and the South German States. He presented it as a patriotic war in defence of the Fatherland.

The results of the Franco-Prussian war

The French lost Alsace, most of Lorraine and the fortresses of Strasbourg and Metz – a cause of lasting grievance for France. She was also forced to pay an indemnity of five billion francs and a German army of occupation was to remain in northern France until it was paid.

With the proclamation of William I as German Emperor in January, 1871, Germany emerged as the dominant continental state, a role which France had enjoyed since the seventeenth century. The Empire, which incorporated the Southern States, was a federal rather than a unitary state.

Two other incidental but significant results of the Franco-Prussian War were (1) that Russia stated, and the Powers accepted, that she would no longer be bound by the Black Sea clauses of the Treaty of Paris (1856), and (2) following the withdrawal of the French garrison, Rome was incorporated as the capital of Italy.

6.4 CHRONOLOGY

1850	**November**	THE PUNCTATION OF OLMÜTZ: the re-establishment of the Germanic Diet at Frankfurt.
1853		Brunswick, Hanover and Oldenburg joined the ZOLLVEREIN (formed 1834), bringing the whole of non-Austrian Germany into the customs union.
1858		WILLIAM I appointed Regent (King of Prussia from 1861).
1859–62		GENERAL VON ROON appointed Minister of War to reform the army; VON MOLTKE becomes Chief of the General Staff. THE STRUGGLE BETWEEN THE LIBERAL MAJORITY IN PARLIAMENT AND THE MONARCHY OVER ARMY REFORM; PRUSSIAN CONSTITUTIONAL CRISIS; WILLIAM I THREATENED ABDICATION.
1862		OTTO VON BISMARCK became Minister-President of Prussia, later Chancellor. Carried on the constitutional struggle with the Prussian parliament for the next four years.
1863	**February**	THE POLISH REVOLT: Bismarck sent von Alvensleben to offer Prussian support to Russia against the rebels.
	July	The Diet of the Germanic Confederation demanded that the two Duchies be taken from Denmark and submitted to the rule of the German Duke of Augustenburg, son of one of the claimants to the succession.
	August	Congress of Princes, summoned by the Austrian Emperor, Francis Joseph, to reform the German Confederation. Bismarck refused to allow William I to attend. THE SCHLESWIG-HOLSTEIN QUESTION: Proclamation of King Frederick VII of Denmark in March 1863 announced the annexation to Denmark of the Duchy of Schleswig. This breached the London Protocol of 1852.
	October	The Diet voted for action against Denmark.
	November	Death of Frederick; succeeded by Christian IX, who signed a new constitution indicating his intention to incorporate Schleswig.

	December	Federal troops entered Holstein.
1864	January	Austria joins Prussia in alliance.
	February	Austrian and Prussian troops invaded Schleswig.
	April	Capture of Duppel forts; German invasion of Denmark.
	April – June	THE LONDON CONFERENCE. War renewed at end of June, resulting in defeat of the Danes. Surrender of the Duchies of Schleswig, Holstein and Lauenburg to Austria and Prussia.
	October	The Peace of Vienna.
1865	August	THE CONVENTION OF GASTEIN: Joint sovereignty of Prussia and Austria over the Duchies. Austria to administer Holstein and Prussia Schleswig. (Lauenburg ceded to Prussia in return for a money payment to Austria.)
	October	BISMARCK MET NAPOLEON AT BIARRITZ: a vague offer of 'compensation' for France in return for a promise of French neutrality in the event of an Austro-Prussian war.
1866	April	BISMARCK, SUPPORTED BY NAPOLEON, CONCLUDED AN OFFENSIVE AND DEFENSIVE ALLIANCE WITH ITALY: Italy to join Prussia if war broke out between Austria and Prussia within three months, with Venetia as the reward for Italy. Bismarck introduced a motion for federal reform into the Frankfurt Diet with the intention that Austria reject it and precipitate a conflict. Prussia and Austria started to mobilize.
		Austria signed a secret treaty with Napoleon III. In return for French neutrality Austria promised to cede Venetia to Napoleon who was then to give it to Italy. The Frankfurt Diet voted against Prussia for violating Holstein territory. The majority of the German States, including Bavaria, Saxony and Holstein, sided with Austria against Prussia.
	June	The Austrian governor of Holstein summoned the Holstein Diet in order to discuss the future of the Duchy. Bismarck denounced this as a violation of the Gastein Convention and ordered Prussian troops into the Duchy.
	June–August	THE SEVEN WEEKS' WAR: complete victory for the Prussians at the battle of SADOWA (Königgrätz) in Bohemia. Napoleon offered to mediate. Bismarck accepted this only on condition that the peace terms should be determined before the armistice was completed. THE PEACE PRELIMINARIES OF NIKOLSBURG: the states of Hesse, Frankfurt, Nassau and Hanover to be incorporated in Prussia: Austria excluded from Germany. German states north of the river Main were to form a North German Confederation under Prussian leadership. The South German States were to remain independent and to be allowed to form a separate confederation. Bismarck took advantage of the fears aroused by French demands for compensation to persuade the South German States (Bavaria, Baden and Würtemberg) to conclude secret military alliances with Prussia, operative in the event of a French attack.
	August	THE TREATY OF PRAGUE: the Germanic Confederation, established in 1815, was dissolved; Austria was excluded from participation in German affairs.
	September	THE BILL OF INDEMNITY: gave retrospective assent to Bismarck's expenditure without the consent of parliament during the constitutional crisis. The majority of the Liberals now rallied to Bismarck, in spite of their previous bitter opposition, because of his achievement of national objectives.
1867	April	THE CREATION OF THE NORTH GERMAN CONFEDERATION. The Luxemburg crisis.
	July	Bismarck brought the four South German States into the Zollverein and established the ZOLLPARLAMENT, customs parliament.
1868–70		THE HOHENZOLLERN CANDIDATURE FOR THE SPANISH THRONE.
1870	June	Leopold of Hohenzollern-Sigmaringen accepted the offer of the Spanish throne.
	6 July	The French foreign minister Gramont made a speech to the French Chamber indicating France would go to war unless the Prussian government withdrew the candidacy.
	9–11	The French ambassador to Berlin, Count Benedetti, requested William I to order Leopold to withdraw.
	12	Prince Charles Anthony, father of Leopold, withdrew the candidacy on behalf of his son. Not content with this diplomatic victory, the French government proceeded to demand guarantees from William I that he should both disavow the candidacy AND promise that it would never be renewed.
	13	THE EMS INTERVIEW: the Prussian king rejected Benedetti's demand; Bismarck amended the EMS TELEGRAM.
	19	French declaration of war on Prussia.

1870–71   THE FRANCO-PRUSSIAN WAR: Bismarck and the armed support of the South German States. Britain secured the neutrality of Belgium. THE BATTLE OF SEDAN (1 September, 1870); THE CAPITULATION OF METZ (27 October); the FALL OF PARIS (28 January, 1871).

1871   18 January   THE FOUNDATION OF THE GERMAN EMPIRE: As a result of the war the new Reich consisted of 25 states. Germany acquired Alsace and eastern Lorraine, including Metz and Strasbourg, and incorporated the States south of the Main in the North German Confederation.

6.5 QUESTION PRACTICE

1   Is it true to say that diplomacy was more important than war in the process of German unification between 1862 and 1871?

Understanding the question

You may point out that the question implies a highly debatable assumption. Whereas it is reasonable to ask whether, say, economic factors were more important than military factors, it is most arguably the case that war and diplomacy are not alternatives, but that war is in fact diplomacy by other means; it is often very difficult to distinguish between them and you should tell the examiner so.

Knowledge required

Bismarck's career and the military and diplomatic events between 1862 and 1871.

Suggested essay plan

1   Rather than opposites, war and diplomacy may be seen as inseparable parts of a nation's dealings with other states; no foreign minister can know when or whether his diplomatic bluffs will be called and may lead to war; this was most certainly the case with Bismarck.

2   Narrative of the first stage of the process of unification: the Polish Revolt and Bismarck's defiance of Prussian liberals in pledging support for the Tsar; the Schleswig-Holstein Question; the 1848 dispute and the 1852 Treaty of London, and the accession of Christian IX of Denmark; Prussia and Austria invaded the Duchies on behalf of the German Confederation and military victory gave Schleswig to Prussia and Holstein to Austria.

3   Narrative of the second stage: Bismarck negotiated with Italy and met Napoleon at Biarritz (1865); Bismarck, Moltke and Roon engineered a dispute with Austria over the administration of the Duchies after the Convention of Gastein; this was partly to score domestic political points against the Prussian liberals; the Austro-Prussian War and the unexpectedly speedy and easy Prussian victory; Prussian military improvements (*e.g.* the Roon Army reforms) and the reasons for her superiority.

4   The establishment of the North German Confederation, under Prussia (and Bismarck's) thumb (1867); its constitution was the nucleus of the Empire to be founded in 1871; the gradual accommodation between the South German states and the Confederation; Bismarck's secret alliances with these states.

5   The domestic rising in Spain and the abdication of Isabella; Bismarck suggested a Hohenzollern candidate for the Spanish throne; French objections to German 'encirclement'; the Ems telegram and the Franco-Prussian War; the easy Prussian victory and the reasons for it (weaponry, logistics, tactics); the foundation of the German Empire.

6   By far the most active diplomacy in all these events occurred immediately before the 1866 and 1870–1 wars, and was directed unsuccessfully towards preventing them. As for Bismarck's own diplomacy, clearly this could not have brought about unification on its own, but it was an important enabling factor. Especially it secured the neutrality of Russia (in alleged gratitude for

Bismarck's support in 1863); and of Italy and France in 1866. Also, by deliberately not humiliating Austria after the 1866 war, Bismarck's diplomacy may, perhaps, have prevented a Habsburg campaign of revenge in 1870.

7   But undoubtedly it was the events on the battlefield which created German unity; diplomacy played only a secondary role.

2   How clear was it before Bismarck's assumption of office in 1862 that Germany would be united under Prussia?

Understanding the question

This question begs a more fundamental one: how clear was it before 1862 that German unification would be achieved at all? If the answer to this is that it *was* clear, then probably it was also clear that it would be achieved under Prussian leadership.

Knowledge required

The internal histories of Prussia and Austria, c. 1848–71; Bismarck and the course of German unification during 1863–71.

Suggested essay plan

1   Bismarck after the event claimed to have had a clear intention of uniting Germany from the time of his appointment as Minister-President in 1862, and also a clear view of how to do it. This seems unlikely; Bismarck was an opportunist rather than a long-term calculator.

2   The only thing that was clear by 1862 was that Germany would not be united under Austrian auspices; the Grossdeutsch solution had been ruled out by Prussia's objection to the inclusion of Austria's Slav lands; Austria would not accept a Kleindeutsch solution; Austria's economic weakness and severe financial deficits further undermined her power in Germany; Austria's last real attempt at dominance was her 1862–3 scheme for a German federal union under Austrian, Prussian and Bavarian leadership; despite the approval of King William of Prussia, Bismarck managed to thwart the proposal, and with it any lingering hopes of an Austrian 'solution'.

3   Since no institutions existed for the 'popular' unification of Germany, it was therefore clear by 1862 that *if* Germany were to be united it could only be under Prussian leadership. Prussia was economically a very strong state, thanks partly to the Zollverein, and this in turn was to enable her to strengthen her military capacity (though not fully before 1862). What she lacked was confidence – the 'humiliation' of Olmütz still rankled – and this is what the leadership of Bismarck supplied.

4   But by 1862 it was still far from clear that Germany *would* be united; the success of the Zollverein perhaps illustrated that formal political union was not economically essential; national sentiment of a cultural sort was in the doldrums at this time – it was far less apparent than in, say, Italy or Poland – and was to be a product rather than a cause of the process of unification; the Liberal Nationalism of the intellectuals and lawyers of 1848 seemed to be in marked decline after the Frankfurt Parliament of 1848; none of the European Powers wanted German unification and the series of wars which created it could hardly have been foreseen in 1862, even by Bismarck.

5   The motive for unification was there in 1862; subsequent successes enabled Bismarck to reconcile the liberal middle

classes (or at least a large proportion of them) to the existing political system in Prussia. Austria, on the other hand, could only suffer from German unification. If it was to come, therefore, it was obviously going to be under Prussian direction.

3 What were the consequences for Germany of the Zollverein?

Understanding the question

You need to consider the economic and political development of the German states c. 1830–70, and decide how far they were prompted or retarded by the Zollverein.

Knowledge required

The creation of the Zollverein and its policies. The economic development of both those states which made up the Zollverein and those excluded from it. The effects of such economic development on the internal politics of the states and on the German unification movement.

Suggested essay plan

1   It is sometimes said that Germany was united by 'coal and iron' not by 'blood and iron'; this view assigns a central role to the Zollverein in the process of German unification; note, though that there have been many examples of the tendency towards economic concentration and integration during the last couple of centuries, but these have not always led on to political unification.

2   The territorial settlement of Germany in 1815; the events leading to the creation of the Zollverein: Prussia abolished internal customs barriers in 1818 in response to pressure from Rhineland industrialists; Prussia concluded several bilateral tariff negotiations with other German states during the 1820s; this increased her status in Germany at a time when Austria was causing annoyance by interfering in the internal police and censorship regulations of the smaller states.

3   In 1834 these bilateral agreements were merged to form the multilateral Zollverein; Metternich attempted to frustrate it but Saxony and Bavaria joined it; note that the states' motives for acceding to it were *economic* and not political; the Southern States would not have joined if they could have forseen its political consequences; most of them were still loyal to Austria culturally and politically.

4   The economic development of the Zollverein to c. 1850; railway building, begun in 1834, had achieved 5,000 miles by 1850; commercial treaties and a large reduction in tariffs in 1845 led to a big increase in trade and manufacturing, especially in Prussia, whose industrial magnates, *e.g.* the Krupps, came to dominate the Zollverein.

5   The attempts of Brunswick, Hanover and Oldenburg and the Hansa towns to remain outside as a rival trade bloc collapsed with the accession of the first three to the Zollverein in 1844, 1851 and 1852 respectively. By this time Prussia's dominance was developing *political* implications. There was the growing imbalance between Prussia and Austria which was accelerated by the financial collapse of 1856–7; it became ever less likely that Austria – whose industries needed protection – could join the Zollverein, whose policies were increasingly free-trade. In 1862 Prussia replaced Austria as France's 'most favoured trading nation', symbolizing the turn-around in fortunes.

6   Industrial leaders from the Ruhr and Rhineland, like Camphausen and Hansemann joined the Liberal Party inside Prussia; this was seen to pose a threat to the old Prussian nobility and the traditional governing class. In the process of unification and subsequent consolidation of Germany, Bismarck made economic concessions to this new group.

7   The Zollverein contributed a great deal to the unification of Germany and to the dominance of Prussia in the new nation-state; at the same time it helped to make the new middle classes a very powerful element in Prussia; to a certain extent Bismarck diverted attention to national German affairs in order to resist the pretension to political power of the new industrial and financial middle class.

4   Document question (the Austro-Prussian War): study the extract below and answer questions (a) to (f) which follow:

Extract

'We had to avoid <u>wounding Austria too severely</u>: we had to avoid leaving behind in her any unnecessary bitterness of feeling or desire for revenge; we ought rather to reserve the possibility of becoming friends again with our adversary of the moment, and in any case to regard the Austrian state as a piece on the European chessboard and the renewal of friendly relations with her as a move open to us. If Austria were severely injured she would become <u>the ally of France and of every other opponent of ours</u>: she would even sacrifice her <u>anti-Russian interests</u> for the sake of revenge on Prussia.

'On the other hand, I could see no future acceptable to us, for the countries constituting the Austrian monarchy, in case the latter were split up by risings of the Hungarians and Slavs or made permanently dependent on these peoples... German Austria we could neither wholly nor partly make use of. The acquisition of provinces like Austrian Silesia and portions of Bohemia could not strengthen the Prussian State; it would not lead to an amalgamation of German Austria with Prussia, and Vienna could not be governed from Berlin as a mere dependency...

'The resistance I was obliged, in accordance with these convictions, to offer to the King's views with regard to the following up of the <u>military successes,</u> and to his inclination <u>to continue the victorious advance,</u> excited him to such a degree that a prolongation of the discussion became impossible; and under the impression that my opinion was rejected, I left the room with the idea of begging the King to allow me, in my capacity as an officer, to rejoin my regiment.

(Bismarck's record of his advice to the King of Prussia at Nikolsburg, 1866.)

Maximum marks

(a) What did Bismarck mean by 'wounding Austria too severely'? (2)

(b) 'The ally of France and of every other opponent of ours'. Why did Bismarck consider France as an opponent of Prussia at this time? (4)

(c) Explain Austria's 'anti-Russian interests'. In what ways might Austria need to sacrifice them to obtain revenge on Prussia for a harsh peace treaty? (6)

(d) What were 'the military successes' and why was the King so anxious 'to continue the victorious advance'? (5)

(e) In the peace treaty eventually made between Austria and Prussia, did the wishes of the King or of Bismarck prevail? What, for the Austrian monarchy, was the constitutional result of the war and the peace? (6)

(f) What light is shed by this passage, especially the second paragraph on Bismarck's long-term aims for the future of Germany? Did the settlement of 1871 conform to these aims? (8)

6.6 READING LIST

Standard textbook reading

J. A. S. Grenville, *Europe Reshaped 1848–78* (Fontana 1976), chapters 7, 8, 14, 15, 17.

David Thomson, *Europe since Napoleon* (Penguin, 1977), chapter 14.

Anthony Wood, *Europe 1815–1945* (Longmans, 1975), chapters 17–19.

Further suggested reading

H. Böhme, *The Foundation of the German Empire* (Oxford, 1971).

J. P. T. Bury (ed.), *The New Cambridge Modern History 1830–70*, volume X (Cambridge University Press, 1964), chapters 19, 22.

M. Howard, *The Franco-Prussian War* (Hart-Davis, 1961).

W. N. Medlicott, *Bismarck and Modern Germany* (English Universities Press, 1965).

O. Pflanze, *Bismarck and the Development of Germany 1848–1871* (Princeton University Press, 1963).

W. M. Simon, *Germany in the Age of Bismarck* (Allen and Unwin, 1968).

A. J. P. Taylor, *Bismarck* (New English Library, 1974); and *The Struggle for Mastery in Europe* (Oxford, 1971).

7 Russia, 1855–1917

7.1 INTRODUCTION

On his accession in 1855, Alexander II was faced with the need to modernize and reform a backward peasant society if Russia was to maintain her status as a great Power, something which was called in question by the defeat in the Crimean War. He introduced a wide range of reforms, of which the most important was the Emancipation Edict. These reforms were intended to strengthen, not weaken, the tsarist autocracy and he refused to create a parliamentary assembly along Western liberal lines.

After his assassination, his successors, Alexander III and Nicholas II, combined very conservative political and social policies with a programme of rapid state-led industrialisation.

The failure to solve the land question at a time of very sharply increasing population and the social changes brought about by rapid industrialization contributed to the outbreak of revolution in 1905 – which also followed Russia's humiliation in the war against Japan. This was succeeded by the period of the State Dumas and further reforms, most notably those sponsored by Stolypin.

Final disenchantment with the regime came in 1917 when the failures of the government, plus seemingly insoluble economic problems, all combined with military defeat and insurrection to bring down the tsarist order.

7.2 STUDY OBJECTIVES

1   The nature of Russian society at the accession of Tsar Alexander II: its backwardness as revealed by the defeat in the Crimean War; the problems facing the autocracy – indebtedness, and the need for modernization and reform to ensure Great Power status.

2   The Emancipation Edict: the background to and reasons for this reform (particularly the Tsar's motives): its terms for serfs and landowners; its failure to solve the question of the land – the continuance of the 'mir' (commune) and burdensome redemption payments for the peasants.

3   The other reforms of Alexander II: the Zemstvo Law, the reform of the judiciary, of municipal government, of the Army and of education; the disappointing reaction of Russian society to the reforms – obstruction by a bureaucracy increasingly staffed with conservative noblemen; Populism; the growth of terrorism – 'Land and Liberty', 'The Will of the People'.

4   The reasons why Alexander's 'revolution from above' was only, at the very most, a half-revolution; his unwillingness to provide a proper liberal regime with a representative parliament – *i.e.* his refusal to introduce a national assembly; the problem of applying Western liberal ideas in Russian society.

5   The impact of the assassination of Alexander II and the reaction under Tsar Alexander III; Pobedonostsev's influence: bureaucratic paternalism and a strenuous upholding of the principles of Russian nationalism, autocracy and of the Orthodox religion – censorship, Russification, pogroms against the Jews; the banning of opposition political groups, curbs on education; the Land Captains.

6   Economic development and industrialization: Sergei Witte and 'state capitalism', funded through foreign loans paid for by heavy taxation; the emphasis on heavy industry, capital goods and the development of the railway system; the social consequences of this programme of forced industrialization.

7   The growth of political opposition to the tsarist order under Nicholas II: the Social Revolutionaries, Liberals and Social Democrats; the growth of Marxist ideas in Russia, with special reference to the early career of Lenin and his ideas on revolution; the significance of the 1903 split of the Social Democratic Party into Mensheviks and Bolsheviks.

8   The Russo-Japanese War and its impact on the Russian government. Note, as background, the great imperial expansion of Russia into Central Asia, Siberia and Manchuria.

9   The causes, course of events and consequences of the 1905 revolution, noting particularly the role in it of the peasants, industrial workers and the national minorities; the massacre of 'Bloody Sunday', the October Manifesto, the soviets and the Fundamental Laws.

10   The constitution, scope and significance of the Dumas; the electoral law of 1907; the career and reforms of Stolypin, particularly his encouragement of private peasant ownership; the reasons for his unpopularity both in reactionary and liberal political circles.

11   Russia at war, 1914–17: the way in which it exposed the underlying weaknesses of the tsarist order; the land problem; inflation; administrative chaos; the personal role of the Tsar and of court circles; the consideration as to whether – without the impact of a disastrous war – Russia might have evolved along the path of evolutionary reform rather than collapsed.

7.3 COMMENTARY

Tsar Alexander II

Russian society and the state in 1855

Russian society: Russia was in 1855 still very much a pre-industrial society. The vast majority of Russians were either landowners' serfs or 'state peasants' (that is 'owned' by the State). These peasants were governed by the nobility and a very small bureaucracy. The tsarist autocracy depended for its power on the nobles' control of the serfs through the *commune*. This institution was the typical village government in Russia and the means whereby land was periodically redistributed to meet the varying needs of the commune's families.

All services to the lord by either labour service or money, or produce rent, were paid communally, as was the poll tax. Furthermore, the serfs *thought* communally as well – *e.g.* the famous phrase: 'We are yours, but the land is *ours*.'

This system discouraged the incentive of individual peasant families, because it offered no guarantee of possession of land from one generation to another. Thus Russian agriculture in 1855 was not increasing greatly in productivity.

This stagnation of agriculture, coupled with poor means of transportation in Russia, meant that Russian grain could not compete in its traditional European markets against the new, highly productive agriculture being introduced in the West. The result was that the income of the landed nobility declined sharply. Indebtedness to the state had reached crisis proportions by 1855 and it appeared to many that serfdom was not providing the benefits to the ruling class which were its justification.

The State: The tsarist regime was also suffering from competition in the provision of raw materials in which, in the eighteenth century, it had had a virtual monopoly, in parallel to its official monopoly within Russia – *e.g.* timber, gold and flax. It experienced severe financial crises in the early nineteenth century. The agrarian economy was not able to sustain the expenditure nor to provide the new technology necessary to maintain Great-Power status in an age of industrialization. As a result, Russia experienced humiliating defeat in the Crimean War and incurred a huge public debt.

By 1855 the economic bases of the traditional autocracy were disappearing. A new system had to be found if the tsarist regime was to preserve its internal political and international standing. The wish for Great-Power status was a constant element in tsarist policy.

The experiment with Liberalism

In the West the political accompaniment of capitalistic economic growth was Liberalism. This had two forms: (1) economic *'laissez-faire'* and (2) parliamentarism and representative government. Both, however, were in direct opposition to the tsarist autocracy.

Since the time of Peter the Great, however, Russia had tended to look to the West for answers to Russia's problems: the nobility often spoke French better than Russian and were often educated along Western lines. It was not surprising that the regime should try again after 1855 to use Western methods of modernization.

The reforms of Alexander II

The accession of a new tsar, Alexander II (1855–81) brought an expectation of change to the system in the direction of progressive Liberalism.

The Statute of Emancipation

It had been axiomatic in Russia that serfdom was a 'necessary evil'. Even Tsar Nicholas I condemned it in principle. Given the economic situation and bearing in mind that he had received a liberal education, Alexander decided in 1857 that emancipation of the serfs was due in the interests of the power as well as the moral prestige of the state and its nobility. The 1861 Statute of Emancipation gave the peasants legal emancipation and also the right to own their own land. But there were notable drawbacks:

(1) The peasants were forced to make very large redemption payments for any land other than their own personal allotments; this was in the interest of the debtor nobility.

(2) The new administrative system, the 'volost' courts etc., was still dominated by the land-holding nobility and was not integrated with the national judicial system.

(3) The system of land tenure remained communal. Thus the most serious obstacle to a liberal/capitalist agricultural system was retained in the new order. This was because the regime would only reform *partly* and not compromise the basis of its tax system and the traditional peasant society – the commune. The result was the continued backwardness of agriculture.

(4) The effect of the Emancipation was further weakened by the counter – attack of the conservatives in the government. After 1861 the agrarian experiment was effectively in the hands of its opponents.

The other reforms

The Emancipation was followed by a series of measures to provide a new state system along liberal lines:

(1) In 1864 the Zemstvos (locally elected assemblies) were set up on an unequal franchise to provide local government services at provincial and county levels.

(2) A liberal system of justice was introduced with a jury system and on the principle of the 'rule of law'. The peasant 'volost' courts were excluded from this system. The police were also reformed.

(3) Education policy was relatively liberalized, universities were made autonomous and censorship was somewhat relaxed. Secondary and elementary schooling was expanded slightly. Education was still seen as socially dangerous, and likely to undermine popular acceptance of the existing order of society.

(4) The Defence Minister, Milyutin, provided the Army with real, in-depth, liberal reforms which prepared it for success in the 1877–8 Russo-Turkish War.

(5) Attempts were made to encourage enterprise by providing credit institutions, such as the State Bank, and by removing barriers to trade *e.g.* the reduction of exports and import duties after 1863.

The impact of the reforms

From 1861, and especially from 1866 with the attempt on Alexander's life, the liberals began to lose credibility and power. This was for several reasons:

(1) The Tsar and the regime were anomalous *as liberals*, because Liberalism called for a parliamentary assembly, which the autocracy denied.

(2) The series of measures after 1861, hailed as 'a revolution from above', really only amounted to half-measures, because Liberalism could not accomplish the task of bolstering up the tsarist regime while modernizing society; and Russian society reacted poorly to the reforms, or at least to many of them. For instance, the increase in political terrorism *after* the reforms dimmed liberal prospects.

(3) The experience of Russian agriculture after Emancipation was varied. In the Ukraine, and in the east, the Emancipation – combined with easier transport – did lead to increased grain production on capitalist lines, which meant that grain exports went up steeply (31% of exports in 1861–5, 47% in 1891–5). But generally speaking, agriculture remained backward, tied down by the communal system and the lack of capitalist enterprise amongst the nobility. The result was the continued impoverishment of a growing peasant population coupled with the bankruptcy of the landed nobility who gave up the land in droves and became officials in the vastly expanded bureaucracy, which increased from 2,000 to 10,000 in the 1870s.

Thus the liberals, such as they were, were swamped by this gentry-bureaucracy.

Russia, 1881–1905

The poor results of the liberal programme, as far as tsarist power and prestige were concerned, meant that – even before Tsar

Alexander II's assassination in 1881 – Liberalism was in many respects a spent force. It is true Loris-Melikov had, in the last year, been used by Alexander to try and provide the necessary reforms and there was even talk of a constitution. But Alexander's death scotched this faint possibility and instead the regimes of Alexander III and Nicholas II embarked on a combination of the repression of political opposition, conservative social policy and a state-led, forced industrialization.

The reaction

The bureaucracy became jealous of any power base but itself:
(1) The Zemstvos came under attack. In 1889 their contact with the peasants, the Justices of the Peace, were abolished and replaced by a Land Commandant who had to be a noble.
(2) The universities were brought under the control of inspectors once more (1884). Censorship was increased.
(3) All other nationalities in the tsarist empire were subjected to Russification. The first pogrom was in 1881.
(4) All political parties in opposition to the regime were banned.

The regime, especially the Ministry of the Interior, followed a very paternalist path in the interests of preserving the traditional social hierarchy:
(1) No attempt was made to abolish communal tenure, in spite of arguments from the Ministry of Finance, because Tolstoy at the Ministry of the Interior felt it was a major agent for social peace and, thus, loyalty to the regime among the peasants.
(2) Education: similar arguments justified the refusal to institute a classless educational system. In secondary education the raising of fees to keep out lower-class pupils in 1887 had the result that there were fewer pupils at this level in 1895 than there had been in 1882. At the elementary level, education was left to the Church. By 1900 only a quarter of the population was literate.

State capitalism

The Ministry of Finance, particularly under Sergei Witte between 1892 and 1903, remained a bastion of Liberalism, of a sort. As in the West, however, development was to be achieved by protectionism and led by the state. Industry was protected by high import duties and encouraged by government demand for the new state railway system, which expanded greatly in the 1890s.

State expenditure and capital investment was paid for partly by a large inflow of foreign loans, made easier after 1897 by putting the Rouble on the Gold Standard. These loans were serviced by revenue from tariffs, but also from the heavy indirect taxes on the urban poor and the peasants, which provided the foreign-currency earner of grain exports.

This hothouse economic programme produced a dramatic expansion of 8% per annum in the 1890s. But it did not stimulate the *internal* market. Demand was state-led and dependent on taking money from the peasants. Thus grain exports were coupled with famines, because the conservative social policy, *viz.* the communes, meant that the agrarian economy could not respond to, or even accommodate, Witte's programme. Therefore from 1900–05 the economy stagnated, as the state ran into financial difficulties. State capitalism left Russia with a much more advanced economic structure, but also with a financially oppressed peasantry and a discontented urban proletariat, the product of a continuing population increase among the peasantry. It was clear by 1905 that something had to change.

The revolution of 1905

Opposition to the autocracy

The growing social turmoil caused by the effects of the modernisation programme found expression in the formation of three, illegal, political parties: (1) the Social Revolutionaries (1901), who represented a new sort of populism, and had redistribution

of the land to the peasants as their first priority; (2) the Liberals (1903), whose main aim was to try to create a liberal democratic constitutional government which could match Russia's newly emerging society; and (3) the Social Democrats (1903), who voiced the growing frustration of the new urban proletariat through the language of Marxism.

Therefore most sectors of Russian society were in opposition to the state: only the noble bureaucrats, the state-dependent industrialists and the Army supported the regime. And even before the Russo-Japanese War broke out in January, 1904, the regime was in a very poor situation. When war resulted in abject defeat, Russia experienced a massive upheaval.

The elements in the 1905 revolution

There were really four uprisings involved:
(1) the rising of the national minorities against Russification, especially in Poland and the Baltic provinces, coupled with demands for political and economic reforms;
(2) the seizure by the peasants of what they saw as their land – *i.e.* the nobles', Church's and state lands – due to the pressures of over-population;
(3) the rising of the urban proletariat – through illegal strikes and demonstrations – against their employers and the autocracy; and
(4) a campaign by the Union of Liberation based on the French banquet campaign of 1848 to force the regime to liberalize.

The events of 1905

On the 9 January workers demonstrating against conditions were fired on by troops outside the Winter Palace. 'Bloody Sunday' was a tremendous blow to the Tsar's prestige and uprisings spread. In February decrees were issued by Nicholas II to try to quell opposition: they promised 'participation' by the people in government. In October there was a general strike by the railwaymen. The regime was now paralysed.

Witte returned from the peace negotiations and took over the government. On 17 October the Tsar issued the 'October Manifesto' which promised an elected Duma with legislative powers and civil liberties. This did not immediately break the opposition, but it did produce a split in the Liberals between the 'Octobrists' and 'Kadets', the former accepting the October Manifesto in good faith, the latter wanting more.

Soviets (workers' councils) were formed in St. Petersburg and Moscow. During October – November there were large-scale peasant riots but workers in St. Petersburg and Moscow alienated their liberal allies by continuing actions against their employers. In December the St. Petersburg soviet was dispersed and Moscow street risings were put down by the regime's forces (by this time, troops were returning from the Far East).

With the issue of the Franchise Law in December, 1905, and the subsequent Fundamental Laws, the government indicated that it had reasserted its control over the situation.

The period of the Dumas

The effect of the 1905 revolution was to create a constitution in Russia and a parliamentary body in the Duma. It also brought about a rethinking of economic and social policy under the ministry of Stolypin. However, the renewed attempt at westernisation through Liberalism again experienced fatal problems so that, under the pressure of another war, the newly-renovated regime again failed.

Constitutional politics

Parliamentarism When the First Duma met in April, 1906 the weakness of the regime had been remedied. The Fundamental Laws already pointed towards a trimming of the Duma's power.

Its radical composition – 30% workers or peasants – ensured its early dissolution. The Second Duma was, however, equally radical; meeting in February, 1907, it was dissolved in June. Only with the Third Duma was there a body which would co-operate with the tsarist order. However, this was based on a restrictive franchise and followed a new bout of repression of opposition.

In practice the Third and Fourth Dumas became only another agency of the huge, sprawling bureaucracy; Russia, by the eve of war in 1914, justified Witte's claim that it was 'ungovernable'. The bureaucracy, the Army, the Council of State, the Duma and the Zemstvos were all competing for power.

Democracy The Electoral Law of 1907 effectively disfranchised the vast bulk of Russians and meant that they remained outside of political decision-making. Government policy towards strikes etc. remained harsh, as seen in the aftermath of the Lena gold-mine strike in 1912. Moreover, the Duma was sometimes more reactionary than the government, as in the continued pursuit of Russification.

The lack of social cohesion before 1914 meant that if the government were in difficulties it could not count on the alienated masses for support. By 1917 even the majority of the Duma, the 'Progressive Front', was calling for radical change.

Economic and social policy

Stolypin's reforms, 1906–11 Stolypin, chairman of the Council of Ministers from 1906, decided that the peasants must be freed from the commune if the 'sober and strong' were to provide the capitalist agriculture and internal market, the absence of which he saw as the main impediment to development. He was probably right.

However, his famous reforms between 1906–11, encouraging peasants to leave the communes, were very disappointing in their results. It now looks as though most peasants were content to continue the communal system and were more intent on increasing their communal landholdings by purchase or by seizure. By 1917 the peasants owned roughly 90% of the land – due to the retreat of the nobility and the need to provide for an ever-increasing peasant population. In many areas the amount of large-scale capitalist production of grain actually decreased as landlords gave up.

Therefore, although the period after 1906 saw a more prosperous peasantry (helped by the state through loans etc.), the central problems remained unresolved. Thus, during the war, peasant farmers retreated from the market and fed their grain to their livestock.

Industry Russia experienced an economic boom from 1906–17 comparable with that of the 1890s. The State again played a leading role with an increased military budget, especially after 1914. Also its inflationary monetary policy to finance the war fuelled the boom in its later stages. The expansion was characterized by ever larger foreign loans. The period saw the rise of huge cartels on the German model. The labour force, not given the same kindly treatment as the peasants, from 1910 became more radical, thus culminating in the St. Petersburg strike of 1914.

Russia in 1917

The war entered in 1914 was the end towards which tsarist economic policy had prepared since the defeat of the Crimean War. At first it caused the nation to unite behind their Tsar. With the stalemate of 1914–16 came disenchantment with the regime, especially given the Rasputin scandal and the terrible financial and bureaucratic mismanagement, itself a result of the ramshackle, anarchic nature of a government apparatus which had never been fully rationalized. This caused overwhelming political pressure from all sides.

Furthermore, the unsolved problem of Russian agriculture crippled the home front in 1917. Due to the chaos caused by war requisitioning and inflation (due to the lack of adequate revenues) and official grain prices, internal transport and trade collapsed and the peasants refused to sell on the black market. The effect of this was the starving masses of St. Petersburg in February, 1917. When the troops, themselves mostly peasants and demoralized by the mismanagement of the war, joined them, the attempt by the tsarist regime to force Russians to maintain its great power status in the new industrial era had come to a close.

7.4 CHRONOLOGY

1855–81		Reign of Tsar Alexander II, the 'Tsar Liberator'.
1861	**March**	THE EMANCIPATION EDICT: all serfs were given personal freedom together with allotments of land for which the original owners were paid by the State in treasury bonds. The peasants were to repay the bonds over a 49-year period. The land was given to the 'mir', village commune, for distribution.
1863–4		THE SECOND POLISH REVOLUTION: this spread into White Russia and Lithuania. Suppressed in May, 1864. Polish autonomy was abolished and Russian administration re-established.
1864		THE ZEMSTVO LAW: the establishment of a system of local self-government. Local boards, Zemstvos – on which the nobility, townsmen and peasants were represented – were empowered to levy taxes for such purposes as primary education and public health. THE REFORM OF THE JUDICIARY: the old system of class courts was abolished, and modernized procedures substituted. The separation of judicial authority from the administrative.
1870		REFORM OF MUNICIPAL GOVERNMENT: the towns given self-government under a council elected by the propertied classes.
1874		ARMY REFORM: the introduction of the principle of universal military liability in place of the former system of taking recruits only from the lower classes.
1875–8		THE EASTERN CRISIS.
1876		The secret society, 'Land and Liberty', spearhead of the Populist movement founded.
1879		Organization of the society 'Will of the People'.
1881	**March**	Assassination of Tsar Alexander II.
1881–94		TSAR ALEXANDER III: the autocratic system affirmed; the influence of the Metropolitan Pobedonostsev.

Chronology cont.		Persecution of religious dissenters and discrimination against the national minorities. Sergei Witte. The industrialization of Russia.
1881		CONCLUSION OF THE ALLIANCE OF THE THREE EMPERORS BETWEEN GERMANY, RUSSIA AND AUSTRIA: renewed in 1884, but in 1887 was replaced by a separate pact between Germany and Russia, the REINSURANCE TREATY.
1891–94		CONCLUSION OF THE FRANCO-RUSSIAN ALLIANCE.
1894–1917		Reign of TSAR NICHOLAS II:
1898		Formation of the SOCIAL DEMOCRATIC PARTY among the industrial workers. Marxism had been introduced into Russia by PLEKHANOV. His moderate programme was challenged by the radical wing under LENIN.
1901		Organization of the SOCIAL REVOLUTIONARY PARTY.
1903		SPLIT IN THE SOCIAL DEMOCRATIC PARTY AT THE PARTY CONGRESS in London: MENSHEVIKS and BOLSHEVIKS. Formation of the Union of Liberation, calling for a liberal constitution.
1904–5		THE RUSSO-JAPANESE WAR, a result of Russian expansion in the Far East. Outbreak of hostilities, 8 February, 1904; fall of Port Arthur, January 1905; the battle of Mukden, 23 February–10 March; naval disaster at Tsushima.
1904	**July**	Assassination of the Minister of the Interior, Plehve.
	November	Zemstvo Congress met at St. Petersburg and demanded the convocation of a representative assembly and the granting of civil liberties.
1905	**January**	BLOODY SUNDAY: the emergence of workers as a factor in the protest movement. A procession of workers led by Father Gapon fired on. Outbreak of strikes.
	March	The Tsar announced his intention to convoke a consultative assembly.
	May	THE UNION OF UNIONS: this brought together the liberal groups in a demand for universal suffrage and parliamentary government.
	June–August	Strikes, agrarian outbreaks, national movements in the border provinces, mutinies in the army and navy. The POTEMKIN incident.
	August	THE TSAR ISSUED A MANIFESTO CREATING THE IMPERIAL DUMA: this assembly to be elected on a restricted franchise with only deliberative powers. The failure of this concession to satisfy popular demands.
	20–30 October	THE GENERAL STRIKE: a spontaneous movement across the country.
	26	The St. Petersburg workers formed the first SOVIET to direct the strike.
	30	THE OCTOBER MANIFESTO: granted Russia a constitution. The Duma to have legislative power. Witte was appointed Prime Minister. The liberal group divided into the CONSTITUTIONAL DEMOCRATIC PARTY and the OCTOBRIST PARTY.
	December	Members of the St. Petersburg Soviet arrested. Insurrection of workers in Moscow. Punitive repression; the 'Black Hundreds'.
1906	**May**	Dismissal of Witte. THE FUNDAMENTAL LAWS issued. The Tsar retained complete control over the executive and armed forces. The legislative power to be divided between the Duma and Imperial Council. The government reserved the right to legislate by decree when the Duma was not in session. THE MEETING OF THE FIRST DUMA: The Constitutional Democrats (the Kadets) formed the largest party. Criticism of the government.
	July	THE DUMA DISSOLVED: Kadet leaders issued the VIBORG MANIFESTO calling upon the country to refuse to pay taxes.
	November	The agrarian reform of STOLYPIN.
1907	**March – June**	THE SECOND DUMA: reactionary groups pressed for a return to the autocratic system, forcing the dissolution of the Duma. New electoral law, greatly increasing the representation of the propertied classes.
1907–12		THE THIRD DUMA: returned a conservative majority; the suppression of revolutionary disorders. Reforms, local administration, justice and health insurance for workers.
1907	**August**	CONCLUSION OF THE ANGLO-RUSSIAN ENTENTE.
1908–9		THE CRISIS WITH AUSTRIA OVER BOSNIA AND HERZEGOVINA.
1911	**September**	The assassination of Stolypin.
1912–16		THE FOURTH DUMA suspended by the Tsar for much of the war. A clear warning of impending revolution given to the Tsar in November 1916.
1912–13		THE BALKAN WARS.
1914	**1 August**	GERMAN DECLARATION OF WAR ON RUSSIA.

7.5 QUESTION PRACTICE

1 **'A disappointing liberal, an inefficient autocrat'. Examine this assessment of Alexander II.**

Understanding the question

The question is about Alexander II and the myth of the 'Tsar Liberator'. It asks you to examine the traditional view expressed in the title. In doing so one must see if it describes Alexander adequately. In addition, one should examine the assumptions behind the question, especially concerning Liberalism in the Russian context, to assess whether or not the judgement in it is fair or not.

Knowledge required

The economic and social structure of Russia in the mid-nineteenth century. Political events from c. 1850–1881. Alexander's personality and policies, especially details of the Great Reforms.

Suggested essay plan

1   The traditional view of Alexander II as the 'Tsar Liberator', 'Russia's White Hope': his reign seen as starting purposefully, losing impetus, then going forward again, only to be cut short by assassination before progress and Liberalism had really begun. The period presented as a chance missed for Russia to enter the modern world; hence the 'Tsar Liberator' becomes 'an inefficient autocrat' and also a 'disappointing liberal'.

2   Examine Alexander's attitude to his own authority: autocratic power must be used to reform the state when necessary – he adopts liberal policies to preserve the Russian autocracy. The state and nobility must be seen to lead, hence the reforms. He justified the emancipation of the serfs by claiming, among other things, that it benefits the nobility. Liberalism was a means to an end, the shoring-up of the power and the prestige of the autocracy (which is identified in Alexander's view with the interests of Russia itself). When unrest continues, Alexander retreats and the liberals are disappointed. But Alexander is not really a Liberal.

3   The connection of 'autocracy' and 'Liberalism' in Russia; the progressive bureaucracy was western educated; the belief in benevolent absolutism; hence the emancipation and other reforms. But Liberalism in the fullest sense is not just about administrative measures, but also about the *form* of the administration – *i.e.* the parliamentary representative form of government. This was rejected by Alexander, who refused, for instance, to contemplate a national assembly. In fact in the circumstances a Russian national assembly would not have been 'liberal' in outlook but a collection of reactionary nobles!

4   A brief comment on the social background: Russia was *not* a capitalist country; the peasants' organization was communal and remained so after the emancipation of the serfs; there was debate about this, but the peasants were not treated as individual farmers; the retention of the commune as an institution meant that modernizing capitalist enterprise was held back.

5   The reaction during the latter part of his reign must be seen in the context of: (*a*) the failure of many of the liberal reforms to increase state power or prestige (for example the adverse reaction to the 1861 emancipation proposals, the acquitting of terrorists by the new juries); (*b*) the increasing conservatism of the bureaucracy in the 1870s; and (*c*) increasing revolutionary activity; note that Russian populism was anti-liberal.

6   At the very end Alexander did try to prevent unrest by structural liberal changes, *i.e.* the Loris-Melikov proposals, but these were not very liberal and not 'constitutional' in the western sense. Moreover, Loris-Melikov found himself alone against the nobles in the bureaucracy.

7   Conclusion: Alexander II was *not* a Liberal but an autocrat; he gave the appearance of inefficiency, but he achieved a considerable amount. To some extent he gave Tsarism a new lease of life. In the last resort his reign was disappointing because the liberal policies he adopted only increased opposition and did not achieve the modernization of Russian society. It is questionable, though, in the circumstances whether this was possible; one must balance Alexander's 'failure' against the inappropriateness of Liberalism in the Russian setting.

2 **How successful was Tsar Alexander II in solving the problems facing Russia during his reign?**

Understanding the question

You should discuss the problems of Alexander II's reign; then assess how far they were solved during his reign, and what effect his policies had after 1881.

Knowledge required

Russia's economic and social structure in the nineteenth century. Alexander's personality and policies. A general knowledge of historical developments in Russia before and after Alexander's reign.

Suggested essay plan

1   The reign of Alexander II, the 'Tsar Liberator', is commonly seen as a progressive period in which Russia *almost* broke out of its medieval backwardness and developed into a modernizing liberal society. He is presented as half-successful; he introduced major reforms in the 1860s, but these were flawed by the reaction of the 1870s. This traditional view needs some qualification. There were the problems as the tsarist regime perceived them *and* the deep-rooted obstacles to any easy transition from a feudal to an industrial society.

2   The major problems facing the regime were bankruptcy of the nobles who governed the serfs; serfdom was no longer profitable and Russian agriculture was unable to compete with the 'new lands' (*e.g.* America) in the grain markets; the bankruptcy of the regime; the general backwardness of Russian society revealed in the defeat of the Crimean War, which itself led to more debts. Alexander II saw a threat to the authority of the Tsar if he did not renovate the Russian state through 'reform from above'.

3   Alexander's solution: the Emancipation; liberalization of the law, education and the economy; the building of railways which, apart from providing improved communications, helped to increase trade and hence revenue; the Zemstvo reforms; the reform of the Army by Milyutin, which led to victory in the Russo-Turkish War.

3   In fact, though, these reforms were flawed: at a deeper level there were problems which went unanswered; the autocratic basis of government remained; the nobility dominated the bureaucracy and resisted radical proposals for reform; in important respects the economic structure of society was little changed – for social and administrative reasons the commune remained and so peasants tended to continue to be subsistence farmers.

4   In the short term Alexander's reforms appeared quite successful in meeting the *perceived* problems, but disappointment during the reign and the crisis of terrorism at the end can be explained by underlying problems which were ignored. Note the poor reaction to the reforms: revolts by the peasantry were more numerous after the Emancipation than before. Liberalism did not really work unless certain crucial Russian institutions were

swept aside. This was the lesson of the period and such radical reform was beyond both the power and the intention of the regime.

5   It could be argued that Alexander II managed to preserve the regime a while longer through reform, but did not really solve the underlying problems. In the longer term the new course he had set would conflict disastrously with Russian traditions and lead to the eventual demise in 1917 of the tsarist order.

3   How far may the period 1881–1914 be regarded as 'wasted years' in the solution of Russia's internal problems?

Understanding the question

You need to assess the effectiveness of the tsarist regime's policies in trying to bring about the modernization of Russian society and its economy. Further, you should be able to discuss alternative policies and their likelihood of success in the Russian setting and to judge whether there was any alternative which would have been less wasteful.

Knowledge required

Tsarist policies 1881–1914. The economic and social structure of Russia during the same period. A general idea of notions of progress and economic development in Europe c. 1900.

Suggested essay plan

1   Russia embarked during the second half of the nineteenth century on a conscious policy of economic development led by the state, which has been described as 'state capitalism'. But this was under an oppressive regime, whose inadequacies were to lead to defeat and revolution in 1917. Much has been written about what could have happened if the regime had been more liberal and progressive. However, these years can only be called 'wasted' if there was a *viable* alternative – given the situation in 1881.

2   Comment briefly on the policies pursued after 1881, particularly Witte's strategy for industrialization, noting that the political order remained autocratic and bureaucratic.

3   The result was that, when Russia went to war and was beaten by Japan in 1904–5, there was a severe political crisis, rebellions among the minority nationalities, peasant riots, upheavals in the capital; this was a direct result of the tsarist policies: an inefficient bureaucracy and Army, starving peasants, an exploited working class, indignant Poles and a revolutionary intelligentsia.

4   However, it can be argued that the alternatives proposed during this period are either impracticable or harmful:

(a) It was too late to revert to a non-developed society.

(b) Russian Populism had been discredited by the apathy of the peasantry.

(c) A Marxist revolutionary solution looked highly implausible in a society which was 80% peasant; in any case both Marx and Lenin thought in terms of an accompanying Western European revolution.

(d) The only acceptable alternative to tsarist repression seemed to be Liberalism, the guarantee of individual and institutional liberties, the guarantee of property and representative government.

5   Note the failure to follow up Alexander II's reforms in the succeeding period to 1905; *e.g.* agriculture was still left under communal control; there was the failure to provide a constitution to regulate government; the Zemstvos were not given adequate authority; there was only 21% literacy in Russia in 1900.

6   Improvements were enacted, though, by the government under Stolypin *after* the 1905 revolution:

(a) the 1906 Land Reform Act;

(b) educational opportunities were greatly increased; and

(c) there was the Duma, a new representative assembly. *However*, Russia remained an autocracy and the new liberal policies can be seen in a number of respects to be very superficial. The Duma was largely ignored by the Tsar after the October Manifesto. The franchise was reduced to produce a more manageable assembly. The fact is that Liberalism had *no real power base* in Russian society. Furthermore, by 1914 only a small percentage of the land had been transferred from communal to individual tenure.

7   Russia had major problems in its economy and society which it could not easily solve. The government policy after 1881 probably made things worse by overstraining the economy. But given their aim of restoring Russia to the status of a great Power, there was arguably no truly viable alternative in the Russian context. In that sense it was unrealistic to class the years 1881–1914 as 'wasted'. It would be more appropriate to observe the tragedy of a country of such potential being unable to shed the shackles of a bygone age before 1881.

4   Document question (Russia – 1905 and after): study extracts I and II below and then answer questions (a) to (g) which follow:

Extract I (concerning the events of 1905)

'After 9 January (Bloody Sunday) the revolution showed that it controlled the consciousness of the working masses. On 14 June, by the rising on board the Potemkin Tavrichesky, the revolution showed that it could become a material force. By the October strike it showed that it could disorganize the enemy, paralyse his will, and reduce him to complete humiliation. Finally by organizing workers' Soviets throughout the country, the revolution showed that it was able to create organs of power.

'...After 17 October, when the conditions for the constitutional deal were already written down, and it seemed that all that was left was to put them into effect, the revolution's further work obviously undermined the very possibility of such a deal between the liberals and the authorities. From then on the proletarian masses, united by the October strike and organized within themselves, put the liberals against the revolution by the very fact of their existence.'

(Trotsky, '1905')

Extract II (concerning developments of 1906–1908)

'Driven into a historical cul-de-sac by the irreconcilable attitudes of the nobility and the bureaucracy, who once more emerged as the unlimited masters of the situation, the bourgeois parties are once more looking for a way out of the economic and political contradictions of their position – in imperialism....

'The so-called "annexation" of Bosnia and Herzegovina was greeted in St. Petersburg and Moscow with the deafening clatter of all the old ironware of patriotism. And the Kadet party, which, of all the bourgeois parties, claimed to be the most opposed to the old order, now stands at the head of militant 'neo-Slavism'....

'...The same government that buried the reputation of its strength in the waters of Tsushima and the battlefields of Mukden; the same government that suffered the terrible sequel of its adventurist policies, now unexpectedly finds itself patriotically trusted by the nation's representatives...it receives the Duma's support for its new adventures in the Far East. More than that: by Right and Left, by the Black Hundreds and the Kadets, it is actually reproached because its foreign policy is not active enough.'

(Trotsky '1905')

	Maximum marks
(a) State briefly what happened on 'Bloody Sunday'.	(3)
(b) What did Trotsky mean by 'the revolution showed that it could become a material force'?	(2)
(c) Identify each of the following: 'Soviets'; 'the Kadet Party'; 'the Black Hundreds'.	(3)

(d) Explain what is meant by
(i) 'the deafening clatter of all the old ironware of patriotism', and
(ii) 'militant neo-Slavism'. (4)

(e) Explain Trotsky's references to 'Tsushima' and 'Mukden'. (4)

(f) Discuss the nature of the 'constitutional deal' and explain what, according to Extract I, were the main difficulties facing 'the proletarian masses' after October, 1905. (6)

(g) How adequate is Trotsky's view in Extract II of the results in Russia (during the years 1906–8) of the 1905 revolution? (9)

7.6 READING LIST

Standard textbook reading

J. A. S. Grenville, *Europe Reshaped 1848–78* (Fontana, 1976), chapter 13.

David Thomson, *Europe since Napoleon* (Penguin, 1977), chapters 15–17.

Anthony Wood, *Europe 1815–1945* (Longmans, 1975), chapters 18, 22.

Further suggested reading

J. P. T. Bury (ed.), *The New Cambridge Modern History 1830–70*, volume X (Cambridge University Press, 1964), chapter 14.

R. Charques, *The Twilight of Imperial Russia* (Oxford, 1965).

M. T. Florinsky, *Russia, a History and an Interpretation*, volume II (Macmillan, 1969); and *The End of the Russian Empire* (Collier Books, 1961).

L. Kochan, *The Making of Modern Russia* (Penguin, 1970).

W. E. Mosse, *Alexander II and the Modernization of Russia* (English Universities Press, 1958).

Hugh Seton-Watson, *The Russian Empire 1801–1917* (Oxford, 1967).

8 The origins of the First World War

8.1 INTRODUCTION

The alliance system created by the German Chancellor Bismarck after the Congress of Berlin in 1878 was intended to forestall the possibility of a war on two fronts against the new German Empire. After his fall from power there was a gradual realignment of the Powers which brought Britain, France and Russia together and, in 1914, to war with the Central Powers.

After 1871 Germany was unquestionably the dominant continental Power; under Kaiser William II she increasingly demanded recognition as a world Power. German expansion and influence, particularly the construction of a large Navy under the guidance of Admiral Tirpitz, led to the growing rivalry of the Powers which characterized the years before the First World War and was illustrated, for instance, in the two Moroccan crises.

The immediate cause of the war, however, lay in the conflict between Slav nationalist aspirations, especially of Serbia, and the Austro-Hungarian Empire. As could be seen in the Bosnian crisis of 1908, the Balkan area was increasingly drawing the Habsburgs and Germany into rivalry and confrontation with Russia.

The assassination of Archduke Francis Ferdinand at Sarajevo on 28 June led to the conflict. The French decision to support Russia and the 'blank cheque' pledge of Germany to Austria-Hungary transformed a local Balkan conflict into a European war.

The question of the responsibility for the outbreak of war continues to be a major debate among historians – a debate in which the role of public opinion, domestic political problems and considerations of military planning have all, in recent years, been given particularly detailed attention.

8.2 STUDY OBJECTIVES

1 The distinction between the long-term and underlying causes of general conflict in 1914 and the immediate occasion of war.

2 The alliance system: as background, Bismarck's network of alliances and its objectives, noting particularly the Dual Alliance and the Reinsurance Treaty; the origins and development of the Triple Alliance and the Triple Entente; an assessment of the individual foreign policy aims of Britain, France, Russia, Italy and Austria-Hungary and of the rivalries of these Powers.

3 German foreign policy from 1890: 'Weltpolitik' (World Policy), the demand for a 'place in the sun' and for colonies, for preponderance in Central and Eastern Europe ('Mitteleuropa'), and for influence in the Near East; the influence of the Kaiser, Bülow, Holstein, Kiderlen-Wächter and Bethmann-Hollweg on the conduct of German policy; the underlying domestic factors of the rapid growth of German population and industrial might.

4 The major pre-war crises, 1914: the First Moroccan Crisis; the Bosnian Crisis; the Second Moroccan Crisis; the Balkan Wars.

5 An understanding of the underlying conflict of interest over the Balkans: the development of nation states and their rivalries; the growth of Panslavism and Pangermanism; the attitude of the Habsburg Empire to the threat of Slav nationalism, particularly Serbian ambitions; the confrontation of Russia on the one hand of Austria-Hungary and Germany on the other over this area.

6 The naval rivalry between Britain and Germany: Tirpitz and the Navy Laws; Fisher and the Dreadnought programme; the impact of this on public opinion in the two countries.

7 Military preparations: the application of science and technology to warfare after 1870; planning and timetabling, involving the growth of permanent general staff; the plans for mobilization and their influence on governments; an examination in particular of the way in which military thinking in Europe was dominated by the implications of the Schlieffen Plan and how the urgent need of both France and Germany for rapid mobilization and for an early offensive by their eastern allies accelerated the tempo of the crisis in 1914.

8 The pre-war mood of public opinion towards the prospect, or possibility, of war; the way in which domestic social and economic pressures (particularly in Germany) intensified international tensions.

9   The Sarajevo assassination and the July Crisis: a detailed grasp of the events leading to war and the way in which a local conflict became a general war; particular attention to be paid to the German 'blank cheque' to Austria, the Austrian ultimatum to Serbia, Russian mobilization; the role of Britain and France and the question of Belgian neutrality.

10   The 'war guilt debate' – the historical controversy over the respective responsibilities of the Powers for the outbreak of war in 1914.

8.3 COMMENTARY

"The 'nightmare coalition'"

The system of alliances which Bismarck created after the Congress of Berlin was intended to preserve peace in Europe and, more particularly, to prevent a war against Germany on two fronts. Such a 'nightmare coalition' of Russia and France would, as he saw it, prove fatal to the new German Empire.

It was just such a war which faced Germany in August 1914. A clear appreciation of its origins must take account of how Europe became divided into the Triple Alliance (Germany, Austria-Hungary and Italy) and the Triple Entente (Great Britain, France and Russia). In the event, from 1915, Italy was to side with the Entente.

The collapse of Bismarck's system

Bismarck's essential objectives had been (1) to isolate France, and (2) to prevent any military conflict, between Austria and Russia over the unsettled countries of the Balkans, such as would force Germany to choose between the Powers.

His complicated system achieved the diplomatic isolation of

France, an understanding with Russia (as in the *Dreikaiserbund* and the *Reinsurance* Treaty) and the maintenance of European peace.

At the same time, this period of the 'armed peace' was one of unprecedented growth in military and naval power and increasing imperial rivalry.

The network of alliances started to disintegrate after Bismarck's fall from power in March 1890. In that year Kaiser William II allowed the Reinsurance Treaty to lapse. The cornerstone of Bismarck's construction was gone. In 1891 an entente, or understanding, was signed between France and Russia. This was followed by a military convention in which the two partners promised to give each other assistance in the event of an attack by Germany.

The diplomatic revolution

In 1904 the Entente Cordiale was established between France and Great Britain, to be followed by the Anglo-Russian understanding of 1907. These were settlements of outstanding colonial rivalries and brought to an end a period in which Anglo-French rivalry (as in the Fashoda crisis of 1898) and Anglo-Russian friction were dominant factors in international affairs.

These were now replaced by a growing fear of the dramatic increase of German power and its imperial demands.

The principal step in the Entente Cordiale was the exchange of French recognition of British dominance in Egypt in return for a French sphere of influence in Morocco. From 1906 onwards this was followed by military discussions between the British and French authorities.

The Anglo-Russian Convention reached agreement over Afghanistan, Tibet and the division of influence of the Powers in Persia.

The emergence of Europe into two rival and increasingly

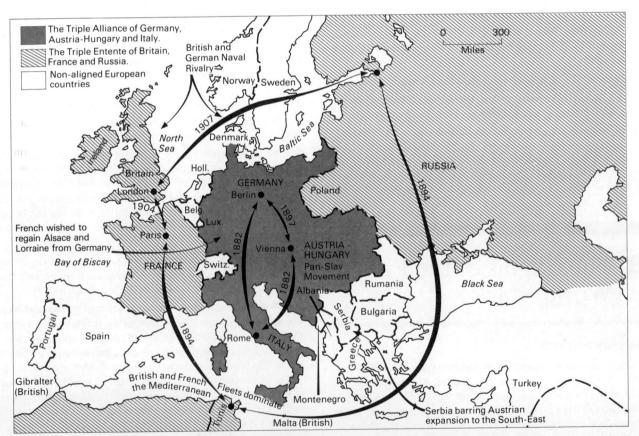

Fig. 4 The origins of the First World War. It is important to understand the reasons for and implications of the emergence of two rival armed camps prior to 1914

hostile camps reduced much of the flexibility of the old balance of power and made it that much harder for statesmen to preserve the peace. In itself this did not make war with Germany inevitable. But it made it more likely.

This was particularly well illustrated in the Moroccan crises of 1905 and 1911, which strengthened the resolve of the other Powers to resist German expansion and demands. The first of these coincided significantly with the weakening of Russia following the Russo-Japanese War.

German world policy

The unification of a majority of Germans under the leadership of Prussia had helped to make Germany into the greatest concentration of power on the continent. Rapid industrialization, concentrated military power, a young and dramatically increasing population (43% increase between 1880 and 1910) had raised it into a position of potential domination over Europe.

With an expanding economy and overseas colonization, she now started to demand recognition as a world power.

The term 'Weltpolitik', or world policy, was first used by Kaiser William in 1896 as a description for the German version of an imperialism which was shared by the other Powers. It was summed up in his statement that 'nothing must henceforth be settled in the world without the intervention of Germany and the German Emperor'.

This concept of expansion found expression in the construction of the Baghdad railway, the launching of a large navy and the quest for colonies. (Already, in the 1880s, Britain and Germany had come into conflict over African and Pacific territories.) Through their naval and colonial ambitions they were seen to menace British imperial interests. By assuming the role of protector of Turkey, Germany was felt to be thwarting the traditional aims of Russia in Asia Minor and in the Balkans.

The naval race

The major instrument for an increasing world role was the construction of a strong navy and this led to growing rivalry with Britain. This was at a time when German economic power was developing, relatively, at the expense of Britain and there was mounting apprehension over Germany's commercial potential.

Admiral von Tirpitz used political persuasion, propaganda and the support in particular of industrialists to gain approval for a German fleet large enough to deter any hostile British action.

At the same time, the British were determined to maintain their wide margin of superiority on which traditionally their supremacy had rested. From 1898 to 1912 successive Navy Laws underwrote this competition, which was seen primarily in terms of the heavy-gun battleship.

This mutual suspicion was underlined when the Haldane Mission of 1912 attempted to reach an understanding with Germany: the German government was willing to offer minor modifications of its naval building programme, but it demanded in return wide colonial concessions and an assurance of British neutrality in the event of war.

Such a promise of neutrality would have meant the destruction of the Anglo-French Entente of 1904 and was quite unacceptable to Great Britain.

The Balkans

Though the changed diplomatic scene operated to bring the Great Powers to war in August 1914, the crisis from which the war immediately originated did not lie in that system itself, in the military and naval competition or in the clash of colonial and economic interests.

The immediate cause sprang from the tension in the Balkans

and the conflict between Slav nationalist aspirations and the Austro-Hungarian Empire.

It was the combination of intense nationalist unrest and instability in the Balkans with the conflicting interests of the Powers, particularly Russia and Austria-Hungary, which made the situation there so explosive.

Austria believed that the recognition of South Slav demands would bring about the collapse of the Empire.

Austria and Russia were the main rivals, though Germany was showing an increasing interest in the Balkans and the Near East (see, for instance, her involvement in Turkey).

Austria-Hungary had a definite interest in preventing the spread of Russian influence through the area, both in those parts immediately affecting her interests, like Serbia, and at Constantinople and the Straits.

Russian nationalists, on the other hand, emphasized their Panslav affinities with the Balkan peoples and stated their determination that the Slavs should not be humiliated by the Austro-Hungarian Empire.

At the same time Russia wished to preserve access to the Mediterranean and, for her own strategic reasons, was anxious to resist any expansion of Austro-Hungarian influence which might make this difficult.

Bosnia and Herzegovina

A major step towards conflict came when Austria annexed Bosnia and Herzegovina in 1908, territory which it had previously only administered.

The Austrian Foreign Minister, Aerenthal, believed that the Empire was close to collapse and that only a policy of annexation could restore Habsburg fortunes. He particularly wanted to humiliate Serbia, which was encouraging Slav nationalism.

He had promised to allow Russia a sphere of influence in the Straits in return for this annexation. But this expansion of Russian influence was blocked by the other Powers.

The issue represented a major diplomatic reverse for Russia, which she was determined not to allow to be repeated.

In May 1909 Germany issued a note to Russia calling upon her to abandon support for the Serbs and to recognize the annexation.

The Balkan Wars

Following Italy's attack on the Turkish Empire for the acquisition of Tripoli, the First Balkan War broke out.

In this, Serbia, Greece, Montenegro and Bulgaria formed a Balkan League to attack Turkey (1912). All wanted parts of Macedonia, but Serbia had wider ambitions. She wanted to provoke a war between Austria-Hungary and Russia which would allow her to incorporate Austria's South Slav provinces.

The Second Balkan War followed in 1913, when Bulgaria attacked Serbia and Greece for refusing to hand over the bulk of Macedonia. Rumania and Turkey seized the opportunity to declare war on Bulgaria.

In the Treaty of Bucharest of August, 1913, Serbia and Greece acquired most of Macedonia. Rumania also gained territory while Turkey recovered part of Thrace. Bulgaria was left with small areas of Thrace and Macedonia.

Sarajevo

The main loser from the Balkan wars was the Habsburg Empire. Serbia was strengthened and had Great Power support: when the Serbs re-entered Albania, both France and Russia backed them.

With this challenge of militant Slav nationalism to the status quo, the Powers moved to defend their own interests. As the Austrian Chief-of-Staff, Conrad, saw it, the loss of the South Slav provinces would reduce the Empire to the status of a small

Power. He hoped to save it by a preventive war against Serbia.

The assassination of the Archduke Francis Ferdinand by Gavrilo Princip, a Bosnian living in Serbia, on the 28 June 1914, led to the culmination of a long-term and mounting conflict of interest between Slav nationalism and the Empire.

The European crisis

The Austrian decision to settle scores with Serbia once and for all was the act that set the European crisis in motion. In contrast with 1908 and 1913, Russia this time stood firmly behind Serbia.

The French decision to back Russia and the unconditional pledge of Germany to Austria, the 'blank cheque', brought the network of alliances and obligations into play. They transformed a local conflict into a European war.

The Austrian decision to proceed against Serbia was made transparently clear in the ultimatum of 23 July, which the British Foreign Secretary, Sir Edward Grey, described as 'the most formidable document I have ever seen addressed by one State to another that was independent'.

Germany felt herself to be increasingly isolated and 'encircled' by the Entente and particularly fearful of the growing strength of Russia. There was a mood among the German High Command of 'now or never'. Germany was unclear whether Britain would go to war.

Her strength, military preparedness and challenge to the interests of the other powers were linked by the summer of 1914 with the determination of her only reliable ally, Austria-Hungary, to arrest her own rapid deterioration as a Power.

The war guilt clause (Article 231) of the Treaty of Versailles attributed responsibility for its outbreak to Germany and her allies. From the start this provoked violent controversy and resentment. It was immediately used to discredit the Versailles settlement and its reparations demands.

Many historians continue to see the conflict as overwhelmingly the result of German ambitions and actions.

Others would support the view put forward by the British Prime Minister, Lloyd George, that the Powers stumbled into war.

Still others lay special emphasis on wider factors such as imperial rivalry, pressure of population, the development of military technology or a 'general will to war'.

Most do not lay sole blame on any one of the Powers. France, Great Britain, Russia, Austria-Hungary and Serbia have all been apportioned responsibility. For example, Britain's failure to make it absolutely clear to Germany that she would support France.

All governments made miscalculations and suffered from misconceptions.

During the later stages there was also the crucial consideration of military momentum as a result of advance planning, *e.g.* the German Schlieffen Plan and a desire not to lose the race of mobilization which statesmen found it impossible to resist.

8.4 CHRONOLOGY

1890	**March**	Dismissal of Bismarck.
	June	REINSURANCE TREATY not renewed.
1891	**August**	FRANCO-RUSSIAN ENTENTE: agreement to consult in the event of threat of aggression to one of the Powers.
1892	**August**	Franco-Russian Military Convention.
1893	**January**	FRANCO-RUSSIAN ALLIANCE SIGNED.
1895	**January**	The Kaiser sent a telegram to Kruger, President of the Transvaal, congratulating him on the defeat of the Jameson raid. Indignation of British public opinion.
1897	**June**	ADMIRAL TIRPITZ nominated State Secretary for the Navy.
1898	**April**	The Reichstag ratified THE FIRST NAVY LAW: German decision to build a battle fleet strong enough to deter the strongest naval power from attacking. THE GERMAN NAVY LEAGUE founded.
	September–November	THE FASHODA CRISIS: on the Nile between Kitchener and Major Marchand. The most acute confrontation between Britain and France in the pre-war period, threatening all-out conflict.
1899	**May–July**	FIRST HAGUE PEACE CONFERENCE: called at the instigation of Tsar Nicholas. A permanent court of arbitration provided.
	October	Start of the Boer War.
1900	**June**	Reichstag ratified SECOND NAVY LAW: fleet of 38 battleships to be built in 20 years.
	June–August	The Boxer Rising in China.
1901	**October–December**	Collapse of Anglo-German alliance negotiations.
1902	**January**	ANGLO-JAPANESE ALLIANCE formed: the end of 'splendid isolation'.
1904	**February**	RUSSO-JAPANESE WAR BEGAN.
	April	ENTENTE CORDIALE BETWEEN FRANCE AND BRITAIN: settlement of colonial differences between the powers, particularly over Egypt and Morocco.
	October	THE DOGGER BANK INCIDENT: acute Anglo-Russian crisis after sinking of a British trawler by the Russian fleet.
	November	Russo-German alliance negotiations broke down.
1905	**March**	KAISER VISITED TANGIER: THE FIRST MOROCCAN CRISIS, testing the strength of the Entente.
	April	Anglo-French military conversations.
	May	Japanese defeat of the Russians at Tsushima.
	July	TREATY OF BJÖRKÖ between Willian II and the Tsar.

	October	DREADNOUGHT constructed.
1906	January–April	ALGECIRAS CONFERENCE: France supported by all the Powers except Germany and Austria. France to have major control in Morocco.
1906	June	THE THIRD NAVY LAW – the Novelle – ratified by the Reichstag.
1907	January	Eyre Crowe's memorandum warning on German foreign policy.
	June	SECOND HAGUE PEACE CONFERENCE BEGAN: British efforts to secure arms limitation thwarted by opposition of other Powers. Germany rejected proposals for compulsory arbitration.
	August	ANGLO-RUSSIAN ENTENTE: settlement of imperial conflicts between the Powers, particularly over Persia, Afghanistan and Tibet.
1908	June	THE FOURTH NAVY LAW.
	October	AUSTRIA-HUNGARY ANNEXED BOSNIA AND HERZEGOVINA: Germany supported Austria-Hungary. Anger of Russian nationalists.
		THE DAILY TELEGRAPH published interview with the Kaiser in which he pictured the German people as hostile to Britain.
1909	February	FRANCO-GERMAN AGREEMENT OVER MOROCCO: Germany recognized France's 'special political interests' in Morocco.
	March	British Navy Bill passed.
1911	May	French occupied Fez in spite of warnings from Germany that they were infringing the Algeciras Act.
	July	THE GERMAN GUNBOAT the 'PANTHER' SENT TO AGADIR: Germany demanded the whole of the French Congo in compensation for their Moroccan rights.
		LLOYD GEORGE WARNED GERMANY IN MANSION HOUSE SPEECH.
	September	The Tripoli War between Italy and Turkey.
	November	THE MOROCCO AGREEMENT SIGNED: Germany agreed to leave France a free hand in Morocco in return for the grant of part of the French Congo.
1912	February	The Kaiser announced major Army and Navy Bills.
		THE HALDANE MISSION TO BERLIN: Germany unwilling to make naval concessions without a promise of British neutrality in the event of war.
	March	THE BALKAN LEAGUE between Serbia and Bulgaria formed.
	October	THE FIRST BALKAN WAR BEGAN: between Bulgaria, Serbia and Greece on the one hand and Turkey on the other.
1913	March	Winston Churchill proposed 'Naval Holiday'.
	June	SECOND BALKAN WAR BEGAN: Bulgaria attacked Serbia and Greece; Rumania and Turkey joined to defeat Bulgaria.
		German Army Bill passed.
		French Army Bill ratified.
	August–October	PEACE OF BUCHAREST – end of the Second Balkan War.
	December	ARRIVAL OF LIMAN VON SANDERS IN CONSTANTINOPLE; Russian protest at German mission.

From assassination to war

1914	28 June	ASSASSINATION OF FRANCIS FERDINAND AND HIS WIFE AT SARAJEVO.
		Mission of Count Hoyos, Chief of Cabinet in the Austro-Hungarian Foreign Ministry; the Kaiser issued 'blank cheque'.
	8 July	Ultimatum to Serbia prepared.
	15 July	Poincaré, the French president and Viviani, Prime Minister and Foreign Minister left for St. Petersburg.
	19 July	Austro-Hungarian Ministerial Council approved an ultimatum to be passed to Serbia on 23 July.
	20 July	Poincaré and Viviani arrived in St. Petersburg.
	21 July	Francis Joseph approved ultimatum.
		Text of ultimatum sent to Berlin.
	23 July	ULTIMATUM SENT TO SERBIA.
	24 July	Austria-Hungary informed France, Russia and Britain of the ultimatum.
		German ambassadors transmitted note in Paris, London and St. Petersburg recommending that the conflict be localised.
		Cambon, French Ambassador in Berlin, proposed a conference.
		Grey, British Foreign Secretary, suggested mediation.
		Russian Council of Ministers considered partial mobilization.

Chronology cont.

25 July	SERBIA REPLIED TO ULTIMATUM.
	Vienna broke off diplomatic relations with Belgrade.
	The Kaiser ordered the return of the Fleet.
	Russian Crown Council approved resolutions of the Ministerial Council.
	Tsar ordered that preparations be made for mobilization.
26 July	Russia asked Germany to exert a moderating influence on Austria-Hungary.
	Grey proposed that a four-Power conference of ambassadors be held in London.
	Austria mobilized on the Russian frontier.
	France took precautionary military measures.
27 July	Austria-Hungary *decided* to declare war on Serbia.
	France accepted Grey's proposals.
	Bethmann-Hollweg, the German Chancellor, rejected the idea of a four-Power conference.
28 July	AUSTRIA-HUNGARY DECLARED WAR ON SERBIA AND SHELLED BELGRADE.
	The Kaiser issued an appeal to Austria to 'halt in Belgrade'.
	The Kaiser appealed to the Tsar's spirit of monarchical solidarity.
	Russia ordered the mobilization of four of her western military districts.
29 July	Vienna refused to enter into negotiations with Belgrade. Tschirschky, German Ambassador in Vienna, presented the Kaiser's 'halt in Belgrade' proposal.
	Germany informed of Russian partial mobilization.
	Germany warned Russia.
	Moltke, German Chief of Staff, demanded general mobilization.
	Bethmann Hollweg made move to keep Britain neutral.
	Grey informed Lichnowsky, German Ambassador in London, that Britain could not remain neutral in the event of a continental war.
	Russian general mobilization ordered, but revoked by the Tsar.
30 July	Austria-Hungary agreed to negotiations with Russia, but refused to delay operations against Serbia.
	Moltke pressed for general mobilization.
	Austria-Hungary order general mobilization for 31 July.
	Russian general mobilization ordered for 31 July.
31 July	Vienna rejected an international conference and ordered general mobilization.
	Russian general mobilization becomes known in Berlin at noon. Kaiser proclaimed 'state of imminent war' an hour later.
	Germany refused to mediate and issues an ultimatum to Russia.
	French Ministerial Council decided to order mobilization for August 1.
1 August	GERMAN ULTIMATUM TO RUSSIA EXPIRED. SHE DECLARED WAR ON RUSSIA AND MOBILIZED. STILL HOPE IN BERLIN THAT BRITAIN MIGHT REMAIN NEUTRAL.
2 August	GERMAN TROOPS OCCUPIED LUXEMBOURG.
	Berlin sent ultimatum to Belgium.
3 August	Germany declared war on France.
	Belgium rejected German demands.
	Italy remained neutral.
	German-Turkish treaty concluded.
	BRITAIN MOBILIZED ARMY; CABINET DECIDED TO ISSUE ULTIMATUM TO GERMANY.
4 August	GERMAN TROOPS INVADED BELGIUM.
	British ultimatum transmitted to Berlin.
	ULTIMATUM EXPIRED AT MIDNIGHT. BRITAIN AND GERMANY AT WAR.

8.5 QUESTION PRACTICE

1   Assess the significance of the Moroccan crises of 1905 and 1911.

Understanding the question

It is essential for you to clarify, above all, the *international* significance of these crises. You must be prepared to describe the effect of the Kaiser's visit to Tangier in 1905 and of the 1911 Agadir incident Eurpe and their consequences on the relations in particular between Germany, France and Great Britain. The word 'significance' in the question indicates that you should be willing to set these events in the wider context of pre-First World War diplomacy. You may also wish to consider the influence of these crises on public opinion within these countries.

Knowledge required

The course of international relations from c. 1890–1914. The Moroccan crises of 1905 and 1911. The immediate outcome of these crises and their longer-term significance.

Suggested essay plan

1   The diplomatic background. These crises should be considered against the background of the Entente Cordiale of 1904, which agreed to terminate Anglo-French colonial rivalries. The French acceded to the British control of Egypt and the British allowed the French to take over Morocco.

2   *The 1905 crisis.* Note that the Kaiser's announcement that he would aid the Moroccans was made at a time when the power of Russia was severely eclipsed by her defeat in the Russo-

Japanese War and when she was hardly in a position to offer France effective support. He hoped to expose the weakness of British support for France – all the evidence shows that his intervention had little to do with German economic interests and a great deal to do with the Entente. At the subsequent Algeciras Conference, though, all the Powers, except Austria, rallied to France's side and the British were so antagonized that they promptly entered into defensive military arrangements with France. J. M. Roberts describes this crisis as marking 'the real end of isolation for England'.

3   *The Agadir incident of 1911.* The arrival of the 'Panther' was seen by Britain and France as a further German challenge to the Entente. Britain also viewed apprehensively the possibility of Germany constructing a naval base close to Gibraltar. The gravity of the crisis was conveyed in the Mansion House speech of the British Chancellor of the Exchequer, David Lloyd George, in which he stated that Britain was not to be treated 'as if she were of no account in the Cabinet of Nations... peace at that price would be a humiliation intolerable for a great country like ours to endure.' This carried particular weight in Berlin since Lloyd George had hitherto shown himself more well-disposed towards Germany than many of his colleagues. In November, 1911 Germany recognized the French position in Morocco in return for the French Congolese territory.

4   The result of both crises was a strengthening of Anglo-French relations, and the second crisis revealed in full the peculiar state of Anglo-German relations in the period of rivalry which preceded the First World War. There was talk of war on both sides but there was nothing concrete at issue to fight about. Germany wanted a spectacular diplomatic victory. Britain, on the other hand, was convinced that if French interests were not supported the European balance of power would be destroyed and British security endangered. Both sides assumed that a North African colonial squabble would involve the European balance of power.

5   Though the Moroccan problem itself was solved, the crises left a legacy of intensified Anglo-German suspicion and hostility. The Agadir crisis wrecked any hope of improved Franco-German relations. Note the effect of these crises on public opinion within Germany – for instance, the Pan-German League and Tirpitz were able to manipulate the national sense of humiliation to press for a larger fleet. The Second Moroccan crisis also encouraged Italy to seek 'compensation' for French gains in Morocco. Italy declared war on Turkey in September, 1911, landing troops in Tripoli. This further assault on Turkish rule encouraged the subsequent outbreak of the First Balkan War.

2   Do you consider that the outbreak of a general war was inevitable after the assassination of Archduke Franz Ferdinand in 1914?

Understanding the question

It queries the inevitability of a *general* war *after* the assassination. It is a specific question and should not be taken simply as an invitation to say whether or not you believe a general war was inevitable, or to enter on a discussion of the forces and rivalries making such a war probable. You should take care to distinguish between the localized Balkan issue and the reasons for the wider involvement of the major Powers in a European conflict, and be clear as to how the one led to the other after Sarajevo.

Knowledge required

The diplomatic background to the First World War. The Balkan problem. The military preparations of the Powers. A detailed grasp of the course of events between the Sarajevo assassination and the outbreak of the First World War.

Suggested essay plan

1   In the view of A. J. P. Taylor, 'It would be wrong to exaggerate the rigidity of the system of alliances or to regard the European war as inevitable. No war is inevitable until it breaks out.' To an extent this is fair comment but it deserves qualification. While the armaments race, Great-Power rivalries and demands for 'preventive war' characterize the pre-war years, one can on the other hand point to widespread anti-war sentiment, the spread of international agencies and a willingness to resolve issues through arbitration. There was no general assumption of the inevitability of a major war, nor was this the reaction of the Powers on receipt of the news of the Sarajevo assassination.

2   The traditional machinery of diplomacy, which had preserved peace in the past and prevented major conflict since the Congress of Berlin (1878), had localized war as recently as during the Balkan Wars. Even if the assassination made an Austro-Serbian war inevitable (and this was not so until the issue of the ultimatum to Serbia), this does not mean that a general war was inevitable.

3   There was, of course, increasing urgency in Vienna that the Slav problem (the threat posed to Habsburg interests by Slav nationalism) be 'solved'. There was also strong evidence that Russia would not accept another humiliation in the Balkans, such as that over Bosnia-Herzegovina in 1908 and that she would not allow the Serbs to be crushed by Austria-Hungary.

4   The extremely severe Austrian ultimatum to Serbia produced a marked Russian reaction and Russia confirmed that she would not allow Serbia to be destroyed. This meant that the Austro-Serbian conflict now threatened a more dangerous Austro-Russian one which, because of the rivalry of the Triple Entente and Triple Alliance, presented a very real *possibility* of a general war.

5   But note that a situation of this sort was by no means unprecedented. War was not inevitable. Britain put forward the suggestion of an international conference to localize the dispute. The German response to this was unfavourable and, on 28 July, Austria formally declared war on Serbia. This step was not irretrievable, for the Austrian forces were not fully prepared for action. Nevertheless, with the crucial 'blank cheque' from the Kaiser, Austria had the guarantee of support for her actions and the German willingness to go to war was the dominant factor in the July crisis. Even this, though, did not make general war inevitable at this stage. The Kaiser favoured the 'halt in Belgrade' scheme: Austria would occupy Belgrade but hold operations there, thus giving a final chance to diplomacy.

6   The question of mobilization: this is a massive and complex operation that, once launched, was very hard to stop. Inescapably it meant that priority is given to military factors. As Professor Michael Howard comments: 'The lesson of 1870 was burnt into the mind of every staff officer in Europe: the nation which loses the mobilization race is likely to lose the war'. He adds: 'In no country could the elaborate plans of the military be substantially modified to meet political requirements.'

7   It was only German mobilization, however, which *inevitably* involved war, because the Schlieffen Plan demanded the rapid invasion of France via a surprise advance through Belgium. Professor L. C. F. Turner has identified the night of 29–30 July as the crucial point. The German Chancellor, Bethmann-Hollweg, had heard that Britain would almost certainly intervene in the war if Germany came into conflict with France. He was, therefore, trying to reverse the trend of German policy and now

restrain Austria. However, when the German Chief of Staff, von Moltke, heard that Austria was only going to mobilize in the Balkans, he recognized that, unless she launched an offensive in Poland against Russia, the German Army in East Prussia would be overwhelmed by the Russians and the Schlieffen strategy gravely compromised. At this point he pressed Austria for immediate mobilization against Russia, promising unqualified German support in a European war. In this manner, because of strategic considerations, Germany pushed forward to make the war inevitable.

8   It is clear that the outbreak of a general war was neither inevitable nor considered to be so as a result of the Sarajevo assassination. However, as the July crisis deepened it became progressively more likely. The *inevitability* of the Austro-Serbian dispute turning into a general European war is much more easily argued in the context of the inflexibility of the mobilization plans and preparations for war, more particularly the German Schlieffen Plan. Once priority was given to military factors the momentum towards war seemed irresistible.

3 'Unfriendly and provocative.' To what extent is this an accurate summary of Germany's policies towards the European Powers between 1890 and 1914?

Understanding the question

It is obvious that other European Powers – particularly Britain, France and Russia – came to see Germany's policies as 'unfriendly and provocative'. You should attempt to separate German motives and intentions from the impression her policies created, and assess the extent to which this impression was justified. Note that the two descriptions are not necessarily inseparable – you may for instance, find her policies 'provocative' but not essentially 'unfriendly'. You should certainly *not* take this question as simply an opportunity to discuss German responsibility for the outbreak of the First World War.

Knowledge required

German policies towards the European Powers and their interests 1890–1914. European diplomatic relations in this period and the attitudes of the other Powers to Germany. The influence of German domestic policy and of her internal problems on her foreign policy.

Suggested essay plan

1   The basic difference between German foreign policy under Bismarck and under William II and his ministers: Bismarck's purpose was to establish Germany as the diplomatic arbiter of Europe – but it was limited, circumscribed and essentially peaceful in character. He had little sympathy with Pan-German ambitions and, for most of his term of office, was markedly unenthusiastic about colonies. Under William II the role of dominant continental Power was no longer enough for Germany. She aspired to the 'place in the sun' of world politics. With 'Weltpolitik', Germany advanced ambitions for recognition as a world Power, with colonies and a large Navy.

2   The 'New Course' in German diplomacy reflected the growth in the power of a nation that had consolidated its position on the Continent and was seeking new outlets for its nationalist and expansionist energies; note the very considerable growth of German industrial potential. However, German imperial ambitions reached their height in the late period of nineteenth-century imperialism when there was little territory left for distribution among the Powers. Seen from one angle, the German leaders simply applied to their own situation imperialistic ideas and ambitions that existed everywhere. Other countries had long been world Powers and were used to rounding off their territories in a suitable way.

3   German demands were accompanied, though, by particularly swashbuckling statements and propaganda, *e.g.* General Bernhardi's book *World Power or Destruction*; and the Pan-German League, which argued that Germany would draw to itself all German-speaking peoples in Europe and emphasized the 'world horizon'. Those Powers which were already established resented what they saw as the reckless and threatening assertion of German power: there was rivalry with Britain, particularly over naval policy; Alsace-Lorraine and Morocco soured Franco-German relations, while Germany's claim to influence in the Near East aroused strong Russian apprehensions.

4   When the actual *content* of German ambitions is examined, they turn out to be lacking in precision. There was a disproportion between the vastness of potential menace, as seen by Britain, France and Russia, and the actual German pursuit of concrete objectives – *e.g.* Germany was throughout this period acquiring major financial interests in the Near East (at this time her share of the Turkish debt rose from 5%–20%). But wild talk from the Kaiser, declaring himself the friend of 300 million Moslems, and the arrival of German military experts in Constantinople gave legitimate interests a flavour of reckless aggression, which was plainly provocative.

5   Over Morocco the Kaiser was anxious to prove that no serious issue in world politics could be decided without him. This was a recurring pattern. In each case there was *something* to be said for German demands. For instance, what gain had Germany to compare with the most recent French acquisitions of Tunis and Morocco? The misfortune of the German situation was that she was in a position where she could not expand without destroying the existing order – that is, expand in line with her growing power, population and economic potential. Her position, therefore, forced her to appear consistently unfriendly in her ambitions. Her presentation of her demands and policies, particularly the interventions of the Kaiser, were diplomatic catastrophes and plainly extremely provocative.

6   Relations with Britain: When Britain launched the Dreadnought in 1906 all previous battleships were rendered obsolete and the navy race became desperate. While for Germany the fleet was a luxury, the British regarded their own fleet as essential for their survival as an imperial world Power. Tirpitz's 'risk theory' amounted to diplomatic blackmail. The effect was inevitably to bring Britain closer to both France and Russia, whereas she might have been a natural ally of Germany. It was not simply with this (and such provocations as the support for Austria during the Bosnian Crisis) that Germany lost the sympathy of the Powers – it appeared to them that she was quite indifferent to their attitude.

7   Note the domestic background to German policy: 'Weltpolitik' represented in part a systematic attempt to mobilize the forces of conservatism in Germany. What Tirpitz called 'this great overseas policy' was intended to silence internal criticism and rally support for the Kaiser and the Reich.

8   There were grounds on which each of the main elements of 'Weltpolitik' might be justified – these elements being the naval expansion, the development of empire in Africa and the commercial and financial penetration of the Near East. It can be argued, though, that for Germany to pursue all three courses at the same time was the worst possible policy. It kept alive the suspicion of the Entente Powers and made them more anxious than ever to stick together.

9   The success of the Triple Entente's policy of containment and rearmament increasingly limited the scope of German

policy and gave rise to increasing fears of 'encirclement'. In the end, the Reich stood alone in 1914 with Austria-Hungary as the only reliable ally. But the outbreak of war was not so much the result of reckless German imperialism as of her desire to have complete security in the face of the other Powers. While her policies from 1890 onwards were undoubtedly ill-judged, tactless and provocative, it is harder to substantiate that they were consistently unfriendly. There were threats but there was also considerable vacillation.

4 Document question (The Outbreak of the First World War): study extracts I and II below and then answer questions (a) to (f) which follow:

Extract I

'The shots of death must ring as a fearful warning in the ears of the Austro-Hungarian government. Franz Ferdinand fell as the victim of his country's Balkan and national policy. He was the defender of clericalism and the mightiest representative of reaction and of so-called Austrian imperialism. He kept Serbia from her goal...the open road to the Adriatic, and so her hate rose as against an enemy. Franz Ferdinand is the victim of a false system which has outlived its time; but however loud may be the death shots of Sarajevo, they will hardly be heard by those who should hear them.

'That alliance (with Austria-Hungary) is not a source of strength but of weakness; the problem of Austria is ever and ever a greater danger to European peace. If we do not wish this danger to be realized in all its hideous reality, we must strive with all our power to bring about friendly relations with France and England.'

(*Vorwärts*, a German Socialist newspaper, July, 1914)

Extract II

'The Austro-Hungarian note was so drawn up as to make war inevitable; the Austro-Hungarian government are fully resolved to have a war with Serbia, and its postponement or prevention would undoubtedly be a great disappointment.'

(The British Ambassador in Vienna to Sir Edward Grey, 27 July, 1914)

Maximum marks

(a) 'The defender of clericalism and the mightiest representative of reaction'. What do you learn of Franz Ferdinand from this description of him? (4)

(b) How had Serbia been kept 'from her goal...the open road to the Adriatic'? (5)

(c) What was 'that alliance' and why might it be regarded as 'a source...of weakness'? (5)

(d) What truth is there in the statement that 'Franz Ferdinand fell as the victim of his country's Balkan and national policy'? (5)

(e) To what 'Austro-Hungarian note' does the Ambassador refer? How far do you agree that it 'was so drawn up as to make war inevitable'? (5)

(f) What evidence is contained in these extracts of German and British attitudes to the prospect of war? To what extent may these extracts be considered reliable sources of information about public opinion at that time? (7)

8.6 READING LIST

Standard textbook reading

J. M. Roberts, *Europe 1880–1945* (Longmans, 1967), chapters 4, 6, 8.

David Thomson, *Europe since Napoleon* (Penguin, 1977), chapters 19–22.

Anthony Wood, *Europe 1815–1945* (Longmans, 1975), chapter 24.

Further suggested reading

Michael Balfour, *The Kaiser and his Times* (Penguin, 1975).

Fritz Fischer, *War of Illusions: German Policies 1911–14* (Chatto, 1975).

O. J. Hale, *The Great Illusion 1900–14* (Harper and Row, 1971).

F. H. Hinsley, *Power and the Pursuit of Peace* (Cambridge University Press, 1963).

L. Lafore, *The Long Fuse* (Weidenfeld and Nicolson, 1965).

A. J. P. Taylor, *The Struggle for Mastery in Europe 1848–1918* (Oxford, 1971).

L. C. F. Turner, *The Origins of the First World War* (Edward Arnold, 1976).

9 The Peace Treaties and resettlement of Europe

9.1 INTRODUCTION

The problems facing the representatives of the victorious Allies, who met in Paris in 1919 to decide the fate of Germany and its defeated partners, were of an unprecedented scale. Not least, four empires – the German, Habsburg, Russian and Ottoman – had collapsed.

Already, before the end of the war, President Woodrow Wilson had issued his celebrated Fourteen Points. His main emphases were on the right of self-determination and the need for a league, later the League of Nations, to preserve peace. He and the other two major figures at the peace conference, Clemenceau and Lloyd George, approached the settlement often from conflicting viewpoints. For instance, Clemenceau wanted to end Germany's status as a Great Power, while Britain required the revival of the German economy to ease her own unemployment.

In Eastern Europe a dangerous power vacuum – of unstable successor states situated between two resentful giants, Germany and the Soviet Union – was confirmed. The settlement in Eastern Europe automatically created grievances. In the Near East the Turks refused to be bound by the Treaty of Sèvres.

The German settlement was the most controversial and aroused widespread criticism for its alleged injustice and severity – the 'war guilt clause', the reparations, the territorial clauses, the loss of her colonies and the drastic reduction in her forces.

Following the invasion of the Ruhr by France and Belgium in 1923, the reparations payments were moderated according to the Dawes Plan, and in 1925 the Locarno Treaties ushered in a period of relative international stability. But the resentments of the revisionist Powers – Germany, Italy and Russia – remained and were to lead to war in the following decade.

9.2 STUDY OBJECTIVES

1 A comparison of the European frontiers of 1914 with those of 1919: particular attention to be paid to Germany and the territories of the Habsburg Empire; a clear grasp of the territorial consequences of the treaties of Versailles, St. Germain, Neuilly, Trianon and the earlier Treaty of Brest-Litovsk.

2   The Turkish settlement: the treaties of Sèvres and Lausanne.

3   The differing approaches of Britain, France, the United States and Italy to the settlement. The individual objectives of the major figures, Lloyd George, Clemenceau and President Woodrow Wilson; the extent to which these influenced the outcome, for example in the questions of the Rhineland and the Italian frontier.

4   The factors which helped to influence the settlement: *(a)* the pacifist mood of public opinion in the aftermath of war; and the desire for retribution; *(b)* the commitments entered into by the Powers during the war (for instance, the Treaty of London, 1915); *(c)* the composition of the Peace Conference, noting particularly the absence of Germany and Russia; and *(d)* the impact of the Russian Revolution of 1917.

5   The underlying principles which helped to shape the treaties. The most important of these were: *(a)* the determination of the Allies to ensure against the possibility of future German aggression and the demands of French security; *(b)* the principles of President Woodrow Wilson – the Fourteen Points, especially national self-determination and the general association of states in the League of Nations; *(c)* the compensation for war losses, reparations; *(d)* the need for the military defence of the Successor States of Central and Eastern Europe (e.g. the incorporation of the Sudetenland into Czechoslovakia); *(e)* the need for economic outlets (e.g. Danzig and the Polish Corridor).

6   Those aspects of the Versailles settlement which were seen by the Germans and others (for instance, J. M. Keynes, *The Economic Consequences of the Peace*) as severe and unjust: the scope of such criticisms and the extent to which they were justified.

7   The Covenant of the League of Nations: the work of the League and its agencies and the attempts made to strengthen it during these years; its weakness as a peacekeeping body.

8   The differing approaches of Britain and France to the question of Germany: their mutual distrust; the Ruhr invasion of 1923.

9   An understanding of the questions of war debts and reparations; the Balfour Note, the Dawes Plan and the Young Report; the attitudes of the various powers to these questions.

10   The Treaty of Locarno and its consequences.

9.3 COMMENTARY

The First World War had proved that Germany was strong enough to defy a coalition of Great Britain, France, Russia and Italy. She had imposed a humiliating settlement on Russia in the Treaty of Brest-Litovsk and only the intervention of the US in 1917 had swung the balance of superior force against her.

The lesson of 1918, in spite of her defeat, was Germany's strength – not her weakness. In population and economic potential she remained the dominant continental Power. It was highly probable that the restrictions imposed upon her by the Treaty of Versailles would, sooner or later, be contested.

The 'Diktat'

There was overwhelming German resentment against the terms of the settlement, which was denounced as a 'Diktat', or dictated arrangement. It was contrary to their expectation that it would be based on President Woodrow Wilson's Fourteen Points. Criticism was above all directed against the following: (1)

Article 231 – the war guilt clause; (2) reparations; (3) the disarmament clauses; (4) the prohibition of German-Austrian unification; (5) the territorial settlement; and (6) the disposal of the German colonies.

Self-determination

In fact the principle of self-determination – the right of peoples to their own self-government and national independence, and which was one of the cardinal points in President Wilson's programme – was applied *against* Germany. This was so, for instance, in the creation of the Polish Corridor.

At the same time, the peace treaty with Austria prohibited the unification of this small state, now shorn of the Habsburg Empire, with Germany. This unification, however, was later achieved by force when Hitler invaded Austria in the Anschluss of 1938.

The wider unpopularity of the settlement

The unpopularity of the settlement for its territorial and economic clauses was echoed abroad. One of the bitterest and most influential denunciations was *The Economic Consequences of the Peace* (1919) by the British Treasury official and economist, J. M. Keynes. He attacked it on both economic and moral grounds and warned of its likely political effects. He was particularly concerned with the likely impact of the reparations payments which were demanded. Though this was disputed, particularly by France, there was soon a fairly widespread view that the settlement was severe and unjust and required modification.

Eastern and Central Europe

A major weakness of the settlement was the situation in Eastern and Central Europe.

The collapse of the Habsburg Empire at the end of the war had left behind new successor states. The British Prime Minister, Lloyd George, pointed out very clearly in his Fontainebleau Memorandum the likely consequences for these states and European peace. They would be extremely vulnerable in the event of a revival of either German or Russian power.

Though the principle of self-determination was allowed to operate, this created new and serious tensions and difficulties. The basic reason for this was that the successor states contained dissatisfied ethnic minorities. For example: Germans and Poles came under Czechoslovak rule and Germans came under Polish rule.

From the start there was instability and rivalry. To give only two instances: Czechoslovakia and Poland were hostile to one another; and Hungary, which had lost substantial territories at the end of the war, was anxious to win them back from the Little Entente States, Yugoslavia, Rumania and Czechoslovakia.

The background to the treaties

It is important to set the peace treaties and their short-comings in the context of the immense problems which faced the statesmen of 1918. No peace treaties of modern times have come under such criticism and abuse.

Woodrow Wilson, Lloyd George, Clemenceau and Orlando were faced with the collapse of four empires, the German, the Habsburg, the Russian and the Turkish. The Paris Peace Conference met against a European background of revolution, chaos and famine. The balance of power had been destroyed. Russia was in the throes of civil war and the US was about to withdraw from political involvement in Europe. Self-determination, whatever the problems it raised – and these were many – was a reality and its consequences had to be confronted.

Criticisms of the settlement should take account of these facts:

Fig. 5 European frontier changes after 1918

the problems of 1918 were on an altogether larger scale than those, for instance, of 1815 and the Vienna Settlement, with which Versailles is frequently and unfavourably compared.

The Decline of Europe

At the same time, Europe was increasingly being challenged overseas. The First World War had speeded up this process.

From being the major overseas investors of capital, Britain, France and Germany had become major debtors to the US. The whole tangle of reparations and war debts is a very important aspect of the history of this period.

The US were now, in a very large way, moving into European

colonial and other overseas markets. Simultaneously, the war accelerated the rise of nationalist movements against European colonial rule.

Britain and France, under economic pressure yet determined to preserve their empires, were to find themselves forced increasingly to make compromises in the face of revived German power.

The Rhineland and the Ruhr invasion

French military leaders at the end of the war had insisted on control of the left bank of the Rhine as a guarantee of security against a future German attack.

In the event, France abandoned this in return for the offer of an Anglo-American Treaty of Guarantee against aggression. This fell through when the US failed to ratify the Versailles Treaty.

France was the leading military power in Europe after 1918, but she had a very much smaller population than Germany and she was without powerful allies, as (1) Britain had refused to renew her wartime alliance, and (2) France's eastern allies, Poland, Czechoslovakia, Rumania and Yugoslavia were weak and divided.

The limitations of French power were very clearly demonstrated when she and Belgium occupied the industrial Ruhr area of Germany to enforce the payments of reparations. Britain and America were extremely hostile to this move. France was subsequently forced to accept considerable revision of the reparations agreement in the Dawes Plan.

These were later to be modified further in Germany's favour in the Young Plan.

Anglo-French relations

The Ruhr episode was only one of a number in which Britain and France revealed a mutual suspicion and division of view which was (1) to prevent a proper understanding between the Powers; (2) to assist Germany in challenging the Versailles system.

The British authorities were wary of French ambitions and anxious to avoid needless entanglement in European affairs.

Secondly, there was a fundamental difference of view over Germany. The major priority of British governments in this period was the revival of trade. And the postwar unemployment seemed to confirm the argument of Keynes that the recovery of the European economy depended on Germany's revival as a trading partner.

Britain was anxious to see the Versailles Treaty revised in Germany's favour if this would help to preserve peace and encourage prosperity.

France, for whom the arguments of national defence were paramount, argued that her security depended on upholding the Treaty as it stood.

The League of Nations

The desire of public opinion to avoid a future war was a very strong one during the interwar years in the democracies.

Many people held traditional power politics, alliances and secret diplomacy to have been largely responsible for the outbreak of war in 1914.

The League of Nations was an attempt to respond to this mood and to provide an international peace-keeping machinery. As an organization it always attracted more trust from public opinion at large than from governments.

From the start it had serious shortcomings: Germany was excluded from it (until 1926); Russia was absent and the US failed to ratify the Treaty of Versailles, of which it was a part.

Germany saw it as a club of the victorious powers, helping to enforce an unpopular and unjust settlement on her. This association of the League with the controversial treaty was a hindrance to its international acceptance as an impartial peacemaker.

Other major weaknesses were (1) the fact that the League had no peace-keeping force at its disposal and (2) the rule that policy decisions of the League Council and Assembly needed the unanimous consent of all states represented. Mussolini's attack on Corfù was an early indication of the League's – later much more apparent – ineffectiveness as a preserver of peace.

One should at the same time bear in mind its successes in dealing with such disputes as Danzig, the Aaland Islands and the welfare work – for instance, among refugees – performed by its agencies.

Locarno

France accepted the international guarantees offered in the Locarno agreements of October 1925. By these, Germany, France and Belgium agreed to recognize their existing frontiers as permanent, including the demilitarized zone of the Rhineland.

Britain and Italy guaranteed this arrangement and the Treaty provided for the settlement of all disputes through the League of Nations.

Locarno offered a temporary illusion of lasting peace and stability, but it had major shortcomings.

It is significant that no such guarantees were offered to Eastern Europe – as France very much wanted.

Germany was admitted to the League, but Locarno did not answer the problem of the revival of German power. To what extent was this compatible with the preservation of peace?

It ended the possibility of military discussions between Britain and France.

Though it inaugurated a period of relative stability, the Locarno settlement meant very different things to the powers concerned:

(1) to the French it offered some guarantee of security;
(2) to the German Foreign Minister, Gustav Stresemann, it was envisaged as a first step towards revision of the treaties; and
(3) to the British it meant a minimum guarantee to France without involving Britain in wider obligations, such as in Eastern Europe.

At this time, in collusion with the Soviet Union, Germany was already evading the disarmament clauses of the Versailles Settlement. Locarno was swept aside when Hitler invaded the Rhineland in 1936.

9.4 CHRONOLOGY

1918	**January**	PRESIDENT WOODROW WILSON'S FOURTEEN POINTS: outlining the peace programme, on the basis of which Germany and Austria sought armistices.
	March	THE TREATY OF BREST-LITOVSK between Russia and the Central Powers. The Russians surrendered the Baltic provinces, the Ukraine, Finland, the Caucasus, White Russia and Poland. THE GERMAN ARMISTICE (11 November 1918) invalidated the treaty.
1919	**January**	THE PARIS PEACE CONFERENCE began: dominated by the 'Big Four', Clemenceau, Woodrow Wilson, Lloyd George and Orlando. A congress of 'allied and associated powers' to determine the peace settlement.
	February	LEAGUE OF NATIONS COVENANT APPROVED: the League was an organization set up to preserve peace and settle disputes by arbitration. President Wilson had called for such a body in his Fourteen Points.
	March	THE COMINTERN FOUNDED at Moscow and the Third International established by the Bolsheviks to promote revolutionary Marxism abroad. THE COUNCIL OF FOUR BEGAN.

	June	THE VERSAILLES TREATY SIGNED: the settlement with Germany; Germany signed under protest and the US refused to ratify the treaty.
	September	THE TREATY OF ST. GERMAIN-EN-LAYE: between the Allies and the Austrian Republic. The independence of Czechoslovakia, Poland, Hungary and Yugoslavia was recognized. Amongst other clauses, Austria was forbidden to unite with Germany and was deprived of about one third of her German-speaking population.
	November	THE TREATY OF NEUILLY WITH BULGARIA: Western Thrace was ceded to Greece and other areas to Yugoslavia.
1920	January	The Treaty of Versailles came into force.
	March	Final US Senate rejection of the Versailles Treaty.
	April	The San Remo Conference: the main lines of the Turkish agreement drawn up. Polish offensive against Russia.
	June	THE TREATY OF TRIANON WITH HUNGARY: old Hungary lost three quarters of its territory and two thirds of its inhabitants.
	July	THE SPA PROTOCOL: Germans arranged reparations payments.
	August	THE TREATY OF SÈVRES: the Sultan's government renounced all claims to non-Turkish territory; the Straits were internationalized. The Poles defeated the Russians at Warsaw.
1921	March	The Russo-Polish Treaty of Riga. Plebiscite in Upper Silesia.
	May	The London Schedule of Reparations Payments.
	November	THE WASHINGTON CONFERENCE BEGAN: discussion of naval limitation and the question of the Pacific and Far East.
1922	February	The Permanent Court of International Arbitration established at the Hague.
	April	THE RAPALLO TREATY: the German and Soviet governments re-established diplomatic relations and pledged cooperation. Both Powers renounced reparations. THE GENOA CONFERENCE: called to consider the Russian problem and general economic questions. It broke down on the insistence by France that Russia recognize its prewar debts.
	August	THE BALFOUR NOTE on war debts: offered to abandon all further claims to reparations provided a general settlement of debts could be arrived at. The US rejected this on the grounds that reparations and inter-allied debts were not connected problems.
1923	January	OCCUPATION OF THE RUHR BY FRANCE AND BELGIUM: followed the German failure to fulfil reparations obligations. Led to passive resistance of German workers and to British and American condemnation.
	July	THE TREATY OF LAUSANNE: the final peace treaty with Turkey following the refusal of Turkish nationalists to be bound by the Treaty of Sèvres and following their victories against the Greeks in Asia Minor. Turkey gave up claims to any territory formerly in the Ottoman Empire occupied by non-Turks.
	August	THE CORFÙ INCIDENT: Greek appeal to the League of Nations against the attack on Corfù by Italy. Greece was made to accept most of the Italian demands.
	September	German passive resistance in the Ruhr ended.
1924	February	Britain recognized the Soviet government.
	April	THE DAWES PLAN: proposed revision of the reparations agreement with Germany; provided for loan to be granted to stabilize the German currency; the plan would help her to meet her treaty obligations in the period 1924–9.
	July	Britain rejected the Draft Treaty for Mutual Assistance. The London reparations conference ADOPTED THE DAWES PLAN.
	October	THE GENEVA PROTOCOL FOR PACIFIC SETTLEMENTS OF INTERNATIONAL DISPUTES: provided for the compulsory arbitration of all disputes and defined the aggressor as the nation unwilling to submit its case to arbitration. The decisive factor in the rejection of this attempt to strengthen the League was the opposition of the British Dominions. (Rejected finally in March, 1925.)
1925	October	THE LOCARNO CONFERENCE AND TREATIES: this most significant treaty confirmed the inviolability of the Franco-German and Belgo-German frontiers and the demilitarized zone of the Rhineland. (This was violated in 1936 with Hitler's reoccupation of the Rhineland.)
1926	September	Germany entered the League of Nations.
1927	May	The World Economic Conference at Geneva. Britain broke off relations with the Soviet Union.
	June	The Geneva Naval Conference: Great Britain, the US and Japan.
1928	August	THE KELLOGG-BRIAND PACT: convention formally renouncing war as an instrument of national policy and providing for the peaceful settlement of disputes.

Chronology cont.

1929   June   THE YOUNG REPORT: proposed a reduction of German reparations by 75% and that payments should be made in the form of annuities. Proposals abandoned, though accepted by Germany, because of the world economic crisis. Hitler declared his opposition to paying any further reparations.

September   The Briand proposal for a European federal union.

October   THE WALL ST. CRASH: heralded the economic slump.

9.5 QUESTION PRACTICE

1   'This is not peace. It is an armistice for 20 years.' Why do you think Marshal Foch so described the Versailles Settlement?

Understanding the question

Foch criticized the Treaty of Versailles in this way because he believed it did not adequately safeguard France against German attack. But you should not regard this question as being specifically about Foch: rather it invites you to consider the Versailles Settlement as an 'armistice' or truce, and to discuss whether its deficiences were in any way responsible for the renewal of war in 1939.

Knowledge required

The main provisions of the Treaty of Versailles and the part played by these provisions in the diplomacy of the 1920s and, especially, of the 1930s.

Suggested essay plan

1   The Treaty of Versailles (signed between Germany and the Western Powers in 1919) resembled an 'armistice' in two respects: (*a*) its short duration (much of it was overthrown even before 1939 – contrast this with the settlements of 1815, 1871 or with that after 1945) and (*b*) when war resumed in 1939–41, Britain, France, the US and the Soviet Union were again pitted against Germany, and, even though Italy and Japan changed sides, the two World Wars can be seen as successive phases of a single conflict.

2   Is it possible to go a step further than this and argue that the Treaty of Versailles in some way *caused* the Second World War? The Treaty has been criticized on two, contradictory, grounds: (*a*) it was not severe enough to prevent Germany from recovering as a military Power – in particular, it did not attempt to undo Bismarck's unification of the country; and (*b*) it was not conciliatory enough to resolve the differences between Germany and the Western governments. Instead, arguments about the Treaty terms poisoned relations between Britain, France and Germany throughout the 1920s and 1930s, and eventually caused a renewal of war.

3   It would be fairer to the peacemakers of 1919 to acknowledge that the Treaty was a compromise between the school of thought based on coercion and that based on conciliation. It was also a compromise between rival French, British and American views. Can it, nonetheless, be argued that the Treaty of Versailles fell between two stools and managed *neither* to coerce Germany *nor* to grant it conditions that it could accept voluntarily?

4   The first of these charges will not stand up. The Treaty in its territorial, financial and, especially, military clauses *were* strong enough to make Germany powerless to wage war. But, under pressure first from Weimar Germany and then from Hitler, the Allies had by 1936 allowed the most important of these clauses to lapse. In particular:

(a) reparations payments were reduced under the Dawes and Young Plans (1924 and 1929) and ended in 1931–2, allowing Germany to expand its heavy industrial capacity in the 1920s with the aid of American loans;

(b) the Allies evacuated the Rhineland by 1930 and permitted Hitler to remilitarize it (in violation of the Treaty) in 1936;

(c) in the 1920s Germany began secretly to rearm, and accelerated this after 1933. In 1935 Hitler announced the reintroduction of conscription and of a German air force, again in violation of the Treaty.

5   Germany, therefore, regained the power to wage war as a result of the *failure to enforce* the Treaty. But what about the second argument? It appears that Germany nearly went to war with Britain and France in 1938 over its treaty frontier with Czechoslovakia, and actually went to war in 1939 over its treaty frontier with Poland. This, however, confuses the *occasion* of war with its deeper causes. The issue in 1939 was not in itself the Versailles Settlement as it affected Danzig and Poland but the determination of Britain and France to prevent a further growth of German power in Europe. Conflict between Hitler and the West would have been likely even if the Treaty of Versailles had placed all the German-inhabited areas of Eastern Europe within Germany's frontiers.

6   The links between the Treaty of Versailles and the outbreak of war in 1939 were, therefore, only very indirect.

2   Why did the Locarno Pact of 1925 fail to live up to the high hopes of its signatories?

Understanding the question

You must be able to assess why Locarno was regarded as such a hopeful settlement for the preservation of peace, what the expectations of the individual signatories were and why these were disappointed by subsequent events.

Knowledge required

The Versailles Settlement, more particularly as it affected Western Europe. Franco-German and Anglo-German relations after 1918, especially the significance of the Ruhr crisis of 1923. The terms of the Locarno Pact and the motivation of the Powers in signing it, noting especially the policy of the German Foreign Minister, Gustav Stresemann. The weaknesses of the Pact and its subsequent collapse.

Suggested essay plan

1   Begin with the postwar tensions arising from the Versailles Settlement: the failure to meet reparations payments led to the Franco-Belgian occupation of the Ruhr and to the revival of the French plan for the separation of the Rhineland from Germany; the French desire for security; the need to achieve reconciliation in the interests of general peace.

2   An assessment of the initiatives to reduce land armaments (the Washington Naval Treaty allowed for limited naval armaments); the French insistence that this disarmament could only follow a security system; the British refusal to accept wide-ranging commitments for universal peace-keeping (*e.g.* the Geneva Protocol); the Locarno Pact was welcomed by Austen Chamberlain as a much more limited (regional as against universal) agreement.

3   The terms of the Locarno Pact: an international guarantee of Western frontiers, with Britain and Italy underwriting this

arrangement; provision for the settlement of disputes through the League of Nations.

4   The differing objectives of the Powers: to the French, Locarno was some guarantee of security against Germany; to the German Foreign Minister it was a step towards the revision of Versailles; to the British, a guarantee to France not involving them in wider obligations, particularly to Eastern Europe.

5   The weaknesses of the Pact: it failed to extend guarantees, as France wanted, to Eastern Europe (an 'Eastern Locarno'); it ended the justification for bilateral military discussions between Britain and France, but did not resolve the problem of French security in the face of the revival of German power.

6   The Locarno Pact did not appear under threat in the period of prosperity between 1925 and 1929; but this changed with the economic collapse, the death of Stresemann and the rapid growth of political extremism, particularly of the Nazi Party. Hitler: his desire to reverse the Versailles Settlement in both the East and the West; his reoccupation of the Rhineland in 1936 (which was his first territorial step to the revision of Versailles) marked the destruction of Locarno.

3   Do you consider that the peace treaties which ended the First World War embodied any general principles?

Understanding the question

You should take care here not to confuse general principles with what were simply prior commitments – for instance, the Treaty of London (1915). You must identify the underlying principles, such as that of self-determination, and explain the extent to which the peace treaties reflected them. Note that the question is not confined to the Treaty of Versailles, and that the scope of your answer should cover the other peace settlements as well.

Knowledge required

The terms of the Treaties. Woodrow Wilson's Fourteen Points. The differing priorities of the peace makers. The extent to which public opinion influenced the peace treaties.

Suggested essay plan

1   Introduction: initially President Woodrow Wilson's Fourteen Points seemed to offer a set of general principles which might form the basis of a reasonable and acceptable overall settlement. In the peace discussions and hard bargaining that followed the war, these were often found to conflict with other principles and to contradict the commitments or interests of the other Powers; the particular difficulty of applying general principles to the situation in Central and Eastern Europe.

2   The principle of self-determination: the new postwar states, such as Yugoslavia and Poland, were based upon this principle; but the frontiers which this would require were in turn modified by other considerations and overriding concerns, such as defence and economic viability; the Sudetenland, and the Polish Corridor; the contradiction of the principle of self-determination in the ban on Austro-German unification.

3   The preservation of peace: President Woodrow Wilson's concept of a general association of states built into the peace treaties; the concept of collective security embodied in the League of Nations was weakened by the exclusion of Germany and the Soviet Union and by the refusal of the American Senate to ratify the agreement; in spite of the League and the sentiments of public opinion in favour of collective security, the discussions and bargaining at Versailles reflected very clearly the preoccupation of statesmen with their individual national interests.

4   French objectives at Versailles: Clemenceau's primary concern was to protect France against future German aggression. He proposed an independent Rhineland, but this idea ran contrary to the principle of national self-determination; instead France obtained the demilitarization of the Rhineland, and its occupation for 15 years and also acquired the Saar coal mines. Note the criticism of French demands; final terms were a diplomatic defeat for France.

5   National ambitions versus the general principle in the Treaty of Sèvres: Greece was compensated for her participation in the war by being given the right to set up a colony at Smyrna on the Turkish mainland (a conflict here with the principle of nationality); Arab lands were transferred from Turkey to Britain and France; Italy was given islands in the Aegean.

6   The general principles advanced at the time of the peace treaties were very considerably modified by power-political and other practical considerations (such as in the refusal to allow Austria to unite with Germany). From the start, Britain and France had showed scepticism over the Fourteen Points; at the same time, the complexity of the situation in Central and Eastern Europe after the collapse of four empires was such as to defy the possibility of the successful application of the principle of national self-determination; though guarantees were given to national minorities, these aroused hostility and led to instability.

4   Document question (The Treaty of Versailles): study extracts I and II below and then answer questions (a) to (f) which follow:

Extract I

'(We are) a government and people seeking no selfish and predatory aims of any kind, pursuing with one mind and one unchanging purpose: <u>to obtain justice for</u> others. We desire neither to destroy Germany nor diminish her boundaries: we seek <u>neither to exalt ourselves nor to enlarge our Empire</u>.'

(D. Lloyd George, 5 January 1918)

Extract II

'It is comparatively easy to patch up a peace which will last for thirty years. What is difficult is to draw up a peace which will not provoke a fresh struggle when those who have had a <u>practical experience of what war means</u> have passed away…. <u>You may strip</u> Germany of her colonies, reduce her armaments to a mere police force and her navy to that of a fifth-rate power; all the same <u>if she feels she has been unjustly treated</u>…she will find <u>means of exacting retribution from her conquerors</u>.'

(D. Lloyd George memorandum, 1919)

Maximum marks

(a) To what is Lloyd George referring when he claims to be seeking 'to obtain justice for others'?   (4)

(b) How far is Lloyd George's assertion that 'we seek neither to exalt ourselves nor enlarge our Empire' borne out by the terms of the Versailles Treaty?   (3)

(c) Explain, with reference to the Versailles Treaty, what Lloyd George is referring to by 'You may strip Germany of her colonies, reduce her armaments to a mere police force and her navy to that of a fifth-rate power.'   (5)

(d) Why, at the peace conference, was the point of view expressed by Lloyd George in Extract II not generally adopted?   (5)

(e) Discuss whether Germany had cause to 'feel she has been unjustly treated'.   (6)

(f) Do you consider Lloyd George right in thinking that 'practical experience of what war means' influenced statesmen to try to preserve the peace settlement during the next 20 years?   (8)

9.6 READING LIST

Standard textbook reading

J. M. Roberts, *Europe 1880–1945* (Longmans, 1967), chapters 9–11.

David Thomson, *Europe since Napoleon* (Penguin, 1977), chapters 24–25.

Anthony Wood, *Europe 1815–1945* (Longmans, 1975), chapters 27–29.

Further suggested reading

J. Joll, *Europe since 1970* (Weidenfeld and Nicolson, 1973).

C. L. Mowat (ed), *The New Cambridge Modern History 1898–1945*, volume XII (Cambridge University Press, 1968), chapter 8.

A. J. Nicholls, *Weimar and the Rise of Hitler* (Macmillan, 1979).

H. Nicolson, *Peacemaking 1919* (Methuen, 1964).

R. Sontag, *A Broken World 1919–39* (Harper and Row, 1971).

E. Wiskemann, *Europe of the Dictators* (Fontana, 1973).

10 Russia, 1917–41

10.1 INTRODUCTION

The revolutions in Russia in 1917 resulted from grave economic and social problems and military failure in the First World War. After the March revolution and the downfall of the Tsar, Nicholas II, the Provisional Government attempted to rule Russia but was obliged to share power with the Soviets (workers' councils). In November, 1917 Lenin and the Bolsheviks seized power, issuing decrees on the paramount issues of land and peace. After elections which only gave the Bolsheviks a minority of the seats, Lenin ordered the closure of the Constituent Assembly and proceeded to consolidate power on the basis of a single-party totalitarian dictatorship.

From the period of the Civil War, in which the Whites were defeated by the Red Army organized by Trotsky, the new Communist order became more and more identified with Russian nationalism. Following the period of 'War Communism', the New Economic Policy and the collapse of revolutionary hopes in Western and Central Europe, Stalin emerged as the dominant figure. He announced the slogan of 'Socialism in One Country'. He pressed ahead with the collectivisation of agriculture and the massive industrialization of the five-year plans. While there was rapid economic development, these programmes were only accomplished with the grossest inhumanity.

Meanwhile, Stalin consolidated his personal power by defeating both the Left and Right Opposition. The murder of Kirov provided him with an excuse for mounting a bloody purge, accompanied by show trials, of the Communist Party, which carried away the Old Bolsheviks and a high proportion of the military command. In foreign policy, tentative moves towards a common front against Germany from the mid-30s were terminated by the Molotov-Ribbentrop Pact. In 1941 Hitler launched his invasion of Russia.

10.2 STUDY OBJECTIVES

1   The failure of Russia during the First World War: the economic and political breakdown in Petrograd; the February Revolution; the reasons for the fall of the Tsar.

2   The Provisional Government: its positive achievements – for instance the eight-hour working day, the freeing of political prisoners, Polish independence; the resistance to the Provisional Government and the reasons for this – for example, the continuation of the war and failure to transfer land to the peasants; the July rising, the influence of Lenin and of Bolshevik propaganda, the Petrograd Soviet, the Kornilov coup.

3   The November Revolution: explanation of the weakness of the Provisional Government; the role of the Petrograd Soviet; the contribution of Trotsky; the Land and Peace decrees.

4   Lenin's contribution: his earlier career as a revolutionary and the development of his ideas; the transfer of political and economic power from the traditional classes to the Bolsheviks; note the position of the Bolsheviks as a minority party and Lenin's decision to disband the Constituent Assembly; a comparison of Bolshevik objectives with the programmes of the other political parties.

5   The Civil War: the terms of the Treaty of Brest-Litovsk and the reasons for its acceptance; the causes of the Allied intervention and its consequences; the aims of the White armies and their leaders; the Red Army and its organization by Trotsky; the explanation of its success; the murder of the Tsar and his family.

6   War Communism and the New Economic Policy (NEP): how, as a party of urban workers, the Bolsheviks attempted to win over the peasantry; the NEP as a recognition of the failure of Bolshevik agrarian policy; the reorganization of industry and agriculture; opposition; the significance of the Kronstadt Mutiny; the suppression of resistance, the CHEKA.

7   The battle for the succession to Lenin; the contrasting ideas of the contestants; the methods used by Stalin to gain power, his role in the Communist Party as General Secretary, his membership of the triumvirate of the Politburo; the extent to which, in organizational powers and ideas, Stalin was the direct successor to Lenin.

8   The first Five-Year Plan and its successors (*a*) as a means of establishing a heavy industrial base for the Soviet Union; (*b*) as 'the second agrarian revolution', a further attempt to do what War Communism had failed to do – to eradicate the Kulaks and establish state or collective farms as the norm of the Russian agrarian economy; and (*c*) the problems of industrial and agrarian discipline; Stakhanovism.

9   An understanding of the policy of 'Socialism in One Country' and the conflict between Stalin and Trotsky's revolutionary programmes.

10   The ways in which divergence and debate in the earlier Soviet system gave way to rigid uniformity where deviation would be punishable by dismissal or death; the liquidation of the Kulaks; the growth of the Stalinist system of dictatorship; the labour camps; the show trials of the 1930s; the judicial murder of the Old Bolsheviks and the army leaders; the degree of control achieved by Stalin over Russian political, economic and cultural life by the end of the 1930s.

11   The preoccupations of Soviet foreign policy in this period; the 'siege mentality' and the attitude towards the Western democracies; relations with Germany: Rapallo; the reaction to the rise of Nazism and Hitler's policies; the reasons for the Molotov-Ribbentrop Pact.

10.3 COMMENTARY

The March revolution

The revolutionary situation in Russia in 1917 resulted from the bankruptcy of the tsarist regime and the grave impact of three years of war. Russia was the first Great Power to suffer real military disaster in the struggle with Germany and its failures led to Nicholas II's downfall. It is questionable in any event, however, how long tsarism as it existed in 1914 could have survived. Military defeat hastened what was most likely to be an inevitable collapse. The war-weariness of the Russian people and the widespread distrust of the pro-German faction in the Imperial Court combined with economic deprivation to produce a series of riots and strikes in Petrograd, leading to a mutiny of the troops garrisoning the capital.

When the revolution broke out in March, 1917, the Bolshevik leaders, who were for the most part at this time in exile, were taken more by surprise than the government itself. An emergency committee of the Duma and the newly-created 'Soviet (Council) of Workers' and Soldiers' Deputies' in Petrograd set up the provisional government under Prince Lvov. This was followed by the abdication of the Tsar and his brother.

The March revolution was not an unwelcome event to the Western Allies to start with. They tended to interpret it as a vote of no confidence in the tsarist management of the war effort. Further, the Provisional Government appeared to share most of the political ideas of the Western Powers – was it not possible to envisage Russia evolving towards a constitutional and democratic order on the Western model? The major significance of 1917 and what followed is that Russia did not follow this path. The Soviet Union developed in a radically different direction.

The Provisional Government and the Soviets

For eight months the Provisional Government attempted to rule Russia. What power it had it was obliged to share with the Soviets and at the end of this period it was overthrown by the Bolsheviks.

In April, 1917, the Germans, knowing that the Bolshevik leaders would undermine the war effort, offered Lenin a safe passage from his revolutionary's exile in Switzerland to Russia in a sealed train. On his return Lenin at once set about exploiting the difficulties of the Provisional Government to his advantage.

In the famous 'April Theses' he announced a political programme which called for peace, the nationalization of the land and the granting of power to the soviets. He underlined what were increasingly seen by the Russian people as the major failures of the Provisional Government – the failure to solve the land question and their continuation of the war with Germany.

The Soviets were the great institutional innovation of the Russian revolution. While the unrepresentative Duma did not appeal to the masses as the basis of a new government, a Soviet was elected in Petrograd in March, and in June a National Congress of soviets was convened. 'Order Number 1' required the election of a committee in each military and naval unit, which was to have charge of arms and which was in effect to decide which orders were to be obeyed. Lenin justifiably spoke of a 'dual power', a sharing of government between government and Soviets. Indeed, shortly after the March revolution the Petrograd Soviet began to act as a shadow government, challenging the acts of the Provisional Government. Within the Soviets at this time the Bolsheviks were only a minority. Both the Social Revolutionaries and the Mensheviks were more numerous. The urban population in Russia at this time was in any case only about a sixth of the total. Russia was still overwhelmingly a peasant economy.

Lenin and the Provisional Government

Lenin judged that if the Bolsheviks offered a programme with sufficient revolutionary appeal they could gain control over the Soviets. While the Provisional Government, which did not have a clear or incisive line of policy, muddled along, discovering increasingly that it could not sustain both war and revolution at the same time, Lenin proceeded to outflank the other parties. He stole popularity from the Social Revolutionaries with his programme of land for the peasants and showed himself more radical than the Mensheviks (who argued that Russia was not ready for a socialist revolution) by calling for an immediate proletarian upheaval. He declared quite unhesitatingly for peace and with this demand he attacked the Provisional Government at its weakest point.

With the last failed military offensive by Russia on the Galician front in July, 1917, the Provisional Government, now under the leadership of the Socialist Revolutionary Alexander Kerensky, began very quickly to lose credibility. In the same month the Bolsheviks made their first, abortive, coup. Meanwhile Kerensky's government showed that it could reach no agreed solution on the central question of land reform. A move by the Right wing – the Kornilov coup – forced him to turn to the Bolsheviks for support. By October the Russian armies were breaking up and the soldiers drifting home; the peasants were taking matters into their own hands across the country and seizing the land. At the same time, the Bolsheviks, strengthened by the defeat of the Kornilov coup, had by now managed to achieve a majority in the Petrograd Soviet. Confusion and administrative chaos had gone too far for the situation to be saved for the Provisional Government.

The Bolshevik revolution

The Bolsheviks seized power on the night of 6 November, 1917. The coup was timed to coincide with the All-Russian Congress of Soviets. This received full power from the Petrograd Soviet, which it in turn delegated to the local Soviets. The main agents of the revolution were the Petrograd Soviet, units of the Army and Navy and the Party organization itself.

The programme formulated by Lenin was fourfold: (1) land to the peasants; (2) distribution of food to the starving; (3) power to the Soviets; and (4) peace with Germany.

The first of these was already spontaneously happening with the peasant seizure of land and the last was brought about by the Treaty of Brest-Litovsk, which involved massive losses of Russian territory and of productive capacity. The second and third were achieved together in that food was distributed only to those willing to grant power to the Soviets.

Soviets sprang up all over Russia, especially in the factories and Army units under the auspices of the Party. The two decrees on land and peace consolidated the revolution at this early stage, winning the initial support of the peasantry for what had started as an urban insurrection. The Congress appointed a Council of People's Commissars to govern the state.

In this new situation power was wielded by the highly-disciplined and organized Party which Lenin had been forging since 1903. In theory Lenin had done what he had promised: he had won all power for the Soviets. In practice the Bolshevik leaders had seized power for themselves, and the concentration of power at the centre – which Soviet Communism involved – was later to be fully revealed in the period of Stalin's rule.

The Consolidation of Power

Though they had won a majority in the Petrograd Soviet, in the local provincial Soviets the Bolsheviks were still outnumbered. The elections for the Constituent Assembly produced in fact a clear majority for the Social Revolutionaries (the Bolsheviks obtained fewer than a quarter of the seats). The Bolshevik military command acted immediately, closing down the assembly. The third All-Russian Congress of Soviets, which the

Bolsheviks were now able to dominate, assumed the functions of the Constituent Assembly.

There was now, quite clearly for all to see, a sharp distinction between what the Bolsheviks were doing and the model of Western parliamentarism and democracy. The Bolsheviks showed themselves willing to seize and consolidate power in defiance of democratic legality, as expressed in the election to the Constituent Assembly.

The authority of the Bolsheviks was further consolidated by the creation of Cheka ('The Extraordinary All-Russian Commission of Struggle against Counter-Revolution, Speculation and Sabotage'), and a month later in January, 1918, the Red Army was founded by Trotsky. Lenin proceeded to lay the basis of a single-party, totalitarian dictatorship principally through the four instruments of the Party (designated 'the Communist Party' in March, 1918), the Soviets, the secret police and the Red Army.

The Russian civil war

The subsequent developments of the Russian revolution (in spite of opposition to this drift of affairs by Trotsky and others) moved increasingly towards identifying Communism with Russian national interests. The period of foreign intervention – the civil war – served particularly to strengthen the power of the Communist government as a national government. It forged the Red Army into a more efficient fighting force for national defence and its triumph over the mixed forces of the counter-revolutionaries, which included British, French, US and Japanese contingents, left the Party and its instruments without serious armed opposition. On the home front the Cheka launched a reign of terror against all elements of 'bourgeois reaction' and eliminated all rivals to the Bolsheviks among the other revolutionary movements.

During the period of so-called 'War Communism' urban workers were subjected to a militarization policy and were enrolled in labour battalions; the peasantry had to submit to forced requisitioning of their produce. Only by such drastic means were the Bolsheviks able to keep the economy functioning at all and to provide the cities and the Red Army with basic rations. By 1920 industrial production had fallen to only 16% of its 1912 level. In this crisis workers' control gave way to strict factory discipline imposed by state-dominated trade unions.

The victory of the Red Army over the Whites in the Civil War was due primarily to the following:
(1) the creation of the Red Army, formed as a volunteer force out of virtually nothing by Trotsky;
(2) a marked lack of coordination among the anti-Bolshevik forces;
(3) the Reds' possession of the strategic advantage of internal lines of communication;
(4) the crucial asset of appearing to be on the Russian people's side – the White armies were associated with reaction and the restoration of the old order;
(5) the conviction amongst the peasantry that a White victory would mean the abrogation of the revolutionary land settlement;
(6) the White generals' association with the intervention by foreign Powers which made the Bolsheviks more and more appear to be the patriotic defenders of the Motherland.

One should remember in this also the very important nationalities' question: the Red victory had an additional significance of being a triumph of the Great Russian over the minor nationalities.

The 'New Economic Policy'

Lenin's revolutionary expectation that the Bolshevik seizure of power would lead to successful upheavals throughout Western Europe was not fulfilled. Furthermore, following the naval mutiny at Kronstadt in March, 1921, Lenin found himself obliged by the seriousness of the economic situation to beat a tactical retreat. He introduced the 'New Economic Policy', which made extensive concessions to private enterprise.

The major aim was to give inducements to get the economy moving again. Conciliation with the peasantry was essential: in Lenin's words 'only agreement with the peasantry can save the socialist revolution in Russia'. By the 'Fundamental Law' of May, 1922, the peasants were given security in land tenure; they were further permitted to sell or lease their land and to hire labour to work it. In industry small private enterprise was allowed an equal freedom.

The critics of this policy, which included Trotsky, pointed to the return of 'bourgeois' habits, the emergence of a class of 'new rich'. But it is important to note that the 'commanding heights' of the economy – heavy industry, foreign trade and the transport system – remained in state hands.

Stalin and 'Socialism in One Country'

After the final failure of Communism in Germany after the First World War, in 1923, the Russian regime reconciled itself to the fact that revolution would not spread throughout Europe in the easily foreseeable future. In the autumn of 1924 Stalin launched the slogan of 'Socialism in One Country'. Russia, he argued, could create Socialism on its own, without relying on foreign help. His policy in fact was to be a mixture of Trotsky's demand for intensive industrialization and his opponents' strategy of subordinating international revolution (which had not taken place) to the national needs of Russia.

This linking of Communism with the historic sentiments of Russian nationalism was to be consolidated by the achievements of the five-year plans, by the cult of Stalin as the great national hero and by the experience of sustained patriotic resistance to the German invasion after 1941.

The collectivization of agriculture

From the mid-1920s the Soviet Union faced a chronic and growing crisis in feeding its urban population. Stalin embarked on a massive and brutal programme to collectivize agriculture against the stubborn resistance of the peasants, more particularly the richer peasants, the Kulaks, whom the New Economic Policy had notably advantaged. Class warfare was encouraged in the countryside. The poorer peasants who had something to gain from collectivization were incited against the richer ones, who had everything to lose from it. By the outbreak of the Second World War about 95% of the Soviet Union's farms had been collectivized.

Apart from the economic aim – the urgent need to achieve greater productivity from the land – there was in Stalin's mind a clear political objective. The conquest and consolidation of political power in a backward country whose peasants clung to private ownership, dictated collectivization. Lenin had hoped that the peasantry might gradually come to accept the modernization of agriculture. But when this expectation failed, and when the NEP did not resolve the crisis of food shortages, Stalin pressed full speed ahead with collectivization regardless of the human consequences.

The five-year plans and industrialization

Lenin's successor also proceeded after 1928 with plans to equip Russia with heavy industry and better transport and to develop new sources of power and industry beyond the Urals. Lenin had not envisaged detailed economic planning in 1917, but under Stalin it became and remained a permanent characteristic of the Communist order. The effect was to produce a highly integrated economy. By 1939, four-fifths of Russian industrial production came from plants built during the previous ten years and, as many of these were situated east of the Urals, they carried modern industrialization into the heart of Asia.

This rigorous drive to transform Russia into a first-class industrial power made necessary a high degree of political centralization. The transition from Marxism to Leninism and from Leninism to Stalinism was brought about by a self-constituted bureaucratic élite which had become independent of the masses in whose name it was supposed to rule. The re-constitution of the familiar Russian tradition of autocratic domination and state control was made possible by the institution of a single-party monopoly.

Stalin's policies involved gross inhumanity – slaughter, deportation, imprisonment and famine on a massive scale. At the same time there was rapid economic progress. It is estimated that in the decade after 1929 the Russian gross national product grew by just under 12% a year. The greatest single failure (as today) was to be found in Russian agriculture. But even there, by paying the collectives far less than the market price and absorbing the difference, the state forced agriculture to swell the accumulation of capital which powered the industrial programme. While the price of collectivization was extremely high and while historians question whether Stalin's brutality was necessary to achieve the pace of Soviet development, collectivization and grain surpluses made it possible to pursue the aims of the five-year plans. Collectivization caused bitter disaffection, not only among the peasantry.

The period of the purges

The assassination of a popular leading Leningrad Communist, Kirov, in 1934 provided Stalin with an excuse for what was to be a bloody purge of the Communist Party. In 1935 Zinoviev and the other Old Bolsheviks were brought to trial and condemned. In the years that followed hundreds of thousands of Party officials and administrators were shot, deported or simply replaced. Half the officer corps of the Army was also removed.

These purges had the evident purpose of, on the one hand, removing all Stalin's earlier rivals and adversaries, and, on the other, of forestalling any potential future opposition. They must be set against the background of the rise of Nazi Germany. Abroad, the well-advertised show trials confirmed conservative-minded statesmen and politicians in their hatred and suspicion of the Soviet Union and of Communist totalitarianism. At the same time the dramatic purge of the Red Army made military leaders in France and Germany sceptical about Russia's value as a possible ally against Germany.

At the same time, sympathizers with the Soviet system in other countries mistakenly saw in the new Soviet Constitution of 1936 an apparently liberal and democratic creation. In fact, though this provided excellent propaganda material for Stalinism abroad, it simply disguised the real totalitarian police-state nature of the regime.

The real significance of the Constitution was that it ended conclusively the era of improvizations. It emphasized the fact that Soviet rule had hardened into its final form. The abolition of the distinctions that had denied full rights to the former possessing classes most clearly demonstrated that they no longer represented any sort of threat to the new Stalinist order, that the power of these classes had been broken.

The Soviet Union and the Western Powers

The attitude of the European Powers to the new phenomenon of Bolshevik Russia passed through various phases. At first the revolution and the terror provoked a violent reaction of fear. Then Russia made her first formal agreement with Germany at the Treaty of Rapallo (1922). During the next decade the Soviet Union made trade agreements with other Western countries and in 1934 she was admitted to the League of Nations.

Her relations with the West were periodically disturbed by the activities of the Comintern (founded in 1919). On the one hand, Stalin seemed intent on 'building Socialism' in Russia alone; on the other, Soviet propaganda continued to argue that world revolution was the inevitable goal of Communism and that no opportunity should be missed to promote that end. Distrust and hostility continued, therefore, to characterize relations with the West.

The years 1936–8, the major years of the purges, were also marked by a new direction in Soviet foreign policy. It seemed unlikely to Stalin that the Western Powers, which had acquiesced in Hitler's invasion of the Rhineland, would be willing to resist German expansion at Russia's expense. This lesson seemed to be underlined by the Spanish Civil War and the Munich Conference. At the same time Stalin was only agreeable to participating in effective collective security against Germany if the Soviet Union was allowed access across Poland, the Baltic states and Rumania. This was something Western governments would not contemplate since it meant the establishment of Russian military domination over Eastern Europe. This was the background to the Nazi-Soviet Non-Aggression Pact of August, 1939.

At the end of December, 1940, Hitler finally took the decision which was essential for his objective of 'Lebensraum' (living space) in the East and he issued his orders for the preparation of the attack on Russia. On 22 June, 1941, German troops attacked the Soviet Union in Operation Barbarossa on a broad front from the Baltic to Rumania.

10.4 CHRONOLOGY

1917	**March**	Outbreak of riots in St. Petersburg.
		The Duma refused to obey an imperial decree ordering its dissolution.
		Abdication of the Tsar Nicholas II; establishing of the PROVINCIAL GOVERNMENT UNDER PRINCE LVOV.
		Power struggle between the Provisional Government and the PETROGRAD (St. Petersburg) SOVIET. Issue of Order No. 1 by the Soviet.
	April	LENIN ARRIVED AT PETROGRAD. His programme: (1) the transfer of power to the Soviets; (2) the cessation of war; (3) the seizure of land by the peasants; and (4) the control of industry by committees of workers.
	July	Bolshevik attempt to seize power in Petrograd; Trotsky arrested; Lenin went into hiding. KERENSKY succeeded Prince Lvov.
	September	KORNILOV attacked the government; in defeating him, Kerensky came under the domination of his Bolshevik allies.

Chronology cont.

	6 November	THE BOLSHEVIK REVOLUTION: the Bolsheviks, led by the military revolutionary committee, the soldiers of the Petrograd garrison and the Red Guards, seized key government offices, stormed the Winter Palace and arrested members of the Provisional Government.
	7	THE SECOND ALL-RUSSIAN CONGRESS OF SOVIETS: the new government assumed the name of the Council of People's Commissars; headed by Lenin and including TROTSKY, as Commissar for Foreign Affairs and STALIN, Commissar for National Minorities. THE LAND DECREE: ordered the immediate distribution of the land among the peasants. Nationalization of the banks; the repudiation of the national debt. Workmen given control over the factories, confiscation of church property.
	25	The elections to the Constituent Assembly returned 420 Social Revolutionaries as against only 225 Bolsheviks. When the Assembly met in Petrograd (January, 1918) it was dispersed at once by the Red troops.
1918	**March**	THE TREATY OF BREST-LITOVSK: between Russia and the Central Powers. The Russians surrendered the Baltic provinces, the Ukraine, Finland, the Caucasus, White Russia and Poland. The government moved the capital from Petrograd to Moscow. Independent governments formed all along the Russian frontier.
1918–20		THE CIVIL WAR: the war with the Cossacks; the struggle for the Ukraine; the war in White Russia and the Baltic region. Allied intervention in northern Russia; the campaigns of Denikin and Wrangel in the Caucasus and southern Russia; the war in Siberia and eastern Russia.
1918	**July**	PROMULGATION OF THE SOVIET CONSTITUTION. MURDER OF NICHOLAS II AND HIS FAMILY.
	August	An attempt by a social Revolutionary to assassinate Lenin inaugurated a systematic reign of terror by the Bolsheviks.
1919	**March**	COMINTERN FOUNDED: for the propagation of revolutionary Marxism abroad.
1920	**April–October**	The Russo-Polish War.
1921	**March**	The Treaty of Riga, defining the frontier between Russia and Poland. THE KRONSTADT MUTINY. THE NEW ECONOMIC POLICY (NEP): a partial restoration of freedom of trade and other changes; declared to be a 'temporary retreat' from communism.
1922	**April**	THE TREATY OF RAPALLO between Germany and Soviet Russia.
	December	THE UNION OF SOVIET SOCIALIST REPUBLICS.
1924	**January**	The death of Lenin; followed by a power struggle between Stalin and Trotsky.
1926	**July–October**	The victory of Stalin over the Leftist opposition bloc led by Trotsky, which called for the discontinuation of the NEP and the speeding-up of world revolution.
1927	**December**	Definitive victory of the Stalin faction over the Trotsky group; the 15th All-Union Congress of the Communist Party condemned all 'deviation from the general Party line'. (Trotsky expelled from Russia in January, 1929.)
1928		NEW SOCIALIST POLICIES; SPEEDY INDUSTRIALIZATION, THE FIVE-YEAR PLANS; THE COLLECTIVIZATION OF AGRICULTURE.
1929	**November**	Expulsion from the Party of Bukharin and other members of the Rightist opposition.
1933		PURGE OF THE COMMUNIST PARTY: about one third of its members expelled.
1934	**September**	The Soviet Union joined the League of Nations, which it had previously denounced. Started to work for collective security and supported France in the scheme for an eastern European pact.
	December	Assassination of Kirov, followed by purges and the trials of many of the older leaders of the Party.
1935	**May**	Conclusion of the Franco-Russian alliance. Alliance between Russia and Czechoslovakia which obliged the Russians to come to the assistance of Czechoslovakia in the event of an attack, provided France decided to act.
	August	Decision at a meeting of the Comintern that Soviet Russia should throw its weight on the side of the democracies against Fascism.
1936	**August**	Zinoviev, Kamenev and others put on trial and executed for alleged plotting with enemy Powers against the regime of Stalin.
	December	Adoption of a new 'democratic' constitution.
1939	**August**	Trade pact concluded with Germany. THE MOLOTOV-RIBBENTROP NON-AGGRESSION PACT: included pledges to maintain neutrality if either country was at war. Secret clauses gave Lithuania and western Poland to Germany and eastern Poland, Estonia, Latvia, Finland and Bessarabia to the Soviet Union.
1941	**22 June**	Operation Barbarossa launched by Hitler against the Soviet Union.

10.5 QUESTION PRACTICE

1 Consider how far Stalin had closed the gap between the Soviet Union and the more advanced countries by 1941.

Understanding the question

This is basically a question on the economic achievements of the Soviet Union before the German invasion. You should consider the problems of economic development in Russia as a backward country and Stalin's solution for them. Given the terminal date, an assessment of Russia's preparedness for war would be in place – as would wider considerations of the social gap between Russia and the advanced countries (which may be taken to be the US and those of Western Europe).

Knowledge required

A general understanding of the structure and history of the Russian economy and society since the 1880s. A detailed grasp of Stalin's policies, particularly the collectivization of agriculture and the five-year plans, and the levels of agricultural and industrial achievement reached. The level of development in the West.

Suggested essay plan

1 The fact that the Soviet Union was able to withstand and finally to defeat the very highly industrialized Nazi Germany is an indication of the level of industrial achievement that Stalin's rule and policies had encouraged. However, although the catching-up process was dramatic it was not complete, and we can still see serious weaknesses in the Russian economy.

2 When Stalin came to power Russia was still a few islands of advanced economy in a sea of backwardness. A *brief* explanation of the collectivization of agriculture, of the five-year plans and on the relationship between them.

3 Industrial achievements: by 1941 Russia was a very major producer in some fields (third in the world league of steel production); coal output had increased fourfold since 1913; in petroleum, Russia was second only to the US; there was great investment in steel plants, dams, electricity, etc.

4 In agriculture the areas cultivated had increased by one third compared with 1913; the grain and beet crops had gone up threefold; milk and meat production had nearly doubled; Russian town dwellers were now considerably better fed, particularly as the growth of industry had reduced the need to export agricultural crops in exchange for machinery. Nevertheless, note the failure of the collective farms to produce the requisite amount of food for the population in a rapidly developing economy.

5 By 1941 the Soviet Union had, in material terms, made a huge stride forward in basic heavy industry, which was the basis for military strength; but to some extent Stalin's totalitarian grasp, the mistakes made by centralized planning and the discouragement of individuality reduced the scale of advance.

6 Consumer goods were in very short supply – in personal consumption and standard of living of the populations there was still a wide divide between the Soviet Union and the West.

7 While the economic gap had appreciably narrowed, it had greatly widened in terms of personal freedom. A new constitution in 1936 had *theoretically* widened the boundaries of freedom: in fact, with the purges and the terror, the drive to uniformity in every sphere, including the cultural, was overwhelming.

8 The invasion of 1941 and the subsequent war effort encouraged a further very considerable development of Soviet industrial might, supported by intense Russian patriotism and investment of American capital. Even so, it is still clear today in its legacy that the Stalinist achievement was a partial one and the narrowing of the gap with the West was also partial: there was the persistent agricultural problem, continuing low levels of consumption and the need for totalitarian controls.

2 Why were there two revolutions in Russia in 1917?

Understanding the question

The question requires you to examine why the March Revolution happened in the first place; but further, why it failed to solve Russia's crisis. You must also be able to explain how a tiny minority radical group, the Bolsheviks, came to dominate the Russian state in the second revolution, with the support of the urban proletariat.

Knowledge required

The structure and history of Russia's economy and society from c. 1880 to 1917, particularly the agricultural problem. The history of radical, revolutionary politics in Russia, especially the Mensheviks and the Bolsheviks. The effects of the First World War on Russia. The period of the Provisional Government.

Suggested essay plan

1 Marxist ideology in the early 20th century argued that a society had to go through *two* revolutions – a liberal bourgeois one, followed much later, at the full development of capitalism, by a socialist revolution led and controlled by 'the dictatorship of the proletariat'. In Russia both revolutions happened in the same year and this fact can only be understood through an analysis of Russian society in the early twentieth century.

2 The contrasts in Russian society: her need to develop the economy in order to be a Great Power; state capitalism and industrialization; the contribution of Sergei Witte; conversely, the problem of the land and the peasantry – capitalist-run estates were the exception; rapid, but localized, industrial development was taking place in a society where the majority of the population were largely self-sufficient peasants. The social structure meant the lack of a strong liberal middle class such as exists in the Western European states; the Tsarist autocracy had stifled the growth of political consciousness along western lines and the social structure seemed not to be capable of producing effective 'reformist' politics.

3 The only real political opposition was in the cities among a Socialist/populist intelligentsia; their ideas were intellectual and radical, and Lenin's ideas were essentially anti-western and anti-liberal; the contrast between the Bolshevik and Menshevik approaches; the Mensheviks attempting to retain the orthodox Marxist view of the two revolutions, in spite of the realities of Russian society.

4 The effects of the First World War: a huge economic boom, but the Government and administration were not flexible enough to cope; the Government broke down under the strains of war in 1917; the failure to feed the towns; financial chaos; the distrust of Tsarist circles; the mismanagement of the war effort; desertion of troops.

5 The March Revolution: a liberal revolution bringing the Provisional Government to power; this had fundamentally no effective new approaches to the problems of Russian society; it did nothing to reclaim the control of local government, which was lapsing into anarchy; it maintained the alliance with the Western Powers as popular support for the war against Germany was rapidly waning; the Provisional Government was forced to share power with the Petrograd Soviet.

6 The Bolsheviks: the acceptance of Lenin's 'April Theses'; the false start in the July Days; the Provisional Government under Kerensky and the Kornilov affair; the Soviet came to be dominated by Bolsheviks as workers were increasingly dis-illusioned and as many local organisers came to see the Bolsheviks as the only sound base on which to build a new form of government.

7   The March Revolution occurred because of the crisis in Russia's economy and society, and because of the strains of the very rapid development to meet the demands of war; but the liberal bourgeois revolution did not work out in practice; there was no substantial liberal middle class and no true capitalist economic structure to form the base of a Western-style political order; the November Revolution occurred as a logical outcome, almost as a reaction, to Western theory and influence; the Bolsheviks alone seemed to offer a 'real' enough solution to the crisis of Russian society, more particularly the land question. Significantly, the Bolshevik take-over was consolidated by the decrees on land and peace.

---

3   'A great revolutionary, but not a great statesman'. How far would you accept this view of Lenin?

Understanding the question

This question appears to ask for a simple comparison of Lenin's achievement in gaining power in 1917 with his apparent lack of decisiveness subsequently and his failure to achieve the 'state' he described in *The State and Revolution*. However, this question asks whether you accept the view of Lenin propounded and it is quite legitimate to argue that it should be rejected because it gives a false perspective on Lenin's aims and understanding of events; the point being that Lenin became a 'statesman' by mistake.

Knowledge required

Lenin's political beliefs and strategy. Russia's political and economic structure and its relation to Western Europe. Bolshevik ideas before the First World War. The events of 1917 and beyond, both in Russia and in Europe. Lenin's policies after 1917 and their consequences.

Suggested essay plan

1   Introduction: the view expressed might appear to be self-evident. The revolutionary who wrote about the idyllic state in *The State and Revolution* became the founding father of one of the most repressive states in the world and failed to achieve his ideals. The end of Lenin's life was a story of failure and tragedy as he saw the growing dominance of a bureaucratic and oppressive state. However, this traditional view of Lenin as 'the revolutionary-who-fails-to-make-a-competent-statesman' is very misleading.

2   This is because: *(a)* the concept of 'statesmanship' used is a traditional one which he would not have accepted for himself; *(b)* he managed to retain power – on a simply pragmatic level he was a very good 'statesman' in that he preserved any government at all in the extremely chaotic circumstances after 1917; *(c)* he never expected to have to be a 'statesman' in the traditional sense at all. His policy after 1917 was a series of expedients, for he was anticipating the spread of revolution.

3   The Marxist concept of revolution and Lenin's expectation in 1917; the notion of 'permanent revolution' (Trotsky); Lenin was willing to make major concessions to Germany in the Treaty of Brest-Litovsk, not least because he believed that Germany would soon be engulfed in revolution.

4   Pure Marxist idealism lay at the heart of Lenin's rationale for 1917, but Lenin was a great revolutionary in practice as well as in theory; his organization of the Bolshevik Party as a party of

professional revolutionaries; his tactics between February and October 1917, especially the 'April Theses'; his disciplining of the Party; his capacity, once in power, for being very ruthless, *viz.* the dissolution of the Constituent Assembly and 'War Communism'; his pragmatism and flexibility in instigating the New Economic Policy.

5   Lenin was never a 'statesman' in the traditional sense; rather he was a pragmatic revolutionary in power; his legacy of strict party discipline and ruthlessness conflicted in its consequences with his idealistic wish for a 'stateless' world. These two strands were inherited respectively in the approaches of Trotsky (idealistic 'world revolution') and the 'Socialism in One Country' totalitarian dictatorship under Stalin.

6   It can be argued that Lenin remained a *revolutionary*. It was only with Stalin that the idea of 'statesman' returned to the Russian scene as the Bolsheviks turned towards 'building Socialism' within the bounds of the traditional state.

---

4   Document question (The Russian Revolution and Lenin): study extracts I and II below and then answer questions (a) to (h) which follow:

Extract I

'The Provisional Government ... should do nothing now which would break our ties with the Allies .... The worst thing that could happen to us would be a separate peace. It would be ruinous for the Russian revolution, ruinous for international democracy .... As to the land question – we regard it as our duty at the present time to prepare the ground for a just solution of that problem by the Constituent Assembly ....'
(Speech by Tseretelli, a minister in the provisional government, June, 1917)

Extract II

'Comrades! .... Kerensky sends troops to suppress the peasants and to defend the landowners. Kerensky has again come to an agreement with the Kornilovist generals and officers who stand for the landowners.
'... If power is (put) in the hands of the Soviets ... there will be in Russia a workers' and peasants' government; it will immediately, without losing a single day, offer a just peace to all the belligerent peoples ....
'... If power is in the hands of the Soviets ... the landowners' lands will immediately be declared the property and heritage of the whole people ....
'... Are you willing to 'be patient' in order that the war may be dragged out longer, the offer of peace postponed, the tearing-up of the secret treaties of the former Tsar with the Russian and Anglo-French capitalists postponed?'
(Lenin, writing in October, 1917)

Maximum marks

(a) Who first led 'the Provisional Government' and who succeeded him?   (2)

(b) Explain Lenin's reference to 'Kornilovist generals'.   (2)

(c) What do you understand by 'the Soviets'?   (2)

(d) Identify 'the former Tsar' and explain briefly the circumstances in which he ceased to reign.   (4)

(e) Summarise the main differences between the policy of the provisional government, as outlined by Tseretelli in Extract I, and the policy proposed by Lenin in Extract II.   (4)

(f) What was the importance of the 'Constituent Assembly' in Russian political affairs in 1917–18?   (5)

(g) Discuss how successful Lenin and his supporters were in obtaining 'a just peace'. (4)

(h) What problems arose concerning Lenin's intention to make the land 'the property and heritage of the whole people', and how far had these problems been overcome by the time of Lenin's death in 1924? (8)

10.6 READING LIST

Standard textbook reading

J. M. Roberts, *Europe 1880–1945* (Longmans, 1967), chapter 13.

David Thomson, *Europe since Napoleon* (Penguin, 1977), chapters 22, 23, 25, 27.

Anthony Wood, *Europe 1815–1945* (Longmans, 1975), chapters 26 and 29.

Further suggested reading

Robert Conquest, *The Great Terror: Stalin's Purge of the Thirties* (Penguin, 1968).

Isaac Deutscher, *Stalin* (Penguin, 1972).

L. Kochan, *The Making of Modern Russia* (Penguin, 1970); and *Russia in Revolution* (Paladin, 1978).

W. Laqueur, *The Fate of the Revolution: Interpretations of Russian History* (Weidenfeld and Nicolson, 1967).

Alec Nove, *An Economic History of the USSR* (Penguin, 1972).

D. W. Treadgold, *Twentieth-Century Russia* (Rand McNally, 1976).

A. B. Ulam, *Lenin and the Bolsheviks* (Fontana, 1969); and *Expansion and Coexistence: the History of Soviet Foreign Policy 1917–67*, (Secker and Warburg, 1968).

11 Mussolini and Italian Fascism

11.1 INTRODUCTION

Fascism marked a revolt against the liberal political ideas of the nineteenth century. It emphasized the power and unity of the state, as against individualism and the division of society into classes. It was militantly nationalistic and rejected pacifism and internationalism. It came to power in Italy against a background of acute economic difficulties, fear of revolution, disappointment at the territorial results of the Peace Settlement and the inability of Liberal governments to resolve these problems.

Mussolini became Prime Minister and Duce (leader) following the march on Rome in October, 1922. In subsequent years he proceeded to suppress political opposition and institute a dictatorship. Other major domestic developments included the setting up of the National Council of Corporations and the conclusion of the Lateran Treaties with the Catholic Church.

In 1935 he issued a major challenge to the international order and the League of Nations with his successful war against Abyssinia. This also drew him towards an alliance with Germany, with which he concluded the Pact of Steel. In 1940 he declared war on Britain and France.

In 1943 he was forced to resign after a coup by King Victor Emmanuel and Marshal Badoglio, and the Fascist Party was dissolved.

11.2 STUDY OBJECTIVES

1 The background of Fascist ideas; Fascism as a European phenomenon with origins in the prewar period; the emphasis on Nationalism; the rejection of parliamentary democracy, Liberalism and the ideas associated with the French Revolution of 1789; its appeal as a means of suppressing Socialism and Communism after the First World War.

2 The Treaty of London and the grievances of Italy at the result of the Treaty of Versailles; the significance of the seizure of Fiume by Gabriele d'Annunzio.

3 Mussolini's background and career before 1919; the formation of the Fascio di Combattimento; the events leading to and including the March on Rome; the 'Squadristi'; the role of the monarchy and the security forces.

4 The political weaknesses in the Italian system which helped to make Mussolini successful in 1922 included: *(i)* the Catholic Church's hostility to the Liberal Italian state: *(ii)* the effects of political management, especially the 'transformism' of Giolitti; *(iii)* the failure of the Italian Left to become a credible parliamentary alternative (the Left was divided between those who were willing to work within the parliamentary system and those who looked for a revolutionary alternative; the Socialist split in 1912 and 1921 on these issues); and *(iv)* the domestic political significance of Italy's entry into the war in 1915 – the intervention crisis as a precedent for 1922.

5 The demand for and appeal of strong leadership; the argument that the Italian parliamentary system had failed; the contribution towards Mussolini's success of the following groups: *(i)* the Italian nationalists and patriots aggrieved by their country's failure to receive territories promised by the Treaty of London; *(ii)* the propertied classes frightened by the seizure of land and the occupation of the factories; and *(iii)* all those who blamed the continuing inflation on the failure of government.

6 The consolidation of Mussolini's power and the means by which he established a one-party state; the new electoral law (1923); the Aventine Secession, following the murder of Matteotti; censorship; abolition of universal suffrage.

7 The economic policies of the Fascist State; the National Council of Corporations.

8 Mussolini's relations with the Roman Catholic Church and Papacy, particularly the significance of the Lateran Treaties (1929).

9 Mussolini's foreign policy: the Corfù incident; the origins, course and international impact of the Abyssinian War; the intervention in the Spanish Civil War; the invasion and conquest of Albania; Mussolini's relations with Germany: the Pact of Steel; the declaration of war on France and Britain.

10 The circumstances of Mussolini's fall from power; the dissolution of the Fascist Party.

11.3 COMMENTARY

The background

Fascism was a European phenomenon. It developed in the disturbed economic and social conditions which were common to most of Europe in the years after the First World War. Its origins, however, lie in the revolt against nineteenth-century liberal ideas and 'bourgeois' society in the pre-war period.

In Italy the main target of this revolt was the dominant political figure, Giovanni Giolitti and his liberal, democratic, reform policies. The younger generation of post-war Italy rejected the cosy and comfortable middle-class values of Giolittian Italy. They demanded action and political excitement instead.

The leaders of the revolt were the Italian nationalists, who argued that Italy could only fulfil her imperial destiny and solve her persistent economic problems through successful war (there was already a strand of extreme nationalism in Italy dating from the struggle for unification). They also rejected Giolitti's compromises with the Socialist Party, which they regarded as the enemy of Italian patriotism. They advocated, in place of Socialist class struggle, the collaboration of all classes under an authoritarian, strong, nationalistic government.

Intervention and the First World War

The revolt against Giolitti most successfully asserted itself in the Intervention Crisis of 1915. The nationalists and their allies on the Right joined other forces violently hostile to Giolitti – the revolutionary Socialists (who included Mussolini at this time), the revolutionary syndicalists, radicals, republicans and democrats – to support Italian entry into the First World War. Giolitti and the majority of the Liberals opted for neutrality. So did the Socialists, thus reinforcing nationalist hostility to Socialism.

Eventually, the neutralist majority was overawed by the Interventionist press campaigns and street demonstrations. In May, 1915 the Italian parliament voted to go to war on the side of Britain, France and Russia, according to the terms of the Treaty of London.

The birth of Italian Fascism

In March, 1919 some of the extreme Interventionist forces coalesced to form the first 'Fascio' under the leadership of Benito Mussolini. The Left-wing origins of Fascism were clearly visible in the political programme of the first 'Fascio', with its emphasis on anti-capitalism, anti-clericalism and republicanism. Indeed, Mussolini launched his movement as a patriotic alternative to the neutralist Socialist Party, but his appeal to the urban working-class failed.

He did succeed, though, in attracting large numbers of ex-servicemen who were disillusioned with their return to civilian life and who were rebuffed by the Socialists. From the start, Fascism was characterized by extreme nationalism and violent anti-Socialism.

The crisis of the liberal state

At the same time as Italy, in common with other European countries in the post-war period, was facing serious economic, social and political problems, the system of government was paralysed by a difficult transition from a restricted franchise to full democracy. As a result of the introduction of universal male suffrage and proportional representation in the general elections of 1919, two strong mass parties, the Socialists and the Catholic Popular Party, emerged and the Liberals lost their parliamentary majority for the first time since the unification.

The Socialists obstinately refused to participate in parliamentary government, and the Popular Party would only do so on stringent conditions. To make matters worse, the Liberals were divided into several warring factions. Between the end of the war and October, 1922, Italy was ruled by no fewer than six different coalitions. It was in any case a country with a tradition of anti-parliamentarism.

Not surprisingly, these weak, unstable governments proved incapable of resolving the grave problems facing Italy, and in their desire for strong, effective government many Italians turned to Fascism.

The three major problems facing the post-war governments were:

1   The 'mutilated victory'

Public opinion was dissatisfied with the gains made by Italy in the Peace Settlement of 1919. The nationalists and the Fascists claimed that Italy had been 'cheated' of her due reward for their war effort and they blamed the weak parliamentary governments for this.

The aspect of the Peace which most outraged Italians was the allocation of the Italian-speaking town of Fiume to Yugoslavia, and in 1919 Gabriele D'Annunzio and his shock troops, the 'arditi', seized the town by force.

The failure of the Italian Government to crush this rebellion further damaged its credibility, demonstrating that it was powerless to prevent Right-wing violence and illegal acts.

2   Economic difficulties

Italian governments also failed to solve the economic problems of the post-war period, the run-down of industry, mass unemployment (swelled by demobilization), high inflation and the land hunger of the peasantry. The immediate post-war period intensified deep-rooted problems in the economy, some of which could be clearly identified at the time of the Unification.

3   The fear of revolution

The years 1918–20 are known as the 'Red Two Years' by Italian historians, for they witnessed a violent upsurge of spontaneous working-class militancy – strikes, street demonstrations, riots and occupations of the factories and land. Though the leadership of the Socialist Party preached revolution, they were unwilling to bring it about.

Nevertheless the aggressive activities of the Socialists gave rise to a widespread belief that there was a real danger of a revolution. As one Socialist warned: 'We shall pay with tears of blood for this fright we have given the bourgeoisie'.

The growth of Fascism, 1920–22

Many who joined the Fascist squads did so because they seemed the only force willing and able to stop Socialism. The most widespread and violent reaction against the Left took place in the Po Valley. There, in the 1920 local elections, the traditional ruling class was unseated by a Socialist victory.

Control of the town halls gave them authority over the local labour exchanges, which the Socialist Peasant Leagues used to raise agricultural wages and force the farmers to accept more labourers then they needed. At the same time, they proclaimed that their ultimate aim was the *collectivization* of the land, thus threatening not only the large landowners but the small peasant farmers as well.

For these social groups it seemed that the revolution was already taking place and in their desperation they turned to the local Fascists for assistance.

Very quickly the Fascist squads were swelled by the unemployed, by students, peasants and all manner of unruly and anti-socialistic elements eager for action and excitement. Between September, 1920 and October, 1922 the Fascist squads literally bullied, beat and burned their way to power in many areas of northern and central Italy. The police tended to turn a blind eye to these activities and in some cases they and the Army actually assisted the squads.

Even Mussolini was surprised by the speed with which 'agrarian fascism' grew. At this time he needed all his resources of energy, his national reputation as an Interventionist, his oratorical gifts and his ownership of the only Fascist daily newspaper, 'Il Popolo d'Italia', to preserve his leadership of the movement.

Power in three stages

In the general elections of May, 1921, Giolitti made a last attempt to restore the authority of the Liberal state by forming an electoral alliance of Liberals, Nationalists and Fascists, but he failed to win a majority. The only beneficiaries of his strategy were the Fascists, who won 35 seats and thus acquired a respectable parliamentary face which they used to win support and sympathy in conservative circles. Even the Vatican showed sympathy for Fascism, taking at face value Mussolini's brazenly opportunistic promise to settle the dispute between Church and State which had existed since the unification.

By October, 1922, the paralysis of parliamentary government was complete and the Fascists exploited the situation by threatening a March on Rome to give Italy a strong, effective government. On the 24th of that month Mussolini, speaking in Naples, laid down the challenge to the government: either solve Italy's problems or make way for the Fascists. Apprehensive that the alternative to capitulation, as the Fascists marched on Rome, would be civil war, the King, Victor Emmanuel III, gave in to this blackmail and appointed Mussolini as Prime Minister.

To consolidate his power, Mussolini persuaded Parliament to introduce a law controlling the press and set up a militia of Fascist 'squadristi' to defend Fascism against the Army should the Army move against him. Most important of all, he obtained approval for a reform of the electoral system, the Acerbo Law of 1923 which stipulated that the party which obtained the largest number of votes in excess of a quarter of the total would win two-thirds of the parliamentary seats. In the 1924 general elections the Fascists won their two-thirds majority, largely through violence and intimidation and because of the divisions of the opposition.

Within a few weeks, however, as a result of the abduction and murder by Fascist thugs of the Socialist leader, Giacomo Matteotti, Mussolini was fighting for his political life. Italian public opinion was outraged by the crime, not least because of Mussolini's alleged involvement.

The opposition parties abandoned parliament to form the Aventine Secession as a moral protest; however, they signally failed to exploit their last chance to overthrow Mussolini. Mussolini's allies – the industrialists, the Church and many liberal politicians – stood by him.

Prompted by the demands of the bosses of the Fascist Party, in January, 1925 Mussolini declared that he accepted 'moral' responsibility for the Matteotti murder and for all other crimes committed during the 'Fascist Revolution'. During the next three years he suppressed the opposition, established a police state and the other institutions of the Fascist dictatorship.

Fascism in power: economic policy and corporatism

Despite the anti-capitalist demands of the early Fascist movement, Fascism in power showed considerable favour to the industrial and agrarian interests, beginning with the *'laissez-faire'* policies of the Finance Minister, De Stefani, appointed in 1922. On the other hand, in accordance with Mussolini's rule that 'economics is subordinate to politics', Fascist economic policy in the 1920s and 1930s was often dictated by prestige motives and the need to prepare for war.

The fixing of the exchange rate at the artificially high level of 90 Lire to the pound was typical of Mussolini's desire to raise Italian prestige abroad. Similarly the policies for autarky, or self-sufficiency in essential products, like the 'Battle for Grain' were pursued in the interests of defence and war. Under Mussolini substantial progress was made in this respect. By 1939 she was able to feed herself.

In the 1930s Fascism gained much respect from foreign statesmen because of its policy of corporatism: the establishment of corporations representing employers, employees and the state to regulate production and industrial relations. This policy was taken to its logical conclusion in 1939 with the abolition of the elected lower house of parliament and its replacement by a Chamber of Fasces and Corporations elected by the Corporations.

In fact the almost total absence of strikes in Fascist Italy was not the result of corporatism, but the effect of police repression, the outlawing of strikes, the dissolution of the free trade unions and their replacement by compulsory Fascist-controlled unions.

Mussolini and the Catholic Church

Mussolini's policy of wooing the Church paid handsome dividends in 1923–4 when the Vatican assisted his rise to power by abandoning the main parliamentary opposition force, the Catholic Popular Party. By 1926 the Vatican and the Fascist regime had developed cordial, even close relations, and over the next three years negotiated secretly to resolve the 60-year-old dispute between Church and State – the 'Roman Question'. On 11 February, 1929, the Vatican and Italy signed the Lateran Pacts which established the State of the Vatican City and restored the Church's power and influence in Italy. Mussolini's prestige was enormously strengthened both at home and abroad by the diplomatic triumph, and in the 1929 elections or 'Plebiscite' the Church ordered Italian Catholics to vote for the single Fascist list. But Fascism's relationship with the Church was not always harmonious: the 1931 crisis over Catholic Action showed the strength of anti-clerical, Fascist discontent with Mussolini's concessions to the Church. Further, though many Catholics approved of Mussolini's economic and corporatist policies and gave enthusiastic support to the Abyssinian War of 1935–6, they drew the line at the introduction of the Racial Laws in 1938. Pope Pius XI saw them as a highly unwelcome indication of the growing influence of Hitler over Mussolini. His successor, Pius XII, strongly disapproved of Italy's entry into the Second World War in 1940 and from this point on the Church sought to dissociate itself from Fascism.

Fascist foreign policy

Fascism was nothing if not violently nationalistic, and this nationalism manifested itself in the 1930s and 1940s in wars of aggression and colonial expansion. In the 1920s, however, Fascist Italy had little scope for aggression or expansion given the dominance of the international situation in the immediate post-war years by Britain and France. Only the 'gun-boat diplomacy' employed by Mussolini in the Corfù crisis of 1923 revealed the true nature of Fascist foreign policy.

Italy's opportunity for aggrandizement came with the Nazi assumption of power in 1933. By 1935 Britain and France were anxious for Italian support against a resurgent, rearming Germany. Mussolini, therefore, decided that they would condone his conquest of Abyssinia, on which he embarked in order to distract attention from the effects of the Great Depression in Italy. The Abyssinian War was both successful and popular with the Italian people, not least because the overwhelming majority resented the economic sanctions taken by the League of Nations.

Hitler was quick to exploit Mussolini's estrangement from Britain and France which resulted from the sanctions policy: he encouraged Italy to involve itself more and more deeply in the Spanish Civil War, thus preventing Mussolini from resisting the German invasion of Austria in 1938.

By 1939 relations between Italy and Germany had developed into a 'Rome–Berlin Axis', though Mussolini already showed signs of resenting Hitler's masterful methods. In March, 1939

Mussolini invaded Albania to 'match' Hitler's occupation of Bohemia. Despite his doubts, Mussolini strengthened his ties with Germany, concluding the Pact of Steel in May, 1939, and finally declaring war on Britain and France a year later.

The fall of Italian Fascism

The War proved to be an unmitigated disaster for Mussolini and Fascism. By the summer of 1943 Italy had lost Greece and all her African colonies, in large part a result of her unpreparedness for war, for which Mussolini must take the blame. The civilian population was demoralized by the defeats, by the food shortages and the effects of the Allied bombing. The last straw was the Allied invasion of Sicily in July, 1943, by which time demoralisation was almost as complete within the ranks of the Fascist Party as among the population at large.

On 24 July the Grand Council of Fascism passed a vote of no confidence in Mussolini and he was dismissed. The Fascist Party was dissolved by order of the King. Fascism was in fact overthrown by a royal coup d'état – the survival of the Monarchy had proved the 'Achilles heel' of the regime. However, it should be noted that Fascism had never been able to obtain such a degree of obedience or totalitarian control as Nazism had in Germany. This can be seen, for instance, in the attitude of the Italian population towards the racial laws.

11.4 CHRONOLOGY

1915	**April**	THE TREATY OF LONDON: concluded with Britain, France and Russia. Clauses promised Italy the South Tyrol, Trentino and Trieste among other territories.
	May	Italian Government denounced the Triple Alliance and mobilized against Austria-Hungary.
1917	**October–December**	THE CAPORETTO CAMPAIGN: the defeat of the Italians by Austro-German advance.
1919	**March**	THE FORMATION OF THE FIRST FASCIO DI COMBATTIMENTO BY BENITO MUSSOLINI (1883–1945): Mussolini was a former socialist and editor of 'Avanti' who had turned interventionist and nationalist.
	April	President Wilson appealed against the Italian territorial claims on the Adriatic. The Italian delegation left the conference.
	September	THE SEIZURE OF FIUME BY NATIONALIST AND WAR HERO, GABRIELE D'ANNUNZIO, following President Wilson's rejection of the Italian claim to the town.
1920	**August**	A general lockout in the metallurgical factories, leading to the occupation of the factories by workers.
	November	Treaty of Rapallo with Yugoslavia: Fiume to be an independent state.
	December	Declaration of war by d'Annunzio; Italian troops bombarded Fiume and forced him to evacuate it.
1921	**January**	Congress of the Socialist Party at Livorno: the Party split into a radical and moderate wing.
	February	Communist and Fascist riots in Florence; the spread of clashes between the two factions.
	June	Fall of the GIOLITTI cabinet, replaced by Bonomi.
1922	**February**	New government headed by Facta.
	May	Fascists drove out Communist city government of Bologna. Conflict extended to all the larger cities.
	August	Fascists seized control of the Milan city government.
	October	Fascist Congress at Naples. Mussolini, having refused a seat in the Cabinet, demanded the resignation of Facta and the formation of a Fascist cabinet. THE MARCH ON ROME: the Fascists occupied Rome, the King having refused Facta's demand for the proclamation of martial law. Mussolini formed the Cabinet of Fascists and Nationalists.
	November	MUSSOLINI GRANTED DICTATORIAL POWERS BY THE KING AND PARLIAMENT: granted until the end of December 1923 to restore order and introduce reforms. Gradual consolidation of Fascist control. Constitution still technically in force.
1923	**January**	A voluntary Fascist Militia authorized by the King.
	August	THE CORFÙ INCIDENT: Greek appeal against Italy to the League of Nations.
	November	THE NEW ELECTORAL LAW: provision that any party securing the largest number of votes in an election (provided it had at least one quarter of the total) should receive two-thirds of the seats.
1924	**April**	In the elections the Fascists received 375 as against the previous 35 seats.
	June	MURDER OF GIACOMO MATTEOTTI, Socialist deputy who had denounced the Fascists. THE AVENTINE SECESSION; majority of non-Fascist deputies left the Chamber.
	July	Rigid press censorship introduced.
	August	Meetings of opposition political groups forbidden.
1928	**May**	A NEW ELECTORAL LAW: universal suffrage abolished; the electorate reduced from 10 to 3 million. Candidates for election to be submitted to voters by the Fascist Grand Council.
1929	**February**	THE LATERAN TREATIES WITH THE PAPACY (ratified June): restored the temporal power of the Pope who was to rule over the Vatican City. A concordat with the Italian Government defined the position of the Church in the Fascist State; state indemnity for the Church.

	April	NATIONAL COUNCIL OF CORPORATIONS ESTABLISHED: to adjust disputes between various groups in the interests of national production. Composed of representatives from the syndicates, employers and government.
1933	July	THE FOUR POWER PACT: between Britain, France Italy and Germany.
1934	July	Attempted Nazi coup in Austria; Mussolini mobilized the army.
	November	Establishment of the Central Corporative Committee.
	December	Outbreak of conflict with Abyssinian troops at Wal Wal.
1935	January	FRANCO-ITALIAN AGREEMENT: in the hope of winning Italian support against Germany, Laval made large concessions to Italian claims in Abyssinia.
	April	THE STRESA CONFERENCE: called by France to consider action against German rearmament and to provide further guarantees for Austrian independence.
	October	ITALIAN INVASION OF ABYSSINIA. League Council declared Italy the aggressor.
	November	The League voted sanctions against Italy; failure of Powers to agree on oil sanction. (1936)
1936	May	Italian army captured Addis Ababa. Annexation of all Abyssinia.
	July	League Council voted to discontinue sanctions. OUTBREAK OF THE SPANISH CIVIL WAR: from the start Mussolini supplied Francoist forces with men and equipment.
	October	Italian-German agreement over Austria: the basis for the Rome-Berlin Axis.
1937	December	THE WITHDRAWAL OF ITALY FROM THE LEAGUE.
1938	March	The German annexation of Austria. No objection offered by Italy.
	October	The Fascist Grand Council abolished the Chamber of Deputies, replacing it with a Chamber of Fasces and Corporations.
1939	April	ITALIAN INVASION AND CONQUEST OF ALBANIA.
	September	Italy declared neutrality at the outbreak of the Second World War.
1940	June	ITALY DECLARED WAR ON FRANCE AND BRITAIN.
	September	Italy, Germany and Japan concluded a three-power pact.
1943	July	MUSSOLINI RESIGNED and was placed under arrest. Replaced by MARSHAL BADOGLIO, who declared the Fascist Party dissolved.
	September	Badoglio accepted terms for Italian surrender. Mussolini rescued from captivity by German troops, proclaimed a Republican Fascist Party.
1945	April	Mussolini captured and executed by Italian anti-Fascist forces.

11.5 QUESTION PRACTICE

1   To what extent is Mussolini's rise to power accounted for by the appeal of Fascism?

Understanding the question

This question invites a full discussion of the reasons why Mussolini came to power in 1922. The word 'appeal' in the question should be interpreted broadly, in terms of the promise of political stability, order and the revival of Italian national prestige as well as the narrower appeal of Fascist ideology.

Knowledge required

The long-term causes of the abandonment of the parliamentary system in Italy and the forces and interests undermining it. The effect of the Versailles Treaty on Italian public opinion. The social instability in post-war Italy, *e.g.* the occupation of the factories. The forces demanding strong government and supporting Mussolini's Fascist bands. The political situation in 1922, particularly the attitude of the King and the Army.

Suggested essay plan

1   Introduction: post-war Europe saw a challenge to the parliamentary system – a particularly strong challenge to a democracy that had been recently established and was seen to be not working well; in Italy discontents with the political system were fuelled by a sense of grievance over the Versailles Treaty as it affected Italy, and by fears of general social disorder and the threat to property.

2   The weakness of the political structure: the Italian State was faced by a hostile Church; democracy was seen by the masses as a middle-class affair concerned with liberal freedoms rather than social betterment; democracy was undermined in 1915 by Italian intervention in the First World War; clear indication of the nature of a state dominated by interest groups.

3   The immediate post-war situation: the 1919 general elections, Liberals losing control of the Liberal state; difficulties in forming a stable and effective government – parliamentary paralysis; the effect on public opinion of the failure of the Versailles Treaty to fulfil earlier promises (*e.g.* the Treaty of London, 1915) to Italy.

4   The seizure of land; the occupation of factories; ex-servicemen form the nucleus of a new party and offer their services to the landed aristocracy and industrialists to suppress the Left; Fascism, which did so much to create disorder, appeared to many to promise the restoration of order; support for Fascism in the armed forces and Royal Family; the fear of Socialism leads the Catholic middle classes to rally to Mussolini.

5   The circumstances of Mussolini's appointment: this was not primarily due to the appeal of Fascism, which was vague and not widely known, but to the appeal of law and order and the fact that there appeared to be no credible alternative; the King who was fearful of being replaced by Aosta, agreed to the appointment of Mussolini in a coalition government; the significance of the March on Rome; official Italy favoured Fascism – a mainly Liberal parliament voted for Mussolini; the alternatives were seen to be military dictatorship or Bolshevism.

2   How did Mussolini consolidate his power in Italy after 1922?

Understanding the question

The question is asking for an examination of the means by which Mussolini transformed Italy from a parliamentary democracy into a Fascist state. It would be proper for you to look at, among other considerations, Church–State relations (as the Catholic Church had been hostile to the Italian State since the latter's creation). Though you may make reference to it, you should not dwell at length on his imperial policy.

Knowledge required

Mussolini's use of emergency powers; the suppression of opposition newspapers and trade unions; and the creation of a special security force. New electoral laws. Control of education and the media. The corporate state, a National Assembly of Corporations replacing Parliament. Church–State relations and the Lateran Treaty.

Suggested essay plan

1   Introduction: Mussolini's assumption of power with the backing or connivance of the King, Army and industry; a *brief* assessment of the Fascist idea of the state.

2   The establishment of the new state on the basis of widespread national support; transformation of the political system: rule by decree replaced by a packed parliament based on the 1923 electoral law giving two-thirds of the seats to the Party getting one-quarter of the votes; the elections were managed by the police and the Blackshirts; there was support for Mussolini from numbers of Liberals and Catholics (the Clerico-fascists); Matteotti's published evidence of electoral malpractices led to his murder; the rigged Parliament further weakened in 1928 when the new electoral law enabled the Fascist Grand Council to decide who should be an MP; Mussolini was then President of the Grand Council and effectively Prime Minister for life; by 1934 this rump Parliament was in turn replaced by a National Assembly of Corporations.

3   The creation of a Fascist society: by 1927 there was the banning of non-Fascist parties, and the suppression of all opposition newspapers; strikes had been outlawed and trade unions of the traditional type had disappeared; there was the OVRA, the secret police; military tribunals had been empowered to deal with political offences; state employees were required to be Fascist Party members. To create a popular basis for a continuing support for Mussolini, the Ministry of Popular Culture was set up; Fascist textbooks had become compulsory and the youth were drafted into Fascist military-style organisations; propaganda in the press, on the radio and in the cinema; the Duce had become a symbol of the new Italy, virile, militaristic and united in purpose.

4   Mussolini's relationship with the traditional institutions; he was usurping more and more of the monarch's customary powers; the Church was won over by the Lateran Treaty, thus ending 60 years of Church–State rupture; the Church was protected by the regime; official Church support was given to the State – *e.g.* to the Fascist list of candidates in the plebiscite of 1929; Mussolini wooed the industrialists with new opportunities arising from the expansion in Africa and the development of the Italian air force.

3   Was Mussolini more than a 'sawdust Caesar'?

Understanding the question

You should concentrate on Mussolini's foreign policy and military achievements and failures and measure these against the grandiose imperial ambitions which he proclaimed for Italy.

Knowledge required

Mussolini's declared imperial ambitions for Italy. Italian foreign policy under Fascism. The Italian campaign and conquest of Abyssinia and Italian intervention in the Spanish Civil War. Mussolini's contribution to the Second World War.

Suggested essay plan

1   Caesar had led his legions successfully against the barbarians; Mussolini's proclaimed ideal was to infuse Italians with martial ardour and to lead them against the contemporary barbarians (the Abyssinians), the effete races (the British and French) and such nations as Albania and Greece, which stood in the way of what he considered to be the natural path of Italian expansion, and the transformation of the Mediterranean into an Italian lake.

2   The Abyssinian campaign: even with relatively advanced military technology (aeroplanes, gas, etc) against half-armed tribesmen, Mussolini did not find it easy to defeat the Abyssinians. He was held up for two months and subsequently controlled the conquered territory by holding it in pockets and terroristic methods. However, the prestige victory over the Abyssinians (which redressed the ignominious defeat of the Italians at the battle of Adowa in 1896) *seemed* gloriously successful to Italians at the time and Mussolini succeeded in rallying public opinion enthusiastically behind him. The defiance of the League had exposed British and French weakness – 52 nations had voted against Italy's action. It was a war which, in Denis Mack Smith's words, 'provided few casualties and many medals'. Mussolini now boasted that Fascist Italy, under strong leadership, had achieved what Liberal Italy had failed to do – to prove itself a major Power.

3   *However*, Mussolini's aping of Roman imperialism gave to Italians a wholly misleading impression of their country's potential strength, military preparedness and of his own capabilities as would-be Caesar. In fact, his megalomania and emphasis on 'Ducismo' proved more and more of a liability to Italy.

4   The Abyssinian War weakened the Italian economy and therefore restricted the likelihood of effective future military preparations; it depleted the war reserves and Italy ran into large budget deficits, in spite of recourse to compulsory loans and capital levies.

5   The impact on his foreign policy: initially he had been concerned to defend Austria against Germany but he abandoned Austria during the Abyssinian War and moved towards forging the 'brutal friendship. with Hitler. Mussolini convinced himself that he was the centre of European diplomatic activity (*e.g.* at Munich) but he was repeatedly upstaged and deceived by Hitler. He showed himself both fundamentally weak and unpredictable – *e.g.* at one moment he vetoed the Anschluss and began to fortify the Brenner. At the next he announced that he was standing behind Germany. After the Anschluss his foreign policy showed a progressive subordination of Italian to German interests.

6   Mussolini declared that a major war was inevitable but he showed little aptitude for preparing for it. For instance (as contrasted with Germany) he did not use the Spanish Civil War to master new tactics or use new weapons. When Italy went to war, in June, 1940, it was not merely militarily unprepared but in a state of marked political turmoil – the consensus which had supported the Abyssinian adventure was rapidly vanishing. 'Ducismo' was losing its hold and Mussolini had plainly not succeeded in militarizing the spirit of the Italian people.

7   Mussolini admitted privately that Italy's unpreparedness for the First World War was as nothing in comparison with her

unreadiness for the Second: equipment was antiquated and there was only enough ammunition for 60 days. Two weeks before his attack on Greece he had demobilized half his forces and then called up a fresh batch of 100,000 raw recruits. The attack proved a humiliating defeat and Italian troops were lucky to hang on to Albania. The subsequent German intervention in Greece and Yugoslavia put an end to Mussolini's ambitions in the Balkans. Note, too, the debacle in North Africa where 250,000 Italian troops were defeated by 30,000 British.

8   The case for describing Mussolini as a 'sawdust Caesar' can be fully substantiated. In his military and foreign policies it would be hard to argue that he was more than this. He *postured* as a great leader intent on grand imperialistic enterprises but he failed to plan effectively. He created large armies which he failed to equip. He proved himself very markedly inferior to Hitler as a dictator.

4   Document question (Mussolini and the Rise of Fascism): study the extract below and then answer questions (a) to (g) which follow:

'When in the now distant March of 1919, speaking through the columns of the "Popolo d'Italia" I summoned to Milan <u>the surviving interventionists who had intervened,</u> and who had followed me ever since the foundation of the "Fasci" of revolutionary action in January 1915, I had in mind no specific doctrinal programme. The only doctrine of which I had practical experience was that of Socialism, from 1903–1904 until <u>the winter of 1914</u> – nearly a decade. My experience was both that of a follower and a leader – but it was not doctrinal experience. My doctrine during that period had been the doctrine of action. A uniform, universally accepted doctrine of Socialism had not existed since 1905, when the revisionist movement, headed by Bernstein, arose in Germany, countered by the formation, in the see-saw of tendencies, of a Left-revolutionary movement which in Italy never quitted the field of phrases, whereas, in the case of <u>Russian Socialism,</u> it became the prelude to Bolshevism.

'The "Popolo d'Italia" described itself in its sub-title as "the daily organ of fighters and <u>producers</u>". The word "producers" was already the expression of a mental trend. Fascism was not the nurseling of a doctrine previously drafted at a desk; it was born of the need of action, and was action, it was not a party but, in the first two years, an anti-party and a movement. The name I gave the organization fixed its character.

'<u>The years preceding the March on Rome</u> cover a period during which the need of action forbade delay and careful doctrinal elaborations. <u>Fighting was going on in the towns and villages.</u> There were discussions but ... there was something more sacred and more important ... death .... Fascists knew how to die. A doctrine – fully elaborated, divided up into chapters and paragraphs with annotations – may have been lacking, but it was replaced by something far more decisive – by a faith. All the same, if with the help of books, articles, resolutions passed at congresses, major and minor speeches, anyone should care to revive the memory of those days, he will find, provided he knows how to seek and select, that the doctrinal foundations were laid while the battle was still raging. Indeed, it was during those years that Fascist thought armed, refined itself, and proceeded ahead with its organization. The problems of the individual and the state; the problems of authority and liberty; <u>political, social, and more especially national problems were discussed;</u> the conflict with liberal, democratic, socialistic, masonic doctrines and with those of the 'Partito Popolare', was carried on at the same time as the punitive expeditions.'

(Mussolini, *Fascism, Doctrine and Institutions*, 1935)

Maximum marks

(a) To whom is Mussolini referring as 'the surviving interventionists who had intervened'?   (2)

(b) Why had Mussolini's experience of Socialism ended in 'the winter of 1914'?   (2)

(c) Which sections of the Italian population were supposed to be included under the term 'producers'?   (2)

(d) What were the 'political, social, and more especially national problems' of Italy in 1919–1922?   (7)

(e) 'Fighting was going on in the towns and villages' of Italy in 1919–1922. Explain how this was important for the Fascists.   (4)

(f) Compare the revolutionary activity of Italian Socialism in 1919–1922 with 'Russian Socialism' in 1917.   (6)

(g) How far do Fascist slogans and Fascist doctrine support this explanation by Mussolini of the nature of Fascism in 'the years preceding the March on Rome'?   (8)

11.6 READING LIST

Standard textbook reading

J. M. Roberts, *Europe 1880–1945* (Longmans, 1967) chapters 13, 15.

David Thomson, *Europe since Napoleon* (Penguin, 1977), chapters 23, 27, 28.

Anthony Wood, *Europe 1815–1945* (Longmans, 1975), chapters 29, 31.

Further suggested reading

A. Cassels, *Fascist Italy* (Routledge, 1969.)

F. Chabod, *A History of Italian Fascism* (Cedric Chivers, 1974).

H. Finer, *Mussolini's Italy* (F. Cass, 1964).

D. Mack Smith, *Italy – A Modern History* (Ann Arbor, 1969).

G. Procacci, *A History of the Italian People* (Penguin, 1973).

Christopher Seton-Watson, *Italy from Liberalism to Fascism* (Methuen, 1967).

S. Woolf, *European Fascism* (Weidenfeld and Nicolson, 1970).

12   The Weimar Republic and National Socialism

12.1 INTRODUCTION

The Weimar Republic was established as a presidential democracy following Germany's collapse in 1918. From the start it faced opposition from both the political Left and Right. It was widely associated in the German mind with the acceptance of the deeply-resented Versailles 'Diktat', the implications of which were humiliatingly re-emphasized in the Ruhr invasion of 1923, by France and Belgium.

    The period between 1924 and 1929 was one of relative stability in Germany. But after the economic collapse of 1929 political extremism gained the upper hand. The assumption of power in 1933 by Hitler, and the groups which supported him, exposed the widespread German opposition to Weimar democracy and the flaws in its constitution. Nazism, which had originated in post-war Bavaria, offered a mixture of violent nationalism, racialism and anti-Communism.

Once in power, and after the Reichstag fire, Hitler proceeded through legislation to set up a totalitarian order under the domination of the Nazi state and its Führer (leader). The process of the Nazi coordination of German society was accompanied by a reign of terror, of which a major element was anti-semitic legislation and persecution. Socialistic social policies were combined with an attempt to secure maximum economic self-sufficiency and with preparations for war.

After the 'Night of the Long Knives' Hitler was granted an oath of unconditional loyalty from the Army and he merged the offices of Chancellor and President. The consolidation of Nazi power within Germany was the prelude to his expansionist foreign policy and military campaigns.

12.2 STUDY OBJECTIVES

1   The end of the First World War; the origins and constitution of the Weimar Republic.

2   The particular difficulties facing the new Republic; its association in the public mind with defeat and the unpopular Versailles Settlement; widespread social and political tensions and uprisings, *e.g.* the Kapp Putsch and the Spartacists; the weaknesses of the constitution; the dependence of Weimar on insecure coalitions; a sound understanding of the various parties and political alignments.

3   The impact of the Ruhr invasion of 1923; the inflationary crisis.

4   The importance of the relatively stable years 1924–9; the Dawes Plan; Gustav Stresemann and Locarno – Stresemann's objectives of understanding with the Western Powers and revision of the Eastern frontiers of Germany.

5   The use made by the extremist parties of the unpopularity of the Versailles 'Diktat' and the effect of the world depression, following the crash of 1929; the origins of the Nazi Party and an explanation of its subsequent very rapid growth.

6   Hitler's own background, personality and ideas; the appeal of his propaganda; the comprehensiveness of the Nazi movement with its appeal to an anti-Communist middle class, with its 'socialism' directed against Germany's traditional rulers and the promise of a restored German greatness; the importance to Nazism of the support from industrialists, bankers and Right-wing politicians.

7   The so-called 'Nazi seizure of power' in 1933; a detailed knowledge of the course of events leading to this and an explanation of how it came about; the Nazi emphasis on 'legality' and 'national revolution'; a consideration of the previous weakening of the Weimar democracy through government by Presidential Decree and the subsequent willingness of the parliamentary parties to vote their own end in the Enabling Law.

8   The process by which Nazi power was consolidated; the Reichstag Fire and the Enabling Act; the attitude to established institutions, *e.g.* the churches; the reign of terror; the racial laws.

9   Nazi economic and social policies.

10   Hitler's relationship with the Army, particularly the events leading to the Röhm purge (the 'Night of the Long Knives'); the Army's attitude to the SA; the Army's oath of unconditional obedience to Hitler in his new role as Führer and Chancellor

on the death of Hindenburg; the policies of conscription, rearmament and the reunification of the German people; the way in which these coincided with the Army's view of Germany's revival as a great Power.

12.3 COMMENTARY

Foundation of the Republic

The Weimar Republic was born out of Germany's defeat in the First World War. A combination of domestic upheaval and President Wilson's demands for the Kaiser's abdication led to its declaration by the new moderate Socialist Chancellor Ebert in November, 1918. The Republic's constitution, drafted in the early months of 1919, established a democratically-elected main chamber, the Reichstag; a second chamber, the Landrat, to represent the interests of the Länder (provinces); a popularly elected president and an electoral system with proportional representation.

Opposition to the Republic

From the start the new Republic was beset by difficulties. The Right-wing elements in German political life, which included the volunteer forces of ex-soldiers, the *Freikorps*, claimed that the new order had been brought about by revolutionaries. It was these, they argued, who had caused the internal collapse of Germany while the Army remained undefeated in the field. This was the legend of the so-called 'stab in the back'.

The Republic also suffered from being forced by the Allies to accept the deeply-resented Treaty of Versailles, the 'Diktat'. In fact a large minority in the Reichstag voted against accepting the peace terms. In its early years Weimar was assaulted by a series of attempted coups from both the extreme Left and the extreme Right, such as the Spartacist revolt, the Kapp putsch and Hitler's abortive revolution in Munich.

In each instance, the Republic was dependent for its survival on the support or at least the neutrality of the Army, though the Kapp putsch was actually defeated by the actions of the German working-class. But the Army was not the only institution left over from Imperial Germany on which the Republic was forced to rely.

Unable or unwilling to carry out a major change of personnel in the administration, the Republic's politicians continued to be served by a Civil Service and judiciary which had little sympathy for the new political system. As J. M. Roberts writes: 'The Right opposed Weimar parliamentarism as an affront to Nationalism; the Left regarded it as a cover for social conservatism.'

The weakness of the Weimar Constitution

The new political order was also flawed in its own constitutional structure. There were dangers in the system of proportional representation, which could lead to a multitude of small parties and instability. This was particularly so when they were faced with the violent nationalism of the Right and the militant emphasis on the unity of the German 'Volk'.

There was potential danger in the popular election of the president, which created the possibility of his appealing to the German people over the head of the Reichstag; and in the presidential right to rule by emergency decree, a right designed and used originally to protect the Republic but one which could be used to undermine the democratic process. Further, in this system the Chancellor had a relative lack of control over his ministers, who were closely tied to the whims of the individual parties.

The main parties lending their support to the Weimar Republic were the Social Democratic Party, the Catholic Centre and the Liberal Democratic parties, though both of the latter drifted to the Right during the latter days of Weimar. The Communists and the National Socialists (Nazis), the extreme parties of the Left and Right, openly proclaimed their wish to overthrow the Weimar Republic. Stable coalitions of those parties willing to participate in a Republican government could only be achieved by either the exclusion of the Left or Right – both of which represented very powerful elements in Weimar society.

The Ruhr invasion and the Great Inflation

This vulnerable political structure and culture was further weakened by the French and Belgian invasion of the Ruhr in 1923, an action designed to secure reparations payments from Germany.

The German currency, already severely inflated by the government's need to pay for the war and its aftermath, spiralled into hyperinflation, due to a renewed crisis of confidence in the Mark and the government's resort to the printing press to finance its policy of passive resistance to the French. The result was the financial ruin of many middle-class Germans, many of whom already felt alienated from the Republic.

Inflation was finally halted with the introduction of a new stop-gap currency, the Rentenmark, and in 1924 an inter-Allied commission worked out the Dawes Plan for a revised arrangement for German reparations payments. This envisaged the economic recovery of Germany, a recovery which was financed largely on short-term loans borrowed from abroad, particularly from the US.

Gustav Stresemann

The German government accepted the Dawes Plan in spite of Right-wing opposition, and the German foreign minister, Stresemann, made further moves towards reconciliation with the West, reaching a high point with the Locarno Treaty of 1925. Stresemann's so-called 'policy of fulfilment' was designed to make concessions which would return Germany to full sovereignty and independence as quickly as possible. But it did not abandon the aims of eventually removing reparations, achieving military equality and revising Germany's eastern frontiers. Though he did not exclude the possible use of force in the East, he was attacked by the Right-wing as far too conciliatory and as a betrayer of German national interests.

The period 1924–29 was one of relative stability for Weimar, but the hostility of the Right towards Stresemann – who was arguably the Republic's most accomplished statesman, who was conspicuously patriotic and who was trying to restore Germany's status and strength within the framework of international relations – was ominous for the future. So, too, was the election in 1925 of the First World War hero Hindenburg to the Presidency. Although he was not the representative of the monarchist cause which some of the voters took him for, he was never a fervent supporter of Weimar and in his immediate circle there were many who were violently anti-Republican.

The move to the Right and the economic collapse

Following the accession of the press magnate Hugenburg to the leadership of the German Nationalists in 1928, the Party made a strong shift to the Right and was in the forefront of the 1929 campaign against the Young Plan, the new Allied scheme for reparations payments. The attack on the Young Plan was combined with a wholesale onslaught on the Republic, which brought into operation an alliance between the Nationalists and the National Socialists. Although, ultimately, the campaign failed to prevent the acceptance of the Young Plan, it was indicative of the scale of the Right-wing opposition to Weimar.

The year 1929 also saw the Wall Street crash, the prelude to the world-wide slump. This led to the recalling of those short-term loans on which Weimar's brief period of relative prosperity had been based. The subsequent depression led to a rapid rise in unemployment, reaching 6 million in 1932. It also resulted in the rapid spread of extremism in German life.

Because of the impossibility of maintaining a stable coalition, German chancellors were forced to rule from 1930 onwards by a combination of ad hoc majorities and presidential decrees. On the Left it was the Communists who gained most from the crisis; on the Right it was the Nazi Party which now – for the first time – became a really decisive force in German politics. In September, 1930, the Nazi Party won 107 seats in the Reichstag, and the Communists 77.

Hitler and the Nazi Party

The Nazi Party had its beginnings, in post-war Bavaria, as originally only one of a number of small radical Right-wing groups. Under the leadership of Adolf Hitler, with his extraordinary oratorical and propagandist talents, this rapidly became a significant force in Bavarian politics. The party suffered a major setback in 1923 with Hitler's abortive coup – an attempt in Munich to mount local forces and march on Berlin (in the year after Mussolini's March on Rome) led to his arrest, imprisonment and a temporary ban on the Party.

From this failure Hitler learnt one major lesson – power was not to be achieved by a coup, but through the democratic and constitutional processes of the Republic. Hitler chose the slower route of a legal accession to power, seeking to destroy Weimar from within rather than from without. This was the path he set out upon when the Party was reconstituted in 1925.

Hitler's ideology

Hitler, born an Austrian of lower-middle-class stock, had developed his basic ideas and prejudices as a layabout in Vienna in the early years of the century. These were a violent German nationalism, ideas of racial superiority, anti-semitism and anti-Marxism. These ideas were later combined with the fixed determination to overthrow the Versailles Settlement as well as the desire to acquire for Germany the massive addition of 'Lebensraum') or 'living-space', in the East.

These themes are all contained in 'Mein Kampf' ('My Struggle') which Hitler wrote in prison after the failed putsch of 1923. His experience as a soldier in the trenches during the First World War only served to strengthen these ideas as well as to develop in him an admiration for militaristic hierarchy.

These ideas were presented to the German people with great political astuteness and had a widespread appeal, particularly among the demoralized lower middle classes with their deep fear of Communism, of the loss of social status and identity and their nostalgic hankering for the return of German greatness.

Nazism was presented as a movement that included the whole German 'Volk', not as a party in the traditional sense. In fact its objectives were the abolition of parties and factions, and the elevation of the power of the Nazi state and its leader over all – the creation of a totalitarian society.

The Party had an anti-capitalist wing which promised to remove economic and social privilege. There were in the development of Nazism significant socialistic elements but Hitler tended to underplay them. He courted the support not only of the masses of the voters but also of the industrial, financial and politically conservative élites, setting out to persuade them that the Nazi Party was the only bastion against Communism. It should be noted that by no means all members of these groups supported

National Socialism, for while they were inclined to endorse its anti-Communist and nationalistic appeal, they were apprehensive of its socially radical elements.

Hitler as chancellor

In the years 1930–32 the Nazi vote grew rapidly. Though he never gained an overall Nazi majority in the Reichstag, it was on the basis of that mass support that Hitler came to power. The Nazi use of the term 'seizure of power' is something of a misnomer. Hitler did not seize power. He was offered it. He had already demanded the Chancellorship in August, 1932 (the previous month the Nazis had become the largest party in the Reichstag with 230 seats) but had been refused by the current chancellor, Von Papen, and the president, Hindenburg.

However, neither Von Papen nor his successor, Schleicher, had been able to form a stable majority in the Reichstag. Von Papen continued negotiations with Hitler behind Schleicher's back and finally persuaded Hindenburg to give Hitler the Chancellorship in the National Socialist/Nationalist government.

As a representative of Germany's conservative élite, Von Papen believed that Hitler's party could be used to give mass popular backing to their attempt to create a conservative-dominated authoritarian state. This would, they hoped, return to Germany's privileged classes the political prominence they had enjoyed before Weimar. Hitler was to be their 'drummer'.

The conservative Right intended to use him to achieve *their* ends. In the event it was to be the other way round. Such a scheme could only operate in the context of the breakdown of parliamentary government and the consequently increased power of the President and the clique surrounding him.

The consolidation of Nazi power

The Nazis' legal take-over and their collaboration with the conservatives in the so-called 'national revolution' was to prove an immense advantage to them when they began to consolidate their power. It provided them with full access to the administrative apparatus of the German state as well as with allies who were willing or eager to see the destruction of that state.

That consolidation or coordination (*Gleichschaltung*) proceeded along a number of lines. The legal basis for the Nazi take-over was established by two pieces of legislation.

First were the Reichstag Fire decrees which Hindenburg issued following the burning-down of the Reichstag by a Communist, possibly with the connivance of the Nazis. These suspended all civil liberties and provided the legal bases whereby the Land governments could be coordinated.

The second piece of legislation followed the March elections, when the Nazis and their nationalist allies gained an overall majority. This was the Enabling Act, which gave Hitler the right to rule by decree for four years and which was the prelude to the declaration of one-party rule. The Act allowed the government to pass laws without consulting the Reichstag.

The reign of terror A reign of terror was the background to the Nazi coordination, with beatings, murders and the establishment of concentration camps for political opponents. Some conservative institutions were fearful of Nazi radicalism. These Hitler bought off with promises which appealed to their self-interest: rearmament for the Army, a Concordat for the Roman Catholic Church, the suppression of the independent labour movement for big business (strikes were forbidden and the trades unions abolished, being supplanted by the Nazi Labour Front).

Hitler appeared to have something to offer for most groups in German society. The Jews, however, were to be ruthlessly excluded from the Aryan Reich. This was signalled by the official boycott of Jewish shops and the laws passed excluding them from public office. Anti-semitism was codified in particular in the Nuremberg Laws of 1935.

Economic policy Nazism wore the mask of benevolent Socialism for very many Germans. Not least, Hitler's job-creation schemes rapidly brought down the unemployment which had done so much to bring him to power. Here he expanded the work of previous governments in the provision of public-works programmes; by 1938 below half a million Germans were unemployed, this fall being also accounted for by the revival in world trade.

The fear of indebtedness was lifted from thousands of German farmers by the hereditary farm law which guaranteed ownership in perpetuity; a new scheme was established to coordinate agricultural production. At the same time, new plans were developed to reduce Germany's expenditure of foreign currency, involving both complicated trading procedures and the encouragement of home production rather than imports. These programmes were designed not only to improve the performance of the German economy; they were also intended to make Germany self-sufficient in food and essential raw materials.

Totalitarian control Nazi rule penetrated every aspect of German life (its control and pervasiveness were markedly more comprehensive than that of Italian Fascism). Schools were purged of opposition elements amongst the staff and were instructed to teach in accordance with Nazi ideology. Membership of the Nazi Hitler Youth became obligatory in 1936 and gradually all other youth groups were dissolved. Marriage and population growth were promoted through propaganda and financial concessions.

Significantly, too, the historic subdivisions of Germany were done away with in 1933 and in fact the Nazi *Gleichschaltung* ended the dominance of Prussia in Germany which had been the legacy of the *Kleindeutschland* unification of Germany under Bismarck.

The Army Hitler still, however, required the support of powerful conservative institutions. His relationship with the Army, *the* institution which could have toppled him, was threatened by the SA, the Nazi stormtroops who, under Röhm, were seeking to usurp the Army's position. They also contained the most distinctively radical elements in the Nazi Party. Furthermore, the independent attitude of Röhm threatened Hitler's position within the Nazi movement itself. To deal with this threat, as well as to settle old scores with other figures, such as Schleicher, Hitler had the SA leaders and others removed in the 'Night of the Long Knives' (30 June, 1934).

The Army, whose attitude had been equivocal towards Hitler, rewarded him for this action by granting him an oath of unconditional loyalty and by accepting his merger of the offices of chancellor and president, thus conferring on himself total dictatorial power.

But the elimination of the SA was not the only reason for the Army's allegiance to Hitler. In the following years he set about sweeping away the hated Treaty of Versailles, He introduced conscription, building up Germany's armaments and, on the basis of this strength, first through diplomacy and then by war, undid the territorial arrangements of the post-war settlement in order to restore Germany to the rank of a world power and to assert her hegemony over Europe.

12.4 CHRONOLOGY

1918	**November**	THE ABDICATION OF KAISER WILLIAM ANNOUNCED IN BERLIN: a Republic proclaimed. The government entrusted to Friedrich Ebert and Philipp Scheidmann. A joint administration of Independent and Majority Socialists took control in Berlin. The struggle between the extreme Left SPARTACISTS and the Majority Socialists.
1919	**January**	THE SPARTACIST REVOLT IN BERLIN, crushed by the provisional government with the aid of the Army. Election of a national assembly to draw up a constitution.
	April	Soviet Republic established in Bavaria; overthrown by forces of the Federal Government.
	June	Proclamation of a Rhineland Republic. UNCONDITIONAL ACCEPTANCE OF THE PEACE TREATY. Germany signed the Treaty of VERSAILLES.
	July	ADOPTION OF THE WEIMAR CONSTITUTION: the President, elected for a seven-year term, was to appoint a Chancellor who in turn chose a cabinet which could command majority support in the Reichstag. By Articles 25 and 48 the President was empowered to suspend constitutional guarantees and dissolve the Reichstag in periods of national emergency. Federal rights were guaranteed. A system of proportional representation ensured that minority parties were represented. Also it meant government by coalition.
1920	**February**	HITLER PROCLAIMED THE 25 POINTS OF THE NATIONAL SOCIALIST PROGRAMME.
	March	THE KAPP PUTSCH: a monarchical coup involving the seizure of government buildings in Berlin. The Government fled to Stuttgart, but the movement collapsed as a result of the general strike of the trades unions.
	May	Germany received reparations bill of 132 billion marks.
	July	Spa Conference: the Germans signed a protocol of disarmament and arranged for reparations payments.
1921	**March**	Allied occupation of Düsseldorf, Duisburg and Ruhrort because of alleged German default in reparations payments.
1922	**April**	Treaty of Rapallo with Russia.
	June	Assassination of WALTHER RATHENAU.
1923	**January**	OCCUPATION OF THE RUHR by French and Belgian troops after Germany had been declared in default. The occupation was condemned by the British and Americans and led to passive resistance among the Ruhr workers. The Government supported this resistance.
	August	New government formed by GUSTAV STRESEMANN, leader of the People's Party.
	September	The government ended passive resistance. Astronomic monetary inflation in Germany.
	November	THE BEER HALL PUTSCH IN MUNICH: General Ludendorff and Adolf Hitler, leader of the growing National Socialist Party, attempted to overthrow the Bavarian Government. The rising was easily put down. Hitler was imprisoned, during which time he wrote MEIN KAMPF.
1924	**April**	THE DAWES PLAN: helped Germany to meet her treaty obligations in the period 1924–9. Gave stability to the German economy.
1925	**April**	FIELD MARSHAL VON HINDBENBURG elected President of the Weimar Republic.
	October	The Reichstag ratified the Locarno treaties.
1926	**September**	Germany entered the League of Nations.
1929	**September**	French forced to evacuate demilitarized Rhineland.
	October	WALL ST. CRASH, signalling the onset of the Great Depression.
1930	**September**	REICHSTAG ELECTIONS: emergence of Hitler's National Socialists as a major party (107 seats as against a previous 12).
1931	**March**	Publication of a project for a German-Austrian customs union; rejected by France and the World Court.
	May	The failure of the Austrian Credit-Anstalt, marking the start of the financial collapse in central Europe. By the beginning of 1932 the number of German unemployed was already more than 6 m.
	July	Hoover moratorium on reparations and war debts.
1932	**May**	BRÜNING resigned; von PAPEN asked by President Hindenburg to form a Ministry responsible to the Executive alone.
	June	The government lifted the ban on the Nazi storm troops. Von Papen coup against the Socialist state government of Prussia.
	July	REICHSTAG ELECTIONS: the Nazis returned more candidates (230) than any other party.

	August	Hitler refused Hindenburg's request that he serve as Vice-Chancellor under Von Papen.
	September	The Reichstag dissolved: the impossibility of securing popular support for Von Papen.
	November	Resignation of Papen.
		Hitler rejected the conditional offer of the chancellorship. His demand for full powers was refused by Hindenburg.
	December	GENERAL VON SCHLEICHER formed a new cabinet.
1933	**January**	Schleicher was forced to resign after his efforts to conciliate the Left and Centre had failed.
	30	HITLER BECAME CHANCELLOR: the Ministry was regarded as a coalition of National Socialists and Nationalists.
	February	THE REICHSTAG fire: the Nazis claimed that it was the first move in a Communist conspiracy. Emergency decrees were issued suspending the constitutional guarantees.
	March	The last Reichstag elections of the Weimar Republic: the Nazi terror.
		THE ENABLING ACT PASSED BY THE REICHSTAG AND REICHSRAT: the Nazi dictatorship firmly established. Suspension of the Weimar Constitution.
	April	The Civil Service Law.
	July	THE NAZI PARTY DECLARED THE ONLY POLITICAL PARTY.
		CONCORDAT WITH THE VATICAN.
1934	**January**	THE REICHSRAT REPRESENTING THE STATES WAS ABOLISHED. Germany became a national rather than a federal state.
	May	Creation of the People's Court, which was set up to try cases of treason. Proceedings to be secret. No appeal except to the Führer.
	October	Constitution of the NAZI LABOUR FRONT.
	June	THE RÖHM PURGE – 'THE NIGHT OF THE LONG KNIVES': directed primarily against representatives of the more radical social revolutionary wing of the Nazi Party which aimed at far-reaching changes in society and the incorporation of the storm troops into the German Army.
	August	LAW CONCERNING THE HEAD OF STATE, COMBINING THE PRESIDENCY AND CHANCELLORSHIP. Death of Hindenburg.
		Plebiscite approved of Hitler's assumption of the presidency and of his taking sole executive power.
1935	**September**	THE NUREMBURG LAWS: all Jews in Germany reduced to second-class citizens; professions closed to the Jews; the banning of intermarriage with them.
1936	**October**	THE FOUR YEAR PLAN: policy of economic self-sufficiency which would make Germany independent, especially in raw materials, in the event of war.
1938	**February**	REORGANIZATION OF THE MILITARY AND DIPLOMATIC COMMAND: VON BLOMBERG, War Minister, and VON FRITSCH, Commander-in-Chief of the Army, were removed.
		Hitler assumed the Ministry of War.

12.5 QUESTION PRACTICE

1   How significant was the support of the German Army and its leaders in the establishment and consolidation of Hitler's power before the outbreak of the Second World War?

Understanding the question

The point is to assess just how much the Army had the power to stand in Hitler's way and whether or not it was the support of the Army which allowed Hitler to achieve his final mastery over Germany. You should give equal attention to both the establishment and consolidation of his power. Note that the question seems to assume that the Army supported Hitler from the start. This can be argued to be a mistaken assumption. One can, on the contrary, approach the question from the opposite angle, examining how Hitler managed to gain its support.

Knowledge required

The relationship between Hitler and the German Army. The course of political events in Germany from 1918–39, with special regard to military involvement in civilian politics. The social basis of Nazi support.

Suggested essay plan

1   Nazism and German militarism are so synonymous in the

popular imagination that the demise of the Weimar Republic and the emergence of the Nazi state *seemed* to follow inevitably, and the Röhm putsch ('the Night of the Long Knives') of 1934 is seen as the culmination of the alliance between the Nazi Party and the Army which carried Hitler to absolute power. However, the relationship between the Army leadership and the Nazis was not so simple. The former were the old aristocratic Right of German society and antipathetic to the new 'radical Right'. It is not a simple story of the Army supporting Hitler from the start, but of Hitler *winning* army loyalty.

2   The initial Army attitude: there was some sympathy with the Nazis and SA (who were ex-Freikorps members) as a force against the Left, particularly the Communists and the trade unions; note that the Weimar Republic was supported by the 1918 Groening-Ebert pact against disorder; the Nazis were also seen, though, as a threat to order (*e.g.* Prussia, where the state was Socialist and the Nazis conflicted with authority).

3   The Army and Hitler's accession to power; the support for Brüning; note that the Army was working to try and block the Nazis from gaining power; in the end Hitler achieved office because of his popularity with the electorate, *i.e. despite* Army attempts to prevent him.

4   Once in power, Hitler worked to secure the trust of the Army leadership; they saw him, as many of the German industrialists

did, as a bastion against the Left; they supported him against the radical wing of the Nazi Party, the SA. The Röhm putsch is a classic example of Hitler's capacity to divide and rule; standing by while the SS despatched the SA leadership and also killed off Schleicher and Bredow, the Army compromised themselves with the regime.

5 The personal oath of the Generals to Hitler; after the Hossbach Memorandum of 1937 Blomberg and Fritsch were dismissed; after this the Army was in Hitler's grip; the only resistance occurred in 1944 (the Stauffenberg plot), when the defeat of Germany loomed.

6 Ironically, then, it is not the Army's initial support but its failure to stop Nazism in 1932–3 which is all-important; the Army leadership's antipathy to Nazism was only overcome by the seeming irresistibility of Hitler's mass popularity with the electorate. Subsequently it is easy to see why the Army supported a leader who promised a re-emergence of the prestige of their institution in German life and who also shared their ambitions for national aggrandizement.

7 It could be argued that it was as much Hitler himself who caused the support of the Army; he used a natural sympathy for his nationalist and anti-Leftist aims (if not for his style) to subjugate an initially reluctant Army to offer unswerving loyalty. It was not so much the Army's support for Hitler that was important as his ability to produce that support in the first place and so to use the Army to enact his designs after 1939.

2 'Hitler's acquisition of power is more easily explained than his retention of it'. Do you agree?

Understanding the question

The question requires you to explain how Hitler first gained power. You must then explain how he avoided becoming simply an instrument for the conservative wing of German political life (a 'front man' for another coalition of the Right) and how he went on to establish and consolidate the Nazi state.

Knowledge required

The previous course of German politics, with particular attention to the years 1929–33: the economic collapse, the rapid expansion of the Nazi Party and the increasingly authoritarian nature of German government before Hitler. The reasons for the appeal of Nazism and the basis of its support. Hitler's achievement of power and the passing of the Enabling Act. The destruction of organized opposition; the key role of the Army, particularly in the 1934 purge. The reason why Hitler's policies were regarded as successful within Germany.

Suggested essay plan

1 Hitler's achievement of office was often thought of as a great tragic accident – due to a gamble, by the conservative upper classes through their representatives, which failed: it was seen as the gaining of power with their connivance in order to resist the Left; as outsmarting them and consolidating his political control through the 'Gleichschaltung'. On the other hand, one should not overstress the contribution of his personality and tactical astuteness at the expense of the underlying reasons why there was a wide basis of support for Nazism in German society.

2 An explanation of the collapse of the Weimar Republic: the economic slump sharply exposing a German lack of faith in democracy; constitutional weaknesses; Brüning and government by decree; growing extremism at the expense of both liberal and conservative parties (*e.g.* the September 1930 elections); Right-wing radicalism in the face of a perceived threat to German

society from Communism; this fear was skilfully used by Hitler, who won the support of bankers and industrialists, who in turn saw him as someone whom they could use to their own ends.

3 The immediate circumstances of Hitler's coming to power; the emphasis on the strength of popular support; note that subsequently the Enabling Act was passed with the support of all the parliamentary parties, except the Socialists and Communists. Note the skilful use of propaganda by the Nazis, particularly Goebbels' contribution.

4 Once in power by legal means, Hitler began to build the Nazi state and he banned, imprisoned and suppressed any opposition; within the Nazi movement the populist wing, the SA, was subordinated (with Army support) in the Röhm putsch, this securing the loyalty of the Army to the regime.

5 In consolidating his control, Hitler depended partly upon the tradition of obedience to authority in German society and increasingly upon the image of a new, powerful and successful Germany. The ways in which he satisfied different groups in society: the working class with employment and social benefits; industrialists with rearmament and the pledge to defend society against Marxism; the Army with conscription and rearmament. He was helped by the economic recovery of the mid-later 1930s. Nazism appeared to offer national unity and purpose after the divisions of Weimar, and the great majority of Germans welcomed the success of Nazi foreign policy.

6 Both Hitler's acquisition and retention of power depended to a considerable extent on his political astuteness; he used the Weimar political system with great skill; once in power he used parliamentary means to suspend the constitution and then built his regime up by subordinating national institutions and organizations to Nazi purposes; his continuance in power rested on very widespread public support and also on his willingness to use methods which traditional German rulers would not normally have countenanced.

3 What grounds were there, by 1929, for believing that democracy had taken root in post-war Germany?

Understanding the question

This requires you to concentrate on the Weimar system of government and its relation to German public opinion. You should try and assess whether or not, before its collapse, it was working – and working *democratically*. You should then make some observations on the cause of the subsequent collapse.

Knowledge required

Details of Weimar politics and a clear idea of the relations between the main interest groups, the industrialists, unions and the government. Economic developments and problems during Weimar, especially reparations, the Great Inflation and its social consequences. The constitution of the Weimar Republic and post-war attitudes towards it in Germany.

Suggested essay plan

1 Superficially things may have appeared stable in 1929; however, the prosperity and Liberalism of Weimar were largely a mirage and there were serious flaws in its democratic character.

2 The Western idea of democracy: the belief in government by the will of the majority of the people, with the rights of all protected by a constitution; also there was the connotation that the people would have the power to make its will effective through parliament.

3 Compare Weimar's constitution with this: show the extent to which it is 'democratic', *viz.* universal suffrage, the protection

of rights, enactment by majority in the Reichstag; control of the executive through a democratically elected President with wide emergency powers over the state; Weimar was *formally* governed by 'vox populi'.

4   Note, however, that for democracy to work there must be a general consensus that the best way of settling differences is by each individual voting for a representative in a political party in an assembly which will then give the decision by vote; this approach of government is usually guaranteed by a protection of civil rights which 'all' accept as sacrosanct. Germany did *not* function in this way before or after 1929.

5   The inflation: the impact of this on the middle classes; those suffering from the economic crisis were alienated from the state, which was felt not to be acting in their interests; the pressure of the large interest groups, not the ballot box, was felt to be deciding government policy; it was widely felt that Germany was operating on the cartel principle (of group interest) and not democratically (in the public interest); the middle classes felt ignored; see how, late in 1929, the question of social insurance created deadlock in the coalition.

6   Weimar politics was 'interest politics': the middle classes in particular were not reconciled to the social revolution and national defeat; all sides were only content so long as the economic prosperity of the Stresemann period continued; but this was a very fragile thing, dependent on loans.

7   Under the Weimar Republic Germany in many respects remained backward-looking – to past glories of the Wilhelmine Reich rather than forward to a democratic future. By 1929 Germany had not truly taken to the democracy foisted on to her after defeat in 1918. Old forms of cartelization and interest politics found new expression in the cobbled-together coalitions of Weimar. A large section of the population was in any case antipathetic to the new state.

4   Document question (The Nazis in 1933): study the extracts below and then answer questions (a) to (h) **which follow:**

JANUARY 30TH, 1933
'...Chief-of-Staff Röhm stands at the window the whole time, watching the door of the Chancellery from which the Leader must emerge....
'The Leader is coming. A few moments later he is with us. He says nothing, and we all remain silent also. His eyes are full of tears. It has come! The Leader is appointed Chancellor. He has already been sworn in by the President of the Reich. The final decision has been made. Germany is at a turning-point in her history....'

'FEBRUARY 3RD, 1933
'I talk over the beginning of the election campaign in detail with the Leader. The struggle is a light one now, since we are able to employ all means of the State. Radio and Press are at our disposal.... The Leader is to speak in all towns having their own broadcasting station. We transmit the broadcast to the entire people and give listeners-in a clear idea of all that occurs at our meetings.
'I am going to introduce the Leader's address, in which I shall try to convey to the hearers the magical atmosphere of our huge demonstrations.'

'FEBRUARY 27th, 1933
'...Suddenly a phone call from Dr. Hanfstaengl: "The Reichstag is on fire!"....There is no doubt but that Communism has made a last attempt to cause disorder.... Goering at once suppresses the entire Communist and Social Democrat Press. Officials of the Communist Party are arrested during the night. The S.A. is warned to stand by for every contingency.'

'MARCH 24th, 1933
'The Leader delivers an address to the German Reichstag. He is in good form.... The Zentrum and even the Staatspartei affirm the law of authorization...only the Socialists vote against it. Now we are also constitutionally masters of the Reich.
'...Now the discussions with the trade unions begin. We shall not have any peace before we have entirely captured these.'

'MARCH 27th, 1933
'Dictate a sharp article against the Jewish horrors propaganda. The proclamation of the boycott already makes the whole clan tremble in their shoes.... The Jewish Press is whimpering with alarm and fear...'

'MAY 1st, 1933
'The great Day of the German Nation has arrived.... An endless continuous stream of men, women and children flows to the Tempelhofer Feld...a million and a half people have assembled there.... Then the Leader speaks...Germany is at stake, its future and the future of our children. A wild frenzy of enthusiasm has seized the crowd.
'...Tomorrow we shall seize the houses of the trade unions. There will hardly be any resistance anywhere. The struggle is going on!'
(Joseph Goebbels, 'My Part in Germany's Fight,' 1935)

Maximum marks

(a) Identify 'the President of the Reich'.   (1)
(b) State briefly what happened to Röhm, and explain why it happened.   (3)
(c) What was *each* of the following: 'the S.A.'; 'the Zentrum'; 'the boycott'?   (3)
(d) Explain the importance of 'the law of authorization'.   (4)
(e) These extracts were written for publication. Comment on the style which Goebbels used.   (3)
(f) Identify *three* groups of people whom, according to these extracts, the Nazis regarded as their enemies.   (3)
(g) What do these extracts reveal of the methods by which the Nazis strengthened their position once they were in power?   (6)
(h) On what grounds would you agree or disagree with the assertion that on 30 January, 1933, Germany was 'at a turning-point in her history'?   (8)

12.6 READING LIST

Standard textbook reading

J. M. Roberts, *Europe 1880–1945* (Longmans, 1967), chapter 13.

David Thomson, *Europe since Napoleon* (Penguin, 1977), chapters 23, 25, 26, 27.

Anthony Wood, *Europe 1815–1945* (Longmans, 1975), chapters 29–30.

Further suggested reading

K. D. Bracher, *The German Dictatorship* (Penguin, 1973).

A. Bullock, *Hitler: a Study in Tyranny* (Penguin, 1968).

J. Fest, *Hitler* (Penguin, 1977).

J. W. Hiden, *The Weimar Republic* (Longmans, 1974).

A. J. Nicholls, *Weimar and the Rise of Hitler* (Macmillan, 1979).

A. J. Ryder, *Twentieth Century Germany from Bismarck to Brandt* (Macmillan, 1973).

W. L. Shirer, *The Rise and Fall of the Third Reich* (Pan, 1968).

E. Wiskemann, *Europe of the Dictators* (Fontana, 1973).

13 The approach of war, 1933–9

13.1 INTRODUCTION

During the 1930s the hopes of the peacemakers of 1919 were destroyed and the League of Nations proved unable to prevent the aggression of Germany, Italy and Japan. The Manchurian crisis revealed the weakness of the League as an institution capable of preserving peace. Its credibility was at an end when it failed to impede Mussolini's conquest of Abyssinia. From 1933 onwards Hitler and Mussolini challenged the existing order.

The origins of the Second World War lie both in their actions and in the other Powers' reaction – or lack of effective reaction – to them. Soon after coming to power in 1933 Hitler withdrew Germany from the disarmament negotiations at Geneva and pulled out of the League of Nations. He announced his determination to reverse the 'humiliation' of the Versailles Treaty and also demanded the conquest of 'Lebensraum' (living-space) for the German people in the East and the subjugation of the Slavs.

The major events leading towards war were the German reoccupation of the Rhineland, the Spanish civil war, the Anschluss, the ceding of the Sudentenland to Germany after the Munich Conference, the subsequent dismemberment of Czechoslovakia and the German-Polish dispute.

There was a very strong public sentiment in favour of peace in these years and a marked willingness of the Western governments – for a range of reasons – to appease the aggressor Powers. British and French policy did much to convince Hitler that they would not resist his demands by military means. The mutual distrust between the Soviet Union and the Western Powers further weakened the prospect of peace.

The Molotov-Ribbentrop Pact, in which Russia and Germany agreed on the partition of Poland, together with the Western guarantee to Poland set the scene for the invasion of September, 1939 and the subsequent wider conflict.

13.2 STUDY OBJECTIVES

1 The significance of the following events in the breakdown of the international order during the 1930s: Japan's attack on Manchuria; the Abyssinian War; Hitler's reoccupation of the Rhineland; the Anschluss; the Czechoslovak crises.

2 The significance of the intervention of Germany, Italy and the Soviet Union in the Spanish Civil War; the influence of the war on subsequent events and on the relations between the Powers; the role of Britain and France in the Spanish Civil War.

3 A detailed grasp of the course of events from the Munich Conference to the outbreak of war in 1939: particularly the German occupation of Bohemia and Moravia; the Molotov-Ribbentrop Pact and the Polish crisis.

4 The various arguments advanced in the West in favour of the appeasement of Germany and Italy; particularly the reasons why there was widespread acceptance that the Germans had good justification for violating the Treaty of Versailles; a clear understanding of the territorial, military and economic effects of the Treaty.

5 How Britain helped to undermine the Versailles system by her failure to threaten action in the Rhineland crisis of 1936, by the Anglo-German Naval Treaty of 1935 (which represented a unilateral breaking of Versailles); Britain's willingness to accept the Anschluss and the sacrifice of Czechoslovakia; an explanation of official British policy during these events.

6 The foreign policy of the Soviet Union in this period: the reasons why she was an unlikely ally for the Western democracies; the influence of the fear of Soviet Communism on Western policies; the belief in some quarters that the real enemy was not Germany but Communism and that a strong and re-armed Germany could be an effective bulwark against this danger; how this affected Anglo-French attitudes, for instance at the Munich Conference (where Russian interests were not considered); the negotiations between Britain, France and the Soviet Union in the weeks preceding the Russo-German pact.

7 The development of relations between Germany and Italy; an explanation of the Pact of Steel.

8 The reasons for the failure of the League of Nations to preserve peace; a consideration both of its weakness as a peace-keeping organization and of its handling of the crises which confronted it, with particular reference to the Abyssinian War; the League and public opinion.

9 Hitler's foreign policy: his aims and ideas of German aggrandizement and racial domination; his tactics for their achievement; the extent to which his actions were determined by long-term calculations and ambitions or by opportunism; the extent to which a widespread acceptance of the German case for revision of the Versailles Treaty worked for him up to and after the Munich Conference.

10 The British and French responses to German policy after the Munich Conference, e.g., the guarantee to Poland; the revolt of public opinion against appeasement and its influence in the immediate pre-war period.

13.3 COMMENTARY

The economic crisis

The world economic crisis of 1929–33 was a major cause of the collapse of the Versailles settlement. The economic breakdown, with a shrinking export trade and mass unemployment, led to political upheaval, hostility between states and the end of peace.

Particularly it encouraged political extremism, both of the Left and the Right. Germany was the most striking example of this. There, massive unemployment reinforced the existing deep resentment against the 'Diktat', reparations and the Weimar Republic. It dramatically boosted the fortunes of Hitler's Nazis.

The collapse of the national economies led countries to be more than ever preoccupied with their domestic social and economic problems. The prospect of costly rearmament, funded by heavy taxation or loans, was more than ever unwelcome to those democratic countries whose most pressing concern was to achieve economic recovery.

This strengthened the argument for appeasement and encouraged the Western Powers (Britain and France) to conciliate Italy and Germany.

Manchuria

The Manchurian affair of 1931–3, which began when Japanese forces seized key points in Manchuria from Chinese garrisons, damaged the authority of the League of Nations and the credibility of the idea of collective security or international peace-keeping.

It weakened the confidence particularly of those in the democratic states who believed that peace could be preserved by such an organization.

Following the Lytton Commission, which went to investigate the situation in the Far East and censured Japan for resorting to force, Japan left the League.

This example of a successful aggressor power defying the League was not lost on the leaders of Germany and Italy.

German demands for equality and rearmament

Hitler's impact on international affairs was felt soon after his accession to power in January 1933.

Germany had insisted on equality of rights in the Geneva disarmament talks. Such demands were regarded by France – which sensed herself to be very vulnerable in the event of any expansion of German power – as quite unacceptable. In October, 1933, Hitler ended German participation in the conference and openly rejected disarmament clauses of the Treaty of Versailles.

In the spring of 1934 France rejected further disarmament discussions. She attempted, without success, to establish the same sort of guarantee for Eastern Europe (an 'Eastern Locarno') as had been achieved in 1925 for the West (see Unit 9, *The Peace Treaties and the Resettlement of Europe*). France signed a pact with Soviet Russia in 1935 but this did not strengthen her position; in some respects it weakened it. It was highly controversial, being violently opposed by Right-wing political circles and there were no subsequent staff talks between the French and Russian military. In 1936 France began to rearm.

The Abyssinian crisis

The consequences of Mussolini's invasion of Abyssinia were no less grave: the League imposed sanctions on Mussolini but not the oil embargo, which could have stopped him. Britain and France tried to put off Mussolini with the Hoare-Laval Pact of December, 1935, by which two-thirds of Abyssinia would have been ceded to Italy.

This brought speedy condemnation in the British Parliament and from supporters of the League. The Government were seen on the one hand to be openly supporting the League and on the other to be doing a deal with the aggressor. Within six months the Italians had conquered Abyssinia. The results for international relations of the Abyssinian crisis included the following:

1   The recently constructed Stresa 'front' of Britain, France and Italy, whose purpose was to defend the international order and more specifically to guard against German aggrandizement, was destroyed.

2   The League of Nations could no longer be considered an effective peace-maker and it exercised no significant influence on subsequent events.

3   Italy was isolated. This led to the Rome-Berlin Axis of October, 1936, the prelude to the war-time alliance of the Fascist Powers.

4   Italy abandoned her resistance to the German domination of Austria. This was shown during the Anschluss (the take-over of Austria) and contrasts with Mussolini's reaction at the time of the murder of Dollfuss and the abortive Nazi putsch in July, 1934.

5   Hitler was given further evidence of British and French unwillingness to resist aggression.

The German reoccupation of the Rhineland

While Britain and France were distracted by the Abyssinian war Hitler struck his first blow against the territorial order. In March, 1936, he reoccupied the Rhineland. In order to justify this he argued that the Franco-Soviet Pact of 1935 contradicted the Treaty of Locarno and threatened Germany.

In fact his invasion and remilitarization violated both the Treaty of Versailles and the Treaty of Locarno.

This action was of very great strategic, as well as political, significance and was the essential prelude to any revision of the eastern boundaries of Germany. By closing the gap in Germany's western frontier, Hitler could block a French offensive in aid of Poland or Czechoslovakia.

It was a major gamble, but the failure of Britain and France to stop Hitler confirmed further his diagnosis of the weakness of the Western democracies while at the same time enhancing his own reputation within Germany.

The Spanish Civil War

Immediately after the outbreak of the civil war in July, 1936, the Spanish Popular Front government asked France for arms. At the same time, General Franco, leader of the military revolt, requested help from Germany and Italy. Within weeks Italy, Germany and the Soviet Union were supplying men and arms.

The French Popular Front government proposed a non-intervention agreement. Of the Powers, only Britain fully respected this while France allowed a small quantity of supplies to reach Spain.

The Spanish Civil War deeply divided public opinion and nowhere more so than in France, which was already at odds over the Franco-Soviet Pact.

The decisive consideration for Britain and France, though, was the danger of a general European war if they became involved in Spain. Such a war neither of them were willing to contemplate. On the contrary, Chamberlain was working to restore relations with Italy and regarded the Spanish conflict as an obstruction to his strategy for European security.

The Nazi expansion

Within a few weeks of coming to power Hitler was addressing his High Command on the need for 'the conquest of new "Lebensraum" (living-space) in the East and ruthless Germanization'. Following his reoccupation of the Rhineland his aggressive intentions became increasingly apparent.

Though he did not have a clearly predetermined timetable, his objectives were constant:
(1) the full revision of the Versailles Treaty in Germany's interest in Europe;
(2) the inclusion of all people of German race within an enlarged Reich; and
(3) the establishment of a massive German preponderance in the East through the conquest of 'Lebensraum'.

The pace of German rearmament indicates that Hitler was not preparing for a general war. For instance, when war broke out in 1939 he had only six weeks' supply of munitions. He adopted a 'Blitzkreig' strategy in the event – a series of short wars. The war which the Western democracies expected and prepared for – a war of attrition similar to that of 1914–18 – was the war Hitler wanted to avoid

The Anschluss

As Hitler accelerated his programme of expansion, the British Prime Minister, Chamberlain, set out to achieve a reconciliation with Italy. Hitler, exploiting the opportunity provided by the decision of Austria's Chancellor Schuschnigg to hold a plebiscite on the future of Austria, improvised invasion plans and German troops occupied Austria in March, 1938.

He proclaimed the annexation of Austria as a province of the German Reich. As an Austrian by birth, the incorporation of his homeland in a greater Germany had been a primary and unchanging aim, stated on the first page of 'Mein Kampf'.

Czechoslovakia

After the invasion of Austria, Czechoslovakia was the obvious next target for Nazi expansion.

She had a strong army and alliances with France and the Soviet Union. She was a barrier to German control of Central

Europe. The Anschluss now made her very vulnerable to economic pressure.

She did not have the undivided loyalty as a state of her inhabitants. There were Hungarian, Slovak, Ruthenian and German minorities. Hitler's justification for now turning his attention to Czechoslovakia was the 3.5 million minority of Germans living in the Sudetenland in Western Czechoslovakia. By her treaty of 1925 France was pledged to aid Czechoslovakia if the latter were attacked. Her own safety was bound up with the fate of Czech fortifications on Germany's opposite frontier.

Faced with German pressure on Czechoslovakia in 1938, she decided that she could not help her ally directly. Britain had made up her mind that Czechoslovakia could not be defended and that no guarantee should be given to France on that account.

Both Powers agreed that the Czechs should be encouraged to make large concessions to the Sudeten German minority.

The arguments for appeasement

The arguments for yielding to Germany's pressure were various:
1   There was the widespread horror of war.
2   There was fairly wide acceptance that the Germans had a good case for violating the Treaty of Versailles. There was the practical criticism of the settlement and the view, advanced by J. M. Keynes and others, that it had been morally unacceptable.
3   There was the belief that a strong Germany could act as a bulwark for the West against the spread of Soviet Communism.
4   Economic considerations – the cost of war preparations and of war. There was no guarantee for instance, that Britain would receive economic assistance in the event of war from the US.
5   British and French policies were also swayed by imperial and strategic considerations. Could the British Empire survive a major war? Italy and Japan, as well as Germany, were now counted as potential enemies. Britain could not, without substantial assistance, confront all three and hope to win.

Munich

On 30 May, 1938, Hitler secretly decided to invade Czechoslovakia in the near future. Military preparations were to be completed by October, 1938. During the last week of September, German troops were massed on Czechoslovakia's borders.

Following appeals from Chamberlain (who had already discussed the issue of the Sudetenland with Hitler at Berchtesgaden and at Godesberg), and from Mussolini and President Roosevelt, invitations were issued by Hitler to a conference at Munich on 29 September.

The agreement signed there provided for the German occupation of the Sudetenland in ten days from October 1. The operation was to be supervised by an international commission. A Four-Power guarantee of the dismembered State would replace the Franco-Czechoslovak treaty once Polish and Hungarian claims in Czechoslovakia had been settled.

Arguments justifying the Munich agreement include: (1) that public opinion supported it; (2) that Britain was unprepared for war and was given an extra year for rearmament; and (3) that it offered a last chance for Hitler – if he went further he would condemn himself in the eyes of world opinion, not least in the US.

To set against this, the arguments of the anti-appeasers were: (1) that the nature of Nazism and Hitler were such that he would go on to further conquests unless a firm stand were taken; (2) that Munich would only encourage him to take further risks; (3) that a major betrayal of Czechoslovakia was involved; and (4) militarily he could have been resisted in September, 1938.

The prelude to war

Hitler's immediate objectives after the Munich agreement were (1) an understanding with Poland on Danzig and the Polish Corridor; (2) a military alliance with Italy and Japan; and (3) the capitulation of what remained of Czechoslovakia.

On 15 March, 1939, Germany occupied Prague and announced the annexation of Moravia and Bohemia. The same day Hungary occupied the Carpatho-Ukraine. Slovakia survived as a vassal state.

The immediate results of this coup were: (1) a decisive hardening of public opinion against Hitler in Britain and France – though the appeasers did not give up hope, they now faced mounting criticism; and (2) Anglo-French guarantees were given to Poland, Greece and Rumania.

Hitler occupied Memel and Mussolini invaded Albania.

Poland

Hitler hoped to win Danzig by peaceful means, but the seizure of Memel convinced Poland that a German attack was imminent and strengthened their resistance.

On 3 April he ordered plans to be prepared for the invasion of Poland by 1 September. It would be attacked he informed his generals, 'at the first suitable opportunity . . . but it must not come to a simultaneous showdown with the West'.

On 23 August, following the failure of Stalin and the Western Powers to reach an understanding, Germany and the Soviet Union signed the Molotov-Ribbentrop Pact. This contained a secret protocol providing for the partition of Poland and

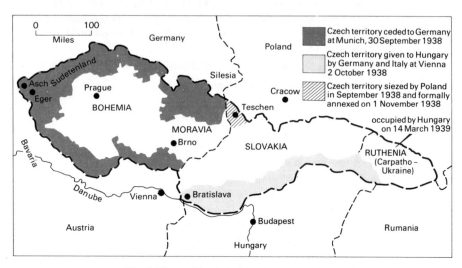

Fig. 6 The partitioning of Czechoslovakia, 1938

designating Nazi and Soviet spheres of influence in Eastern Europe.

With this crucial promise of benevolent neutrality Hitler was freed – for the moment – from the threat of war on two fronts.

13.4 CHRONOLOGY

1933	**January**	HITLER BECAME GERMAN CHANCELLOR.
	February	Japan left the League of Nations following the reception of the Lytton Report by the League on Manchuria.
	October	GERMANY LEFT THE DISARMAMENT CONFERENCE AT GENEVA AND THE LEAGUE OF NATIONS.
1934	**January**	NON-AGGRESSION PACT BETWEEN GERMANY AND POLAND - the first break in the French alliance system.
	July	THE MURDER OF DOLLFUSS: attempted Nazi takeover in Austria.
	September	The Soviet Union joined the League of Nations, reflecting her fear of German aggrandizement.
1935	**March**	HITLER ANNOUNCED CONSCRIPTION AND THE CREATION OF A MILITARY AIR FORCE.
	April	THE STRESA AGREEMENTS between Britain, France and Italy; in the face of growing German power.
	May	THE FRANCO-SOVIET PACT: concluded for five years. Each party promised the other aid in the event of unprovoked aggression.
	June	THE ANGLO-GERMAN naval agreement. Franco-Italian military convention.
	October	THE ITALIAN INVASION OF ABYSSINIA: the League declared that Italy had resorted to war in disregard of her obligations under Article XII; voted to impose sanctions.
	December	THE HOARE-LAVAL PLAN.
1936	**March**	THE GERMAN REOCCUPATION OF THE RHINELAND: Hitler denounced the Locarno agreements, taking advantage of the Abyssinian crisis and pleading danger to Germany from the Franco-Soviet Pact.
	July	THE SPANISH CIVIL WAR BEGAN: became the focus of Fascist and anti-Fascist intervention.
	August	The French proposed a non-intervention agreement for the Spanish civil war.
	November	ROME-BERLIN AXIS announced. Germany and Japan signed the Anti-Comintern Pact.
1937	**May**	NEVILLE CHAMBERLAIN became British Prime Minister.
	July	Japanese invasion of China.
	November	THE HOSSBACH CONFERENCE. Italy joined the Anti-Comintern Pact.
	December	Italy left the League.
1938	**March**	GERMAN OCCUPATION OF AUSTRIA AND THE ANNEXATION (ANSCHLUSS).
	April	Anglo-Italian agreement signed (implemented in November 1938).
	September	HITLER AND CHAMBERLAIN MET AT BERCHTESGADEN: Hitler stated his demand for the German annexation of the German areas of Czechoslovakia. Anglo-French plan for the Czech cession of the Sudetenland to Germany. HITLER AND CHAMBERLAIN MET AT GODESBERG: further demands by Hitler; for the immediate surrender of the Sudetenland without the dismantling of military or industrial establishments. MUNICH CONFERENCE AND AGREEMENT: Chamberlain and Hitler signed the Anglo-German Declaration. The transfer of the frontier region and provision for Polish and Hungarian claims on Czechoslovakia. The Polish ultimatum to Czechoslovakia over Teschen (occupied by Polish forces in October).
	November	Southern Slovakia and part of Ruthenia given to Hungary.
	December	Italy denounced the Rome agreements of 1935.
1939	**March**	GERMANY OCCUPIED BOHEMIA AND MORAVIA. Hungary occupied Carpatho-Ukraine. Lithuania ceded Memel to Germany. Madrid surrendered to General Franco. Anglo-French staff talks began. PROVISIONAL ANGLO-FRENCH GUARANTEE TO POLAND.
	April	ITALY INVADED ALBANIA. ANGLO-FRENCH GUARANTEES TO GREECE AND RUMANIA. Hitler denounced both the Anglo-German naval agreement and the German-Polish Pact.
	May	Provisional Anglo-Turkish pact announced. GERMAN-ITALIAN PACT OF STEEL SIGNED'. Hitler addressed his generals on Danzig and Poland.

July	Scare over Danzig.
4 August	Danzig-Polish customs crisis.
12	Anglo-French-Soviet military talks began in Moscow.
23	MOLOTOV-RIBBENTROP NON-AGRESSION PACT SIGNED IN MOSCOW: the Nazi-Soviet Pact included pledges to maintain neutrality if either country was at war. Secret clauses gave Lithuania and western Poland to Germany and eastern Poland, Estonia, Latvia, Finland and Bessarabia to the Soviet Union.
	Mussolini suggested that Britain should persuade Poland to surrender Danzig to Germany.
24	Danish-Polish customs talks broken off.
25	ANGLO-POLISH AGREEMENT SIGNED.
28	Hitler demanded Danzig, the Polish Corridor and parts of Silesia.
29	Poland persuaded by Britain and France to postpone full mobilization.
	Mussolini urged Hitler to accept the British proposals as a basis for settlement.
30	Polish mobilization announced.
31	Poland informed Germany that she was considering the British proposals of direct negotiation.
	Mussolini proposed conference to discuss Versailles grievances.
1 September	GERMANY INVADED POLAND.
	Britain and France warned Germany.
2	French moves in support of an Italian idea for a conference.
	Protests in the House of Commons over the government's appeasement of Germany.
3	BRITISH AND FRENCH ISSUED ULTIMATA TO GERMANY – DECLARATION OF WAR.

13.5 QUESTION PRACTICE

1   How far is it true that World War II was essentially the result of Hitler's aggressive policies?

Understanding the question

As with any 'how far' question, this one invites you to show precisely *in what ways* Hitler's policies contributed to the coming of war. You should consider how these policies were 'aggressive'. You should refer to the outbreak of the *world* war in 1941, as well as of the *European* war in 1939.

Knowledge required

Hitler's underlying ideas and objectives. The relations between Germany and the Western Powers, 1933–39. The relations between Germany, the USSR and the US, 1939–41.

Suggested essay plan

1   A war is possible only if *two* sides are willing to use force. Both Hitler and the Western Powers must be examined.

2   Hitler's objectives, as set out in 'Mein Kampf' and elsewhere, fall into two main groups:

(a) The regaining of Germany's freedom of manoeuvre as a great Power by breaking out of the constraints imposed by the Treaty of Versailles. By 1937 this had largely been achieved through: (i) (above all) successful rearmament; (ii) the remilitarization of the Rhineland in March, 1936, allowing the construction of defences for Germany's western industrial areas; and (iii) the breach in the unity of World War I victors caused by the Abyssinian crisis and the formation of the Rome–Berlin Axis. By 1937 war was, therefore, *possible* because Germany was again in a position to fight one. In allowing this to happen, the Western Powers perhaps had a 'passive' responsibility for the war but this must be carefully distinguished from Hitler's 'active' one.

(b) Hitler's first set of objectives was not 'aggressive', in the sense of entailing threats against and attacks on Germany's neighbours. But he also desired 'Lebensraum' or 'living-space' – *i.e.* the conquest of an empire in the USSR and Eastern Europe. This was aggressive by any definition. Yet, when Hitler began to expand outside Germany's frontiers in 1938–9, his intentions appeared at first to be confined to bringing all the Germans of Eastern Europe within the Third Reich. Hence the absorption of Austria and the Sudetenland in 1938; and the crisis over Danzig and Poland in 1939.

3   Britain and France. The second precondition for war was the shift on the Western side from 'appeasement' (notably the Munich Settlement of September, 1938) to a policy of resistance to Hitler (the British guarantee to Poland in March, 1939). The British government was *not* prepared to fight over the issue of incorporating the Eastern European Germans in the Third Reich. But by March, 1939, it was convinced that Hitler also threatened the *non-Germans* of Eastern Europe (evidenced by the occupation of Prague, and of reported German threats to Rumania and Poland). Rather than permit German domination of Eastern Europe, the British were willing to risk war by issuing guarantees to Poland and (later) Rumania. When Hitler invaded Poland in defiance of the guarantee, in September, 1939, Britain and France declared war on Germany. The West, therefore, took the final step – but note paragraph 4 below.

4   Unlike his position on Czechoslovakia in 1938, Hitler was determined to settle the Polish crisis by force rather than compromise. He gambled (because of the Nazi–Soviet Pact) that Britain and France would back down rather than honour the guarantee. The issue in 1939 was not just Danzig and the German minority in Poland but whether Germany should dominate Eastern Europe.

5   The events of 1939 were not the end of the story. In 1941 Hitler invaded Russia, as he had always intended to do. This was even more clearly aggressive than his actions in 1938–9. In December, 1941, he declared war on the US. It is true that the US had been greatly assisting both Britain and the USSR in their war against Germany (*e.g.* Lend-Lease; destroyers for convoy duty, etc.) but these American actions in turn were in response to Hitler's policies in the preceding years.

2   How far was the destruction of the peace treaties the mainspring of German foreign policy after 1933?

Understanding the question

You are being asked for an evaluation of Hitler's foreign policy.

The German leader was insistent on the need to revise the treaties in Germany's interest and the achievement of equal status among the Powers; but he was also driven by an ambition for a greatly expanded Reich and for racial domination in the East. Assess the relative significance of these as driving forces in his foreign policy.

Knowledge required

The clauses of the peace treaties, particularly Versailles, and the reasons for their unpopularity in Germany. Hitler's foreign policy objectives. The course of this policy after 1933 and the relation of the major developments (*e.g.* the reoccupation of the Rhineland) to the revision of the treaties and to Hitler's wider aims.

Suggested essay plan

1   German foreign policy in the 1930s demanded equality of status, a revision of the treaties and the incorporation of previously German territories and populations and the unification with Austria; but Hitler's ultimate objectives were far more ambitious, embracing the domination of Central and Eastern Europe and the achievement of 'Lebensraum' (living-space) for the Third Reich.

2   The challenge to the peace treaties included: rearmament, the reoccupation and remilitarization of the Rhineland, following the Anglo-German naval treaty which broke the Versailles Treaty; the Anschluss (forbidden by the Austrian Peace Treaty); the Sudetenland (incorporated in Czechoslovakia by the Peace Treaty for strategic and economic reasons); the Munich Conference, being the apogee of Hitler's foreign policy success without war; after this the wider nature of his objectives became more apparent to the Powers and to public opinion.

3   The moves to achieve German security: Hitler's intervention in Spain was partly directed towards establishing a state well disposed to Germany on France's frontiers to reduce the threat of France; the Rome-Berlin Axis and Anti-Comintern Pact; the Nazi-Soviet Pact ensuring that Germany would not be faced with the possibility of a two-front war when she attacked Poland.

4   The expansion of the Reich: in his speech of 5 November, 1937, Hitler expounded his overall purpose of expanding Germany into European Russia. His foreign policy followed this aim: the occupation of the rump of Czechoslovakia in March, 1939; the partition of Poland after the Polish campaign; Operation Barbarossa in 1941; plans for the establishment of German colonists in Russia and the driving back of Russian peoples into Asia.

5   If the destruction of the peace treaties had been *the* mainspring of German foreign policy in the Nazi period, it is arguable that this process could have been completed without war. But this objective, was only part of, and means towards achieving, a far larger design; though the justification offered by Hitler was in terms of righting 'wrongs' done to Germany in the peace treaties, the essential driving force was the far more comprehensive design.

3   Examine the extent to which the policies of Nazi Germany, Fascist Italy and Soviet Russia were responsible for the outbreak of war in 1939.

Understanding the question

This requires you to analyse the contribution of the policies of the three Powers to the outbreak of the war and to weigh their respective responsibility. Your answer must obviously also indicate an awareness of the role of the other Powers. The question should not be taken as an invitation to give a narrative account of the origins of the war.

Knowledge required

A comprehensive understanding of the course of events leading to war, and the relations between the Powers. German foreign policy, particularly from 1936 onwards. Italian foreign policy. Soviet policies, particularly the Nazi-Soviet Pact.

Suggested essay plan

1   Introduction: the impact of the rise of the dictators; the revisionist Powers, Germany and Italy, eager to reverse the consequences of the First World War; and willing to resort to aggression and intervene in the affairs of other states; contrast this with the policies of the Western Powers; responsibility means not simply the immediate responsibility, but also the contribution of these policies to a warlike climate in international relations.

2   Fascist Italy: Mussolini's advocacy of violence and war as a means of reviving Italian greatness; his invasion of Abyssinia gave evidence of the willingness of Italy to use force to achieve her foreign policy objectives and destroyed the credibility of the League of Nations as a preserver of peace; Italy's new relationship with Germany and her breach with France and Britain; intervention in Spain; in 1938 Mussolini acted as an 'honest broker' at the Munich Conference; in 1939 his role might be seen as one of trying to restrain Hitler; Mussolini's overall responsibility is to be seen more in his role as a pioneer to Hitler's defiance of the Western Powers and as distracting them from the greater threat of German aggrandizement.

3   Nazi Germany: initially Hitler's policy was regarded with some international sympathy as redressing the wrongs of Versailles; but one should emphasize the aggressiveness of the Nazi creed and the far more extensive aims of Hitler's foreign policy; Germany's intervention in the Spanish civil war, such actions as the bombing of Guernica and Madrid intensifying the apprehensions of the Western Powers; German intentions more clearly revealed in the Czech crises, more particularly after the Nazi occupation of Bohemia and Moravia, which led public opinion in Britain to turn against appeasement.

4   Hitler's policy towards Poland led inevitably, after Britain's guarantee to her, towards war; consider the argument that, even without the Nazi-Soviet Pact, his determination and vision of a new Reich would have forced him inexorably on.

5   Soviet Russia: Soviet involvement in Spain intensified British and French suspicions and was seen as part of that worldwide drive to establish Communism, which those distrustful of Stalin's Russia saw as her ultimate foreign policy aim; this weakened the prospect of collective security; Russian policy was essentially defensive; the Nazi-Soviet Pact, which seemed to buy security extending Russian frontiers westwards into Poland; responsibility lay too with Britian and France in failing to secure a pact with Russia which would make the guarantee to Poland a military reality; the Pact, however, was of crucial assistance to Hitler in pursuing his objective of further German aggrandizement.

6   What made war certain was the Polish guarantee, but only to the extent that Britain and France were forced into war; the overwhelming responsibility was Germany's, given the scale of Hitler's ambitions and his unwillingness to limit them; wars between Germany and Poland and Germany and the Soviet Union were inevitable in these circumstances; so was war with the Western Powers unless they were willing to tolerate the creation of a German Europe; while Italian and Soviet policies,

as well as the actions of Britain and France, helped to contribute towards the outbreak of war, they were not ultimately the decisive factor.

4   Document question (Diplomacy – 1939): study the extract below and then answer questions (a) **to** (f) **which follow:**

Extract

'Duce,

'For some time Germany and Russia have been meditating upon the possibility of placing their mutual political relations upon a new basis. The need to arrive at concrete results in this sense has been strengthened by:

'1   The conditions of the world political situation in general.

'2   The continued procrastination of the Japanese Cabinet in taking up a clear stand. Japan was ready for an alliance against Russia in which Germany – and in my view Italy – could only be interested in the present circumstances as a secondary consideration. She was not agreeable however to assuming any clear obligations regarding England – a decisive question from the German side, and I think also from Italy's . . . .

'3   The relations between Germany and Poland have been unsatisfactory since the spring, and in recent weeks have become simply intolerable, not through the fault of the Reich, but principally because of British action . . . These reasons have induced me to hasten on a conclusion of the Russian-German talks. I have not yet informed you, Duce, in detail on this question. But now in recent weeks the disposition of the Kremlin to engage in an exchange of relations with Germany – a disposition produced from the moment of the dismissal of Litvinov – has been increasingly marked, and has now made it possible for me, after having reached a preliminary clarification, to send my Foreign Minister to Moscow to draw up a treaty which is far and away the most extensive non-aggression pact in existence today, and the text of which will be made public. The pact is unconditional, and establishes in addition the commitment to consult on all questions which interest Germany and Russia. I can also inform you, Duce, that, given these undertakings, the benevolent attitude of Russia is assured.'

(Letter from Hitler to Mussolini, August, 1939)

Maximum marks

(a)   Explain why 'relations between Germany and Poland have been unsatisfactory.'   (6)

(b)   (i)   Who was Litvinov and who replaced him?   (2)
     (ii)   Why did Hitler regard this as an important change?   (2)

(c)   Explain why Japan 'was ready for an alliance against Russia'.   (5)

(d)   What was Hitler's purpose in informing Mussolini of the imminent agreement between Germany and Russia?   (4)

(e)   Mussolini replied to this letter: 'As far as Russia is concerned, I completely approve.' Why could the publication of such sentiments in Italy have constituted a political embarrassment to him?   (4)

(f)   Discuss the view that the information revealed in this letter meant that the outbreak of a European war in the near future was inevitable?   (8)

13.6   READING LIST

Standard textbook reading

J. M. Roberts, *Europe 1880–1945* (Longmans, 1967), chapter 15.

David Thomson, *Europe since Napoleon* (Penguin, 1977), chapters 26–28.

Anthony Wood, *Europe 1815–1945* (Longmans, 1975), chapters 29–30.

Further suggested reading

A. Bullock, *Hitler: a study in Tyranny* (Penguin, 1968).

E. H. Carr, *International Relations between the Two World Wars* (Macmillan, 1947).

F. H. Hinsley, *Power and the Pursuit of Peace* (Cambridge University Press, 1963).

C. L. Mowat (ed.), *The New Cambridge Modern History 1898–1945*, volume XII (Cambridge University Press, 1968), chapter 23.

R. Sontag, *A Broken World 1919–39* (Harper and Row, 1971).

A. J. P. Taylor, *The Origins of the Second World War* (Penguin, 1963).

C. Thorne, *The Approach of War 1938–39* (Macmillan, 1976).

E. Wiskemann, *Europe of the Dictators* (Fontana, 1973).

14   The Cold War and Europe

14.1   INTRODUCTION

The term 'Cold War' describes the mounting tension between East and West which dominated the international scene during the years after the Second World War. There had already been considerable disagreements between the Soviet Union and the Western Powers during the war and there was a legacy of acute distrust which went back to the Russian revolution of November, 1917 and the subsequent civil war.

In August, 1941 Roosevelt and Churchill had drawn up the Atlantic Charter to lay the basis for an enduring peace. In this they declared that boundaries would be drawn up in accordance with the principles of national self-determination. However, the Soviet Union informed the West that they expected to keep those territories which they had won during their pact with the Nazis. At the same time, American troops remained in Europe, particularly as occupation forces in Germany and Austria, and American economic dominance was overwhelming.

The new juxtaposition of the superpowers brought them into direct confrontation. This was ideological as well as being territorial and was concerned with considerations of security. The conflict of interest was first sharply apparent over Poland and then Germany. Berlin was a particular focus of East-West hostility and Germany was divided between the Western and Soviet blocs.

President Truman announced a policy of 'containment' of Communism and offered extensive military and economic aid to Turkey and Greece. This was followed by the setting-up of the North Atlantic Treaty Organization. The conflict intensified with the Sovietization of Eastern Europe and the Russian acquisition of the atom bomb. The Cold War widened into a global conflict with the outbreak of the Korean War.

14.2   STUDY OBJECTIVES

1   As background: an understanding of relations between the Soviet Union and the Western Powers during the pre-Second

World War period; Western attitudes to the Russian Revolution; the involvement of the Powers in the Russian Civil War; the Soviet view of the West, and Stalin's policies.

2   The significance of the Atlantic Charter; the Tehran, Yalta, San Francisco and Potsdam conferences; the significance in particular of the decisions over the future of Poland and Germany; a clear grasp of the territorial changes involved and the conflicting views of the Allied Powers.

3   The division of Germany and Berlin; the key role of Germany in the post-war period, noting particularly the intensification of conflict between East and West over the Berlin blockade and airlift; the question of German rearmament; the emergence of the Federal Republic of West Germany, and the Berlin Wall; the policies of Chancellor Adenauer; how the concept of a united Germany was abandoned by both East and West as a result of the considerations of military security.

4   The extent to which President Roosevelt's death and his replacement by President Truman affected US policy towards the Soviet Union; the contrasting views of President Roosevelt and Stalin on the post-war world order; Russian considerations of national security; the doctrine of containment; the role of the atom bomb in East-West relations; the impact of the Korean War.

5   The Marshall Plan (offered initially to Eastern Europe as well as to the nations of Western Europe); the Brussels Treaty; NATO; the purpose and significance of each of these and its influence on the growing divergence of East and West.

6   The ideological conflict; the development of the notion of a world divided into two blocs, Communist and anti-Communist; the extent to which the Cold War was a conflict of ideas; the growth of anti-Communism in the West; the significance of the Communist parties of France and Italy in Western calculations.

7   The Sovietization of Eastern Europe and the consolidation of Russian power; the process of transformation from 'Peoples' Democracies' to satellite states; Comecon; Cominform and the Warsaw Pact; the significance of the rift between Yugoslavia and the Soviet Union; developments in Eastern Europe after the death of Stalin, particularly the Hungarian Revolution.

14.3 COMMENTARY

The background

The postwar division of Europe into Eastern and Western Blocs dates from the wartime agreements of the Allies. As early as January, 1943, they agreed to demand the unconditional surrender of Germany. This ensured the creation of a 'power vacuum' in Central Europe which would inevitably be filled by Germany's conquerors. An agreement was reached between Stalin and Churchill in October, 1944, in which the two leaders divided the Balkans into spheres of influence. As a result of the 1943 armistice, Britain and the US had already brought Italy firmly into their sphere of influence.

The confrontation between the Soviet Union and the Western Powers which developed into the Cold War can be traced through an almost continuous series of wartime diplomatic meetings, during which major differences emerged between the Powers. At the same time it should be understood in the context of the original deep ideological hostility and mutual suspicion between Russia and the West which followed the revolution of 1917 and Allied intervention in the Russian Civil War.

There was the conference of foreign ministers in Moscow in October–November, 1943, followed by the summit meeting at Tehran, which brought the three great wartime leaders, Churchill, Stalin and Roosevelt together for the first time.

Particularly crucial for subsequent developments were the conferences of Yalta and Potsdam. Unlike in 1919, the victors could not agree on the terms of the treaty to be signed with their main enemy, Germany.

The Soviet Union and Europe

The Soviet Union wanted certain clearly-defined benefits from the peace: reparations for the damage done to her by the German invasion, security against any possible repetition of it, the recognition of her status as a Great Power and security for the future of the Communist system.

Between 1941 and 1945, in addition to about 20 million lives, one third of the productive resources of the Soviet Union had been destroyed; reparations or aid were very seriously needed. She set out to safeguard herself from any recurrence of German invasion by military alliances, territorial annexations and control over the satellite buffer states of Eastern Europe. Above all, Stalin was determined to incorporate into the Soviet Union those territories which had been taken away from Russia after the First World War. Russia took advantage of victory to recover territories – extending from the Baltic to the Black Sea – which had been part of Tsarist Russia before 1917.

These included the areas through which Germany had attacked her in both wars. This was particularly true of the large area of Poland which she now recovered, encouraging the Poles to make up for the loss by moving westwards at the expense of Germany. The dominant motive for this annexation was security.

The Sovietization of Eastern Europe

It was soon clear that Stalin thought of the so-called 'Peoples' Democracies', which were set up in Eastern Europe in the wake of the Red Army at the end of the war, as interim regimes at best.

Originally Stalin had hoped to obtain American aid and at the same time to control and exploit Eastern Europe to the Soviet Union's own economic and political advantage. Failing in the former, he concentrated on the latter. One after another, the Eastern European countries succumbed to Communist plots and manipulation which brought them more and more under Russian control: in Poland, Rumania, Bulgaria and Hungary non-Communist politicians were eliminated and the peasant parties outmanoeuvred or taken over. By 1947 all four countries had virtually become Russian satellites, and in February, 1948, Czechoslovakia fell victim to a Communist coup. Yugoslavia and Finland were left out. Tito's independence and defiance of Stalin led to the expulsion of the Yugoslav state from the Communist Bloc and also to purges in other Eastern European countries.

During the interwar years Russia had resented the role of Eastern Europe (as envisaged by Churchill and others) as a buffer against the spread of Bolshevik ideas. Now this area became a 'cordon sanitaire' against the penetration of Western and 'cosmopolitan' ideas and influence in Russia. Whether a state had been an ally of Nazi Germany or a victim, the process was the same. After a period of coalition government in which the Soviet occupying forces saw to it that Communists gained leading positions, all the satellites were sooner or later transformed into one-party states with governments dependent on and subservient to the Soviet Union.

The US and Europe

At the end of the Second World War the US had more than doubled its industrial production and was determined to reorganize the world to make it free for trade. In August, 1941, Churchill and Roosevelt had met to draft the Atlantic Charter, a general statement of principles on which a post-war settlement might be based.

Both powers disavowed any national gain or the implementing of any territorial changes contrary to the wishes of the inhabitants. All peoples should have the right to choose their own forms of government in a new world of economic co-operation. In January, 1942, these were embodied in the Declaration of the United Nations, which all governments at war with Germany signed.

In contrast to the very specific Russian objectives, the post-war aims of the US in world affairs appear more theoretical and less tangible.

It is clear that President Roosevelt expected the Soviet Union to try to establish its control over the states of Eastern Europe and that this could well have unpleasant consequences for the people of Poland and the other states involved. He seems to have believed that the fundamental interests of the US and the USSR could be reconciled without too much conflict or confrontation. He expected, or at least hoped, that the United Nations Organization would develop into a functioning world peace system.

If he had taken a different line in 1945, leaving considerable American forces in Europe in an attempt to stand up actively to Soviet pressure, he would have run into powerful opposition at home. During the Yalta Conference Roosevelt told Stalin that within two years of the war's end there would be no American troops left in Europe. The US army was reduced from about 5 millions in May, 1945, to less than half a million in March 1946. While the US possessed the atom bomb, the Soviet Union had a very marked superiority of conventional forces in Europe.

The question of Poland

After Roosevelt's death on 12 April, 1945, tension between Russia and her allies very noticeably increased. His successor, Harry Truman, right from the start of his presidency took a much firmer line with the Russians.

In January, 1945, the 'puppet' Provisional Government of Poland at Lublin had been recognized, in spite of the existence of the Polish government-in-exile in London. At Potsdam, in July, Churchill complained bitterly to Stalin that the Russians were surrounding parts of Eastern Europe with an 'iron fence'.

During the summer of 1945 American Lend-Lease aid to her allies, including Russia, was abruptly terminated. Stalin's request for an American loan, originally made early in 1945, was finally turned down in March 1946. It was the Soviet defiance of the American wish for liberal democratic regimes and particularly their attitude towards Poland that provoked the tougher attitude on the part of Roosevelt's successors. As early as the summer of 1945, when the Soviet Union was plainly unwilling to allow a Western-type democratic regime to be established in Poland, Britain and America were protesting at the breach of the Yalta agreement.

The question of Germany

At the end of the war the US, Britain and the USSR were agreed that Germany should be kept under strict control and that she should remain a single unit. In fact, by 1949, two new German states had come into existence. Germany, as well as Europe, was partitioned as a result of the Cold War.

The reasons for this included the inability of the victors to agree on other aspects of the German problem, the very great political, economic and strategic importance of Germany itself (neither East nor West could allow the other side to monopolize it), and the external circumstances of the Cold War, of which the Korean War was the most important.

Reparations. The major victors in 1945 were agreed that Germany should be disarmed and de-Nazified, divided *administratively* into zones of occupation but treated *economically* as a single unit. It was intended that she should pay for imported necessities out of current production.

The idea of dismembering Germany, which had been put forward during the war, was quietly forgotten. But she lost East Prussia to the USSR and all other territories beyond the Oder and the Western Neisse rivers, which were left under Polish administration. The Potsdam decision meant that Germany lost nearly a quarter of her pre-1938 territory.

The policy of economic unity was contradicted by the problem of reparations.

At the Potsdam Conference the Russians, whose need for reparations was very great, secured agreement for the removal of goods from their zone of occupation, but nothing was settled with the Western Allies about the *extent* of these claims. The Allies were similarly to be entitled to dismantle and remove property. They soon found, however, that this process left them with an obligation to provide their own zones with imported food which had to be paid for by their own taxpayers since German production was unable to foot the bill. Soon, therefore, the West called for the restoration of German industrial capacity and production.

Berlin. Berlin had been excluded from the zonal system. It lay deep inside the Soviet zone of occupation but it was placed, in joint occupation, under a separate joint allied authority, the Kommandatura. From the start Berlin was a crucial point of confrontation in the Cold War. During 1947–8, steps were being taken by the Western Powers to establish a West German state. which contradicted the idea of a unitary Germany. In order to bring pressure to bear on the West, Stalin ordered the blockading of Berlin and the Allies – having considered sending an armed convoy to force its way along the road from the British zone – decided instead to keep the city going by the airlift of essential supplies. The Soviet blockade was lifted after 318 days, but the problem of Berlin remained as one of the most vexed issues in East-West relations.

The German Federal Republic. After elections in August, 1949, the German Federal Republic came into existence with its capital in Bonn and with Konrad Adenauer as Chancellor. Adenauer postponed any serious pursuit of German reunification and concentrated on integrating the new state within Western Europe. A year later the question of German rearmament became a live issue, following the outbreak of the war in Korea in June 1950, and the US announcement that a substantial West German contribution was essential to the defence of Western Europe.(See Unit 15, *The Recovery of Western Europe.*)

The Cold War and 'containment'

The principal features of the Cold War were clearly apparent in 1946 in the 'Iron Curtain' speech of Sir Winston Churchill at Fulton, Missouri, and in that of President Truman's Secretary of State, J. F. Byrnes, at Stuttgart in the same year in which he announced the merging of the British and American zones (Bizonia). These clearly indicated that the wartime alliance had been superseded by East-West confrontation and that the US now regarded Western Europe as an essential sphere of influence.

Though President Truman had to accept the exclusion of western influence from Central and Eastern Europe, he warned the Soviet Union – the Truman Doctrine of March, 1947 – that further pressure would meet with American resistance. This may be taken as the first fully comprehensive statement from either side of the existence of the Cold War. This announcement was precipitated by the economic weakness of Britain in 1946–7, which forced her to give up her military and economic support to Turkey and Greece.

The Truman Doctrine spelled out the policy of 'containment', which was designed to curb Russian expansionism and to resist the spread of Communism. This was further underlined by the strengthening of the West European economies by Marshall Aid and the creation of NATO.

NATO was an association of 11 states which declared that an armed attack on any one of them in Europe or in North America would be regarded as an attack on them all. In such an event each would go to the help of the ally attacked. Its creation meant the formal collapse of the World II alliance.

The European policy of resisting Communist expansion was transformed into a worldwide confrontation by the outbreak of the Korean War.

The Cold War embodied both a conflict of interests between the superpowers and their allies (some historians see Stalin's policies essentially in terms of Russian nationalism) and a confrontation of ideas, Western democratic and Communist.

Both sides exaggerated the other's intentions. On the one hand, the Soviet Union accused the United States of trying to establish anti-Soviet regimes in countries on the borders of the USSR; on the other, the US appeared to take with very great seriousness the theoretical Communist aim of spreading revolution to the whole world and regarded every increase in Soviet influence and strength as a step towards that end. The American fear of the Soviet Union was matched by an often hysterical fear of Communist subversion at home. This was intensified by the ending of the Western monopoly of nuclear weapons in 1949 and the atomic arms race.

14.4 CHRONOLOGY

1941	**August**	THE ATLANTIC CHARTER: PRESIDENT ROOSEVELT and WINSTON CHURCHILL issued a statement of fundamental principles for the post-war world, emphasizing opposition to any territorial changes contrary to the wishes of those immediately concerned. Sovereign rights and self-government should be restored to those who had been deprived of them.
1943	**October**	Moscow Conference of British, US and Soviet Foreign Ministers: agreement to set up an international organization for peace and security.
	November	THE TEHRAN CONFERENCE: the first inter-allied summit attended by Stalin as well as Roosevelt and Churchill. Discussion of a landing in France and of cooperation in the peace settlement.
1945	**February**	THE YALTA CONFERENCE: Stalin agreed to enter the war against Japan – accepted the Curzon Line as the Soviet-Polish frontier; conceded that Poland would have the Oder-Neisse Line as a western boundary and agreed to Poland having free elections as soon as possible.
	May	TERMS OF UNCONDITIONAL GERMAN SURRENDER SIGNED: supreme authority vested in the Allied Control Council of Great Britain, France, the US and the Soviet Union. Each of these powers administered its own occupation zone. Berlin was divided into four sectors.
	June	THE SAN FRANCISCO CONFERENCE: completion of the Charter for the United Nations Organization.
	August	THE POTSDAM CONFERENCE: discussion of allied control in Germany, reparations, disarmament, and of the dissolution of Nazi institutions; the Oder-Neisse Line and Russian intervention against Japan also discussed.
1946		Economic fusion of the US and British Zones of Germany, (Bizonia).
1947	**March–April**	MOSCOW CONFERENCE OF FOREIGN MINISTERS: showed considerable disagreement on the German question. Conflict between Russian demands for reparations, and the Western policy of making Germany economically self-supporting.
	July	THE MARSHALL PLAN: a programme of European recovery proposed by George Marshall, US Secretary of State. A committee was set up to draft a EUROPEAN RECOVERY PROGRAMME. The Soviet Union and its satellites refused to participate in the programme of European reconstruction.
	October	THE COMMUNIST INFORMATION BUREAU (Cominform): established by the Communist parties of the Soviet Union, Yugoslavia, Bulgaria, Hungary, Poland, France, Italy and Czechoslovakia to coordinate the activities of the European Communist parties.
1948	**March**	Soviet delegates walked out of the Allied Control Council after charging the Western Powers with undermining the quadripartite administration of Germany. THE BRUSSELS TREATY: signed by Great Britain, France, Belgium, the Netherlands and Luxembourg. It constituted a 50 year alliance against attack in Europe and provided for economic, social and military cooperation.
	June	YUGOSLAVIA EXPELLED FROM THE COMINFORM: for alleged doctrinal reasons and hostility to the Soviet Union.
	July	Disagreement between the Soviet Union and the West over the latter's programme of economic and currency reforms brought complete Soviet stoppage of rail and road traffic between Berlin and the West – THE BERLIN AIRLIFT. COMMUNIST COUP AND PURGE IN CZECHOSLOVAKIA.
1949	**January**	THE COUNCIL FOR MUTUAL ECONOMIC ASSISTANCE (COMECON): originally intended by Stalin to enforce an economic boycott against Yugoslavia.
	April	THE NORTH ATLANTIC TREATY ORGANIZATION (NATO): between Great Britain, Belgium, Canada, Denmark, France, Iceland, Italy, the Netherlands, Norway, Portugal and the US – for mutual assistance in the event of attack.

		The three Western Powers agreed on an OCCUPATION STATUTE for Western Germany, which assured the Germans of considerable self-government, while reserving far-reaching powers to the occupation authorities.
	May	The Western Parliamentary Council adopted the BASIC LAW for the Federal Republic of Germany.
	September	KONRAD ADENAUER elected Chancellor of the Federal Republic.
		Soviet Union renounced its treaty of friendship with Yugoslavia.
1950	**June**	OUTBREAK OF THE KOREAN WAR.
1952	**February**	GREECE AND TURKEY JOIN NATO.
	May	THE EUROPEAN DEFENCE COMMUNITY: an attempt to create a supranational European army with common institutions. Rejected by the French Assembly, 1954.
1953	**March–August**	The death of Stalin.
		The USSR proposed that a Big Four Conference be held within six months on a German peace treaty. Moscow suggested that East and West Germany met first to set up a provisional all-German regime.
1955	**May**	The Federal Republic of (Western) Germany gained sovereign status.
		WEST GERMANY JOINED NATO.
		THE WARSAW PACT: set up to provide for the establishment of a unified military command with headquarters in Moscow – 'The Eastern European Mutual Assistance Treaty'.'
1956	**October–November**	THE HUNGARIAN UPRISING.

14.5 QUESTION PRACTICE

1 Why, after the Second World War, did Germany occupy so important a place in the quarrels between West and East?

Understanding the question

A straightforward question which needs to be answered, however, at both a general and a specific level.

Knowledge required

Developments in Germany, c. 1945–54, and their effect on East-West relations.

Suggested essay plan

1 Start by presenting a paradox: during 1941–5 the Allies were all agreed that Germany must be completely defeated. There was also substantial Soviet–US agreement about how Germany should be treated after the war: it should lose territory, be disarmed, economically weakened, and possibly split up (*viz.* the 1944 Morgenthau Plan). Yet, by 1948–9, the future of Germany caused the most serious US–Soviet crisis of the early Cold War. Why?

2 The general reasons for Germany's importance were *(a)* its geographical position – Western and Soviet troops met along the Elbe, and West Berlin remained as a Western enclave that was highly vulnerable; and *(b)* Germany's potential, shown in both World Wars, as an extremely formidable military and industrial Power, and the evidence of the inter-war years that it could rapidly recover from defeat. Neither the Western Powers, especially France, nor the USSR, which had just lost 20 million dead, wanted Germany to recover as an *independent* Power. But if it were to recover as a *dependent* Power it might give a decisive advantage to whichever of the post-war blocs it was attached to. The wartime solution to this dilemma was to prevent Germany from recovering economically and militarily at all. This, however, proved very difficult to implement in the post-war years. Because of all these difficulties, there was no precise agreement over a peace treaty with Germany even when East-West relations were good. Once the Cold War had begun, all chance of a peace treaty with Germany was ended.

3 The *specific* course of events: Germany was not the issue which started the Cold War. American–Soviet tension began over Poland and Eastern Europe in 1944–5. It intensified over the crises in the Near East in 1946–7 (*i.e.*, Iran, Turkey and Greece). In 1945–6 the main dispute in Germany was over the USSR's desire for reparations (and especially its desire to take these from the Western occupation zones). This however, was relatively minor. Once tension had begun elsewhere it became important to prevent Left- or Right-wing extremism winning support in Western Germany. By 1947 industrial recovery in Germany had become essential to prevent an economic crisis and possible social revolution in Western Europe as a whole. Hence the development of a Western commitment to a unified and prosperous West Germany (*e.g.*, J. F. Byrnes's Stuttgart speech, September, 1946; the economic fusion of the British and US occupation zones, January, 1947; the Marshall Plan, announced in June, 1947; the London conference (February–June, 1948) agreed on a West German currency and state; the June, 1948, currency reform; the West German constitution adopted, May, 1949).

This re-emergence of a German state (with two-thirds of the German population and the main industrial areas) was what the USSR had wished all along to prevent. Hence Stalin responded by: *(a)* tightening his control of Eastern Europe, and *(b)* the Berlin blockade, a direct response to the currency reform but frustrated by the Berlin airlift. The final phase came as a result of the USSR acquiring an atomic bomb and the invasion of South Korea. The US now pressed for Western Europe's non-nuclear defences to be strengthened by the creation of a West German Army and West Germany's admission into NATO (provided by the Paris agreements of October, 1954). In May, 1955 the USSR responded with the creation of the Warsaw Pact.

4 *Conclusion.* The Cold War did not start over Germany but, once it had begun, there was an irreconcilable conflict between the Western desire to see a united, prosperous (and eventually rearmed) West Germany and the Soviet desire to prevent this.

2 Show how and why a 'cold war' developed after 1945. What were its consequences?

Understanding the question

The question asks you to explain (and not just to narrate) the transformation of the relationship between the United States and the Soviet Union from an alliance against Germany (and

Japan) during 1941–5 into open hostility, coupled with the formation of opposed alliances and the beginning of an arms race.

Knowledge required

Soviet and US foreign policy, 1944–50. The outlines of European political history over the same period.

Suggested essay plan

1   At the most general level, the Cold War was made probable by the emergence in 1945 of the US and the USSR as by far the two most powerful nations in the world. The war had caused the defeat or enfeeblement of all the other Powers. It also made the US and the USSR more anxious to eliminate threats to their 'security' than they had been before. But they had little in common once Germany was defeated (the US, unlike the USSR, was a private enterprise democracy without vulnerable land frontiers or a history of foreign invasion), and they disagreed radically in their desires for a post-war settlement.

2   In addition to these general points, it is necessary to explain the circumstances and timing of the 'outbreak' of the Cold War, and to study the motives of both sides.
(*a*) *The USSR:* Stalin's maximum aims are not known. His minimum aims certainly included: (*i*) preventing Germany from ever recovering as a military threat to the USSR; (*ii*) establishing a security screen in Eastern Europe, which would probably require reliable Communist dictatorships in the countries occupied by the Red Army in 1944–5; and (*iii*) similar buffer zones in the Far East (*e.g.* Manchuria) and possibly the Middle East (*e.g.* northern Iran).
(*b*) *The USA:* The US leaders were initially close to Stalin's view about Germany (*e.g.* the Morgenthau Plan, September, 1944). For reasons of idealism (*e.g.* the Atlantic Charter, 1941) and domestic circumstances (*e.g.* the sizeable Polish vote), however, they criticized and tried to limit the establishment of Communist dictatorships in Eastern Europe (notably in the Yalta agreement over Poland). But they could do little more than slow this process down. More important was the evidence in 1946–7 that appeared to show a desire by the USSR for unlimited expansion *outside* Eastern Europe (*e.g.* the Iran crisis, March, 1946; Turkey, August, 1946; and the Greek civil war). To this the US leaders responded by: (*i*) announcing the policy of 'containing' Communist expansion (*e.g.* the Truman Doctrine, March, 1947, and aid to Greece and Turkey); and (*ii*) giving aid to prevent economic crisis and social disorder in Western Europe (*i.e.* the Marshall Plan, announced June, 1947). The US leaders were encouraged by (*i*) historical analogy – *e.g.* the failure of appeasement in the 1930s; and (*ii*) geopolitics – the belief that the US would be threatened if Western Europe's industrial resources fell under Soviet control.

3   But the Marshall Plan required the economic recovery of West Germany (accompanied by the formation of the West German state). This appeared to threaten the Soviet Union and started a process of 'escalation' (*e.g.* the Berlin blockade of 1948–9) causing the US to enter a defensive alliance with the Western Europeans (the North Atlantic Treaty, April, 1949). After the USSR acquired an atomic bomb (1949) and the invasion of South Korea (1950), the US pressed for the rearmament of West Germany (agreed in 1954). By then Europe was more stable, and Asia had become the main theatre of East-West conflict.

4   *Consequences :*
(*a*) The ceasefire lines of 1945 became the boundary between two Germanies and two Europes with very different political and economic systems.

(*b*) Suspicion and rivalry between the US and USSR both within and outside Europe, which still continues.
(*c*) The Cold War and the division of Germany made it possible to heal the old France/Germany conflict. But while East-West tension continues both East and Western European states remain dependent on the US and the USSR, and have little diplomatic manoeuvre.
(*d*) Note, finally, that outside Greece the Cold War in Europe did not become 'hot'.

3   To what extent was the Cold War 'a conflict between two irreconcilable ideologies'?

Understanding the question

This is a difficult question which requires you to identify the ideological elements of the Soviet-American conflict and to balance them against other possible explanations of that conflict.

Knowledge required

Soviet-American relations, c. 1943–50. The domestic political and economic systems of the US and the USSR, and the fundamental assumptions of their leaderships.

Suggested essay plan

1   Officially Marxist-Leninist, the USSR after World War II was a command economy under a single-party dictatorship; the US was a liberal democracy with a predominantly market economy. The ideological element in the Cold War was the belief of each side in the superiority of its domestic system, and its willingness, if necessary by force, to preserve and/or expand the geographical area covered by that system. This has to be set against the alternative explanations of the Cold War as a conflict of economic interests or (more convincingly) as a conflict between rival states which would have taken place whatever the domestic political system.

2   The US and the USSR had had different domestic systems for many years. But during the inter-war years they did not fear attack from each other – unlike the period after 1945; and in the years 1941–5 they were preoccupied with the danger from Germany and Japan. By 1945 Germany and Japan were defeated, and other Powers (Britain, France, China) were so weakened that the US and the USSR were far stronger than any other states except each other. If *both* states had been democratic, or *both* Communist, conflict might still well have occurred. But the conflict was much harder to control because of the ideological suspicion each side felt for the other, which encouraged each to interpret the other's actions in the worst possible light. To explain the *timing* and *circumstances* of the conflict, however, it is necessary to be far more precise.

3   *The USSR:* There was the widespread belief in the US by 1950 that the USSR wished to impose Communist dictatorships throughout the world. But (*a*) many Communist movements – *e.g.* in China, Yugoslavia, and Greece – were not under Soviet control; and (*b*) the USSR *was* responsible for the communization of the areas of Eastern Europe occupied by the Red Army. But this can be explained as the result of a devastating war with Germany. Stalin wanted a security screen of reliable governments on his border, and Communist control of the governments was the best way to make them reliable. Ideology was, therefore, at the service of security. It *may* be, however, that the USSR felt its security also required the Communist control of Western Europe. This is still unknown.

4   *The USA:* The American government pressed in 1945 for non-communists to be included in the East European governments. But what persuaded the Americans of a need for

'containment' was the evidence that the Soviet Union wished to expand *outside* Eastern Europe (*e.g.* Iran and Turkey, 1946; Greece, 1947). If Communist governments came to power in Western Europe, democracy and private enterprise would be endangered. But the US would itself feel under threat if the USSR controlled Western Europe's industrial resources. To prevent this, the US supported the economic and military recovery of Western Germany after 1947: this was precisely the threat which the Soviet Union had wished to guard against. From this point on, suspicion on the two sides was self-reinforcing.

4　The motives of security and ideology on the two sides were inextricably interlinked but, during the years 1946–8, ideological differences came to seem more threatening than before because of the elimination of the other Powers. Soviet behaviour in Iran, Greece and Turkey in 1946–7, and US policy towards West Germany in 1947–8, seem to have persuaded each side that the other endangered its existence.

4　Document question (the Cold War):
study extracts I, II and III below and then answer questions (a) to (g) which follow:

Extract I

'From Stettin in the Baltic to Trieste in the Adriatic <u>an iron curtain</u> has descended across the continent. Behind that line lie all the capitals of the ancient states of Central and Eastern Europe.

'...Athens alone...is free to decide its future at an election under British, American and French observation. The <u>Russian-dominated</u> Polish government has been encouraged to make enormous and wrongful inroads upon Germany....<u>The Communist parties...in all these Eastern States of Europe, have been raised to pre-eminence and power far beyond their numbers and are seeking everywhere to obtain totalitarian control...and, so far, except in Czechoslovakia, there is no true democracy.</u>'
(Winston Churchill, Speech at Fulton, Missouri, 5 March, 1946)

Extract II

'Since the war Turkey has sought additional financial assistance from Great Britain and the United States for the purpose of effecting that modernization necessary for the maintenance of its national integrity. That integrity is essential to the preservation of order in the Middle East...

'To ensure the peaceful development of nations, free from coercion, the United States has taken a leading part in establishing the United Nations. The United Nations is designed to make possible lasting freedom and independence for all its members. We shall not realise our objectives, however, unless we are willing to help free peoples to maintain their free institutions and their

Fig. 7
The division
of Europe

national integrity against aggressive movements that seek to impose upon them totalitarian regimes. This is no more than a frank recognition that totalitarian regimes imposed upon the peoples, by direct or indirect aggression, undermine the foundations of international peace and hence the security of the United States.'

(President Truman to Congress, 12 March, 1947)

Extract III

'The American arguments for assisting Turkey base themselves on the existence of a threat to the integrity of Turkish territory – though no one and nothing actually threatens Turkish integrity. This 'assistance' is evidently aimed at putting this country also under US control. Some American commentators admit this quite openly...that an American alliance with Turkey would give the USA a strategic position incomparably more advantageous than any other, from which power could be wielded over the Middle East....

'We are now witnessing a fresh intrusion of the USA into the affairs of other states...But the American leaders...fail to reckon with the fact that the old methods of the colonizers and die-hard politicians have out-lived their time and are doomed to failure.'

(Article in *Soviet News*, 15 March, 1947)

Maximum marks

(a) What did Churchill mean by 'an iron curtain'? (2)

(b) Explain why Churchill considered the Polish government to be 'Russian-dominated' and how this situation had come about (4)

(c) Explain what Churchill meant by 'The Communist parties...in all these Eastern States of Europe, have been raised to pre-eminence and power far beyond their numbers and are seeking everywhere to obtain totalitarian control'. (4)

(d) Explain the particular circumstances which led Truman to enunciate his 'doctrine' (Extract II). (4)

(e) Why did the writer of Extract III refer to 'the old methods of the colonizers and die-hard politicians'? (4)

(f) Compare the reasons advanced in Extract II with those advanced in Extract III for United States policy towards Turkey. (5)

(g) What developments in Europe in 1947 and 1948 led to an even greater rift between the Communists and the Western democracies? Which of these developments would have made Churchill revise the last assertion in Extract I, 'except in Czechoslovakia, there is no true democracy'? (7)

14.6 READING LIST

Standard textbook reading

H. Stuart Hughes, *Contemporary Europe: A History* (Prentice-Hall, 1976), chapters 13, 14, 15, 17, 19.

David Thomson, *Europe since Napoleon* (Penguin, 1977), chapters 29–30.

Further suggested reading

M. Crouzet, *The European Renaissance since 1945* (Thames and Hudson, 1970).

Isaac Deutscher, *Stalin* (Penguin, 1972).

André Fontaine, *History of the Cold War*, volumes 1 and 2 (Vintage Books, 1970).

Alfred Grosser, *Germany in our Time* (Pall Mall, 1971).

L. J. Halle, *The Cold War as History* (Harper and Row, 1975).

J. Joll, *Europe since 1870* (Weidenfeld and Nicolson, 1973).

W. Laqueur, *Europe since Hitler* (Weidenfeld and Nicolson/Penguin, 1970).

R. C. Mowat, *Ruin and Resurgence 1939-65* (Blandford, 1966).

15 The recovery of Western Europe

15.1 INTRODUCTION

Most of the countries of Western Europe faced a painful task of reconstruction after 1945; economic revival was the primary requisite after the devastation of the Second World War. Following the war, American aid provided statesmen in most of the non-Communist states with vital experience in economic cooperation. The Marshall Plan required recipients to coordinate their economic policies in order to use the aid to the best advantage.

During the war leaders of the Resistance movement in many countries had voiced the desire for a post-war reorganization of the European states system, for the reduction of national autonomy and even for a federal Europe. The Cold War threat to Europe and the problem of how to contain the revival of German power offered further encouragement to ideas of greater European integration.

The first really major step in this direction was the Schuman Plan, which led to the setting-up of the Coal and Steel Community. This was followed by the abortive attempt to create a European army, the European Defence Community. In 1957 the nations of the Six pooled their nuclear energy resources in Euratom and established the European Economic Community, which became the Nine in 1973 and is still growing.

15.2 STUDY OBJECTIVES

1   Western Europe in 1945: the political, economic and social consequences of the Second World War – particular attention to be paid to Germany, France and Italy; the scale of destruction and the factors favouring economic reconstruction; the significance of the movement of populations from Eastern Europe; the restoration of democracy.

2   The historical antecedents of the post-war ideas for European unity, *e.g.* the influence of the resistance movement against Hitler.

3   The economic arguments in favour of integration, such as the need for large markets, competitiveness and technological advance.

4   The political arguments in favour of integration: the influence of the loss of faith in Nationalism as a result of the Second World War; the decline of European power in the world; the fear of the Soviet Union and the spread of Communism; the desire to contain Germany and to prevent any future European war; American encouragement for moves towards German unification.

5   The difference between the federalist and functionalist approaches to European integration; the contribution of leading European statesmen to the European movement, particularly Konrad Adenauer and Alcide de Gasperi; the particular role of Jean Monnet.

6 The Marshall Plan or European Recovery Programme; its aims, scope and results; the OEEC; Benelux; the Council of Europe.

7 The revival of Germany: the transition from occupation to Federal Republic; the 'Economic Miracle'.

8 The Schuman Plan and the European Coal and Steel Community; the various reasons for the Plan and the institutions and organization and institutions of the ECSC.

9 The question of European defence: the Brussels Treaty and Nato; the Pleven Plan and the debate over the European Defence Community; the reasons for the defeat of the EDC and the role of public opinion in this debate.

10 The Messina Conference and the Rome Treaties; the institutions and working of the EEC; the role in particular of the Commission, the Council of Ministers and the European parliament.

11 The effects of the EEC on the development of Western European industry and agriculture.

12 General de Gaulle and the EEC: the Fouchet Plan, the veto of British entry in 1963 and 1967 and the reasons for this; the Gaullist view of Europe in the world and of France in Europe; the Franco-German Treaty (1963); the boycott of Community institutions (1965).

13 The enlargement of the Community (1973).

15.3 COMMENTARY

By 'integration' we mean the process of political and economic unification of the nation-states of Western Europe since the end of the Second World War in 1945. 'Integration' is to be distinguished from 'cooperation' by the fact that the participants in integration must delegate part of their national sovereignty to a a body with supranational powers.

The background

The idea of the unification of Europe was not new. Thinkers from the sixteenth century onwards had put forward schemes for a more or less federal European government. There had been Pan-European movements between the two world wars and the demand for European union was very strongly advanced by the resistance movement against the Nazis.

Common opposition to the Hitler regime had brought resistance fighters and exiled governments of different nationalities closer together. This movement condemned the system of nation-states and nationalism as the cause of Europe's war. As resistance activists expressed it in July 1944: 'Federal union alone can ensure the preservation of liberty and civilization on the Continent of Europe, bring about economic recovery and enable the German people to play a peaceful role in European affairs.'

Factors favouring integration

There were a number of reasons why European statesmen were persuaded to move in the direction of European integration after 1945. The major ones were:

1 The discrediting of Nationalism and the fact that, during the war, the nation-states of Europe had not been able to offer a minimum of security to their inhabitants against the aggressor.

2 The impact of the Cold War and the need for European defence.

3 Europe had ceased to be the dominant force in world affairs and was now replaced by the two superpowers, the US and the USSR. The idea grew of giving to Europe the necessary strength to preserve its independence and identity in the world.

4 The need for large markets and the coordination of economic efforts in a modern technological economy if European countries were going to be able to compete, for instance with the US.

5 American support for a strong Europe capable of resisting Soviet aggression – the US consistently exerted pressure in favour of integration.

6 A means of solving 'the German question', by incorporating her in a larger unit on the basis of reconciliation with France.

7 A general disposition towards European union on the part of the Catholic parties.

Federalism and Functionalism

The federalist movement, which derived such encouragement from Resistance circles, called for European unity through the transfer of political sovereignty to a new central authority. In practice, though, at the end of the war the occupying armies handed over political control to restored national governments.

There was also the functionalist approach, represented most notably in the plans of M. Jean Monnet. He had had long experience of international affairs, having been Deputy Secretary-General of the League of Nations. At the end of the war he became the first head of the French Economic Planning Commission. He was to be one of the most influential figures in the movement towards European integration.

He argued that a new international order could be built, not as the federalists hoped, by an out-and-out attack on the principle of state sovereignty and independence, but by the establishment of specialized agencies with real power to carry out certain functions, which would attract authority away from national governments. In the first instance the integration was to be *economic*, though the longer-term objective was political. The hope was that, just as the *Zollverein* had helped to pave the way for German unification during the nineteenth century, so too the economic integration of Europe would prepare the way for political cohesion. There was in any case a widespread recognition, going far beyond those who were enthusiastic for European unification, that the economic and political problems of the Western European governments could not be solved within a purely national framework.

The Marshall Plan

A particularly strong impetus towards European unity came from the United States, as a result of the growing Cold War conflict between East and West, with the Truman Doctrine and the Marshall Plan.

The background to this was the collapse of British power and a desperate economic situation in Europe. The Cold War was the context within which the European integration movement was really reborn. The American Secretary of State, George Marshall, put forward a programme of massive economic aid to Europe. This was prompted, not least, by the political risks as the Americans saw them; an extremely weakened Europe would be vulnerable to Soviet pressure both from without and from within. They viewed with apprehension the existence of very large Communist parties in both France and Italy.

By April, 1948, the Organization for European Economic Cooperation had been established to administer the aid programme. Through the plan more than 11 billion dollars of aid were made available by 1951 to the principal Western European countries. With this assistance their economies were able to recover, by that time, their pre-war level. Once this process of substantial economic reconstruction had been set in motion it also became possible for statesmen to think purposefully in terms of Western European economic integration. This represented a change from the immediate post-war period, when the various governments had been hard-pressed simply to rebuild their devastated economies.

The offer of economic aid was made to the whole of Europe, including the Soviet Union. Before Moscow forced them to withdraw, Poland and Czechoslovakia had accepted the Marshall Plan with enthusiasm.

The OEEC succeeded in three principal tasks: (1) co-operating in the distribution of Marshall Aid; (2) freeing trade between its members from tariffs and other barriers; (3) creating the European Payments Union for monetary transfers. It was, however, an inter-governmental organization with no supra-national powers of decision. Awareness of its limitations and the success of the integration of Belgium, the Netherlands and Luxemburg (Benelux) helped to encourage the movement towards more effective integration.

NATO and the Council of Europe

The two most significant Western organizations to be set up in 1949 were NATO (The North Atlantic Treaty Organization) and the Council of Europe. NATO committed its members to consult together on the means of planning their common defence and to 'take such action as it deems necessary' in case of attack.

In fact the Western European countries had begun to establish a system of defensive military alliances before the escalation of the Cold War conflict in 1947. Both the Franco-British Treaty of Dunkirk and the Treaty of Brussels, signed by Britain, France and the Benelux countries were, however, signed partly with fears of a revived Germany in mind.

The crisis that led the West to expand the Brussels Treaty into NATO was the Berlin blockade of June, 1948–May, 1949.

The European Coal and Steel Community

In May, 1950, the French Foreign Minister, Robert Schuman, following the suggestion of Jean Monnet, proposed a European Coal and Steel Community.

One motive behind this initiative was that French political leaders wished to ensure that German strength would never dominate Europe again. In particular, France wanted to maintain some control over the Ruhr industrial area. The pooling of coal and steel production would help to solve this problem.

There were other major considerations. In his announcement of the plan, M. Schuman declared: 'The pooling of coal and steel production will immediately provide for the establishment of common bases for economic development as a first step in the federation of Europe and will change the destinies of those regions which have long been devoted to the manufacture of munitions of war, of which they have been the most constant victims'.

The Coal and Steel Community (ECSC) was formally set up in April, 1951 and was a supranational organization. The High Authority of the Community had real power to enforce its decisions on member states, even though its functions were limited to the control of the coal and steel industries.

The essential feature of the Schuman Plan was that six European governments were prepared to accept voluntary limitations on their national sovereignty in a vital part of their economic life with the political aim of progressing towards a united Europe. One of the main purposes of the ECSC was to serve as a pioneer for other movements towards integration.

The West German Chancellor, Konrad Adenauer, welcomed the Plan – among other reasons as a means of rehabilitating Germany and giving her greater status. He was intent on integrating West Germany into the West as a priority over German reunification. In Italy, Alcide de Gasperi was equally anxious to find a European framework which would offer his country a stable economic background against which to resolve its internal problems. The Benelux countries were already committed to a customs union and were looking beyond this to a closer economic union.

The European Defence Community and the European Political Community 1950–54

The outbreak of the Korean War led to American fears that the Soviet Union might move against Western Europe. The next area in which integration was proposed was that of defence.

The Pleven Plan was a French proposal for solving the problem of fitting Germany into the Western Alliance in response to the American pressure for German rearmament.

The idea of a strong independent German national army was unacceptable to very many Europeans, not least to large numbers of Germans. If German rearmament was essential – as the US argued it was – then it was logical for the armed forces to be controlled by a similar system. The federalists went further, to argue that a European army could only be directed by a European government.

On the proposal of de Gasperi and Robert Schuman, an ad hoc assembly, formed by an enlargement of the Common Assembly of the ECSC, was directed to draw up a political statute for a European Political Community.

From 1950–54, the arguments for and against the European Defence Community (EDC) dominated the political life of France and Germany.

In the first place it raised the fundamental question as to whether Germany should be rearmed at all. It was only five years since the Allies had resolved to abolish German militarism for ever and the German people had to a large extent supported the decision that the new Federal Republic would have no armed forces. There was also a marked resistance in France to the idea of the French Army merging its identity in a common European army.

The refusal of the French Assembly to ratify the EDC led to a serious setback for the European integration movement, but arrangements were speedily made for the rearmament of West Germany in a new and acceptable form. The outcome was the agreement of October, 1954, to enlarge the Brussels Treaty organization into a Western European Union into which Germany and Italy would both be admitted. They were also to become members of NATO and the function of the WEU was essentially to ensure that Germany did not create armed forces larger than 12 divisions and that she respected her pledge not to manufacture atomic, biological or chemical weapons. Britain also undertook to keep four army divisions and her tactical air force on the Continent.

The Treaty of Rome

The movement towards further European integration was revived with the Messina Conference of June, 1955. In October of that year. M. Monnet announced the formation of the Action Committee for the United States of Europe. These laid the basis for the Treaty of Rome by which the six members of the ECSC agreed to go ahead with the integration of their economies on a much broader scale, which was to include agriculture. This was signed in March, 1957, and at the same time the Six authorized the establishment of Euratom.

The provisions of the Treaty of Rome contained articles of two categories: those designed to achieve (1) 'positive' and (2) 'negative' integration.

The first included such aims as the development of common European policies for industry and technology and the second involved all measures aimed at removing tariff barriers and obstacles to the free movement of labour and capital.

The major Community institutions which came into existence

were: (1) the executive Commission; (2) the Council of Ministers; (3) the European Parliament, with largely consultative powers; and (4) the Court of Justice.

The EEC emerged as more than a traditional inter-governmental organization but less than a full federal structure. The Commission represented the unity of the Community, but final decisions were taken by the representatives of the individual member states in the Council of Ministers.

The enlargement of the Community

The EEC was helped by various factors to establish itself. One of these was the continuing rapid pace of economic growth and prosperity, to which it contributed. Another was the continued support of the US.

To start with, the British – who had seen themselves primarily as a World Power with overriding overseas commitments – reacted to the EEC by proposing a wider free-trade area of which Britain, the Community and the other OEEC countries could be a part. Negotiations for this broke down and Britain, under pressure from the Scandinavian countries, decided to go ahead with the construction of a smaller free-trade area. This led in 1960 to the setting up of EFTA with seven members – Britain, Denmark, Norway, Sweden, Switzerland, Portugal and Eire.

A year later the British Prime Minister, Mr. Macmillan, announced his government's decision to apply for membership of the Community. This application was vetoed by General de Gaulle in January, 1963. De Gaulle had moved from his previous hostility to the Community to the view that the Six was not only a useful economic group but one which could serve French political interests. This lay behind his initiative, the Fouchet proposals. There is little doubt that his veto was motivated primarily by political considerations he wished to exclude a potential rival from the Community, and to resist Anglo Saxon dominance in Europe.

The final and successful negotiations for the enlargement of the Community were opened in June, 1970, during the presidency of de Gaulle's successor, M. Pompidou. All the major issues of the British case were settled within a year. In January, 1973, the Six became the Nine with the accession of the UK, Denmark and Eire.

15.4 CHRONOLOGY

1946	**September**	WINSTON CHURCHILL'S ZÜRICH SPEECH, calling for Franco-German reconciliation within a 'United States of Europe'.
1947	**July**	ANNOUNCEMENT OF MARSHALL AID: to stimulate European recovery. A committee set up to draft a European Recovery Programme; European Payments Union set up to distribute this.
	October	Economic union of Belgium, the Netherlands and Luxembourg – BENELUX.
1948	**March**	The Brussels Treaty: Benelux, France and England.
	April	The OEEC (Organization for European Economic Cooperation) formed.
1949	**April**	NATO (North Atlantic Treaty Organization) formed.
	May	The Statute of the Council of Europe signed, consisting of the Brussels Treaty powers plus Sweden, Denmark, Eire, Norway and Italy, followed by Iceland, Greece, Turkey, West Germany and Austria.
1950	**May**	THE SCHUMAN PLAN: proposal to place French and German coal and steel under a common authority.
	October	THE EUROPEAN DEFENCE COMMUNITY: devised and presented by M. Pleven. Recommendation of a European Ministry of Defence and the integration of Germany into the defence of Western Europe rather than the rearmament of West Germany as an independent unit. After four years of debate it was rejected by the French National Assembly.
1951	**April**	THE EUROPEAN COAL AND STEEL COMMUNITY (ECSC) TREATY signed by the Six – Benelux, France, Germany and Italy – to become operational in 1953. This Treaty set up the first common European authority, the ECSC High Authority, subject to democratic control through an assembly and to the rule of law through the Court of Justice.
1954	**October**	Western European Union formed: the Six plus Britain.
1955	**June**	MESSINA CONFERENCE of the foreign ministers of the Six. Set up a committee under M. Spaak to study ways in which 'a fresh advance towards the building of Europe could be achieved'.
1957	**March**	ROME TREATIES SIGNED, setting up the European Economic Community and Euratom.
1958	**January**	THE ROME TREATIES CAME INTO EFFECT.
1959	**May**	EUROPEAN FREE TRADE ASSOCIATION (EFTA) formed by Austria, Denmark, Norway, Sweden, Switzerland, Portugal and the UK.
1962	**March**	The UK applied for membership of EEC.
1963	**January**	Signing of Franco-German Friendship Treaty. DE GAULLE VETOED BRITISH ENTRY INTO EEC.
1965	**July**	French boycott of Community institutions for seven months in opposition to the Commission's proposal that all import duties and levies be paid to the Community budget, and that the powers of the European Parliament be increased.
1967	**May**	The UK, Eire and Denmark submitted applications for membership.
	July	COMMUNITY EXECUTIVES MERGED INTO A SINGLE COMMISSION.
	December	Council reached deadlock after de Gaulle's refusal to accept UK entry.

1969	April	The resignation of General de Gaulle.
	December	The Hague Summit – the Six agreed to enlarge the Community.
1971	June	Agreement reached on Britain's entry to the EEC.
1973	January	DENMARK, EIRE AND THE UK JOINED THE COMMUNITY. The Six became the Nine.

15.5 QUESTION PRACTICE

1   Would you agree that the Marshall Plan saved Europe from certain economic collapse?

Understanding the question

A precise answer to this question cannot be given but you can explain the economic crisis which the Marshall Plan was intended to solve, and speculate on what might have happened without it.

Knowledge required

The European economic crisis of 1947, and its political background. The Marshall Plan and its effects.

Suggested essay plan

1   The nature of the Western European economic crisis of 1947: (a) shortages of food and raw materials (especially coal), due to a delayed recovery of production, was worsened by a bad harvest and a harsh winter; this led to inflation, and labour unrest; (b) the financial crisis (the 'dollar gap') – Europe's export industries had been dislocated, and it had difficulty in earning foreign currency, especially as its pre-war foreign investments had mostly been lost; in particular, most European countries ran very large balance-of-payments deficits with the USA, above all because of their imports from it of the capital goods and machinery needed for reconstruction (Germany had been the other big pre-war exporter of machinery but industrial production in West Germany was one-third of the pre-war level); the danger was that the European countries would halt their economic expansion programmes due to a shortage of dollars, thus leading to further social unrest and to political crises which would benefit the Left in Germany and elsewhere, in turn further damaging business confidence.

2   The Marshall Plan and its effects: the Plan was announced in principle in Secretary of State Marshall's Harvard speech, in June, 1947. The distribution of the US aid was worked out by the European nations (including West Germany but *not* members of the Soviet bloc) in the Organization for European Economic Recovery (OEEC). The aid was used for: (a) the finance of European imports from the US, especially capital goods; (b) investment in basic industries in Europe. In its first year the Plan accounted for 4% of the national income of the 14 European recipients. It allowed: (a) a continuing, very fast industrial growth (especially in West Germany); (b) the restoration of the trade balance between Western Europe and the US by 1950–1.

3   The contribution of the Plan was: (a) economic – the removal of the bottleneck of the dollar gap, and the revival of German machinery production for supply to the rest of Europe (other measures, such as the German currency reform of 1948, helped here); and (b) political – the solution to the problem of German recovery. Germany's neighbours (especially France) needed what it could produce but feared that Germany's economic revival would enable it again to become a military threat. The Plan overcame this obstacle because: (a) Germany's recovery would take place only as part of a recovery of Western Europe as a whole and hence would, in relative terms, be less threatening; (b) Germany's neighbours would not get reparations but would be compensated for this by US aid; and (c) German recovery would be less dangerous if within the political framework of a continuing presence of US troops in Europe and of French

alliances with Britain (*e.g.* the Dunkirk and Brussels treaties, 1947 and 1948) and with the USA (the North Atlantic Treaty, 1949).

4   The Marshall Plan therefore not only solved the problem of the 'dollar gap' but also permitted German economic recovery to take place without the controversies over reparations and French security that followed the First World War (though, while reassuring France, it gravely alarmed the USSR).

2   Examine why it was possible to establish a European Economic Community but not a European Defence Community.

Understanding the question

A straightforward question requiring you to compare the failure of the proposed EDC (1950–4) with the successful establishment of the EEC in 1955–8.

Knowledge required

The history of the European integration movement, c. 1950–8. Attitudes towards it of the various national governments, notably Britain, France, West Germany and the US.

Suggested essay plan

1   Set the context: the early 1950s were the most favourable historical moment for moves to pool sovereignty in Western Europe. Before that point there was no West German state, and wartime memories were still too bitter; also, by the late 1950s, traditional national consciousness was reasserting itself (especially in France after 1958 and De Gaulle's assumption of power). In 1950 two initiatives were launched by the French government: (a) the Schuman Plan, which was to lead on to the European Coal and Steel Community, and later to the EEC; and (b) the Pleven Plan, for the European Defence Community, which did not succeed. Agreement, therefore, was reached over economic questions but not defence, which was much closer to the essence of national sovereignty.

2   Both the Schuman and the Pleven plans were French attempts to reduce the risks of German recovery: French reasons were (a) economic, in the case of the ECSC – *e.g.* the need to assure France supplies of German coking coal and to keep some international control over the Ruhr; and (b) military, in the case of the EDC – *e.g.*, the US government, as a result of the Korean War, had pressed for the creation of West German armed forces, and the Pleven Plan, intended as an *alternative* to this, proposed a European force which would include German units but with a multi-national command. In the emergency caused by the Korean War, the Plan was approved by the French parliament; the EDC treaty was signed in 1952 and approved by most of the parliaments of the ECSC 'Six' but defeated by the French parliament in August, 1954 by a combination of Communist votes with those of nationalists hostile to German rearmament.

3   Why did the French reject the EDC treaty? Reasons include (a) the lukewarm attitude of the French Premier, Mendès-France; (b) after the death of Stalin in 1953 and the resulting slackening of East-West tension, any strengthening of West European defences seemed less urgent; and (c) the reluctance of Britain to participate in the EDC as a possible counterweight to Germany. Note, however, that the October, 1954 Paris accords (which the French parliament approved) *did* provide for a West German army but this was to be balanced by a British commit-

ment to keep troops indefinitely on the Continent. This settlement of the defence question made further integration easier.

4 The EEC: this was accepted in principle by the ECSC 'Six' at Messina in June, 1955 and confirmed by the Treaty of Rome, March, 1957, which all six parliaments ratified. There were four main reasons for this success:

(a) the project was deliberately *not* presented as leading on to political unification but rather took the form of a broadening out of the successful ECSC, which had benefited the European coal and steel industries while not infringing intolerably on national sovereignty, and whose institutions were copied by the EEC;

(b) agreement was probably assisted by the revival of the Cold War (Hungary, October–November, 1956);

(c) Britain's absence was less serious than for the EDC, and made negotiations simpler; and

(d) a successful balance was struck between Germany's desire for larger export markets and the interests of France: France obtained (i) European assistance for the French civil nuclear programme, through Euratom, (ii) European aid for France's colonies, and (iii) special transitional arrangements to protect French industry.

5 *Conclusion.* The EDC was a much more ambitious project than the EEC, which built on the earlier success of the ECSC. The French attitude was the key: France was given adequate reassurances over the EEC but not over the EDC.

3 Examine the historical origins of the European Economic Community.

Understanding the question

An open-ended question, which requires you to balance long- against short-term explanations.

Knowledge required

The historical antecedents of European unification. The stages of European integration, c. 1950–8.

Suggested essay plan

1 (a) The origins of 'European' consciousness are very old; there were the precedents of the Roman Empire, Charlemagne, medieval Christendom at war with the Sarcens and later with the Turks, and the nineteenth-century cooperation of the Powers in the 'Concert of Europe'. Since the sixteenth century, however, there had been the rise of rival allegiances – the Reformation, the spread of national consciousness in Europe, and ideological divisions caused by the French and Russian revolutions and by the spread of Fascism. (b) On the economic side, there had been the spectacular increase in intra-European trade due to the industrial revolution and the railways before 1914; there was evidence also (*e.g.* from the German Zollverein) that a customs union could promote economic growth and political unity (cf. also the US). The growth, however, of intra-European trade was much weaker between the wars. (c) The 'European' movement: advocates of surrendering some or all national sovereignty to pan-European institutions. This grew stronger between the wars and won over some statesmen (*e.g.* Briand and Herriot in France).

2 These long-term trends are less important than these specific circumstances after 1945:

(a) The strengthening of the European movement by the Second World War; many of the post-war statesmen most concerned were in opposition or exile during the war (*e.g.* Monnet, Schuman, Adenauer), and there was also the wartime Resistance. Favourable circumstances for the spread of the pan-European movement included: (i) the rival economic power of the US seemed more formidable than ever; (ii) East European Communist parties were now backed by the Red Army; and (iii) there was increasing hostility to the colonial Powers outside Europe. All of these, among other reasons, made continuance of European internal political and economic conflicts appear suicidal.

(b) In spite of this, there was deep resentment against Germany in the former German-occupied states of Western Europe. The timing of the unification movement is explained by the British/US rehabilitation of West Germany due to the Cold War – *i.e.* the economic recovery under the Marshall Plan; the West German constitution (1949) and West German rearmament (agreed by 1954). The biggest obstacle to European integration was Franco-German hostility. This was overcome by: (i) the European movement; (ii) the need to strengthen Western defences against the USSR; (iii) the presence in Germany of British and US forces, which was a reassurance to France against possible renewed German aggression. But it was the acceptance of the Schuman Plan (1950) – the French proposal for a European Coal and Steel Community designed to assure France's coking coal supplies and to keep German heavy industry under some international control – which proved to be the crucial breakthrough; the Plan's function was to reassure France about the consequences of German economic recovery.

3 The foundation of the EEC was a smaller step than that of the ECSC, and needs less explanation. 'European' enthusiasts in the governments of the Six were seeking a new line of advance after the 1954 failure of the European Defence Community, and chose to broaden the methods of cooperation of the ECSC to other branches of industry, Purely economic motives were by now increasingly important rather than political ones, though agreement helped by the revival of the Cold War (*e.g.* Hungary, October-November, 1956; the launch of the Sputnik, October 1957). There was wide acceptance of the view that the removal of international barriers would encourage economic stability and growth, and there was the German desire for larger markets.

4 *Conclusion.* The EEC was made possible by the ECSC, which was a solution to the problems of German recovery. The EEC was made possible also by economic growth, which it accelerated, and by the European movement, which paradoxically declined after 1958.

15.6 READING LIST

Standard textbook reading

H. Stuart Hughes, *Contemporary Europe: A History* (Prentice-Hall, 1976), chapters 15, 17, 19, 20.

David Thomson, *Europe since Napoleon* (Penguin, 1975), chapter 30.

Further suggested reading

Brian Crozier, *De Gaulle: the Statesman* (Methuen, 1974).

A. Grosser, *Germany in our Time* (Pall Mall, 1971).

J. Joll, *Europe since 1870* (Weidefeld and Nicolson, 1973).

R. C. Mowat, *Ruin and Resurgence 1939–45* (Blandford, 1966).

A. J. Ryder, *Twentieth Century Germany from Bismarck to Brandt* (Macmillan, 1973).

D. W. Urwin, *Western Europe since 1945* (Longmans, 1968).

R. Vaughan, *Twentieth Century Europe: Paths to Unity* (Croom Helm, 1979).

Part II
Britain, 1815–1951
Introduction

16 Lord Liverpool's Administration, 1815–27

During the Napoleonic War a new Britain had come into being. Few nations have experienced so rapid and disturbing a process of change as occurred in Britain in the two decades before 1815. During the eighteenth century the population almost doubled, open fields were enclosed, farming was rationalised and many yeomen were driven from the land.

Industrial changes after 1780 meant the advent of the machine and steam power, great new industrial cities, the rise of coal, iron and cotton and the decline of the old woollen and cottage industries. Changes of such magnitude meant anxiety and misery for many but they also meant great opportunities and profits for others.

The impact of war

War had intensified this process and had accelerated the pace of change. In agriculture it had hastened the enclosure movement and encouraged much investment in land that would not normally be profitable. The coal, iron and textile industries received enormous stimulation from the demands for war goods and uniforms. European trade was severely disrupted for over 20 years and new markets, for example, in South America, had to be found for British trade. The financial burden of the war was enormous (the National Debt grew from £240 million to £860 million) and the introduction of income tax and higher indirect taxes meant that all classes suffered.

The war on the whole was profitable for producers, *i.e.* farmers and industrialists, but oppressive for the lower orders who were forced to work long and hard in the new mills and factories, live in squalid, overcrowded conditions in the new industrial towns, bear the brunt of higher indirect taxes and be severely repressed in the name of 'Church and King' if they objected.

Perspectives

During the war conflicts of interest were hidden as all energies were devoted to defeating France, but what would happen when the pressures and the artificial demands of war were removed? Would the lower orders continue to accept unquestioningly their new way of life? Would the new industrial middle class be satisfied to be ruled by an old-fashioned aristocracy that looked back to the agricultural world of the eighteenth century rather than forward to the industrial and manufacturing world of the nineteenth? How would the old ruling order respond if challenged on such key issues as 'cash, corn and Catholics'? Would they resist, adapt or reform?

By 1815 therefore, the world of the eighteenth century had been shattered. The agricultural and industrial revolutions had destroyed the balance of society and new classes had appeared for whom there was no place in the established order. Dissatisfaction was aggravated and intensified by the French Revolution, with its radical ideals of civil and religious liberty. It was exacerbated also by the repressive policy of the Younger Pitt's government in the 1790s, and by the disruptive demands of a war that dragged on intermittently for over 20 years.

16.1 INTRODUCTION

The main problem facing Lord Liverpool's government was the widespread distress and discontent of the post-war world. Protests were treated as unsympathetically as in the 1790s. Informers and *agents provocateurs* were employed by a government which feared that 'English Jacobins' would imitate the French Revolution and attempt violently to overturn the existing political order. Consequently, the military was used to break up peaceful meetings and laws were introduced to deter working-class expression through speech, meeting or writing. Only after the return of prosperity in 1822 did the government relax its vigilance and introduce some limited, but welcome, reforms.

16.2 STUDY OBJECTIVES

1 Lord Liverpool, his political talents and philosophy: how could the man whom Disraeli called the 'arch-mediocrity' retain power for longer than any Prime Minister since? Liverpool managed to display consummate political skill, despite his irritable temper; Asa Briggs' assessment was that Liverpool was 'a mediator of men and ideas'.

2 The composition of the ministry and the changes of 1822. Was there a reactionary period 1815–22 (associated with Castlereagh) followed by a liberal awakening 1822–7 (associated with Canning)? What was the effect on policy-making of Peel's and Robinson's displacement of Sidmouth and Vansittart, Canning's succession to Castlereagh, Huskisson's promotion to the Board of Trade, and the adhesion of the Grenville party?

3 The division of the ministers into so-called 'liberal' Tories (*e.g.* Canning and Peel) and 'high' Tories (*e.g.* Eldon and Wellington); the difficulties in the way of such a neat classification – for instance, the 'high' Tory Castlereagh took the 'liberal' side on Catholic emancipation, while Peel championed the Protestants. Did the 'liberals' and the 'highs' appeal to different classes in the country? Did the distinction correspond to the chronological break in 1822 or not?

4 Support for the ministry in parliament was very strong among the independently-minded country gentlemen who, though willing to defeat the government on specifics (*e.g.* the 1816 Property Tax), could not forgive the Whigs for Fox's support of the French Revolution, and so always supported the government on issues of confidence. The Whigs were further weakened by internal divisions, poor leadership and the implacable hostility of George IV. A general mood of reaction in the nation, dating from the 1790s, also helped the Tories.

5 Castlereagh's diplomacy at Vienna and the subsequent rejection of the Holy Alliance (1820 State Paper): Canning pursued a similar policy but in a more flamboyant and publicity-seeking style – *e.g.* the French invasion of Spain, the Greek Revolt and the Latin-American revolts. The bitter cabinet divisions (1825–6) as Wellington led the Tory rebels against Canning's isolationist foreign policy.

6 The switch from a protectionist economic policy (*e.g.* 1815 Corn Law) to Huskisson's and Robinson's freer trade policies – tariff reductions, the relaxation of the Navigation Code and of the Combination Laws; the new policies associated with 'dear money' economics (deflation), following the resumption of cash payments in 1819, and with retrenchment; the 1828 Corn Law began the slow process towards the repeal of 1846.

7   The move towards the more efficient control of law and order, following the fiasco at *Peterloo*; Peel's penal reforms and the development of the Metropolitan Police; law-and-order considerations forced the grant of Catholic emancipation (1829).

8   Political and personal motives combined to persuade the 'high' Tories not to serve under Canning (1827); the reasons for the fall of Goderich's ministry; the reasons for the fall of Wellington (1830) and of the 50-year-old Tory regime itself; the revolt of the Ultra-Tories (*e.g.* Knatchbull) against the economic (and especially the currency) policy and Catholic emancipation.

16.3 COMMENTARY

The return to peace

The coming of peace in 1815 did not bring the stability and prosperity that the nation desperately hoped for. To the landed classes peace brought falling prices, lower profits and the inability to pay the large investment mortgages taken out during the war. The result was often bankruptcy and widespread distress on the land. As foreign corn became available once more, English corn prices collapsed. The response of a parliament dominated by the landed classes was to bring in the Corn Law of 1815, which immediately antagonized the two other emerging groups in the state – the middle and lower orders.

The wartime demand for coal, iron, armaments and profitable government contracts ceased. Trade with Europe stagnated as many European nations had brought in protective tariffs. As a result industrial production had to be scaled down, priorities rearranged and new markets discovered. Unemployment in agriculture, industry and trade rose alarmingly. Demobilization aggravated the problem.

You must ask yourself to what extent were these economic, trading and financial problems due to the effects of war – or were the changes wrought by the industrial and agricultural revolutions bound to have serious and disruptive consequences in any event?

The radical protest

Each class in society, therefore, was damaged by the return to peace but in the general ruin and confusion the lower orders suffered most. As their economic plight deepened they became willing listeners to the growing band of radicals such as William Cobbett, Major Cartwright and 'Orator' Hunt, who demanded fundamental reforms in Church and state. Find out what you can about the ideas of the Radicals and the changing nature of working-class protest during the period 1790–1830. Note that the government particularly resented the radical tendency to link social distress to politics.

Distress and discontent, 1815–20

Whatever the nature of the protest, the discontent that burst forth from 1815 was so menacing that law and order could be maintained only by the use of the armed forces.

At Ely and Littleport county agricultural labourers, faced with unemployment and high bread prices, marched under banners inscribed 'Bread or Blood' and rioted throughout the area. In the industrial towns of the Midlands and the North unemployed workers smashed machinery. London did not escape disorder and, after the Spa Fields meeting in 1816, *Habeas Corpus* was suspended. In 1817 fresh disturbances occurred in the provinces. There was an attempt at armed insurrection at Pentrich in Derbyshire.

The Tories won the general election of 1818 – mainly because the disaffected had no vote and the opposition Whigs failed to offer an alternative to the Tory policy of rigorously maintaining law and order.

The revival of *Habeas Corpus* set the Radicals free. More disorder followed, culminating in the 'Peterloo Massacre' at St. Peter's Fields, Manchester, in 1819. The Regent congratulated the magistrates and Yeomanry for their prompt and decisive action but the nation was outraged. Early disorders seemed to paralyse the Whigs but *Peterloo* encouraged them to challenge Tory policy and put themselves at the head of the constitutional movement for parliamentary reform.

Undaunted, the government passed the Six Acts in the autumn of 1819. This further suppression of English liberties led to the Cato Street conspiracy in 1820. This attempt to murder the Cabinet proved to be the final act of lawlessness of these disturbed and unhappy years.

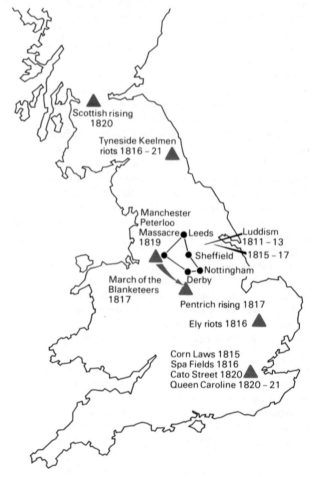

Fig. 8 Social unrest in Britain 1815–21

The policies of transition 1820–22

Problems in the economy and trade brought about by the abrupt change from war to peace had by now largely been solved. Despite the continued efforts of the radical journalists and the attacks of the unstamped press on Tory policy, their cause declined as employment picked up–apparently confirming Cobbett's dictum: 'I defy you to agitate a fellow with a full stomach.'

You must know the chronology of distress during these years and the details of major events like Peterloo and Pentrich. Why did the Whigs provide such ineffective opposition?

The cabinet had, nevertheless, one final popular agitation to survive when, on the death of George III in 1820, Caroline, the estranged wife of the Prince Regent, returned from six years' exile to claim her position as Queen. George's attempt to divorce her precipitated a crisis.

The importance of the event is not in the unsavoury revelations that emerged about the Queen's conduct but its political effects. The stock of the monarchy was further reduced; a humiliated government was forced to bow to the pressure of public opinion. On the other hand note how Whigs and Radicals, who had supported the Queen, gained in popularity and thus was forged an understanding between Whigs, Radicals and the people which was to bear fruit in the reform crises of 1830.

Government policy 1815–20

The Lord Liverpool government may have weathered the immediate post-war storm but the Radicals attacked its policy as cruel, repressive and uncaring. Shelley, in his sonnet *England in 1819*, called the Prime Minister and his colleagues 'Rulers who neither see, nor feel, nor know.' They appeared to equate reform with revolution and blindly crushed the symptoms of economic distress without considering the underlying causes.

Furthermore, the Radicals argued that the government actually added to the general distress by introducing the Corn Laws, abolishing income tax and returning Britain to the gold standard. In general they reacted to general unrest not by reconsidering their policies but by political repression. Is the radical case too one-sided in its approach?

In defence of the policy of the Lord Liverpool administration, one could argue that the first duty of any government is to maintain law and order. Also, it was well known that the means at the disposal of the government were unequal to the task. They had to show a determination to govern or the result might have been catastrophic. Having seen the effects of the revolution in France, the ruling classes, as their every letter to the Home Office appeared to confirm, feared violence and imminent revolution in England.

Firmness, then, they thought was essential. As to the economic distress, the government believed that – in an age of *laissez-faire* – intervention in economic affairs was undesirable and, that because of the absence of information, the lack of a trained government bureaucracy and the limited administrative and financial resources available, it would in any case be ineffective. They introduced the Corn Law because agriculture was the main factor of production and still the major employer in the country. It offered the best chance of helping the whole nation out of depression.

According to conservative apologists, the government was not unduly repressive in its handling of the situation nor was it greatly obsessed with disorder and treason. Ministers were alarmed but generally pragmatic in their response. Both centrally and locally they did what they could to alleviate distress and subscribed generously to soup kitchens and relief funds.

On the whole, it is claimed that government reaction was largely what might reasonably be expected. According to Professor Gash, 'they were often urged to do more; they could hardly have done less.' This explanation may appear to you to excuse too much. You must, however, be able to argue effectively both the 'repressive' and 'pragmatic' explanations of government policy.

How near to revolution?

What is remarkable about the years 1815–20 is that a people so oppressed by economic exploitation, unemployment, hunger and political repression, and so used to violence in their everyday lives, did not act more violently than they did. Make a list of points explaining why a revolution did not occur. Halévy maintained that Methodism prevented the middle-class leaders of post-war disaffection from going too far. E. P. Thompson agrees that there was a real possibility of revolution during the period but suggests that there were not enough Methodists to influence events more than marginally, although paradoxically some radical leaders did benefit from Methodist instruction in reading, writing and public speaking. More important than the influence of religion may be the government's unwavering determination to maintain law and order. Would you agree that the potential leaders of a revolution, such as Cobbett, Cartwright and Hunt, had neither the will nor the means to bring one about?

Reconstruction and reform, 1822–27

The suicide of Castlereagh gave Lord Liverpool the opportunity to reconstitute the ministry in 1822. Younger men in the Party were given key posts. Peel became Home Secretary; Huskisson, the President of the Board of Trade; Robinson, the Chancellor of the Exchequer, and Canning, Leader of the House of Commons and Foreign Secretary. These men were all capable administrators and, as they all had middle-class origins, were much more likely to be in tune with the needs of a developing industrial society.

They also realised that, after the alarms of the period 1815–20, a wider basis for popular support had to be found. Basically, it was the return of prosperity, rather than any preconceived policy by Liverpool, that gave them the opportunity to introduce some modest reforms. Note that the reforms of the Liberal-Tories were mainly economic and legal. Their political outlook did not change – it was simply not challenged in the 'twenties.

You must know the details of the work of Peel and Huskisson. The most important contribution of 'Prosperity' Robinson was that he consistently balanced the budgets and cut taxation. Until his death in 1827, Canning gave the ministry an aura of liberalism by his open rejection of the Holy Alliance and his confident vindication of nationalism abroad. Note that some very important reforms were carried out by the Tories *after* 1827. The Corn Laws were modified and a sliding scale was introduced to give some flexibility. Dissenters were given civil and religious liberty by the Repeal of the Test and Corporation Acts in 1828, but O'Connell had to wrest Catholic emancipation from a reluctant government in 1829.

These concessions on cash, corn and Catholics, however, merely sowed dissent in the ranks of the Tories. The challenge to their most fundamental beliefs, especially in the granting of Catholic emancipation, prompted charges of betrayal against Peel and Wellington. These post-1827 reforms under Wellington were basically concessions. Pressured from outside parliament, and, after Canning's death, divided within, these constitutional changes finally brought about the destruction of 'Liverpool Toryism.'

An enlightened administration?

Armed with the details of the major reforms of the ministry and your knowledge of the policies of governments immediately before and after 1822, you should now be able to tackle the central question of this section: how liberal were the liberal-Tories? They were, indeed, responsible for some important reforms.

Huskisson's free-trade measures, which upset the protectionists, were liberal in outlook. He dismantled the Navigation Acts and put the Corn Laws on a sliding scale. His policies were strongly supported by Canning's foreign policy, in which the promotion of trade was always a basic principle. Peel's improvement of prison conditions, his reduction of the number of capital offences, his reform of the Criminal Code, and his establishment of the Metropolitan Police Force in 1829 – these were reforms that would not have been contemplated by Sidmouth and Eldon. Certainly these reforms appeared liberal compared to what had gone before but less so when compared to the Whig reforms of the 1830s.

There remain however, certain important factors to be borne in mind. Lord Liverpool did not consciously encourage reform in the 'twenties: all the new appointments had served during the repressive period. In most cases the process of reform had already begun or was being investigated. The 'liberal-Tories' were not innovatory. Before and after 1822 they had sought to foster commerce and trade, to protect property and to strengthen law and order. Apart from the Corn Laws – which they believed to be vital for the national interest – their economic measures were progressive.

There was, however, no bold commitment to improvement: the reforms that occurred did not attempt to alter the fundamental constitutional structure. Parliamentary reform was never seriously considered and pressing problems of a new urban industrial society, such as public health, poverty, education and factory reform, were ignored. The 'liberal-Tories' *could* have attempted reforms similar to those of the Whigs in the 1830s but did not.

Finally, consider the argument that there is a real continuity in government policy from 1815–27 and that 'the essence of the Liverpool system was opposition to constitutional changes combined with innovation in administrative and social policy.'

16.4 CHRONOLOGY

1815	**January**	CORN LAW (SECOND): prohibited the import of foreign corn until the price of English corn reached 80s. a quarter.
1816	**March**	REPEAL OF THE PROPERTY (*i.e.* INCOME) TAX: necessitated the introduction of more indirect taxes which, proportionally, fell heaviest on the poor. LUDDISM: machine-breaking in the East Midlands, Yorkshire and Lancashire.
	May	ELY AND LITTLEPORT RIOTS: the seizure of Littleport by agricultural labourers; dispersed by military force.
	December	SPA FIELDS RIOT, London: organized by the Spenceans and led by Thistlewood with the object of seizing the Tower; they were dispersed at the Royal Exchange.
1817	**January**	A London crowd attacked the Prince-Regent's coach.
	February	HABEAS CORPUS SUSPENDED.
	March	Act against Blasphemous and Seditious Libels. MARCH OF THE BLANKETEERS from Manchester to petition the Regent for relief.
	June	PENTRICH REVOLT: organized by J. Brandreth partly owing to the influence of a government spy 'Oliver'; easily suppressed by a small military force near Nottingham.
1819	**May**	'BULLION COMMITTEE', chaired by Peel, recommended a return to the gold standard.
	16 August	'PETERLOO MASSACRE': meeting in St. Peter's Field, Manchester, for parliamentary reform; magistrates sent a body of Yeomanry into the 80,000 crowd to arrest 'Orator' Hunt; they panicked; the Hussars charged the crowd; 11 were killed and 400 injured. The government introduced THE SIX ACTS: these aimed to prevent delay in the administration of justice, seditious libels and training in the use of arms; they gave power to seize arms, imposed a 'newspaper stamp' tax and proscribed seditious meetings. They were not strictly enforced. Factory Act: forbade the employment of children under nine in cotton mills.
1820	**January**	Death of George III, accession of George IV.
	February	CATO STREET CONSPIRACY: a plot formed by Thistlewood to assassinate the cabinet. Conspirators were seized in a stable in Cato St. in north west London. Thistlewood and four others were executed.
	August	THE QUEEN'S TRIAL. George IV attempted to divorce Caroline, his estranged wife; public sympathy for the Queen (as George was unpopular) led to widespread disturbances in London.
1821	**May**	RETURN TO THE GOLD STANDARD, whereby the Bank of England gave gold for notes. This was a deflationary measure, causing some distress.
1822	**August**	Castlereagh committed suicide. RECONSTRUCTION OF THE MINISTRY: Liverpool gave Canning, Peel, Huskisson and Robinson key posts in the government.
1823	**June**	HUSKISSON'S RECIPROCAL TRADE TREATIES substantially increased British trade. Revision of the Navigation Acts: lifted restrictions on foreign trading vessels. PEEL'S REVISION OF THE CRIMINAL CODE: modified the severity of the criminal code and improved prison administration.
1824	**June**	REPEAL OF THE COMBINATION ACTS of 1799 and 1800 which had prevented liberty of association among working men. Organized by F. Place and carried by J. Hume in Parliament.
1825	**December**	NEW COMBINATION ACT limiting freedom passed in 1824 after a wave of strikes. Financial crisis: wild speculation led to the collapse of 60 banks in seven weeks.
1826	**July**	General election. Trade recession.
1827	**April**	CANNING BECAME PRIME MINISTER on the disablement, by illness, of Lord Liverpool. Unity of the Tory Party broken.

16.5 QUESTION PRACTICE

1   Compare and examine the attitude of Lord Liverpool's government to social and economic problems before and after 1822.

Understanding the question

The most difficult problem of interpretation concerning Liverpool's government is the extent to which the ministerial changes of 1822 affected the content of policy-making. This question requires you to adjudicate on this matter with respect to economic and social policy.

Knowledge required

The social and economic condition of England 1815–30. The economic and social policies of Liverpool, Vansittart, Huskisson and Robinson.

Suggested essay plan

1   Superficially, the social and economic policies (especially the reduction of tariffs in the 1820s) seem to fit the picture of a reactionary government turning 'liberal' after 1822. Robinson replaced Vansittart as Chancellor, and Huskisson became President of the Board of Trade. On the other hand, Robinson and Wallace had previously been at the Board of Trade, and Vansittart was not intrinsically hostile to free trade.

2   The causes of post-war distress 1815–21; loss of wartime demands on the economy, and the disruption to major industries (*e.g.* iron); unemployment was exacerbated by the return of ex-soldiers from the continent; prices were still high from wartime inflation.

3   The causes of agricultural distress 1819–22; good harvests reduced the price of corn; much of the wartime cultivation had to be abandoned because it was on land not inherently suitable for wheat crops; many farmers with mortgages were also damaged by the rise in the value of money after 1819, *i.e.* by deflation.

4   Social and political consequences of the economic situation 1815–22; severe industrial unrest culminated in *Peterloo*; a back-bench agricultural revolt threatened the government's position in parliament (1819–22). All government policies before 1822 were determined by the need to minimise urban distress and agrarian political discontent.

5   The main threat to law and order was thought to be the possibility of inadequate food supplies; until c. 1821 ministers thought that home production must always be the basis of food supply; this belief led to the 1815 Corn Law to protect agriculture.

6   In 1821 ministers perceived that, given the exhaustion of suitable arable land, Britain could not always be self-sufficient in food; it would therefore be necessary to import food more regularly in order to keep foreign farmers in production; moreover, irregular imports were disrupting the gold standard; so, despite the political dangers, ministers began the cautious move towards free trade in corn, which resulted in the 1828 Corn Law, and finally in the 1846 repeal.

7   Until c. 1822 ministerial freedom of action was limited by the lack of any financial surplus; the 1816 loss of property tax meant that Vansittart had to live from 'hand to mouth'. An upswing in the business cycle (1821–6) increased government revenue, which enabled Robinson and Huskisson to reduce indirect taxation. Such 'freer trade' made them appear very liberal but Vansittart would probably have done the same if he had still been in office.

8   Anyway, Huskisson's 'free trade' policies were cautious and based more on bilateral and reciprocal agreements with other countries than on unilateral action; but he did significantly relax the Navigation Laws and also amended the Combination Laws.

9   A desire for retrenchment, or public spending cuts, prevailed throughout the period and was a consequence of the unprecedentedly large national debt (£860 million in 1815).

10   The real division in policy concerned attitudes to the currency; here the turning-point came in 1819, with the triumph of the bullionists; 1822 made no significant difference to this, most crucial, issue.

2   Would you agree that Lord Liverpool's government 1822–7 was a 'Liberal-Tory' administration?

Understanding the question

Do not waste too much time defining 'Liberal-Toryism'. You are mainly required to examine the ideology of the Liverpool administration in its post-1822 phase, and also the nature of its political appeal.

Knowledge required

The policies and philosophy of the Liberal Tories. The political support for them in the country at large. An assessment of the 'liberality' of the reforms.

Suggested essay plan

1   The implications of the word 'liberal' in the 1820s; usually used by enemies as a term of abuse; 'Liberal-Toryism' was not a half-way house between Tory and Whig but rather a set of political attitudes and instincts (especially free trade and *laissez-faire*) which were growing within both the Tory and Whig parties, and which would eventually develop, through 'Peelism', into Gladstonian Liberalism. Quote Peel, 1820, as saying that 'the tone of England is more *liberal* than that of the government'.

2   Obviously, Liverpool's government remained 'Tory' in the sense of wishing to maintain the unreformed constitution and to suppress radical discontent by strengthening the forces of law and order (*e.g.* Canning's support for the Manchester magistrates after *Peterloo*); they all wished to maintain the established Church but differed as to whether this could best be done by

resisting or conceding Catholic claims. The prominence of Liberal-Tories like Canning after 1822 did not help the 'Catholic' cause, as Liverpool kept this an 'open' question in cabinet.

3   Nevertheless, in 1822–7 Liverpool's government undoubtedly divided into 'liberal' and 'high' Tory camps; by 1825–6 the cabinet was fundamentally split, with some ministers barely on speaking terms. Liverpool's personal support ensured the dominance of such 'liberals' as Canning, Huskisson and Peel, while Wellington, Vansittart (Bexley) and Eldon attempted to undermine policy from within the government. The main sources of dispute were commercial and monetary policy, the recognition of the South American colonies and attitudes to the Greek revolt.

4   Some important 'liberal' (but non-controversial) reforms were carried out by Peel at the Home Office and Huskisson at the Board of Trade. Canning, by his flamboyant statements on foreign affairs, gave the ministry a 'liberal' aura. Compared to the period 1815–20, these reforms did appear enlightened but they were limited. The government refused to tackle the big constitutional, political and social questions of the day – *e.g.* the Poor Law, factory and parliamentary reform etc. Compared to the Whig reforms which follow they appear restrained.

5   It is sometimes thought (see W. R. Brock and Asa Briggs) that 'liberal' Tories appealed to mercantile and manufacturing interests (Canning and Huskisson were both MPs for Liverpool) as against the agricultural bias of the 'high' Tories. In fact, liberal Tories were as much attached to the landed establishment as anyone, and the fact is that – except in the case of corn – free trade was not then popular with commercial interests. If Liverpool's government after 1822 had any class-based appeal, it was rather to the 'rentier' interests who benefited from their 'sound money' policies.

6   And yet Canning was willing to compromise on policy (as over Greece), and Huskisson partially retreated on the Combination Acts. It is probable, therefore, that the divisions over Liberal Toryism can largely be explained in simple personality terms, especially the 'high' Tory distrust of Canning. Their policy of conservatism, however 'liberal', was still fundamentally a policy of preserving the essential basis of aristocratic power and the constitution inherited from the eighteenth century.

3   How near did England come to revolution between 1815 and 1832?

Understanding the question

This question requires, not only your assessment of the strength, unity and potential for violence of the radical movements during the years 1815–32 but also of the readiness and efficiency of the forces of law and order.

Knowledge required

The social and political history of the years 1815–32. The structure of government and the organization of repression.

Suggested essay plan

1   Professor Gash has called the years 1815–48 the most prolonged period of social discontent in English history; certainly, it came *nearer* to revolution during these years of industrial transformation and dislocation than at any other time – but 'how near' remains open to question.

2   Economic and social factors making for discontent: the growth of population and pressure on food supplies, led to high prices and resentment against the 1815 Corn Law ('the most naked piece of class legislation in English history', according to Lord Blake); economic fluctuations caused cyclical unemployment and the permanent *under*-employment of skilled workers

(hand-loom weavers, stockingers, frame-work knitters) displaced by technological advances; cholera, overcrowding and insanitary conditions in the new industrial cities.

3   The course of political defiance, 1815–21; Spa Fields, the Blanketeers, Luddism, Peterloo, the Queen Caroline riots, the Cato Street Conspiracy; the recurrence of violence 1830–32 'Captain Swing' riots in Nottingham, Bristol, etc. The chronology of violence suggests that it was closely connected to the economic depression – a fact which perhaps suggests that it was a response to short-term difficulties, not an endemic aspect of society.

4   As Lenin (an expert on the subject) argued: deterioration plus activity does not equal revolution. Three preconditions are required for revolution: (a) an active revolutionary group ready to seize power; (b) a weakness, or inability to enforce law and order; (c) a loss of nerve, or a crisis in the affairs of the ruling order. These preconditions did not exist.

5   There is more reason for believing that the radical leadership of working-class protest (Cobbett, Hunt, Place, etc.) was fundamentally cautious, believing in 'moral' rather than 'physical' force and in courteous appeals to parliament rather than direct action.

6   The Establishment was more vulnerable in 1829–32 than in 1815–21 because it was bitterly divided over the Reform Bill crisis, and because working-class malcontents found allies in the middle class who were demanding political representation (Attwood, Prentice, Baines and the Political Unions). On the other hand, the forces of law and order had been strengthened during Peel's period at the Home Office; London even had its Metropolitan Police – a much more effective peace-keeping force than the yeomanry – and the government's nerve never failed.

7   Conclusion: protest during the years 1815–32 was less truly revolutionary than Chartism later, if only because it was *not* directed against the State as such; resentment was directed against landlords, fundholders, master-manufacturers, or at particular laws and inventions (*e.g.* threshing-machines), and against corruption in the state (what Cobbett called 'the Thing') but not against the State itself; once the latter was cleansed, it was thought that it would once again protect the common people. It was the 1832 Reform Act which – defining for the first time ever full citizenship (*i.e.* the possession of the vote) in material, socio-economic terms (the £10 householder, the 40s freeholder) – directed working-class anger against the State itself and thereby turned it into a genuinely revolutionary movement.

4   Document question (Lord Liverpool's government, 1815–27): Study the extract below and then answer questions (a) to (g) which follow:

line

1   'As I lay asleep in Italy
     There came a voice from over the Sea,
     And with great power it forth led me
     To walk in the visions of Poesy.

5   'I met Murder on the way –
     He had a mask like Castlereagh –
     Very smooth he looked, yet grim;
     Seven blood-hounds followed him;

     'All were fat; and well they might
10   Be in admirable plight,
     For one by one, and two by two,
     He tossed them human hearts to chew
     Which from his wide cloak he drew.

'Next came Fraud, and he had on,
15   Like Eldon, an ermined gown;
His big tears, for he wept well,
Turned to mill-stones as they fell.

'And the little children, who
Round his feet played to and fro,
20   Thinking every tear a gem,
Had their brains knocked out by them.

'Clothed with the Bible, as with light,
And the shadows of the night,
Like Sidmouth, next, Hypocrisy
25   On a crocodile rode by.

'And many more destructions played
In this ghastly masquerade,
All disguised, even to the eyes,
Like Bishops, lawyers, peers, or spies.

30   'Last came Anarchy: he rode
On a white horse, splashed with blood;
He was pale even to the lips,
Like Death in the Apocalypse....

'Let the charged artillery drive
35   Till the dead air seems alive
With the clash of clanging wheels,
And the tramp of horses' heels.

'Let the fixed bayonet
Gleam with sharp desire to wet
40   Its bright point in English blood
Looking keen as one for food.

'And let Panic, who outspeeds
The career of armed steeds
Pass, a disregarded shade
45   Through your phalanx undismayed....

'Men of England, heirs of Glory,
Heroes of unwritten story,
Nurslings of one mighty Mother,
Hopes of her, and one another;

50   'Rise like Lions after slumber
In unvanquishable number,
Shake your chains to earth like dew
Which in sleep had fallen on you –
Ye are many – they are few.'

(Stanzas from Shelley's 'The Mask of Anarchy', written on the occasion of 'the Massacre at Manchester,' 1819)

Maximum marks

(a) What office did Castlereagh hold in Lord Liverpool's government? For what reasons and with what justification could lines 5–6 ('I met Murder on the way – He had a mask like Castlereagh') be applied to the way he carried out the duties of that particular office?   (4)

(b) For what other reason did Shelley associate Castlereagh personally with 'Murder'?   (2)

(c) What posts in Liverpool's government were held by Eldon and Sidmouth respectively and why did Shelley attack them so bitterly in lines 15–25?   (6)

(d) Explain the reference to 'spies' in line 29.   (3)

(e) Describe and account for the importance attached to the incident referred to in lines 34–41.   (6)

(f) What were the reasons for the 'Panic' referred to by Shelley in line 42?   (4)

(g) Name the man who might be said to have attempted, a year after this was written, to act in the spirit of lines 50–54, and why did Englishmen in general *not* 'Rise like Lions after slumber' in the 1820s?   (6)

16.6 READING LIST

Standard textbook reading

A. Briggs, *Age of Improvement* (Longmans, 1959), chapter 4.

R. K. Webb, *Modern England* (Allen and Unwin, 1980 edn.), chapter 4.

A. Wood, *Nineteenth-Century Britain* (Longmans, 1960), chapters 3, 5.

Further suggested reading

W. R. Brock, *Lord Liverpool and Liberal Toryism 1820–27* (Cass, 1967).

J. E. Cookson, *Lord Liverpool's Administration: the Crucial Years* (Scottish Academic Press, 1975).

B. Hilton, *Corn, Cash, Commerce: The Economic Policies of the Tory Government 1815–30* (OUP, 1977).

A. Mitchell, *The Whigs in Opposition 1815–30* (OUP, 1967).

W. Reitzel ed., *The Autobiography of William Cobbett* (Faber and Faber, 1967).

M. I. Thomis and P. Holt, *Threats of Revolution in Britain 1789–1840* (Macmillan, 1977).

E. P. Thompson, *The Making of the English Working Class* (Gollancz, 1963).

17   The Whig reforms, 1830–41

17.1 INTRODUCTION

This is a period of active reform and political reorganization. The most important change, the Reform Act of 1832, was effected through the mobilization of public opinion by assorted radicals and political pressure groups – such as the Birmingham Political Union.

The Whigs, elected in 1830 on a reform platform, saw their task as remedying the main grievances of the people. Although the measure they passed in 1832 was essentially conservative in nature, their action probably saved England from revolution.

Encouraged by their own belief in progress, and helped by the Benthamite doctrine of utilitarianism, they proceeded to introduce several major reforms before their reforming zeal petered out in 1836.

17.2 STUDY OBJECTIVES

1   The Tories won the 1830 election but Wellington failed to appease the ultra-Tories or Canningites; the government was defeated on the Civil List issue and resigned. Lord Grey became Prime Minister against a background of political, social and economic unrest. The Whigs were strengthened by the adhesion of the Canningites and Radicals. The ministry was predominantly aristocratic but was determined to solve the problem of reform.

2   Note the content of the Reform Bill, the struggle to pass the Bill, the influence of economic distress, the publicity of the radicals, the pressure of the Political Unions (mark especially Attwood and the BPU), and the NUWC. The riots and disorder at Nottingham, Bristol, etc. (1831) and 'the days of May' (1832) – 'to stop the Duke' etc. The object of the Whigs was to appease the middle classes – 'the wealth and intelligence of the country'. The Tories resisted fiercely; Peel was 'unwilling to open a door'.

3   The terms of the Reform Act – were they conservative and modest? Or a revolutionary enactment? There was little change in the composition of the House of Commons, few concessions to Radicals, and corruption continued. The Crown, the Church, the House of Lords, the aristocracy – all survived the trauma. However, a first breach had been made – the Act had completed the religious changes begun in the late 1820s. The Reform Act was 'great' because of the struggle for its passing rather than the significance of its contents.

4   The Radicals demanded further reforms; the Whigs were reluctant but continued – at least up to 1836. The cause of reform was helped by utilitarian doctrine and Benthamite procedures – *i.e.*, the Royal Commission, legislation and central (supervised) control. The importance of Chadwick in this 'revolution in government'.

5   Whig legislation: the abolition of slavery was finally forced through by the pressure of the Dissenters and the evangelicals; £20 million was allocated to assuage the West Indian lobby – which was less vociferous now that profits of the sugar trade were in decline. The Benthamites, however, were unable to get state control of education: Benthamite intrusion was fiercely resisted by the Anglicans and the Dissenters. The Benthamites had to be content, therefore, with an annual grant of £20,000. The campaign for a Factory Act – by Tory landowners (*i.e.*, Oastler), the Radicals, evangelicals and factory operatives – was successful but disappointingly utilitarian in character. The creation of a Factory Inspectorate prepared the way for later, more effective, legislation.

6   The 1834 Poor Law Amendment Act; the deficiencies of the old system; the middle-class attack on Speenhamland. The Act, which was Benthamite utilitarian in conception and execution, finished independent, parochial administration by amalgamating parishes into Unions, managed by elected Boards of Guardians; the results of the Act, the attitudes of the working class and of the Radicals; the Anti-Poor Law movement.

7   The 1835 Municipal Corporations Act: this was a corollary of the 1832 Reform Act. It swept away corrupt municipalities and gave 'shopocracy' the vote. The Act was largely permissive and so postponed municipal reforms until the 1870s. The Act was 'poison for the Tories' as it made the boroughs Whig strongholds.

8   The Radicals were disappointed as there was no Church disestablishment. However, the civil registration of births, marriages and deaths was introduced, tithes were abolished, and the non-denominational University of London was founded. Working-class leaders, however, were irritated at the suppression of the trade unions and the lack of further constitutional reforms. The middle classes were irritated by the continuance of the Corn Laws. The failure of the Whigs to tackle these questions led to the rise of Chartism and of the Anti-Corn Law League in the later 1830s when Whig reforms had dried up.

9   The Whigs were divided and weakened after 1834 by the rifts with the Radicals and the O'Connellites, the weakness of Melbourne's leadership, the inability to solve financial problems and the constancy of the Irish difficulties. Note the re-emergence of the Tories under Peel; the Tamworth manifesto offered a new Conservative creed for the nineteenth century. Peel's qualities of leadership contrasted with those of Melbourne; the 'Bedchamber Question'. The Whigs continued in office but not in effective power; the 1841 election ended Whig rule.

10   The Whig reforms were creditable: they laid the foundation of a modern, administrative state but the governments had their failures. They failed in Ireland despite their reforms (*viz.* Education, Poor Law, Municipal Corporations, tithes); they failed in finance (this was less excusable); and they failed to satisfy working-class aspirations. However, they had appeased the middle classes, they had tackled fundamental problems and had made sure that England's progress to democracy would be an orderly one.

17.3 COMMENTARY

The need for reform

Lord Grey's government faced a difficult situation in 1830. His Cabinet contained men of both parties but was predominantly made up of the great landed Whig aristocrats. (It was said that his Cabinet surpassed in acreage any previous Cabinet.) It dealt severely with the 'Captain Swing' riots before, ironically, turning its attention to reform. In fact, Wellington's government had reluctantly begun the reform of the constitution in 1828. The question now, unlike then, was not whether reform should be allowed but how much should be granted.

The unreformed system

Prior to 1832 the unreformed electoral system was a confusing and antiquated mixture of local customs, practices and abuses. The industrial revolution and population growth of the late eighteenth century had rendered the system obsolete and unrepresentative. The most obvious defects were: the existence of nomination (or 'rotten') boroughs; the widely mixed urban franchises; the unrepresented industrial cities; regional under- or over-representation; the limited numbers who had the vote (only 160,000 out of 16 million); the few elections that were contested and the almost complete domination of the system by the landed élite. You must be able to give examples of each of the above defects.

The Tories had the major, though not exclusive, share of the unreformed system and they (with the Church and the Army) were its fiercest defenders. They argued in its defence that (a) it had stood the test of time, (b) it allowed all classes and interests representation and (c) that it was a deliberative assembly, not a congress of delegates. Even the much-attacked 'rotten' boroughs, it was alleged, had given men of ability (like Pitt, Burke, Fox, Canning and Peel) a means of entry into the Commons. Finally, the Tories maintained that reform would undermine the habits of obedience, deference and respect for property.

In short, they favoured a system based on the ruling aristocracy, the landed gentry, the primacy of the countryside and the Anglican Church. You must now find out what those who wished to change the system wanted.

The reform crisis

The actual pace of change, however, was to be dictated as much by events within parliament as outside it. Hence the crucial importance of the break-up of the Tory Party after Liverpool, and the first adjustments to the constitution that followed. The religious bastion of Anglicanism had now been breached.

The radical campaign which had begun in the 1760s with Wilkes, and had continued under Wyvill, Cartwright, Cobbett and Burdett, at last appeared to be achieving success. You should

● Rotten Boroughs

● Large Towns with no MPs

Representation in the Commons 1830 – 32

	County Seats		Borough Seats		University Seats		Totals	
	1830	1832	1830	1832	1830	1832	1830	1832
England	82	144	403	323	4	4	489	471
Wales	12	15	12	14	–	–	24	29
Scotland	30	30	15	23	–	–	45	53
Ireland	64	64	35	39	1	2	100	105
Totals	188	253	465	399	5	6	658	658

Fig. 9 Representation in the unreformed Parliament, 1830

consider the question of why reform was delayed until 1832 and what combination of circumstances made it possible then. (Note that the reform movement was made up of ultra-Tory discontent after Catholic emancipation, the economic distress of 1830, the outpourings of the press on reform and the activities of the Political Unions – especially Attwood's BPU.)

Wellington's determination to oppose reform forced the Whigs' hand. Lord Grey, realizing that the price of power and general law and order was a Reform Bill, and knowing that he could rely on support across party lines, determined to act boldly. He entrusted the drafting of the Bill to a committee of four –

which included the two most prominent reformers in the party, Lord John Russell and Lord Durham.

The passing of the Bill

The political complexion of Grey's cabinet assured the propertied classes that reform would not mean revolution: the object of the first Bill introduced by Russell was primarily to rally middle-class support around the prevailing aristocratic system. With the Tories incredulous at the proposed changes – 168 seats were to disappear – the struggle to pass the Reform Bill began.

Some historians argue that over the next year England was

in a near-revolutionary state. The public, roused by the Political Unions, demanded 'the Bill, the whole Bill and nothing but the Bill.' When it was jeopardised they rioted, threatened a run on the Bank of England, and burned the castle at Nottingham, the Bishop's Palace at Bristol and the gaol at Derby. There were threats of a refusal to pay taxes and fears of mutiny in the Army. The ingredients for a revolution seemed present as the Radicals and Political Unions tried to impress on the government that a failure to grant reform might precipitate a revolution. How did Lord Grey respond and why was he able to avert this threat?

The Whig case

The Reform Act attacked the most obvious defects of the un-reformed system: the 'rotten' boroughs, unequal franchises, unrepresented cities, regional under- and over-representation. Could it be argued that some of the changes reinforced the landed hold on power? (Check on this aspect and pay particular attention to the Chandos clause.)

The Whigs, of course, maintained that the Bill was a moderate, conservative, final measure. They could not afford to lose the support of too many aristocrats or Tory partisans. Lord Brougham stressed that only half a million of the propertied and educated class – 'the wealth and intelligence of the country, the glory of the British name' – would get the vote. Macaulay neatly summed up their fundamental objective: 'that we may exclude those whom it is necessary to exclude, we must admit those whom it is safe to admit.' According to W. R. Brock, 'the poachers were to be turned into gamekeepers.'

The strength of the Whigs' case was that they were offering a practical remedy for a felt grievance, the most that parliament would tolerate and the least that the country at large would accept. They also maintained that it would preserve aristocratic rule, and the events of the decades that followed showed that their political judgement was sound.

The Tory case

The Tories rejected all these arguments and insisted that the Reform Act would destroy the balance of the constitution. According to Peel, their chief spokesman, royal power and patronage would be undermined and the Church and House of Lords threatened. The Act, he insisted could not be final. Once the principle of change was accepted more reforms would inevitably follow. Hence he demurred, 'I am unwilling to open a door I see no prospect of being able to close'. Wellington foresaw the decline of the landed interest and a fresh onslaught on the Corn Laws. Alexander Baring, MP for Thetford, prophesied that 'the field of coal would beat the field of barley.' In general the Tories rejected Whig arguments about 'finality' and 'moderation' and viewed with alarm the prospect of a House of Commons filled with merchants, tradesmen and radicals.

Effects of reform

About effects of the Act, there are two schools of thought. One tries to minimize the importance of the Act because, for example, the worst fears of the Tories were not realized. It is argued that the Act lessened but did not eliminate bribery, that most of the basic radical demands were ignored and the working classes were denied the vote. Furthermore, the middle classes, who gained the franchise, failed to make use of it to vote into parliament the financiers, manufacturers and bankers who might have represented them.

In fact, the new House of Commons looked very like the unreformed chamber – 150 MPs still related to members of the Lords; 400 had no profession; 71 were lawyers; 75 had Church patronage; 64 were Army or Navy officers. Only 49 merchants were elected. According to Greville, 'the reformed parliament was very much like every other parliament.' You can discover further arguments for minimizing the effects of the Act if you study its impact on the powers of the Crown, Church and House of Lords. Would you agree that each seemed to survive this testing ordeal in the 'thirties and emerged with its powers largely unimpaired?

On the other hand, the Act did effect certain important changes. It got rid of the most notorious 'rotten' boroughs, made the system more representative and gave a substantial number of the middle classes the vote. The manner of its passing was also significant; the Commons had asserted its power over the monarchy and the House of Lords. The force of public opinion and public pressure had been acknowledged. The Act stimulated more interest in both local and national politics and, as more elections were contested, both parties had to reconsider their political outlook and their attitude to the new voters.

Do these changes, however, amount to a revolution? Perhaps the fundamental significance of the Reform Act was that it sanctioned change in the constitution when the time was ripe, showed how this change could be achieved and peaceably opened the door to democracy.

Further reforms, 1833–41

The governments of Grey and Melbourne proceeded to tackle the problems posed by the new industrial society. The predominantly landed parliament accepted the passing of a series of acts which only appeared to sharpen class divisions and emphasize class consciousness. Consequently, reforms dried up after 1836 as the Whigs tired of attempting to satisfy the parliamentary Radicals and their working-class allies.

Nevertheless, by then they had many important reforms to their credit. Slavery in the colonies was abolished in 1833 and the first government subsidy was given to education in the same year. A Factory Act remedied the worst abuses in most textile mills in 1833 and, more important in the long run, appointed four inspectors to supervise the working of the Act. The old Poor Law was amended in 1834. Local government was reformed by the Municipal Corporations Act of 1835 and finally some minor concessions were made to Dissenters in 1836.

Study the details of these measures and note how the influence of Benthamite utilitarianism runs through most of the reforms. Benthamism provided a procedure for reform: investigate, legislate, centralize and, using 'feedback' information (mainly supplied by the new government inspectors), amend where necessary. This procedure was a tremendous asset to reformers and more especially to Edwin Chadwick, who was the driving force behind the most important of the Whig reforms.

The Benthamites did not have it all their own way, however, for they were continuously challenged by humanitarians and Evangelicals such as Lord Shaftesbury. There was a clear conflict of outlook between these two groups in the 1830s and 1840s. You must examine their distinctive ideas and see where they differed, clashed and overlapped in their attitude to the above reforms.

The new Poor Law

The most important of the Whig reforms was the Poor Law Amendment Act of 1834. This clearly revealed the utilitarian spirit of the age; Chadwick played a major part in its formulation and application. The new Act rested firmly on Benthamite convictions – the greatest happiness of the greatest number – and recommendations – the workhouse test and the principle of 'less eligibility.'

Find out what you can about the arguments used to justify this reform, the motives of Chadwick, the terms of the Act and its effects in the decade that followed. Evaluate the arguments of those who opposed the Act. You should also have some knowledge of the activities of the anti-Poor Law movement. Was the Act a brave, rational approach to a problem which, by the introduction of the Speenhamland system in 1796, had undermined the will of the labourer to work and as a result encouraged idleness and pauperism? Or was it, as many humanitarians maintained, 'a vast engine of social degradation' which punished people for being poor and unemployed through no fault of their own? The choice for the poor now seemed to be work, workhouse or starve.

Achievements

Although the Whigs were by 1835 clearly in decline, they had many achievements to their credit. True, they failed in Ireland and in finance but they had seen the need to make a place in the constitution for the middle classes and they had, at least, sought to grapple with the conditions of life and work in the new industrial society. Because of the conflict of interests involved, many of the reforms were compromises, as in factory reform, or permissive, as in municipal reform. They did, however, try to solve the problems posed by the old and muddled local arrangements by promoting a system of administration that was national, standardized and efficient.

Reform before 1830 in many ways was a matter of expediency; now, under the pressure of utilitarianism and social needs, it was becoming more a matter of policy. The pattern of government had been changed in a decade and the foundations of an administrative revolution laid.

Failures

But perhaps the significance of the work of the ministry lay less in what it achieved than what it failed to achieve. Many people came to recognize that they were not the beneficiaries of reform. Why were the radicals and their working-class followers so bitterly disappointed? Why did Oastler call the Whigs 'bloody, base and brutal'?

It would appear at first sight that the middle classes should have been happy with these reforms which, on the whole, favoured their utilitarian and entrepreneurial outlook. Yet they were also disappointed. Find out why. What were the immediate results of this widespread dissatisfaction with the Whig reforms?

Decline

By 1840 the Whigs were in disarray. Death and retirement had deprived the Party of its ablest leaders. They failed to balance the budget, to agree on a policy for Ireland, or to deal with the intransigence of the House of Lords. Under the disinterested Melbourne, the Party withered away, allowed Peel to emerge as the most positive politician in the Commons and thereby, win decisively the general election of 1841.

17.4 CHRONOLOGY

1830	**September**	'SWING' RIOTS: agricultural labourers burned hayricks and smashed threshing machines in southern agricultural counties.
	June	Accession of William IV. 'JULY REVOLUTION' in Paris: ultra-conservative government of Charles X overthrown; the liberal Louis-Philippe was made King.
	November	Lord Grey became Prime Minister. Union of Canningites and Whigs.
1831	**March**	FIRST REFORM BILL passed by the House of Commons. Defeated in Committee.
	April	General Election. Reform fever; demand for 'the Bill, the whole Bill and nothing but the Bill'.
	June	SECOND REFORM BILL introduced. Passed by the House of Commons.
	October	Rejected by the House of Lords. REFORM RIOTS in Nottingham and Bristol.
	December	THIRD REFORM BILL introduced.
1832	**May**	Amended by Lords. Grey resigned. KING PROMISED TO CREATE ENOUGH WHIG PEERS to get the Bill through the Lords.
	June	GREAT REFORM ACT passed. It redistributed 143 seats and extended the franchise to the £10-householder in the towns.
1833	**August**	ABOLITION OF SLAVERY: ended slavery in the British colonies. £20 million given in compensation. FACTORY ACT prohibited the employment of children under nine, granted an eight-hour day to young persons under 13, and appointed inspectors. EDUCATION ACT: first government grant of £20,000 to the two religious societies (Anglican and Non-conformist) which controlled education. The grant was increased and inspection introduced in 1833.
	September	OXFORD MOVEMENT founded: publication of TRACTS FOR THE TIMES.
1834	**March**	Trial and transportation of the Tolpuddle Labourers.
	July	Lord Melbourne became Prime Minister.
	August	POOR LAW AMENDMENT ACT abolished the Speenhamland system and instituted a new system of poor relief based on the workhouse.
	November	Robert Peel became Prime Minister.
	December	Tamworth Manifesto: Peel's electoral address to his constituents at Tamworth. Clear assertion of Conservative principles.
1835	**March**	ECCLESIASTICAL COMMISSION established to investigate the question of Church reform.
	April	Lord Melbourne became Prime Minister.

1836	September	MUNICIPAL CORPORATIONS ACT instituted a more uniform system of local government in the towns. It diminished corruption and abolished municipal privileges.
1836	August	REGISTRATION ACT: births, marriages and deaths to be recorded. DISSENTING MARRIAGES ACT: marriage in Church, chapel or before a registrar allowed. Removed a major Non-conformist grievance. Tithe Commutation Act. Stamp duty on newspapers reduced from 4d to 1d.
1837	June	Death of William IV. Accession of Queen Victoria.
1838	May	PUBLICATION OF 'THE PEOPLE'S CHARTER'.
	September	FORMATION OF THE ANTI-CORN LAW LEAGUE.
1839	May	Lord Melbourne resigned on the Jamaica Bill. 'Bedchamber Crisis': Peel refused to take office unless the Whig ladies of the Royal Household resigned. The Queen refused his request and Melbourne was restored as Prime Minister.
1840	January	PENNY POST established.
	10 February	Marriage of Queen Victoria to Prince Albert of Saxe-Coburg Gotha.
1841	May	Government defeated in the Commons.
	June	General election. Conservative victory.
	September	Robert Peel became Prime Minister.

17.5 QUESTION PRACTICE

1 How far is it true to say that the Whig governments of the 1830s began the 'Age of Reform'?

Understanding the question

This calls for a discussion of whether or not the governments of Grey and Melbourne embarked upon a major programme in ways which marked their achievements as different from those of previous ministries, and began a new era in British politics.

Knowledge required

The major legislative enactments of the Whig governments concerned. Some knowledge of the work of the governments of the 1820s and their attitude towards reform. Some knowledge of the progress of reform later in the nineteenth century.

Suggested essay plan

1 It would be an obvious mistake to say that there were no reforms before the Whig return to office in 1830. For example, the first Factory Act in 1802, and further modest measures of factory regulation in 1820, 1825 and 1830. Peel created the new Metropolitan Police in 1829; the Combination Acts (which restricted trades unions) were partially repealed in 1824–5. Small beginnings were made in parliamentary reform – two members from the corrupt borough of Grampound were transferred to Yorkshire. In 1828–9 the repeal of the Test and Corporation Acts and Catholic emancipation brought significant extensions of religious equality.

2 The first five years of the Whig ministries saw an unmistakable quickening of the pace of reform: the Great Reform Act brought the first large instalment of parliamentary reform and removed much of the more discreditable features of the old electoral system. This measure seemed to be much more sweeping and thorough in the Britain of 1832 than it may do to later observers. In 1833 slavery was abolished throughout the British Empire, there was the appointment of the first factory inspectors and, in 1834, the Poor Law was reformed. In 1835 the Municipal Corporation Act provided a new basis for local government by reforming the government of existing boroughs and setting up municipal institutions in urban areas previously without them.

3 An important distinction may be made between the reforms enacted before and after 1830. In the 1820s the Tory governments were willing to make some improvements to the existing system but shied away from major fundamental reforms; the repeal of the Combination Acts was the work of Place and Hume; even the creation of the Metropolitan Police was given the appearance of a government response to a study on policing by a House of Commons Select Committee, though in fact the committee was guided by Peel. In general the Tories were willing to consider 'judicious improvements' but no sweeping measures of reform.

4 Where important constitutional reforms did take place in the 1820s they were forced on the government rather than welcomed by it; the repeal of the Test and Corporation Acts in 1828 was proposed by an opposition member, Lord John Russell; the granting of Catholic emancipation in 1829, though opposed by Peel, was conceded because of the Clare by-election in Ireland and the threat of serious trouble there.

5 On the other hand, the Whig reforms of the 1830s saw a different spirit. The reforms which the government itself devised and enacted were much more sweeping. Instead of the disfranchisement of individual boroughs, a broad plan of parliamentary reform was implemented; the abolition of slavery covered *all* British possessions; the Poor Law system throughout England and Wales was investigated and altered, and town governments were adjusted on a national basis. The government embraced major reforms as an absolutely essential part of its work.

6 Some measures provided an important foundation for future reforms – *e.g.* inspectors appointed under the 1833 Factory Act provided a model for later inspectorates of schools, mines and lunatic asylums; reformed town councils provided a basis for future growth in the work of local government; the registration of births, deaths and marriages provided new statistical resources for future governments, and this was only part of the increased information made available. The Poor Law Commission and the factory inspectors, for instance, had the duty to lay regular reports before parliament, and the orderly accumulation of evidence about social problems was an important prerequisite for further reforms in future years.

7 In conclusion, though the Tory governments of the 1820s enacted some reforms, this was done so gingerly and sometimes so reluctantly that the much more thorough-going work of the Whigs entitles them to be regarded as the real pioneers of the 'Age of Reform'.

2   How far did the 1832 Reform Act fulfil the aims of its creators?

Understanding the question

This question cannot be answered properly by a narrative account of the reform crisis of 1831–2. It requires first of all an examination of the objectives which the framers of the measure had in mind, and then a consideration as to how far the nature and effects of the 1832 Act matched those objectives.

Knowledge required

The nature of the leadership in Grey's government. The pre-1832 electoral system. The main features of the 1832 Reform Act. The post-1832 electoral system.

Suggested essay plan

1   The Grey Cabinet was composed mainly of aristocrats – including a number of important ministers, *e.g.* Melbourne and Palmerston, who had recently served in Tory governments. They did not believe in democracy and there is no reason to expect that they might have done in the Britain of 1832. The cabinet committee which drafted the Bill included Lord Durham, Lord John Russell, Sir James Graham, Bart. and Lord Duncannon.

2   The reform of Parliament was one of the ministry's main aims. Grey and Russell were advocates of reform for many years before 1830; they wished to preserve the predominance of the aristocracy, to broaden the basis of support for the constitution, and to eliminate, or greatly reduce, aspects of the electoral system which brought the representative system into disrepute. In particular, they wished to conciliate the 'middle classes', though this term was not used with any great precision. They aimed for 'a better representation of the property and intelligence of the country'. In this context the word 'intelligence' was used in the older sense of 'information' (surviving in the modern term of 'intelligence corps'). This meant improving the electoral influence of the propertied and educated sectors of society. To do this they sought a modest extension of the franchise, together with provisions which (1) would make the franchise more uniform throughout the country; (2) would make elections simpler and cheaper, and (3) would reduce corruption.

3   Before 1832 both the franchise and distribution of MPs had changed little for very many years; the counties returned only about a quarter of MPs, though a minority of the population and the wealth of the country was in the towns; the borough franchise was very varied, usually depending on the provisions of the charters governing individual boroughs. In some – Preston, for example – the franchise was quite wide; in others – like Portsmouth – only members of the corporation had the vote. There were also many other varieties of borough electorate; some ancient boroughs – such as Old Sarum – had decayed but still returned MPs; in other boroughs bribery was an established part of the electoral scene; newer urban areas – *e.g.* Manchester and Leeds – were without representation.

4   The government's reform proposals included: the removal of MPs from boroughs too small or decayed to justify their retention, giving MPs to new towns, increasing the number of county MPs and introducing standard electoral qualifications. In the counties, 40-shilling freeholders kept the vote, but the Act also enfranchised certain groups of copyholders, leaseholders and tenants. In the boroughs, the main category of voters was to be those who owned or occupied property of a rateable value of £10 or more – the £10 householders.

5   The results of the 1832 Act in practice represented partial fulfilment of its creators' aims; the redistribution of seats went some way to correct the earlier imbalance between county and borough members; the important newer towns received their own MPs. Many smaller boroughs lost their MPs or now returned one instead of two. Over 300,000 new voters were added to the half million existing before 1832.

6   Corruption did not disappear, but the number of very corrupt constituencies diminished with the disappearance of many of the smaller ones. Some of the new post-1832 voters proved as willing to accept bribes as their predecessors; more than 50 seats were still under the control of individual patrons, most of them aristocrats.

7   The Whigs largely succeeded, in the immediately following years at least, in broadening the support for the electoral system. When the Chartist movement developed a few years later, many of the most strenuous reformers of 1831–2 had been conciliated by the reform of 1832, and were now strong defenders of the reformed system against the Chartists. The electoral changes of 1832 greatly strengthened the existing order, and the aristocracy continued after 1832 to enjoy a predominance very similar to that in pre-1832 years.

3   Why did the Whig hold on power decline in the years after their great victory in 1832?

Understanding the question

After the triumphant passing of the Great Reform Act in 1832, the Grey government scored a sweeping electoral victory in the first reformed general election in December, 1832 but, by 1841, they were defeated and out of office. The question requires an explanation of this transition.

Knowledge required

The results of the general elections of 1832, 1835, 1837 and 1841. An appreciation of the difficulties encountered by the Whigs in holding their followers together and maintaining their popularity. The strong recovery of the Conservatives during the 1830s and the reasons for it.

Suggested essay plan

1   By the end of 1832 the old Tory Party had been crushingly defeated in the battle over the Reform Act and reduced to a small minority in the new House of Commons. The victorious government went on to enact important reform measures (1833–5), including the abolition of slavery, and factory, poor law and municipal reform.

2   However, Whig parliamentary supporters were not a coherent body: they ranged from an important bloc of aristocratic Whigs to Radicals and Irish nationalists under Daniel O'Connell. Even the Radicals were not united and included different groups unable to work together even on agreed policies. It was very difficult for the government to hold together these varied segments, and impossible to prevent a great deal of bickering and public disunity.

3   The general election of 1835 saw Conservatives under Peel gain about 100 seats – though not enough to keep Peel's minority government of 1834–5 in office. In this, and in the following election in 1837, losses among Whig supporters were particularly marked within the radical wing of the 'coalition'. When, after 1835, Whig reform measures ceased, Radicals blamed their losses on their link with the supine Whigs, while Whigs attributed their increasing weakness in parliament to public hostility against the militant pronouncements of some radical groups.

4   Some of the reforms enacted by the Whigs also aroused opposition. The new Poor Law brought much popular hostility and opposition to the elements of centralization and interference in local affairs included in the 1834 changes. Also, the Whig

government reform of the Church of England, especially the control of its property and income, was disliked by many Anglicans at a time when the Church was a very influential body. Attempts to impose unwelcome reforms on the established Church of Ireland, especially to divert that Church's revenues to social purposes, lost the government important supporters – two senior cabinet ministers, Stanley and Graham. Grey himself resigned in 1834, mainly for the same reason.

5   Meanwhile, under Peel's skilful leadership, the opposition, now taking the name of the Conservative Party, experienced a distinct recovery. For the general election of 1835 Peel issued the celebrated Tamworth Manifesto, offering the electorate the concept of an attractive Party which, while determined to conserve the vital foundations of British society, was not a reactionary body but a Party itself willing to contribute useful reforms. Peel was careful to present a public image of conspicuous moderation, while Conservative propaganda stressed the need to defend the reformed establishment against the excesses of radicalism and the nationalist Irish. A closer alliance between the Whigs and the Irish, under the Lichfield House compact (1835), did nothing to prevent this impression gaining ground.

6   Other elements in Whig policy also contributed to the growing support for the Conservatives: the moderation of Whig Home Secretaries towards the early activities of the Chartist Movement; this may have been wise but the restraint in dealing with democratic agitation was unpopular with the reformed electorate, which showed no desire to extend its privileges. The deepening economic depression from 1837, plus the widespread distress it entailed, exposed the Whigs to heavy criticism. They seemed incapable of devising a financial policy to solve these problems, and their last budgets saw a series of deficits in the national accounts.

7   The Whig government, therefore, declined in strength for a variety of reasons: the basic disunity of their wide range of followers was a serious weakness, while some of their policies brought opposition in their train. They lost important adherents, and successive general elections cut away the great majority they had won in 1832. The 'Bedchamber Crisis' of 1839 showed how precarious their hold on power had become. On the other hand, thanks largely to the dexterous leadership of Peel, the opposition was better able to appeal to influential public opinion, and by 1841 was well placed to eject the Whigs and win a sweeping electoral victory.

4   Document question (The Problem of the Poor Rate in the Early Nineteenth Century): study the extract below and then answer questions (a) **to** (g) **which follow:**

'We, the undersigned magistrates of Suffolk, beg leave respectfully to submit to the Committee for the Revision of the Poor Laws, the following observations.

'In the present alarming state of the poor rates it must be obvious that there must have been some very unexpected and extraordinary alterations when these rates are swelling to an amount not only unprecedented but beyond what any actual change in the situation of the agricultural population might appear to warrant. We trust it will appear that one cause among many others has added very materially to the burden of the assessment; and that it is indefensible as being unjust, impolitic and cruel.

'The circumstances to which we allude is a practice which has prevailed of giving reduced and insufficient wages to labourers in husbandry and sending them to the poor rate for the remainder of the sum necessary for their support. A practice like this becomes an assessment, not so much for the relief of the poor as of their employers. If the poor rates were levied solely on those who

employed labourers, the evil, though great, would be less oppressive. But when it is recollected that the small occupier who cultivates his little farm by his own labour and that of his family; that the tradesman, the mechanic and (where cottages are rated, or where a little land is attached to them) even the labourer is compelled to pay to this assessment, the hardship and partiality of the practice is most evident. Further, the professional man, the shopkeeper, the artisan all are taxed for the payment of labour from which they derive no immediate benefit and in the profits of which they have no participation.

'The practice tends to debase the industrious labourer to the class of the pauper: it habituates him to the reception of poor relief; it teaches him to look to the rates for his usual maintenance instead of applying to it reluctantly in sickness or old age.

'For the evils thus detailed, the existing laws furnish no relief. If the labourer whose earnings are insufficient for his support, applies to the magistrate, the magistrate, having no power to fix the rate of wages, MUST, however reluctantly, ORDER relief from the poor rate; and, as this order is final and conclusive, the several classes aforementioned as aggrieved by this unequal assessment, are precluded from the benefit of appeal against the overseers account and left without remedy against this glaring act of injustice and oppression.'

(From a Memorial of the Magistrates of the County of Suffolk to a Select Committee of the House of Commons on the Poor Laws, 1817)

Maximum marks

(a) What system of poor relief is described in the extract? Where and when is it supposed to have originated?   (4)
(b) What other systems of poor relief were also in use in rural areas?   (4)
(c) Under what legislation had a magistrate 'final and conclusive' power to order relief from the poor rate?   (2)
(d) Explain the term 'the overseers'   (2)
(e) Comment on the nature and validity of the grievances detailed and on their description as 'unjust, impolitic and cruel'.   (6)
(f) What measures were eventually taken to teach the labourer not 'to look to the rates for his usual maintenance instead of applying to it reluctantly in sickness or old age'   (6)
(g) Examine the public response to the measures eventually taken in removing the grievances complained of by the Suffolk magistrates in this memorial.   (7)

17.6 READING LIST

Standard textbook reading

A. Briggs, *Age of Improvement* (Longmans, 1959), chapter 5.

R. K. Webb, *Modern England* (Allen and Unwin, 1980 edn.), chapter 6.

A. Wood, *Nineteenth-Century Britain* (Longmans, 1960), chapter 5.

Further suggested reading

M. Brock, *The Great Reform Act* (Hutchinson, 1973).

G. B. A. M. Finlayson, *England in the Eighteen-thirties* (Arnold, 1969).

N. Gash, *Reaction and Reconstruction in English Politics 1832–52* (OUP, 1965).

A. Llewellyn, *The Decade of Reform: The 1830s* (David and Charles, 1972).

J. D. Marshall, *The Old Poor Law 1795–1834* (Macmillan, 1974).

M. E. Rose, *The Relief of Poverty 1834–1914* (Macmillan, 1972).

D. Southgate, *The Passing of the Whigs 1832–86* (Macmillan, 1962).

J. T. Ward (ed.), *Popular Movements c. 1830–1850* (Macmillan, 1970).

18 Chartism and the Anti-Corn Law League

18.1 INTRODUCTION

During the late thirties England was plunged into depression. The condition of the lower classes was miserable. The manufacturing middle classes were also affected by the slump and condemned the Corn Laws as an interference with free competition and free trade.

Both classes founded movements to express their discontent and to challenge the prevailing political structure. The Victorian age opened, therefore, with the threat of revolution in the air – a threat the ruling orders could not afford to ignore.

18.2 STUDY OBJECTIVES

1   Chartism was the product of industrialization but was also a part of the radical tradition dating back to Major Cartwright's *Take Your Choice* (1776); it represented the fundamental belief that economic exploitation and political subservience could be righted by parliamentary means.

2   Background to Chartism: the emergence of the provinces; the desperation of the hungry 1840s; born out of working-class hunger and anger – 'a knife and fork question' (Harney); Chartism as a phase in the 'social tension' chart. There was also anger at exclusion from 1832, the crushing of the trade unions and the introduction of the New Poor Law. The unstamped press publicized these discontents and provided an organizational network.

3   Institutional background: the LWMA and articulate, artisan, London radicalism; later joined by Attwood and BPU (idea of petition). Reinforced by popular, local, radical activity – especially that of O'Connor (and his newspaper *The Northern Star*) in Leeds and the West Riding.

4   The local nature of the Chartist movement: it was strongest in centres of old decaying industries – *e.g.* textiles, wool, stockings, in Stockport, Leeds, etc. Also London, Birmingham, Tyneside, South Wales, Glasgow. The handloom weavers were the backbone of the movement, which was weak in the South-East and in agricultural areas. The movement was class conscious – 'fustian

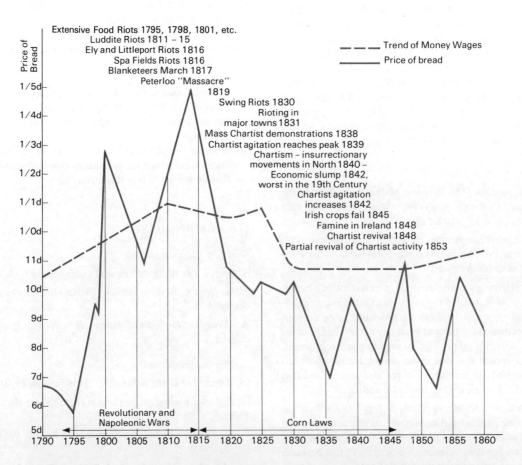

Fig. 10 This diagram represents an index of social tension in Britain between 1790–1860 in relation to the trend of money wages and the price of bread

jackets, blistered hands and unshorn chins'; it was singularly working-class (it rejected the overtures of the ACLL and of the Complete Suffrage Union). Note the Irish contribution – especially that of O'Connor and *The Northern Star*. It was a disparate movement held together by its leaders and by rallies, by its organization and its press.

5　Strategy: the debate in the London Convention, 1839; 'moral or physical force', petition or rebellion? The rejection of the petition, the Newport Rising, the Chartist arrests, the collapse (1840). This was mainly due to a lack of organization and by government determination to stand firm. The excellent use by the government of the Army (General Napier), of the railways and the electric telegraph to maintain control. The Chartists reorganized and founded the National Charter Association (1840). The rejection of the second petition (1842), the Plug Plot fiasco, the disintegration of the leadership as each leader pursued his own objective. O'Connor offered his land scheme; Lovett offered his on education; others Owenism, religion etc.

6　The decline of the Chartist movement as economic conditions improved; Kennington Common (1848), the rejection of the third petition, the loss of mass appeal. Was failure due to the defects in the movement's leadership? To defective strategy? To localism? To dependence on hunger? Or was it that the government and the middle class were determined to resist and that Chartism was doomed from the outset – *i.e.*, that failure was due more to middle-class strength than to working-class weakness? The significance of the movement.

7　Background to the Anti-Corn Law League: the trade depression post-1836; middle-class dissatisfaction with the results of the 1832 Reform Act; their irritation with their continued social and political inequality. The failure of the Whigs to repeal The Corn Laws. The ACLL was eminently a Manchester and middle-class movement, comprising industrialists, manufacturers, etc. together with humanitarians and some radicals. They believed that the landed classes were idle, feudal and immoral (John Bright); for them the battle was not simply about a customs duty but was about a challenge to the social and political order.

8　Motives: free trade and enlightened self-interest. Argued that, apart from economic benefits, free trade would bring international interdependence and peace. (The Chartists were sceptical and refused overtures to join.) The ACLL's strategy was for peaceful persuasion by the use of the press, by propaganda, by pamphlet; there was also the annual motion in parliament, the buying of freeholds, etc. The contribution of Cobden and Bright, the excellence of the central, well-funded organization in Manchester – but the movement was struggling after 1842.

9　The impact of the Irish famine: yet, the ACLL regarded the repeal of 1846 as *their* triumph; they felt that they had convinced Peel and the nation of the justice of their case. These claims were exaggerated, as were their predictions (and the protectionists' fears) about the results. Agriculture was not ruined, corn prices did not collapse, the aristocracy did not relinquish power; neither did mid-Victorian prosperity rest solely on Corn Law repeal and the alleged benefits of free trade.

18.3　COMMENTARY

A divided society

By the mid-forties England had attained industrial maturity, symbolized by the advent of the railways and proudly celebrated in the Great Exhibition of 1851. She had become the workshop of the world. Most Englishmen welcomed the age of improvement and gloried in the rapid and startling changes that occurred.

Others feared the consequences of the changes and the class, material and social divisions resulting from them.

In his novel *Sybil or the Two Nations*, published in 1845, Disraeli captured brilliantly the mood and contrasts of this new England, and the gulf that had emerged between the rich and poor. It is important that you understand why England had become such a divided society in the 1830s. Consider also in what ways the economic crisis of the late 'thirties quickened the pace of politics and encouraged the working and middle classes to express their discontent by forming two of the most important political pressure groups of the nineteenth century.

Chartism: the causes

Chartism, said Carlyle, was 'bitter discontent grown fierce and mad'. It was an attempt to remedy the social and economic evils of the 1830s by political methods and thereby bring about the regeneration of English society. It also expressed the desire of the working classes for social justice. The major demand of the Chartists was for parliamentary reform.

The six points of the People's Charter – adult manhood suffrage, vote by ballot, annual parliaments, equal electoral districts, payment for MPs and the abolition of property qualifications for MPs – already had a chequered radical history which stretched back to the 1760s. In this respect Chartism did not create, but gave expression to, a working-class movement that had been growing for many years. From its inception it was a volatile mixture of social and economic as well as political grievances.

According to Professor Briggs, Chartism was born out of 'hunger and anger'. It was a gut reaction to the economic de-

Fig. 11 Centres of Chartist activity

pression of the late 'thirties and an angry response to the Reform Act of 1832, to the collapse of Owenite Unionism and to the introduction of the New Poor Law. It encouraged every person or group with a grievance, mission or political demand to enrol. Its strength derived from its ability to encompass all the dissatisfactions of a discontented Britain.

Make a list of the different groups that belonged to the Chartist movement and be able to discuss the aims and objectives of each of the leaders.

Nature

The institutional origins of the People's Charter are to be found in the London Working Men's Association – to which most prominent Radicals of the day belonged. The document itself was drawn up in May, 1838, by William Lovett (with the advice of Francis Place) and, with Attwood's help, was launched nationally in 1839. Originally, therefore, Chartism was a movement dominated by the educated and politically experienced sector of the working class.

Gradually, however, as the depression deepened and the unemployed joined in greater numbers, the rights of labour were added to their original demands for the rights of man. Feargus O'Connor did most to emphasize this new development as he stressed the proletarian origins of the mass of the Chartists, their 'fustian jackets, blistered hands and unshorn chins'. To him Chartism was fundamentally an expression of the struggle between the 'the haves and have nots'.

Local diversity

It is very difficult to understand Chartism without a full understanding of its regional and local diversity. According to Professor Briggs, Chartism was strongest where economic hardship was greatest, *i.e.* in the decaying centres of the older industries or in new single-industry towns like Stockport. It was predominantly an urban and industrial phenomenon.

The impact in agricultural areas was feeble. The movement included superior craftsmen such as carpenters and printers, as well as factory operatives, handloom weavers, miners and nailmakers. There was, then, widespread, if locally concentrated, support.

It is important to note that each area had its own grievances, its own leaders and its own priorities. In Manchester it was an expression of class conflict between cotton masters and men. Across the Pennines in Leeds it was galvanized by O'Connor into an aggressive physical-force movement for all those handloom weavers displaced by machinery. In Birmingham, a centre of small workshops, it was, according to Engels, more radical than Chartist. In Tyneside it had strong support from the aristocracy of labour – the better-off artisans.

National unity, then, was more apparent than real. Some modern historians argue that it is more accurate ot talk about 'Chartists' rather than 'Chartism.'

Find out what you can about the conflicting aims and personal rivalries of the Chartist leaders, and the local nature and differing demands of their supporters.

Events

It was Feargus O'Connor who gave the movement whatever appearance of national unity it possessed. His fiery oratory, perfected in the anti-Poor Law movement, backed by his radical newspaper *The Northern Star*, won him the support of the handloom weavers of the West Riding. In 1838 he made his bid for leadership. His appeal to physical force thrilled his discontented followers but alienated the more moderate moral-force leaders such as Lovett and Attwood.

At the People's Convention (an ominous title to those who remembered the French Revolution), held in London in 1839, a compromise solution to this difference of approach was agreed in the slogan 'peaceably if we may, forcibly if we must'. All the plans of the first phase failed – the petition, the 'national holiday', Frost's Newport rising. The Convention eventually broke up late in 1839 in disarray; most of the leaders were soon imprisoned.

A new organization – the National Charter Association – was formed in 1841 but more disappointments followed in 1842 with the rejection of the second petition and the failure of the 'Plug-Plot' strikes in Lancashire and Staffordshire. Chartism was visibly weakening. Undaunted, O'Connor founded a Chartist Co-operative Land Society in 1843 but by 1848 this was also in difficulty and was wound up in 1851.

You should make notes on the major events (listed above) and be able to explain their importance in the history of the movement.

1848

Inspired by the recent revolutions in Europe, the Chartists decided, in April, 1848, to make a final effort. It was agreed to hold a monster meeting in London at Kennington Common. The proposed march to parliament was banned by a determined government; in the Commons a third petition was derisively rejected. Effectively this marked the end of the movement, which was dispersed rather than destroyed. It struggled on into the 1850s. Some aberrations, like Christian Chartism and Teetotal Chartism, continued to be popular but the great days of Chartism – 'the revolution that never was' – ended in the year of revolutions!

Failure

Clearly then, Chartism failed. This is not surprising as the Chartists set themselves an almost impossible task. Their main problem was how to attain a revolutionary goal by constitutional means. Essentially, the movement failed to obtain middle-class (or parliamentary) support for the Six Points.

Satisfied by the extension of the franchise in 1832, or active in the Anti-Corn League, the middle classes either ignored, shunned or condemned Chartism. The leadership itself was divided and often confused in its aims. Most projects, including the petitions, were poorly organized. O'Connor may have given Chartism what unity it possessed but his egocentric attitude, his bitter denunciations of alternative approaches (such as Temperance Chartism and the Complete Suffrage Union) and his physical-force speeches alienated many sections of the movement. His calculated revolutionary bluff (the trump card of the masses) failed to move a government that was determined to stand firm.

The government used troops, the railways and the new telegraph system to contain early Chartist meetings and marches; and in General Napier they had made a masterly choice to control events in the North. Secure in the knowledge of the undivided support of English property-owners, the government handled the movement intelligently and effectively

Moreover, because Chartism was heavily dependent on economic slump for its vital driving force, the return to prosperity under Peel fatally weakened it. The irrelevance of some sources of Chartism (such as the Land Plan), and the elimination of others, all contributed to its decline. In this respect the Kennington Park rally and the rejection of the petition in 1848 were in many ways an anti-climax after years of marching, drilling, demonstrating and talking about physical force.

Chartism, then, had major weaknesses – as Kitson Clark maintains – but these were not the fundamental causes of its failure. Against a strong government and the middle classes – both totally determined to preserve law and order and thereby save England from revolution – Chartism, well or badly led, was bound to fail.

Significance

The movement was not, however, without significance. It gave a severe shock to the established order and made clear to the government and to the middle class the extent of the 'condition of England' question. It improved the morale of the working class, and provided excitement and a sense of community. It showed the more intelligent leaders how, and how not, to agitate and indicated the importance of middle-class support.

If one sees Chartism primarily as an *economic* movement, then Peel's efforts to 'make England a cheap country for living' and the return to prosperity in the 'forties constituted a kind of Chartist success. In this respect one could agree with John Stuart Mill's verdict that 'Chartism was the victory of the vanquished'.

The Anti-Corn Law League

The problem

Most of the newly-enfranchised middle classes were satisfied with the Reform Act of 1832 – though some were still unhappy at their social inferiority and the continued domination of parliament by the landed aristocracy. The Corn Laws still rankled but while England was prosperous, up to 1836, resentment by the middle classes was held in check and their energies were concentrated on schemes such as currency and education reform.

When trade collapsed in 1836 and their manufacturing profits were threatened, they turned their attention once more to the campaign for free trade. If they could get rid of the Corn Laws and the remaining tariffs, the prospect was an endlessly expanding market for their goods. Cobden argued persuasively that Corn Law repeal and free trade would give manufacturers more outlets for their production, tackle the 'condition of England' question by expanding employment and lowering the price of bread, make British agriculture more efficient and productive by exposing it to foreign competition and, perhaps most

important of all, promote international peace through contact and trade. Bright, meanwhile, broadened the basis of support by introducing the moral element. He stigmatized the landlords as parasites collecting unearned rents and oppressing the agricultural labourers with Game Laws and other remnants of feudal power. Supporters of the League were largely, but not exclusively, the manufacturing and trading classes but were joined by Dissenters and humanitarians and all those who believed that cheap bread was a fundamental need. The Chartists remained suspicious.

Economic or political?

The most significant aspect of this efficiently organized and brilliantly led campaign, however, was psychological. The Corn Laws were the symbol of the privileged position of the old aristocratic order. Destroy the Corn Laws and the landed order would also collapse. Success would bring victory for their interests and their class. For this reason, Morley said, 'it was not a battle about a customs duty but a struggle for political influence and social equality'. 'Repeal,' Croker said, 'would bring the overthrow of the existing social and political system.'

Find out what you can about the ways and means adopted by the League in its efforts to gain the repeal. You should also discover some biographical facts about Cobden and Bright and be able to assess their contribution to the movement.

Repeal

Masterly as their campaign was, it did not effect repeal and had, in fact, begun to falter after 1842 as harvests improved and bread prices fell. Encouraged by these events and stung by ACLL abuse, the Protectionists also began to fight back. Despite some ACLL support in parliament (including Charles Villiers' annual motion for reform), the ruling order stood fast under Peel. Unlike the Chartists, spokesmen of the ACLL made it clear that their constitutional agitation had to succeed by persuasion or not at all. Yet by 1845 it was clear that they had failed to win over the Commons.

However, as N. McCord says, although 'the legend of the

Price of Wheat per quarter 1811–1864 (yearly average)

	s.	d.		s.	d.		s.	d.
1811	95	3	1829	66	3	1847	69	5
1812	126	6	1830	64	3	1848	50	6
1813	109	9	1831	66	4	1849	44	6
1814	74	4	1832	58	8	1850	40	4
1815	65	7	1833	52	11	1851	38	7
1816	78	6	1834	46	2	1852	41	0
1817	96	11	1835	39	4	1853	53	3
1818	86	3	1836	48	9	1854	72	7
1819	74	6	1837	55	10	1855	74	9
1820	67	10	1838	64	4	1856	69	2
1821	56	1	1839	70	6	1857	56	5
1822	44	7	1840	66	4	1858	44	3
1823	53	4	1841	64	5	1859	43	10
1824	63	11	1842	57	5	1860	53	4
1825	68	6	1843	50	2	1861	55	5
1826	58	8	1844	51	3	1862	55	6
1827	58	6	1845	50	9	1863	44	9
1828	60	5	1846	54	9	1864	40	5

Fig. 12 Table showing the average price of wheat per quarter between 1811–64

success of the ACLL dies hard' it was 'Peel in parliament' that repealed the Corn Laws. He did it in direct response to the Irish famine and furthermore insisted that he did it in the national, as opposed to middle-class, interest.

The propaganda of the League, however, provided sound arguments and had prepared society for the acceptance of repeal. Russell declared the Whigs in favour in 1845 (the Edinburgh letter) and by that date Peel had already been convinced by Cobden's arguments. All that was needed was an occasion and that was provided by the Irish famine. Cobden generously acknowledged Peel's role, while an elated Bright declared that 'the middle class have effectively vindicated their right to rule.'

Effects

The possible effects of repeal had been exaggerated by both Protectionists and Free Traders. English farming did not collapse. Rationalised under the threat of foreign competition, it made use of new technology and entered a golden era of 'high farming' and prosperity. Prices were stable in the 1850s (wheat averaging 52s. a qr.).

The real effects came in the 1870s after the revolution in sea transport and the opening of the vast American prairies, which completely undermined British ability to compete in corn growing. The Protectionists had abandoned their opposition to Free trade long before – their great spokesman Disraeli having declared in 1852: 'Protection is not only dead but damned.'

18.4 CHRONOLOGY

1836	**June**	LONDON WORKINGMEN'S ASSOCIATION founded by William Lovett.
1837	**May**	Birmingham Political Union reorganized; now publicizing an ultra-radical political programme initiated by Thomas Attwood.
	November	NORTHERN STAR published, with Feargus O'Connor as its owner; soon selling 50 000 copies per week.
1838	**May**	THE PEOPLE'S CHARTER PUBLISHED.
	September	MANCHESTER ANTI-CORN LAW ASSOCIATION FOUNDED.
1839	**February**	CHARTIST CONVENTION in London.
	March	ANTI-CORN LAW LEAGUE FOUNDED: led by Richard Cobden and John Bright. ANTI-CORN LAW CIRCULAR published: 15 000 circulation.
	May	FIRST CHARTIST PETITION: 1.28 million signatures; rejected by 235 votes to 46.
	November	FROST'S NEWPORT RISING: Chartists from the valley marched on Newport; they were scattered, with casualties, by troops. The leaders were transported to Australia.
1840	**July**	NATIONAL CHARTER ASSOCIATION founded.
1841	**September**	Richard Cobden was elected MP for Stockport.
1842	**January**	COMPLETE SUFFRAGE UNION founded in Birmingham by Joseph Sturge.
	May	SECOND CHARTIST PETITION: contained 3.3 million signatures; rejected by the House of Commons – 287 votes to 49.
	August	PLUG-PLOT STRIKES in Lancashire and Staffordshire. Five hundred Anti-Corn Law League delegates marched to Westminster to lobby MPs.
1843		THE LEAGUE: ACLL newspaper published.
1844	**August**	Cobden and O'Connor debate the Corn Laws at Northampton.
1845	**April**	CHARTIST LAND PLAN: a co-operative venture. Units of four acres distributed; three estates established before the Company crashed in 1848.
	October	Partial failure of the Irish potato crop.
	November	'Edinburgh Letter': Russell accepted the need to repeal the Corn Laws.
1846	**June**	CORN LAW REPEAL. Repeal was to be effected over three years. Farmers were to receive some government aid as a palliative.
1847	**July**	Feargus O'Connor elected MP for Nottingham.
1848	**March**	Riots in London, Glasgow and Manchester prompted by rising unemployment and the European revolutions.
	April	Chartist Convention in London.
	10 April	CHARTIST MEETING AT KENNINGTON COMMON. O'Connor was persuaded to abandon the march; the government enrolled 150 000 special constables. The Third Chartist Petition (2 million signatures, many forgeries) rejected.
	May	Chartist National Assembly. Chartist riots in Northern England.
1851	**May**	The Great Exhibition (Crystal Palace) opened in Hyde Park.
	August	Chartist Land Company wound up.
1855	**August**	Death of Feargus O'Connor; he was buried at Kensal Green Cemetery, London. An estimated 50 000 people attended his funeral.
1858	**February**	Chartist Conference agreed to cooperate with moderate Radicals.

18.5 QUESTION PRACTICE

1 How far was the Anti-Corn Law League responsible for the repeal of the Corn Laws?

Understanding the question

It is very important that a question like this, which demands analysis and discussion, should *not* be answered by a simple narrative account of the League. Much to be preferred is a response carefully confined to an assessment of the League's responsibility.

Knowledge required

The working of the political system in early Victorian Britain. The nature and activities of the Anti-Corn Law League. An appreciation of other forces which contributed to the repeal of the Corn Laws.

Suggested essay plan

1 The Anti-Corn Law League was one of the most famous political pressure groups in British history. The contribution of Cobden and Bright is well known; the contrast between the failure of the Chartist movement and the success of the League is often noted.

2 It is natural to assume a clear connection between the work of the League and repeal; the organization and expenditure involved in the League's campaigns was scarcely to be seen in any earlier political pressure group. After a shaky start (1838–41), it quickly established itself on a footing of continuous activity – with a headquarters' organization, regionally distributed agents, a propaganda machine and formidable public speakers; all contributed to an impression of power. Moreover, after 1846 the Leaguers were keen to emphasize their role in bringing about repeal. Note Peel's remarkable tribute to Cobden in 1846, attaching to him the principal responsibility for the coming of free trade in corn.

3 It is, however, possible to suggest that these evaluations of the League's achievements are exaggerated. Consideration of the post-1832 system will suggest difficulty for radical agitation to grasp real power. The Corn Laws could only be repealed by an Act of parliament, and it is not clear that the League was ever in a position to control the legislature. In practice the electoral system was still dominated by aristocratic Whigs and Conservatives. It is difficult, perhaps impossible, to measure the effects of the League's propaganda campaign: property and influence remained of crucial importance in determining political power, and political agitation never succeeded in shifting that balance of power to any great extent. The House of Commons only had a small minority of MPs closely connected with the League and, in comparison the serried ranks of landed Whigs and Conservatives, were formidable.

4 Moreover, other factors unconnected with the work of the League were moving in the direction of repeal; economists advocated free trade, and the political establishment had already begun the process before the League was founded. The tariff system had already been considerably reduced, including modifications of the Corn Laws in 1823. Although the Conservative victory in 1841 owed something to the defence of those laws, Peel revised the Corn Law in 1842, reducing the level of protection; his budgets contained many tariff reductions. By 1846 the Corn Laws stood out as a survival of the policy of protection which in other areas had been substantially discarded. The success of Peel's policy of free trade (1842–5) in restoring prosperity provided a cogent argument for going further. Peel himself had privately decided against the Corn Laws long before the League's final agitation. The Irish famine provided him with an opportunity to take a step he had been contemplating for several years.

5 Of course, the League's work certainly had some success: its greatest achievement was to hasten the conversion of the Whig opposition, which meant that by late 1845 the only possible alternative government was committed to repeal. Whig votes were an important component in the Commons majority which voted for repeal in 1845, enabling Peel to overcome dissentients in his own Party on the issue. The threat of continued, even heightened, agitation by the League may have persuaded the House of Lords to accept repeal. The League, however, played no part in drafting the term of the repeal measure – it played a supporting rather than a leading role in 1846.

6 Conclusion: the primary responsibility for the repeal of the Corn Laws lay with Sir Robert Peel and his government. The League played a part in preparing public opinion and in converting the Whigs but, despite all the noise which they made, the Leaguers were never in a position to exert decisive political influence.

2 'Social and economic distress gave rise to Chartism, the advent of prosperity killed it'. Discuss.

Understanding the question

This calls for a discussion of how far Chartism can be seen simply as a response to social conditions, with hard times producing popular pressure for radical reform while easier conditions led to political apathy or acceptance of the existing order.

Knowledge required

The nature and history of the Chartist Movement. The shifts in economic conditions during and after the main period of Chartist activity.

Suggested essay plan

1 The coincidence in the timing of economic depression and the peaks of Chartist activity is sufficiently obvious to demonstrate the existence of an important connection between the two. Many areas suffered during the depression which became obvious in 1837 and deepened. These years saw the growth of Chartism into a widespread movement – the first Chartist convention, the first petition and the extremism involved in the Newport Rising of 1839 and other smaller outbreaks. The 'Plug Plot', sparked off by wage reductions resulting from the depression, led to Chartist involvement in widespread strikes; hope of transforming them into a national strike for a People's Charter. The last great flare-up of Chartism, culminating in the Kennington Common fiasco, also owed much to economic troubles in that year. In contrast, more prosperous years, the mid-40s, did not see such strenuous mass support for the Charter. It was not surprising, then, that Chartism was often seen as a 'knife-and-fork' question, or that subsequently historians saw the movement as a barometer of early Victorian economic and social troubles.

2 Although there was a close connection between Chartism and distress, things were not as simple as the link might suggest; it is important to remember that Chartism was an episode in the lives of many people, rather than the full story of those lives. Many Chartists were active reformers long before the Charter was drafted and many remained active reformers long after 1848. The connection with immediate economic and social conditions does much to explain the extent and fluctuations of mass support for Chartism but cannot provide a complete picture of the nature of the movement.

3 None of the six points was a *new* radical demand: most of them were advocated in the later eighteenth century, and indeed earlier. The concept of universal manhood suffrage had a long history; the secret ballot was also not a new idea, having been

advocated in the earlier 1830s by Radicals like MP George Grote, who had introduced in parliament several motions in its support. Also, the committee which drafted the 1832 Reform Bill gave the ballot serious consideration. The Charter, therefore, brought together an important group of long-standing radical proposals and welded them into a single programme. This was hardly a response to immediate social conditions.

4   Similarly, many of those who rallied to the Chartist cause had already been active in earlier political campaigning – including the struggle over parliamentary reform in 1831–2, the campaign for factory reform and the protests against the new Poor Law. For many, Chartism was only one phase in a long record of radical political activity. It was not, then, the case that all Chartists were motivated by the immediate pressures of the economic depression.

5   This concept of Chartism cannot do justice to the considerable intellectual and ideological elements within the movement. The original members of the LMWA who played a key role in the drafting and launching of the Charter were not men suffering distress but men convinced of the need for fundamental change on rational grounds. Other prominent Chartists, *e.g.* George Julian Harney, were democrats on intellectual rather than immediate 'bread-and-butter' considerations.

6   Although Chartism as a mass movement disappeared as economic prosperity grew, it is too much to say that prosperity killed it, for not all Chartists abandoned their cause. In many parts of the country small groups of dedicated Chartists held together for many years after the failure of 1848; some played a part in the parliamentary reform campaigns of the 1860s. What prosperity did was to eliminate *mass* popular support, which in the late 1830s and 1840s had intermittently enlisted under the Chartist banner. A hard core of dedicated radicals remained faithful after 1848 but were never able to recruit such an extensive following as distress and suffering had brought to the Chartist cause during the worst depressions of the early Victorian years.

7   Conclusion: the quotation in the question contains a good deal of truth but is less satisfactory as a *total* explanation of the rise and decline of Chartism.

3   Did the Chartist Movement have any chances of achieving its aims?

Understanding the question

It is sufficiently obvious that the Chartists did not succeed in placing their six points on the statute book. The question requires discussion of whether the movement ever could have been victorious, given the context in which it had to operate.

Knowledge required

An appreciation of early Victorian society and the distribution of power within it. The nature and the resources of the Chartist movement and the difficulties facing it.

Suggested essay plan

1   At first sight it may seem obvious that a movement which could enlist such mass support must have had some chance of success. At the peaks of its activity Chartism could claim millions of supporters, and the dedication of many cannot be questioned.

2   However, an examination of the nature of society in which Chartism operated casts doubt upon this first impression. In the early Victorian years most people lived in small scattered rural communities, which presented any radical crusade with serious problems of organization and communication. Not until 1851 did the census show a bare majority living in towns, many of these places being quite small; though industrial and urban society was growing fast, Britain was still very far from being a predominantly industrialized and urban society. In 1851 the biggest single occupational groups were still agricultural labourers and domestic servants, both being groups which were very difficult to enlist in radical political activities.

3   The majority of Chartists were poor, some desperately so; many had little or no education. While this could provide material for mass demonstrations in hard times, it did not provide a promising foundation for sustained organization. Mass support fluctuated according to social conditions, eager to protest in bad years but likely to drift away when better times came.

4   Democracy was now an accepted doctrine; it is important to realize how drastic the demands of the Charter were in the Britain of 1840. Society was profoundly unequal at that time, and always had been. The seemingly modest provisions of the 1832 Reform Act appeared startlingly sweeping at the time. How much more radical, then, was the People's Charter, which demanded a fundamental change in the organization of British society and its ruling elite.

5   Sometimes revolutionary change can be introduced because of the weakness or lack of nerve of those in power at the time but there was never any chance of this in early Victorian Britain, where the aristocratic ruling élite possessed a serene confidence in its right to rule. Britain's rulers occupied strong positions in the economic, political and social life of the country and Chartism was never able to shake this predominance. The armed forces remained loyal and could be used to restrain or repulse Chartist pressures. The Whig government during the years 1838–40 was confident enough to adopt a moderate and restrained line of resistance to Chartism, avoiding executions and, thereby, martyrs. When alarmed by Chartist activities, they were provoked into displays of strength rather than thoughts of surrender. The House of Commons was confident enough to reject repeated Chartist petitions, with many MPs not even bothering to attend the debates.

6   Moreover the Whig reforms of the earlier 1830s – the Reform Act of 1832, the Poor Law and the Municipal Corporation Act – reinforced support for the existing order by conciliating many men of influence previously opposed to the unreformed system. The leaders in the reform agitation were by 1839 councillors, aldermen, mayors or Poor Law guardians, possessing the vote and unwilling to see their privileges diluted by the coming of democracy under Chartist auspices. Instead of joining the Chartists these men were now inclined to join or lead the special constables, or issue anti-Chartist proclamations from the reformed town councils.

7   When to these factors we add the disunity always present in Chartist ranks – *e.g.* the various sub-divisions such as 'moral-force' or 'physical-force' Chartism, Christian Chartists and Temperance Chartists – we see it to be an umbrella movement covering a wide variety of differing local groups of reformers. So we may well conclude that Chartism had little real chance of success. And yet, it is well to remember that some successful revolutions have been the work of small groups of dedicated men. Such successes, however, do not occur against resistance so strongly entrenched as that of the aristocratic rulers of early Victorian Britain.

4   Document question (Thomas Attwood introduces the Charter): study the extract below and answer the questions (a) to (g) which follow:

'MR. ATTWOOD was thankful for the indulgence extended to him, and would only trespass a few minutes longer upon the attention of the House. But he wished to say a few words in explanation of his own peculiar situation. Although he most cordially supported the petition, was ready to support every word contained in it, and was determined to use every means in his power in order to carry it out into a law, he must say, that many reports had gone abroad, in regard to arguments said to have been used in support of the petition on different occasions, which he distinctly disavowed. He never, in the whole course of his life, recommended any means, or inculcated any doctrine except peace, law, order, loyalty, and union, and always in good faith, not holding one face out of doors, and another in that House; but always in the same manner, and in the same feeling, fairly and openly doing all that he could as a man, a patriot, and a Christian, to work out the principles which he maintained; and to support the views of the petitioners. He washed his hands of any idea, of any appeal to physical force. He deprecated all such notions – he repudiated all talk of arms – he wished for no arms but the will of the people, legally, fairly, and constitutionally expressed–and if the people would only adopt his views, and respond to his voice– if they would send up similar petitions from every parish in England, and go on using every argument which justice, reason, and wisdom dictate, they would create such an action in the public mind, which would again act upon Members of that House– that giving due allowance for the prevalence of generous feeling among English gentlemen and the English people, if the people would act in that manner – if they proceeded wisely and discreetly, washing their hands of all insolence and violence– he was confident they would ultimately secure the attentive consideration of that House. Having said so much, he should now read the prayer of the petition . . . .'

(Hansard 3/XVLIII, 222)

Maximum marks

(a) Where was Attwood's political base and what cause first brought him into politics? (2)
(b) In what year was the speech made? (2)
(c) Comment on the background and nature of the petition. (4)
(d) Attwood rejected physical force. Who was the leader of physical-force Chartism? Outline (briefly) his contribution to the movement. (5)
(e) Did the Chartists subsequently proceed 'wisely and discreetly'? (5)
(f) What, in essence, was 'the prayer of the petition'? (3)
(g) Why did Chartism fail to 'ultimately secure the attentive consideration of that House'? (6)

18.6 READING LIST

Standard textbook reading

A. Briggs, *Age of Improvement* (Longmans, 1959), chapter 6.

R. K. Webb, *Modern England* (Allen and Unwin, 1980 edn.), chapters 5, 6.

A. Wood, *Nineteenth-Century Britain* (Longmans, 1960), chapter 8.

Further suggested reading

A. Briggs (ed.), *Chartist Studies* (Macmillan, 1959).

G. D. H. Cole, *Chartist Portraits* (Macmillan, 1941).

P. Hollis (ed.), *Pressure from Without in Early Victorian England* (Arnold, 1974).

D. Jones, *Chartism and the Chartists* (Lane, 1975).

N. McCord, *The Anti-Corn Law League* (Allen and Unwin, 1958).

D. Read, *Cobden and Bright* (Arnold, 1967).

E. Royle, *Chartism* (Longmans, 1980).

19 Peel and the Conservative Party

19.1 INTRODUCTION

Peel was a great, if not the greatest, Prime Minister of the nineteenth century. He was also one of the most capable Victorian statesmen. He dominated early Victorian England more completely than Gladstone or Disraeli dominated the later years. He reconstructed the Tory Party after the Reform Act of 1832 but was always inclined to put nation before party in matters of policy.

He demonstrated his leadership by helping to modernize Britain during his 1841–46 administration but then split the party he had built by repealing the Corn Laws in 1846. Ironically, this enabled his opponent, Disraeli, to become leader of the Tories and left his ablest lieutenant, Gladstone, without a party.

19.2 STUDY OBJECTIVES

1 Peel's family background, personality, and upbringing; a shy, withdrawn boy who got his own back on the world by coming top in examinations ('double First' at Oxford).

2 Peel's success as Irish Secretary; his businesslike approach to politics and methods of dealing with Irish discontent (Ribbonism, etc).

3 His role as chief 'protestant' politician and the development of a 'Peelite' group within the Liverpool party.

4 Peel's role in the resumption of cash payments (and his later determination, culminating in the 1844 Bank Act) to safeguard that settlement. His absorption of political economy and cautious move towards an income tax (finally realized in 1842) and freer trade.

5 Peel's penal reforms as Home Secretary (not original, note the influence of Bentham, Romilly and Mackintosh) and the Metropolitan Police; more 'liberal' than Sidmouth and Eldon, but less so than Russell and Brougham.

6 Wellington's government of 1828–30 and events leading to Catholic emancipation; Ultra-Tory revolt against Peel, and his basic disgust with the Tory Party thereafter; quit Oxford seat.

7 Peel's opposition to the 1832 Reform Act (he was unwilling to open a door which would not be easily closed) but refusal to form an anti-reform ministry made the passage of reform inevitable.

8 Peel's 1834–5 administration and the Tamworth Manifesto led to a resurgence of the Tory Party (or the new Conservative Party). Its principle was attachment to the Crown, Church and Aristocracy, coupled with cautious, non-organic, reforms.

9 As the leader of the opposition–he was moderate, often sustaining the Whig government against its own radical supporters–Peel built up a reputation for supreme honesty and statesmanship, above petty party squabbles.

10 Extra-parliamentary speeches (mainly to respectable artisans and manufacturers) and a surge of Tory support in the counties and the boroughs led to a national swing and victory in the 1841 election. Peel's remarks on corn were ambiguous, trying to distance himself from supporting protection – but protection, nevertheless, was instrumental in bringing the Tories to success.

11 Though he recognized his debt to the Party, Peel was unwilling to allow it to dictate policy. He regarded himself as a minister of the Crown whose duty it was to govern in the national interest. The main problem was finance; he quickly demonstrated his administrative superiority in the field by the brilliance of his budgets.

12 Peel's radical financial policies and decision to reimpose income tax meant a review of the tariff structure, which in turn brought up consideration of the Corn Laws. These were modified in 1842; protectionist MPs made their displeasure felt. Farmers founded the Anti-League and Central Agricultural Protection Society in 1844. The split between Peel and the Party widened in 1845 over the Maynooth grant.

13 The social condition of England in the 1840s (*e.g.* the suffering in Paisley and Stockport horrified Peel); his policies to relieve distress (direct for indirect taxation) aimed to do this through increased prosperity; this culminated in the Corn Law repeal, for which the Irish Famine was probably used as a mere pretext. He rejected the back-to-the-land paternalism of the Young England and Ten Hours movements because of the effects it would have on productivity.

14 The political condition of England in the 1840s (Chartism and the Anti-corn Law League spearheading working- and middle-class attacks – albeit separate – on the aristocracy). The Corn Law repeal was partly to appease the League, which might otherwise have swept the board at the next election; and partly to persuade the Chartists that not all statesmen were indifferent to their discontents (*i.e.* it was a symbolic gesture); but the repeal was primarily to save the aristocracy from its mistaken, and to Peel suicidal, attachment to a Corn Law which was ineffectual and, in the long run, untenable. For him repeal was essentially a preserving measure.

15 Out of office, Peel was still regarded as the greatest statesman of the day; he was the darling of the mercantile, manufacturing and professional classes of the North, and was popular with the workers because of cheap bread and national economic recovery. The failure of Chartism vindicated the work of his ministry.

16 Peel's posthumous political influence; the slow peregrination of his closest disciples (Oxford high-Church 'Peelites') into the Liberal Party, together with his ideals of little England, free trade, and self-help individualism. An alternative view is that Peelism went into the development of Conservatism.

19.3 COMMENTARY

Peel's background

By the time of the Great Exhibition in 1851 the world of *Peterloo* and the Six Acts seemed only an uncomfortable memory. This was largely due to the successful policies of Robert Peel. *The Times*, in Peel's obituary in 1850, remarked: 'Peel has been our chief guide from the confusions and darkness that hung around the beginning of the century to the comparatively quiet haven in which we are now embayed'.

Though he started his career as a Tory, he ended as a liberal-Conservative and became a good guide primarily because of his ability to adapt to changing circumstances and changing needs. He was the son of a successful and enlightened cotton manufacturer who left him a wealthy inheritance and an understanding of the outlook and standards of the business class. By education (Harrow and Oxford; he was the first undergraduate to receive a double First in Classics and Mathematics), and temperament he became associated with the ruling class. It was to prove an uneasy relationship.

Cash and reform, 1819–27

Entering Parliament in 1809, Peel was, from 1812 to 1818, the Chief Secretary for Ireland where he established a reputation for firm but honest and impartial administration. He became MP for Oxford University in 1817 and, in 1819, was chairman of the 'Bullion Committee', which advocated a return to the gold standard – a move to check inflation and put the currency back on a sound footing.

This deflationary measure, however, was bound to upset the landed classes who, on the strength of high prices, had during the war years invested large sums at high mortgage repayment rates in agricultural improvements. Attwood, in 1820, argued that 'Peel's Act' and the ending of cheap money was at the root of the economic difficulties of farmers in the 1820s. From 1822 Peel, as Home Secretary, was busy with a series of legal and criminal reforms. In 1829 he crowned this work by founding the London Metropolitan Police Force.

Catholic emancipation

The problem which concerned Peel in the 1820s, however, was Catholic emancipation, which he firmly resisted until 1829. Events in Ireland, organized by O'Connell and the Catholic Association, left Peel with the choice of two evils in 1828 – emancipation or rebellion. Peel wished to resign, but Wellington persuaded him to stay on to carry the Bill.

As the MP for Oxford, a bastion of the High Church, he was immediately branded a traitor and apostate by angry, resentful Anglicans and Protestants. In the circumstances it is not surprising that he opposed the Reform Bill and his strong defence of the established Church after 1832 was partly an attempt to reinstate himself with the Party. His break with Wellington during the reform crisis gave him more political independence. Hence it was no surprise when, after Melbourne's dismissal, Peel was called back from Italy by the King to form his first government in December 1834. Look carefully at Peel's decisions on 'cash and Catholics' and try to discover why these issues were so important to the Tories. Do you consider that Peel betrayed the Party on these issues?

The Tamworth Manifesto, 1834

Although Peel lost the election he gave the Tory Party what it desperately needed in the 1830s – leadership and direction. In the Tamworth Manifesto (which N. Gash says 'caused a prodigious sensation') he outlined his plans and policy 'for the Party which from now on will be called Conservative.'

In seven pages of print he stated his policy – a policy designed to appeal to all moderate men interested in maintaining order and good (*i.e.* non-partisan) government. He accepted the 1832 Reform Act and the possibility of future moderate reform of proven abuses. Peel was trying to turn the Tory Party of one particular class into the Conservative Party of the nation. He gradually built up Party strength in the 'thirties.

In the interests of good government, he refrained from factious opposition. He supported the Whigs when he believed their measures were just and for the national good.

When Peel became Prime Minister for the second time, in 1841, his control of the Party was absolute. During the 'thirties he had proved his ability and potential for leadership; yet he remained unpopular with at least half the Party. Many Tories did not like his middle-class background, his commercial ideas, his liberal outlook and his occasional attempts to put policy before the Party. His relationships, both Party and private, were often strained and uneasy (note the details of the 'Bedchamber Question' in 1839.) Many regarded him as cold and aloof – 'an iceberg with a slight thaw on the surface.'

Peel's second ministry, 1841–46

By 1841 the Whigs were in total disarray. Facing a deficit of

£2 500 000, they seemed utterly bereft of political enthusiasm or ideas. Peel won a clear majority of 80 seats at the general election and gathered a strong team around him – Wellington, Aberdeen, Stanley and, later, Gladstone. In an England in the depths of depression the suffering, and the concerned, proposed many different remedies such as Chartism, Corn Law repeal and co-operation.

Note that Peel made the surprising – and radical – decision that a bold policy of fiscal and financial reform was needed to resolve England's economic problems. 'We must,' he wrote to Croker, 'make this country a cheap country for living.'

Free trade budgets

Peel first turned his attention to the Corn Laws and revised the sliding scale of 1828. For prohibition, he now wanted to sub-stitute 'fair protection'. Corn imports were allowed below 73s. a quarter but, to the Tories, this was unpalatable medicine and precipitated a ripple of apprehension among Protectionist back-benchers. Cobden, on the other hand, saw it as a tinkering measure – 'a bitter insult to a suffering nation.'

It was to tariff revision, however, that Peel looked for rapid improvement in the country's economic position. By lowering tariffs (currently paid on over 1200 articles) he hoped to stimulate trade and manufacturing in Britain.

It is essential that you know the details of what came to be known as Peel's great 'free-trade' budgets of 1842 and 1845. Basically, a sliding scale of tariff duties and selective remissions was introduced during the ministry and, by 1847, over £8 000 000 in government customs and excise revenue had been remitted. In 1841 Peel re-introduced income tax to make good the current deficit left by the Whigs and then continued it to balance his future tariff concessions.

Despite the remission of £8 000 000 in tariffs, Peel's govern-ment was in surplus by 1846. This was a magnificent achievement, fully vindicating Peel's bold, fiscal approach. Admittedly, he was aided by a run of good harvests and a railway-building boom which encouraged investment, industry and employment. But it was an ambitious programme which deserved success.

To support these free-trade measures he introduced a Bank Charter Act in 1844. This completed the work he had begun in 1819, prepared the way for the supremacy of the Bank of England and restored confidence in the pound. Opinion is divided on the success of this Act but all agree that the Companies Act that followed (which enacted that all companies must now register and display an annual balance sheet), encouraged business and public confidence, and stimulated investment.

Make a careful study of Peel's fiscal and financial measures and then consider whether you would agree that Peel 'laid the foundations of mid-Victorian prosperity.' (Don't forget that you must first discover the economic basis of mid-Victorian pros-perity before you can decide on Peel's contribution).

Social reforms

Peel, once characterized as 'a businessman who brought business methods to government', is justly famous for his financial and administrative measures. He was a brilliant administrator and financier, rather than a social reformer.

As was shown by his monetarist approach to the 'condition of England' question when he took office in 1841, he did not believe that the state could solve problems by intervention. Yet his ministry was not devoid of social reform achieved by state intervention: a Mines Act was passed in 1842, a Factory Act in 1844, and – after Chadwick's investigations into the sanitary condition of the labouring poor had prompted the setting up of a Royal Commission – a Public Health Act under Russell.

The complicated and detailed administrative work of Peel's ministry was helped enormously by the 'administrative re-volution in government' which occurred in the 1830s and 1840s. During these years the foundation of the Civil Service was laid, more efficient procedures were devised to speed up parliamentary legislation and more dedicated administrators appointed to supervise all aspects of government work.

Ireland

Peel cleverly withstood O'Connell's attempts to repeal the Act of Union but his later Irish policy created apprehension in the Conservative Party. He aimed to persuade reasonable, middle-class Catholics to forsake O'Connell's repeal agitation and supp-ort the Union. In an attempt to bridge the sectarian gap between Protestants and Catholics, he founded three secular national colleges in Belfast, Cork and Galway, only to have them attacked by both denominations as 'godless'.

The greatest stir in England, however, was caused by his decision in 1845 to increase the annual grant to the Catholic Training Seminary at Maynooth from £9 000 to £26 000. Emancipation may have been granted but not, it appears, toleration. The Commons was inundated with hundreds of hostile petitions. Gladstone resigned, but Peel stubbornly forced through the Commons what he regarded as a just and beneficial measure – a statesmanlike but not very popular action.

The Irish famine

It was Ireland, in fact, that brought about Peel's downfall. As soon as the news of the Irish famine came through in 1845, Peel thought about repealing the Corn Laws. It was an idea that had been maturing in his mind for some time and derived from the inescapable logic of his financial and fiscal measures.

Although Wellington argued that 'it was potatoes that put Peel in his damned fright', it was not Ireland – according to Professor Gash – but the condition of England that provided the underlying motive for repeal, *i.e.* the success of his free trade budgets; the belief that agricultural prosperity depended on efficient techniques (not protection) and that continued middle-class antagonism was undermining the position of the landed aristocracy and undermining, therefore, the political stability of the nation.

Peel, then, was prepared to repeal the Corn Laws for what he believed to be the best interests of the whole nation. He seems to have underrated the strength of opposition among Tory back-benchers. Although a successful leader, he was not a popular one; many Tories had come to resent his liberal brand of Con-servatism. It is important that you understand the apprehensions of Tories during this period. Why did they consider the Corn Laws to be so important?

Corn Law repeal

Peel's Cabinet was divided. Russell's 'Edinburgh Letter', in which he declared for free trade, encouraged Peel to resign. Russell, however, was unable to form a government and 'passed the poisoned chalice back to Sir Robert'. Peel generously refused to leave the Queen in extremity and agreed to come back. The Party was irretrievably split. The rebels – led by Lord George Bentinck and, more important, in the Commons by Disraeli – attacked Peel with bitter and personally wounding invective.

With the aid of the Whigs and a third of the Conservatives, Peel forced the Bill through, only to be defeated the same night on a Coercion Bill for Ireland. Bentinck and Disraeli maintained that Peel had betrayed his Party on the three great issues of the day – cash, Catholics and corn.

Find out what you can about the motives of the Protectionists and pay particular attention to the under-rated contribution of

Bentinck. Would you agree that Peel could perhaps, be accused of desertion but not betrayal?

Kitson Clark maintains that Peel had simply put nation before Party in an effort to solve one of England's bitterest social issues. Furthermore, it could be argued that the terms of repeal demonstrated that he wished to be the surgeon, not the executioner, of the landed interests.

Repeal was to be spread over three years, accompanied by a £2 000 000 drainage loan, a revision of Poor Law rates and a cut in import duties on seeds and fertilizers. Peel wanted to encourage 'high farming' rather than attempt to guarantee 'high prices' by maintaining the Corn Laws. Above all, by repeal he felt that he was defusing the bitter anti-aristocratic feeling that permeated the middle classes. Finally, you should consider the argument that Peel, unlike the Protectionist Tories, believed that retaining the Corn Laws was more dangerous than repealing them.

Assessment

According to Professor Gash, Peel's lifework was to fashion a compromise between the system he inherited and the necessities of a new industrializing society. It is most important that you understand that this system did not posit two rival parties advocating alternative policies: it was designed primarily to enable the king's government to carry out national policy. Peel's primary aim then was to be a loyal minister of the Crown and to secure efficient, stable government. He was prepared to do whatever was necessary to achieve this.

The growth of the party system and its prejudices after 1832 was a dilemma for Peel because, as an old-fashioned royal minister, he favoured government by a centre group in the Commons – an alliance of moderates which would exclude the extremists of both parties. Note that it was a combination of Whigs and moderate Tories together that ensured that Catholic emancipation and Corn Law repeal were carried; Peel for his part supported moderate Whig reforms in the 1830s.

His Conservatism, therefore, was not a party label for eventually he was prepared to break the Party in the national interest. As he said in 1846: 'If I fall I shall have the satisfaction of reflecting that I do not fall because I have shown subservience to a party [and] preferred the interests of a party to the general interests of the community.' Peel put policy before Party and, by insisting on flexibility in the face of national problems, he helped to preserve the prevailing system (and aristocratic power) in a period of distress and political challenge.

Peel's relations with his Party were, therefore, perpetually strained. As its leader it was his task to convince the Party of his brand of Conservatism and the value of 'centrist' politics. He failed. It was his merit as a statesman that he was prepared to adapt his policies to the need for change. It was his defect as a party leader that he did so in a manner which seemed high-handed and autocratic.

It would be wrong, however, to think that his political career ended in failure. He split the Party but retained the loyalty of its ablest members – Gladstone, Graham, Cardwell, Aberdeen. His repeal of the Corn Laws gave him immense national popularity, his fiscal reforms solved the depression of the 'hungry forties' and the combination of the two probably saved England from revolution in 1848. The 'forties were, indeed, 'Peel's decade.' More than any other, he had shown himself to be 'the statesman of the industrial revolution.'

19.4 CHRONOLOGY

1788	**February**	Born; son of Sir Robert Peel, a Lancashire cotton spinner.
1812	**August**	Chief Secretary for Ireland until 1818; established the Royal Irish Constabulary.
1817	**June**	Became MP for Oxford University.
1819	**May**	Became CHAIRMAN OF THE 'BULLION COMMITTEE'; he recommended a return to the gold standard.
1822	**November**	Became HOME SECRETARY; modified the severity of the criminal code.
1827	**April**	Peel refused office under Canning (because he opposed Canning's policy of Catholic emancipation).
1828	**January**	As Home Secretary under Wellington, Peel founded the Metropolitan Police Force.
1829	**March**	Peel INTRODUCED CATHOLIC EMANCIPATION; the break-up of the old Tory Party.
1831	**March**	Peel opposed the Great Reform Bill during its two year passage through parliament.
1834	**December**	Peel's First Ministry; lasted until April, 1835. TAMWORTH MANIFESTO; Peel's address to his constituents at Tamworth; an electioneering document on a grand scale; it advocated 'a careful review of institutions, civil and ecclesiastical . . . the correction of proved abuses, and the redress of real grievances'.
1835	**April**	Peel was defeated, on the issue of Irish Church revenues, by an alliance of O'Connell and the Whigs – the 'Lichfield House Compact'.
1839	**May**	THE BEDCHAMBER QUESTION, Peel refused to take office on Melbourne's resignation unless Whig ladies of the Royal Household, in attendance on the Queen, resigned.
1841	**September**	Prime Minister; as PM he was 'masterful and autocratic'.
1842	**March**	FIRST FREE TRADE BUDGET; reduced the import duty on 700 articles. Income Tax of 7d in the pound was introduced to offset the fall in income.
	April	MODIFICATION OF THE CORN LAW further eased the restrictions on the entry of grain during periods of dearth. MINES ACT; women, and children under ten years of age, forbidden to work underground.
	July	BANK CHARTER ACT: this was the logical development of Peel's work on the 'Bullion Committee' of 1819. COMPANY ACT: companies now had to register with the Board of Trade and produce an annual balance sheet. This gave some public accountability and encouraged investment.

	August	RAILWAY ACT: instituted 'parliamentary train' – third class passengers to be carried at 1d per mile. This Act was basically a victory for *laissez-faire*. The railway lobby resisted attempts at further government control. FACTORY ACT: children aged eight to 13 to work a six-and-a-half-hour day. Maximum 12-hour working day for women. Dangerous machinery to be fenced.
1845	**February**	MAYNOOTH GRANT: despite strong opposition Peel increased the annual grant to this College, which trained Irish Catholic priests. Gladstone resigned. Queen's Colleges founded: Belfast, Cork and Galway added to the existing Anglican college – Trinity, Dublin; this was opposed by Anglicans and condemned as 'godless' because of the absence of religious teaching.
	October	IRISH FAMINE: outbreak of potato blight in autumn.
	November	'EDINBURGH LETTER'. Lord John Russell's letter to his constituents in London accepting the necessity for repeal of the Corn Laws.
	December	THE TIMES stated the government's intention to repeal the Corn Laws. Peel resigned; Russell failed to form a government.
1846	**June**	CORN LAW REPEAL. Conservative Party split. Government defeated on Coercion Bill for Ireland by a combination of Whigs, Irish MPs and Protectionists. Peel resigned.
1850	**2 July**	Peel died (from injuries sustained after a fall from his horse).

19.5 QUESTION PRACTICE

1 Examine the role of Peel in contemporary British politics.

Understanding the question

This is a searching question. In considering his political role you should understand the way Peel functioned as a politician, what he stood for in contemporary estimation, and what effect he had on the development of politics in his lifetime.

Knowledge required

All aspects of British political history in the first half of the nineteenth century.

Suggested essay plan

1   Peel is often credited with founding the Conservative Party (Gash, Blake), in the sense of a nationalist, pragmatic, unideological, executive Party, above sectional interest. More plausibly, he can be seen as the founder of the Liberal Party, into which most of his supporters (Gladstone, Cardwell, etc.) drifted, taking with them his 'little England' internationalism (see Aberdeen's foreign policy and hostility to Palmerston), free trade, self-help and individualist ethos. Whichever version is correct Peel's was a formative and creative role in politics.

2   Until 1828, Peel's role was to defend the protestant establishment, which made him the darling of the Ultra-Tories. The latter badly needed a political leader with talent.

3   After 1828 Peel never filled this sort of role, being not quite trusted. Many saw him in the role of weather vane, judging the correct moment to implement reversals of policy (see Bagehot, 'he was converted at the conversion of the average man'); others simply saw him in the role of traitor.

4   As Home Secretary Peel grew into the role of the supreme executive politician. He could boast that 'he never proposed any measure that he did not carry.' This was most obvious during 1841–6, when he dominated all departments and even introduced some of Goulburn's budgets. A tireless administrator, he commanded all aspects of government business.

5   After a brief negative role from 1830–3, Peel bounced back with the Tamworth Manifesto; this was vacuous enough – defending the virtues of the status quo and opposing extremism – but it inaugurated his image (emphasized in several extra-parliamentary speeches, mainly to master manufacturers and respectable working-men's organizations) as a champion of Crown, Church, and Aristocracy.

6   His 'responsible' behaviour as leader of the opposition in the late 1830s, sometimes even supporting the Whig ministers against their own supporters, contributed to his perhaps rather spurious image as a supremely honest politician.

7   His role in office was to solve the condition of-England question through radical finance; confirm England as a progressive industrial nation by promoting economic recovery; remove the dissension between the social classes by Corn Law repeal, thereby saving the landed aristocracy from being overthrown and England from revolution.

8   Despite Peel's aspiration to be an executive politician, Disraeli's judgement is correct: that Peel was the 'greatest parliamentarian that ever lived'. Though no orator, he was a superb debater, deploying arguments and stastics with aplomb, and always gave an impression of solid reasonableness which belied a really rather doctrinaire and stubborn personality.

9   The role he played with most conviction, however, was that of political martyr (1846–50). His opponents found this insufferable, but disciples were inspired.

2 What were the motives of those who brought about Peel's downfall?

Understanding the question

This question calls for an analysis of the political crisis of 1845–6. You must consider the motives of those individuals and groups that opposed Peel's actions in those years.

Knowledge required

Details of the events of 1845–6. The structure and beliefs of the Tory Party in parliament and the country.

Suggested essay plan

1   A narrative of the events of 1844–6; back-bench revolts against Peel on the sugar duties and the ten-hour day (1844); agitation over agricultural relief and Maynooth (1845); the Irish famine and the Corn Law repeal (1846); a vote on Irish coercion toppled the government.

2   The role of Russell and of the Whig opposition; did he deliberately hand 'a poisoned chalice' back to Peel in December

1845? Outline the reasons for supposing that Russell's attempts to form a ministry (frustrated by a quarrel between Grey and Palmerston) were genuine.

3   The motives of the Tory malcontents: their long-standing dislike of Peel's dear-money and deflationary policies; and their suspicion that he might betray them on corn, as he had on the Catholics. The Corn Laws were to them a guarantee of high rents and a vital symbol of status and power; both threatened by the Manchester manufacturer. Also landlords urged on by Richmond's Anti-League of mainly tenant farmers.

4   Peel would probably have survived, albeit with difficulty, but for the personal venom and persistence of Disraeli and Bentinck. Bentinck's motives easy to fathom: an honest, un-political man, he felt that he had been 'sold'. He also blamed Peel for having augmented the cares which led to the death of Canning in 1827; this added a personal bitterness to his criticism.

5   Disraeli's motives more ambiguous: what will clearly not do is to say that he was paying Peel out for not being given office in 1841. It was the convention for MPs to offer their services, as Disraeli had done then; moreover, Peel's government had no more consistent or loyal back-bench supporter than Disraeli throughout 1842–3. It was 1844 that marked the turning-point.

6   Probably Disraeli sincerely believed what he alleged – that Peel had been intellectually seduced by the utilitarians, econo-mists, sophists and calculators. Disraeli's 'Young England' (Smythe, Manners, etc.), which called for a return to organic, agrarian, paternalist, protectionism, struck a widespread response in Tory hearts ('the finest brute vote in creation') during 1844–6 – as the Ten Hours and Anti-Poor Law movements showed.

7   But it is obvious too that Disraeli deeply disliked Peel personally, and also that he felt snubbed by him, not in 1841, but subsequently. Peel openly despised the Party that 'spent their days in hunting and shooting and eating and drinking', and *never* took it into his confidence. Disraeli was a Party man before all things, and longed to participate in the great game of politics. Peel, with his executive mentality, quite starved Disraeli's political appetite, and reaped the consequences.

3   'The paradox is that on the issues of corn, cash and Catholics, Peel was at odds with his Party.' Discuss.

Understanding the question

Unquestionably Peel thwarted his supporters' wishes on these issues. You must decide whether he was 'at odds with' them all along, or whether he acted reluctantly and under the pressure of circumstances.

Knowledge required

Peel's policies with regard to the gold standard and the Corn Laws. Their economic implications for different sections of the community. The nature of the Catholic question and reasons for its political importance. Attitudes of the Tory backbenchers.

Suggested essay plan

1   The nature and background of the Catholic question: anti-Catholicism was the most heartfelt instinct of the English squire-archy since the seventeenth-century; even 'the immortal Mr Pitt' had offended on this issue, so it was Peel's staunch protestantism that mainly accounted for his swift rise to promin-ence in the Tory party – a rise otherwise inexplicable as he was a manufacturer's son who rarely courted support.

2   Peel's refusal to serve under the 'Catholic' Canning in 1827 reassured the Party. But Peel had also been Irish Secretary and

Home Secretary, and so had a pragmatic approach to law and order. Returning to office in 1828, he quickly perceived that the Catholic Association threatened the civil peace, and his reaction to O'Connell's victory in the Clare bye-election was swift and decisive. So on this issue, Peel was probably not 'at odds with' the Party, but saw the need to 'about-turn' in a crisis.

3   The currency issue: the decision to return to gold in 1819 ('Peel's Bill') led to an increase in the value of money and a decline in the price of primary and secondary produce. This benefited investors ('fundholders') and exporters, but damaged debtors and producers for the home market. Many agricultural Tories were heavily mortgaged (having expanded cultivation in war-time) and so suffered on both counts. By 1822 a hostile agri-cultural lobby had formed, and monetary discontent (Knatch-bull, Chandos) contributed to Wellington's and Peel's fall in 1830.

4   Peel, being devoted to sound money, was always at odds with most of the Party (see the March, 1833 division on Attwood's distress motion); the 1844 Bank Act, intended to 'safeguard' the gold standard, was even more deflationary.

5   Peel's intentions on corn were more uncertain. Theoreti-cally he favoured free trade; politically he wished to leave it alone. The requirements of government forced him gradually towards repeal. Especially he feared that protection was jeopardising food supplies, disrupting the currency, increasing manufacturing unemployment, and creating public strife (spearheaded by the Anti-Corn Law League).

6   Anyway, he thought agriculturists could survive repeal by capitalizing, and so did many back-benchers. The *real* pressure for protection came from farmers, who often forced their land-lords' hands. When they lined up behind Bentinck and Disraeli the protectionists made it clear that what they were against was Benthamism, Whiggery, Peelism and consensus politics. Hence Disraeli's peroration: 'Let men stand by the principle by which they rise, right or wrong...Above all maintain the line of demarcation between the parties...' The real charge against Peel was not free trade but 'betrayal'.

4   Document question (Peel and Conservatism): study Extracts I, II and III below and then answer questions (a) to (h) **which follow:**

Extract I

'To the electors of the Borough of Tamworth.
Gentlemen,
On the 26th of November last...I received from His Majesty a summons wholly unforeseen and unexpected by me...(to assist) His Majesty in the formation of a new Government....
'I never will admit that I have been, either before or after the Reform Bill, the defender of abuses or the enemy of judicious reforms. I appeal with confidence, in denial of the charge, to the active part I took in the great questions of the currency...(and) ...in the consolidation and amendment of the Criminal Law....
With respect to the Reform Bill itself...I consider the Reform Bill a final and irrevocable settlement of a great constitutional question....
'Then as to the spirit of the Reform Bill and the willingness to adopt and enforce it as a rule of government. If by adopting the spirit of the Reform Bill it be meant that we are to live in a per-petual vortex of agitation...I will not undertake to adopt it. But if the spirit of the Reform Bill implies merely a careful review of institutions...combining with the firm maintenance of estab-lished rights, the correction of proved abuses and the redress of real grievances; in that case, I can, for myself and colleagues, undertake to act in such a spirit and with such intentions.....
'Our object will be—the maintenance of peace—the honourable

fulfilment of...all existing agreements with Foreign Powers...
(and) the just and impartial consideration of what is due to all
interests—agricultural, manufacturing and commercial.'

(From Peel's election address, 1834)

Extract II

'Hush!' said Mr Tadpole. 'The time has gone by for Tory governments; what the country requires is a sound Conservative
government.'
'A sound Conservative government,' said Taper, musingly, 'I
understand: Tory men and Whig measures.'

(From Disraeli's novel 'Coningsby' published 1844)

Extract III

'A Conservative government is an organized hypocrisy.

*(From Disraeli's speech in the House of Commons, 10th March
1845)*

*Maximum
marks*

(a) Explain the circumstances in which Peel received the
    'wholly unforeseen and unexpected' summons (3)

(b) What part did Peel play in 'the great questions of the
    currency' (2)

(c) Explain Peel's reference to 'the active part' he took in
    'the consolidation and amendment of the Criminal
    Law'. (3)

(d) What comment might a hostile critic make on Peel's
    claim that he had 'never', before 1834, been a 'defender
    of abuses or the enemy of judicious reforms'. (4)

(e) Why did Peel feel it necessary in 1834 to declare that he
    considered the Reform Bill 'a final and irrevocable
    settlement' (2)

(f) What 'proved abuses' did Peel 'correct' and what 'real
    grievances' did he 'redress' in the years 1841–46? (6)

(g) Explain what 'Tadpole' and 'Taper' understood as the
    differences between 'Tory', 'Conservative' and 'Whig'
    in Extract II (5)

(h) What connection can you trace between Peel's promise
    to give 'just and impartial consideration of what is due to
    all interests—agricultural, manufacturing and com-
    mercial' and Disraeli's reference to 'organised hypocrisy'
    in Extract III? (6)

19.6 READING LIST

Standard textbook reading

A. Briggs, *Age of Improvement* (Longmans, 1959), chapter 6.

R. K. Webb, *Modern England* (Allen and Unwin, 1980 edn.),
chapter 6.

A. Wood, *Nineteenth-Century Britain* (Longmans, 1960), chapter 9.

Further suggested reading

R. Blake, *The Conservative Party from Peel to Churchill* (Fontana,
1972).

N. Gash, *Peel* (Longmans, 1972).

N. Gash, *Politics in the Age of Peel* (Harvester, 1977).

G. I. T. Machin, *Politics and the Churches in Great Britain 1832–
68* (OUP, 1978).

R. M. Stewart, *The Foundations of the Conservative Party 1830–
67* (Longmans, 1978).

R. M. Stewart, *The Politics of Protection: Lord Derby and the
Protectionist Party 1841–52* (Cambridge 1971).

20 British foreign policy, 1815–65

20.1 INTRODUCTION

The year 1815 marked the end of the long struggle with Revolutionary and Napoleonic France. At Vienna a new, more stable
world had to be fashioned from a Europe ravaged and changed
by almost 23 years of war. Britain favoured a balance of power
that would contain French expansionist policies in Europe and
restrain Russian imperialism in the Near and Far East. Initially
Castlereagh was prepared to work with Metternich to achieve
these ends but, under Canning, British foreign policy gradually
became more independent and isolationist as European reaction
became more pronounced.

Palmerston typified British power and confidence in mid-
century and consequently he followed a jingoistic policy favoured
by the mass of the British electorate. Many did not like his diplomatic approach or political style but his colourful assertion of
British power made him 'the most English minister'–loved at
home and hated abroad. Though he remained the archetypal
John Bull, nevertheless, his actions demonstrated his awareness
of the complexity of European problems and his determination
to follow the interests of Britain in them all.

20.2 STUDY OBJECTIVES

1 The changed patterns of European international relations
after 1815. Allied reasons for fighting France differed; their
attitudes to the Vienna settlement also differed but all were
determined that France would not threaten Europe again.
Castlereagh favoured the Quadruple Alliance for this purpose,
the Tsar a Holy Alliance. The importance of Nationalism and
Liberalism as factors in post-1815 European affairs; Britain's
attitude to these new developments.

2 The basic principles of British foreign policy–maritime and
commercial; politically and territorially satisfied but economically insatiable; aggressively free trading; 'nation of shopkeepers'.
Worked for a European balance dividing France and Russia
(England's two most likely enemies). Castlereagh tried to maintain a balance by the Congress system, Canning by independent
action and Palmerston by a mixture of the two. In the Near East,
British policy was to check Russian expansion and defend our
trade. Britain appeared to favour Liberalism, especially when it
coincided with strategic or commercial considerations–*e.g.*
Belgium or the Spanish-American colonies.

3 Castlereagh (1812–22): his contribution to European diplomacy at Vienna; the Quadruple Alliance, and his attitude to the
Holy Alliance–a 'sublime piece of mysticism and nonsense'; the
Congresses; his attitude to collective intervention; his European
outlook, belief in discussion, mediatorial role. Castlereagh had
no time for Nationalism, Liberalism or public opinion. He failed
to explain his aims, was accused of associating with European
despots, and was, therefore, unpopular at home. Yet he laid
down the basic principles of British foreign policy in his State
Paper (May, 1820). His approach was pragmatic–refusing to act
on questions of 'abstract and speculative principles'.

4 Canning (1822–27): was he more opportunist? Nationalist
in approach, and less European-minded, he rejected diplomacy
by conference–believing in the principle 'every nation for itself'.
Was he a supporter of Liberalism? Or only when the Navy could

Fig. 13 Europe after 1815 (study unit 1 to gain an overview of the situation in Europe in 1815).

be used to maximum effect? He wished to make foreign policy exciting and popular. He was unable to intervene when France invaded Spain but warned her against attempts to regain the Spanish-American colonies. More positive aid to the Portuguese constitutionalists. Canning's surprising intrusion into the Eastern Question – he worked with Russia to effect a solution; Navarino (1827). After Canning's death, Wellington returned to traditional ways – the 'Greek affair no substitute for Turkish power'. In domestic issues, Canning was conservative. Is his liberal reputation exaggerated? Did he digress from basic principles as laid down by Castlereagh?

5   Palmerston (1830–65): was he a disciple of Canning or a shrewd mixer of diplomatic and nationalist bluster? He was popular at home and disliked abroad. Did he follow the basic principles or was he inconsistent? Was he a champion of nationalism, of liberalism, of public opinion or was he narrowly insular? His statement. 'We have no eternal allies and no perpetual enemies'. suggests that he was basically an opportunist. He supported Belgium (fear of French expansion), Spanish and Portuguese constitutionalists, and Italian unification (but he refused to intervene in 1848). His attitude to Mehemet Ali and the Eastern Question; his pursuit of Britain's commercial interests – even to forcibly opening China to western merchants (*e.g.* the two Chinese Wars). His use of the Navy, and his brinkmanship (*e.g.* Don Pacifico): his relations with Queen Victoria and Albert. Palmerston's failure to see that times had changed after the Crimean War; his error in encouraging the Poles in 1863 and the Danes (over Schleswig-Holstein) in 1864.

6   How closely did British Foreign Secretaries of this period

follow the basic principles? To what extent did they stand for intervention or non-intervention? Did their differences lie less in ideology than in temperament, less in Party than in whether they were activists or passivists? Were they pragmatic opportunists who always had a clear idea of Britain's needs – and tailored their policies and actions to suit?

20.3 COMMENTARY

There was a noticeable continuity and consistency underlying the principles of British foreign policy from 1815 to 1865. All British Foreign Secretaries saw their main task as the upholding of what they believed to be British interests. Constrained as they were by king, cabinet, parliament and people, however, they did not have a completely free hand in policy-making.

Commerce and trade

Britain's most vital interests arose out of her position as a great maritime and commercial power. Politically and territorially she was satisfied; commercially her appetite was rapacious. Her pursuit of markets – as in China – was vigorous and, on occasions, aggressive. Generally, merchants could rely on the support of the government which, incidentally did not ask other nations for preferential treatment; all that was needed to guarantee British success was free trade and open competition.

The Navy

To protect and extend this worldwide trading network a powerful Navy was vital. A strong Navy, furthermore – by facilitating the acquisition of bases from Latin-America to China, from which, without occupation, Britain could control trade incursions by foreign rivals – enabled her to adopt a policy of 'informal empire'.

Despite Admiralty conservatism and policies of economic retrenchment, the government accepted responsibility for keeping the sea lanes open and, to a lesser extent, for keeping rivals out of British spheres of influence. This, then, was the period of the 'Pax Britannica' when, virtually unchallenged, the British Navy ruled the seas.

The balance of power

To British Foreign Secretaries, foreign policy signified Europe; Britain was quite capable of dealing with minor imperial problems on her own. In Europe she was suspicious of France and Russia – both aggressively expansionist powers – the former in Europe, the latter in the Near and Far East. Believing in a balance of power, Britain decided that a strong central bloc, comprising Austro-Hungary, Prussia and Italy, was essential to keep her two most likely enemies apart.

Secure as an offshore island, and believing in a policy of non-intervention, Britain could not, however, afford to be isolated. What she wanted was a system of limited liability – preferring to add her weight to European affairs only in times of crisis. For Britain, therefore, the balance of power meant the independence of each nation and the predominance of none.

Other areas

Britain also desired neutral or friendly relations with the Netherlands, Portugal and the United States – all-important for strategic or commercial reasons. Her main interest in the Near East was to check Russian expansion and defend her own Mediterranean trade – a policy that imposed the unpopular task of supporting the declining autocratic power of the Ottoman Empire – and in the Far East the British aim was to secure the northern borders of India (our major piece of 'formal Empire') against Russian encroachments. Britain was also determined to gain an economic foothold in China before Russia could do so.

In so far as there is any general statement of the principles of British foreign policy, it is to be found in Castlereagh's State Paper of May, 1820. Note that although the methods and style of each Foreign Secretary (even the lesser ones) might have changed, the principles, from 1815 to 1865, remained unaltered.

Viscount Castlereagh, 1812–22

As victor, Britain made no European territorial claims at the Congress of Vienna, but Castlereagh cleverly acquired a string of islands and ports to add to her 'informal empire'. His main object at Vienna was to curb the power of France and restore Europe once more to peaceful normalcy.

Though aware of Britain's interests, Castlereagh had a European outlook. He believed that European problems could be solved in congress by discussion and negotiation. He had little time for the new forces of Nationalism, Liberalism or public opinion and was more concerned with implementing policies which he regarded as possible and immediately relevant. Thus, he dismissed the Czar's scheme for a Holy Alliance as 'a piece of sublime mysticism and nonsense' and instead persuaded the Allies to continue with the Quadruple Alliance.

Many contemporaries could not distinguish between the two alliances; and Castlereagh, because of his association with European despots, was branded a reactionary. But Castlereagh was no hardened reactionary: he perceived that Britain could secure her interests by working with the other Powers. His

mistake was that he neglected to explain this policy to cabinet, parliament or people. Note, however, that the congress system had no permanent centre, organization or secretariat. Consider, therefore, whether it is acceptable to refer to it as a 'system'?

The congresses

At *Aix-la-Chapelle* in 1818 it was agreed that the Allied army of occupation should be withdrawn from France and that she should rejoin the European community of Powers. But among those powers the publication of Metternich's *Carlsbad Decrees* and the subsequent revolutions in Spain and Naples quickly exposed divergencies. The Holy Alliance favoured collective intervention, which Castlereagh firmly opposed. As a result, Britain sent an observer instead of a delegate to *Troppau* in 1820 and to *Laibach* in 1821. Another congress, to consider the Greek revolt, was planned for *Verona* in 1822, but Castlereagh committed suicide before it was convened.

Assessment of Castlereagh's policy

Castlereagh's policy after 1815, built on the personal contacts he established during the Napoleonic War, was one of the most personal ever pursued by a British Foreign Secretary. He had no obvious successor. While in office he intelligently exploited Britain's position of strength and worked well with Austria to restrain Russia and France. His fault was that he tended to see political affairs too much in terms of the *status quo* and refused to acknowledge that fundamental new forces were affecting events in Europe.

In his obituary, the *Scotsman* stated: 'The name of Castlereagh will long be connected with tyranny abroad and all that is slavish and oppressive at home.'

On the other hand, C. J. Bartlett reminds us that Castlereagh worked unfailingly to abolish the slave trade and supported Catholic emancipation. He maintains, therefore, that Castlereagh, who has never been accused of neglecting British interests, was not a reactionary and that there was much in his foreign policy that was 'distinctive and original'. Where do you think the truth lies?

George Canning, 1822–27

The change from Castlereagh to Canning produced no break in the continuity of foreign policy. But 'where Castlereagh made discreet signals, Canning waved a flag in conversations, despatches and speeches'. Canning openly rejected the congress system and was much more nationalist in outlook: 'each nation for itself and God for us all.'

As he was not on intimate terms with the rulers of Europe – few of whom knew or trusted him – his more independent approach was to be expected. More brilliant than Castlereagh, he saw the weakness of a system which did not encourage the support of parliament and people. On the whole, Canning followed and developed the principles outlined by Castlereagh in his State Paper of 1820.

Spain, Portugal and the New World

During his period in office Canning was confronted with four major problems. In his handling of these he smashed the already frail structure of the Holy Alliance. Powerless to do anything when France invaded Spain in 1823, he was, through Britain's control of the seas, the master in the dispute about the Spanish American colonies. In 1825, he recognized the independence of a number of these with the now famous justification: 'I resolved that, if France had Spain, it should not be Spain with the Indies. I called the New World into existence to redress the balance of the Old.' In the old world of Portugal, where constitutionalists and reactionaries struggled for control, Canning also showed his sympathy for the constitutionalists by sending the Navy to help them in 1824 and 1826.

The Greek Revolt

When the Greeks rebelled in 1821 Canning was mindful of Castlereagh's advice: 'Barbarous as it is, Turkey forms in the system of Europe a necessary evil.' He had two choices: to support Turkey and oppose Russia (traditional) or work with Russia to ensure a peaceful solution. To Metternich's dismay he chose the latter: the Petersburg Protocol advocating that Greece become an autonomous state under Turkish suzerainty. An Allied fleet was sent to enforce a blockade on the unwilling Turks in 1827 and the battle of Navarino followed. The news of the destruction of the Egypto-Turkish fleet caused a sensation when it became known. Despite Navarino, however, his policy eventually proved successful: the Greeks got their independence at the London Conference of 1830.

Assessment of Canning's policy

When making an assessment of foreign affairs under Castlereagh and Canning, the most important fact to bear in mind is that there was no fundamental change of policy when Canning took over in 1822. Canning merely accelerated existing trends. His policy differed from Castlereagh's in method rather than principle.

According to P. J. Rolo, Canning also fostered a legend – the legend of his own liberalism. Indeed a close examination of the motives behind his policies demonstrates that liberalism to Canning was something to exploit rather than to follow. Much of his domestic policy was reactionary: until his dying day he remained an opponent of parliamentary reform.

Nevertheless his style and his use of the press to publicize his policies made him popular. If the interests of Britain are taken as a criterion he must be judged a most successful Foreign Secretary.

Comparing Canning with Castlereagh

When comparing the two one should remember that both their reputations are exaggerated – Canning being less, and Castlereagh more, liberal than is usually stated. Both were Tories, both served under Lord Liverpool, both were primarily concerned with British interests and both were moving away from the congress system.

Canning, however, was more in tune with nineteenth-century developments and, unlike Castlereagh, cleverly turned everything into a publicity campaign. The public enjoyed his 'liberal' victories but his independent approach led Metternich to complain that the ministry of Canning 'marked the end of an era in the history of England and Europe'.

Lord Palmerston

Castlereagh had been a great European figure; Canning had been much more nationalist. So also was Palmerston, who added a lively sense of Britain's honour and prestige to Canning's policy of liberal-nationalism. Like Castlereagh, Palmerston was also prepared to use negotiation and conferences to solve European problems. A master of improvisation, he chose his allies (and his policies) according to the situation – now absolutist, now liberal: 'We have no eternal allies and no perpetual enemies. Our interests are eternal and perpetual and those interests it is our duty to follow.'

The cornerstone of his policy was national interest. He was careful to avoid any firm European commitments. A. J. P. Taylor maintains that, in general, he tried to use Britain's strength to preserve the European balance of power. In the ideological struggle between absolutism and Liberalism Palmerston carefully adhered to a policy of neutralism. Palmerston, then, did not deviate from the basic principles laid down by his predecessors. Firmly believing that Britain should be 'the arbiter of Europe', he wanted to have a free hand in dealing with problems as they arose. He also hoped that British opinions would be listened to and respected.

Remember that behind Palmerston's bluff 'John Bull' exterior there was an experienced and confident diplomat (who could speak several languages), a skilled drafter of despatches and a shrewd assessor of international relations.

1830–41

Palmerston's career can best be understood if divided into three sections. His period of greatest success was from 1830 to 1841, when he triumphantly resolved major international problems in Belgium, the Iberian Peninsula, the Ottoman Empire and China. You must be familiar with the details of these important events but, more important, be able to explain the reasons for the policies that Palmerston adopted in all of them and why he was successful.

1846–51

Palmerston's middle period in office revealed mixed fortunes but no obvious failures. After an uncertain beginning, with the affair of the Spanish marriages in 1846, he resisted the attempt to partition Switzerland in 1847 and, for all his talk of liberalism, he maintained a rigorous neutrality during the 1848 revolutions. His use of the Navy over Kossuth and Don Pacifico eventually induced the Queen to send him a memorandum deploring his unconstitutional methods. When he recognized the *coup d'état* of Louis Napoleon the following year, Russell took the opportunity to dismiss him.

1855–65

Palmerston returned as Prime Minister in 1855 to a changed international scene. After the Crimean War (see *Wood* chapter XI) the European balance constructed at Vienna was finally destroyed. Industry and science had made new and powerful European armies controlled by dedicated and unscrupulous nationalists such as Cavour and Bismarck. Palmerston and Russell were out of place in this new diplomatic environment and, predictably, made mistakes. Where the Navy was supreme, however, they were still able to take a strong and successful line – as in China and Italy. Finally they misjudged two European crises – the Polish revolt in 1863 and the Schleswig-Holstein affair in 1864 – events, at that time, well outside the range of British power.

Assessment of Palmerston's policy

Palmerston undoubtedly strengthened the influence of Britain in Europe but his dictatorial and aggressive methods made him unpopular with her wartime allies. Conversely, he enjoyed great popularity at home where, according to Jasper Ridley, the people had become 'more Palmerstonian than Palmerston himself'. Abroad, his bullying methods upset some. But generally he possessed a shrewd understanding of European diplomats and events and fashioned his policies accordingly.

Perhaps one should conclude with his own vindication of 'bluster' (used against Russia during the Mehemet Ali crisis): 'I am all for making a clatter against Russia – depend upon it, that is the best way to save you from the necessity of making war against her.'

20.4 CHRONOLOGY

1814	November	CONGRESS OF VIENNA.
1815	June	Napoleon defeated at Waterloo.
	September	HOLY ALLIANCE signed. Rulers of Russia, Austria and Prussia agreed to prevent war and work for European unity on a Christian basis. Castlereagh refused to join.
	November	QUADRUPLE ALLIANCE made between Britain, Russia, Austria and Prussia. These nations agreed to work together in a concert to keep the peace. This treaty was largely the work of Castlereagh.
1818	September	CONGRESS OF AIX-LA-CHAPELLE. Allied armies of occupation withdrawn from France.
1820		Revolts in Spain, Naples and the Spanish-American colonies.
	May	CASTLEREAGH'S STATE PAPER: outlined the principles of British foreign policy.
	October	CONGRESS OF TROPPAU. Castlereagh for non-intervention in European revolts but the other Powers issued a protocol sanctioning the use of force to maintain the *status quo*.
1821	January	CONGRESS OF LAIBACH to discuss the revolt in Naples. Britain sent an observer.
	April	Greek revolt against the Ottoman Empire.
1822	August	Death of Castlereagh. Canning became Foreign Secretary.
	October	CONGRESS OF VERONA to consider the Greek revolt. Britain sent observer.
1823	April	FRENCH TROOPS INVADE SPAIN. Ferdinand VII restored.
	December	Monroe Doctrine of 'America for the Americans' enunciated by President Monroe. Any interference by European Powers in the American hemisphere would be deemed an unfriendly act by the US.
1824	December	Independence of Mexico, Columbia and Buenos Aires recognized by Canning.
1825		Independence of Chile, Bolivia and Peru recognized by Canning.
1826	April	PROTOCOL OF ST PETERSBURG with the new Tsar Nicholas. Agreement to mediate in Greek-Turkish War.
	December	Combined naval and army force, to support Portugal against Spain, despatched by Canning.
1827	July	TREATY OF LONDON ratified the Protocol. France and Prussia became signatories.
	August	Canning died.
	October	BATTLE OF NAVARINO. Turkish-Egyptian fleet sunk by an Allied fleet under Admiral Codrington. Greeks now assured of independence.
1828	April	Russo-Turkish War.
1830	July	Revolution in France: Charles X replaced by Louis Philippe.
	August	STRUGGLE FOR BELGIAN INDEPENDENCE which Britain favoured, but feared that French help might become French ownership. Palmerston's diplomatic skill, and Louis Philippe's good sense, saw it to a successful conclusion in 1839.
	November	Palmerston became Foreign Secretary.
1831	November	WAR BETWEEN OTTOMAN SULTAN AND MEHEMET ALI. Turkey defeated but Russian intervention saved her.
1833	July	TREATY OF UNKIAR SKELESSI between Turkey and Russia; they agreed to give each other mutual help. Palmerston feared a Russian protectorate over Turkey.
1834	April	NEW QUADRUPLE ALLIANCE – Britain, France, Spain and Portugal. Palmerston strongly supported the constitutionalist girl queens in Spain and Portugal. Helped to expel their reactionary uncles.
1839	June	WAR BETWEEN OTTOMAN SULTAN AND MEHEMET ALI renewed. Mehemet Ali was victorious and the major Powers, alarmed at the possible destruction of the Ottoman Empire, intervened.
	November	CHINESE OPIUM WAR. Chinese authorities tried to stamp out the illicit opium trade with British India. Disputes led to violence, demands for compensation, and eventually, war.
1841	July	TREATY OF LONDON. Unkiar Skelessi revoked. The great Powers agreed to maintain the integrity of Turkey.
	September	Peel and the Conservative Party in office.
1842	August	Treaty of Nanking, in which Britain received Hong Kong, ended Chinese War.
1846	July	Palmerston became Foreign Secretary.
		THE SWISS SONDERBUND. Attempted secession by seven Catholic Cantons from the Swiss Federation. Palmerston prevented European intervention on their behalf. Protestant Cantons suppressed the Sonderbund (November 1847).
	August	Affair of the Spanish Marriages: this broke friendly relations between France and Britain.

Chronology cont.

1848		Year of revolutions. French monarchy overthrown (February). Metternich overthrown (March). Palmerston remained neutral but supported the Sultan when he refused to hand over the Hungarian leader, Kossuth, to Austria.
1850	**June**	DON PACIFICO. Palmerston's 'Civis Romanus sum' speech.
1851	**August**	QUEEN'S MEMORANDUM: commented unfavourably on Palmerston's individualistic methods.
	December	LOUIS NAPOLEON'S COUP D'ETAT; recognized by Palmerston without Cabinet approval. Palmerston forced to resign.
1854	**March**	CRIMEAN WAR BEGAN. Britain and France declared war on Russia.
1855	**February**	Palmerston became PM; Clarendon became Foreign Secretary.
1856	**March**	CRIMEAN WAR ENDED. Paris Peace Conference.
	October	Chinese wars (second and third) ended in October, 1860 by the TREATY OF PEKIN.
1857	**May**	Indian Mutiny; ended April, 1859.
1858	**January**	ORSINI 'BOMB PLOT': attempt to assassinate Louis Napoleon. Palmerston's 'Conspiracy to Murder' Bill rejected and he resigned.
1859	**June**	Palmerston became PM; Russell became Foreign Secretary.
1860	**May**	Garibaldi's campaign for Italian unification in Sicily and Naples. Palmerston and Russell gave moral support by maintaining a benevolent neutrality.
1861	**February**	American Civil War began. Britain remained neutral despite difficulties over the TRENT in November, 1861, and the ALABAMA in June, 1863.
1863	**February**	POLISH REVOLT. Russell led the Poles to expect British help against Russia but sent them none.
1864	**February**	WAR OVER SCHLESWIG-HOLSTEIN: Prussia and Austria invaded Denmark. Russell sympathized with the Danes and led them to expect British help; encouraged them to resist. No help was sent.

20.5 QUESTION PRACTICE

1 **'National interest was always his overriding objective.' Is this an accurate assessment of Castlereagh's foreign policy after 1815?**

Understanding the question

The question demands an analysis of the principles underlying Castlereagh's foreign policy. Remember that his contemporaries accused him of allowing general European concerns to override British interests. You may choose to assess the accuracy of the above statement in the light of this popular criticism.

Knowledge required

A thorough understanding of the principles of British foreign policy. A detailed knowledge of Castlereagh's foreign policy from 1815–22.

Suggested essay plan

1 Introduction: Castlereagh became Foreign Secretary in 1812. His outlook was coloured by his experiences as Chief Secretary for Ireland in the 1790s. His ability was respected but he was not revered. A combination of the Irish experience of rebellion and the European experience of the French Revolution made him a determined advocate of peace and maintenance of the *status quo*. As leader of the House of Commons he was acknowledged as the main driving force of Liverpool's administration; consequently he was blamed for its repressive policy at home and its association with foreign despots abroad.

2 The continuity of ideas underlying British foreign policy during this period. Peace, non-intervention, the balance of power, economic expansion. The way in which the objectives approached may have differed, but the principles were permanent. The basis of these principles was that of Britain's national interests and this was clearly expounded by Castlereagh in his State Paper of May, 1820.

3 At Vienna, Castlereagh accepted that the aim of the Congress was European peace and a new balance of power after the upheavals of the Napoleonic War. Castlereagh was generous to France; he was also agreeable to national sentiment being ignored for the sake of shaping the new Europe. Territorially satisfied, he added to Britain's 'informal empire' by acquiring a string of islands across the globe. When faced with the Tsar's apparently interventionist Holy Alliance, he refused to join and offered instead the continuation of the more practical Quadruple Alliance. To maintain the ideals of European peace, stability and the balance of power he was willing to try the Congress system – but to him the system was always one of limited liability. Britain still favoured the policy of a free hand, not wishing to become too entangled in European affairs – an involvement which, above all, could have detrimental economic effects.

4 Increasingly, Castlereagh was out of step with the reactionary interpretation of the Vienna Settlement, which encouraged intervention, and wished to stifle Liberalism and emergent Nationalism. Revolts in Spain and Naples were a setback to Castlereagh's attempts to curtail collective action. To Castlereagh the Congress 'system' was not a system for the government of other states. He began to withdraw – only sending an observer to Troppau and Laibach. This concern prompted him to produce his State Paper of May, 1820, in which he reaffirmed the fundamental principles of British foreign policy.

5 Did Castlereagh follow national interests in these events, or was he the dupe of foreign reactionaries, the pawn of Metternich? An examination of these policies demonstrates clearly Castlereagh's proper regard for the interests of his country – a stable France, the control of the seas, economic expansion (note his opposition to possible Franco-Spanish intervention in the South American colonies), the balance of power, etc. His withdrawal from the congress system indicated his increasing dissatisfaction with its activities. European-minded he may have been, but his actions indicated clear limitations to the primacy of European considerations in his policies.

6 Castlereagh's support for the international method of maintaining peace was not popular in an era of nascent nationalism. Yet he believed that Britain's interests could best be secured by working in concert with the European Powers. His failure was in neglecting to explain this to the cabinet, parliament or people.

What Castlereagh sought to achieve by cooperation and personal contact with the European Powers Canning tried to accomplish by more independent methods–but even he failed to prevent France from intervening in Spain. Castlereagh may have favoured a system of international cooperation but he rightly believed that European peace and security were as much British interests as European ones.

2   To what extent can we describe Canning's foreign policy as liberal?

Understanding the question

Canning's foreign policy has always been regarded as liberal–the key question is *how* liberal? The question implies that his reputation for liberalism may have been exaggerated, that there may be more continuity in the aims of British foreign policy than is sometimes suggested.

Knowledge required

An understanding of the aims and objectives of Castlereagh's foreign policy from 1815–21. A detailed knowledge of Canning's foreign policy and actions in Spain, Portugal, the New World and the Near East.

Suggested essay plan

1   Castlereagh is often said to have represented the more repressive side of the Younger Pitt's character; equally Canning may have inherited the more progressive ideals of his master. It is useful to remember that Canning began life as a Whig but soon settled comfortably into the role of a Pittite Tory. Like Pitt, he favoured Catholic emancipation but displayed few liberal tendencies in other domestic affairs. He was an enemy of Jacobinism, of parliamentary reform and of Corn Law repeal. He also supported the repressive policy of Liverpool's administration from 1815–22.

2   It was in foreign policy that Canning gained his reputation as a liberal. This was primarily based on his rejection of the Holy Alliance, together with Metternich's attempt to rule Europe by the congress system–which Canning openly denounced. Metternich's consequent dislike of Canning simply added to the latter's reputation as a progressive.

3   How liberal was Canning? He condemned the French invasion of Spain in 1823 and expressed the hope that the Spanish constitutionalists would eventually triumph. He warned about French intervention in Portugal or the rebellious Spanish American colonies. These sentiments may have been liberal, but they also admirably expressed British foreign policy objectives–particularly with regard to protecting our burgeoning trade with South America. His 'New World' speech (a piece of retrospective bravado giving political recognition to the 'New World') was made as much to support and protect a rapidly expanding market for British goods, as to uphold the cause of Liberalism. What he wanted was the preservation of Portuguese independence and the denial of the Spanish colonies to France.

4   Unable to act over the French invasion of Spain, Canning was determined to maintain Britain's interest in the Iberian Peninsula by supporting Portugal–note the strategic importance to Britain of its Atlantic coastline and harbours. He sent a naval squadron to the Tagus in 1824, followed in 1826 by a detachment of British troops to Lisbon to help the constitutionalists and to prevent further instability or foreign interference. In the years up to 1826, therefore, he had successfully defended British interests abroad, increased his own reputation and won the admiration of many liberals in Europe.

5   The Greek revolt presented Canning with his greatest problem. British foreign policy demanded that she support Turkey as a barrier against Russian expansion. However, British public opinion was very pro-Greek (*e.g.* Byron, Philhellenism). The main danger was a Russo-Turkish war which Canning worked hard to avert–the Petersburg Protocol. Canning rejected the congress approach by reaching an understanding with Russia that Greece be allowed autonomy under Turkish suzerainty. An allied fleet was sent to force this solution on a reluctant Sultan. Navarino was the result but Canning was dead before the news reached London and the Russo-Turkish war, which he had striven to avoid, followed. Nevertheless, the Greeks got their independence in 1830 but Canning had left the congress system in ruins.

6   The liberals claimed that Canning's innovative policy was in the Near East. They claimed that his attempt to work with Russia to effect a solution to the Eastern Question was bold and imaginative. It is unlikely, though, that Canning was doing this solely to further the cause of Greek nationalism. There was British trade in the Mediterranean to consider and the strength of the Ottoman Empire as a barrier to Russia. Others maintained that his Near Eastern policy was a return to working with the other Powers (the concert system), excluding Metternich, whom he personally disliked, from the negotiations.

7   Canning's policies over Spain, Portugal, the New World and Greece certainly appeared progressive. His main motives were, however, not the furtherance of European Liberalism but the implementation of the basic principles of British foreign policy (and the flashy pursuit of personal power and admiration). He was not an unqualified supporter of Liberalism but his political style seemed to mark him as the symbol of resistance to European despotism. While acknowledging the more overtly progressive nature of his policies (compared, say, to Castlereagh) and his dedication to ending the slave trade, his reputation for liberalism in foreign affairs is probably exaggerated. As Professor Gash says: 'under his flamboyant liberal guise, Canning was an acute, cautious and conservative foreign minister.'

3   To what extent could Palmerston be described as a disciple of Canning?

Understanding the question

You should begin with a consideration of the aims and style of Canning's foreign policy. To what extent did Palmerston continue Canning's approach in style and method? Is there enough similarity to justify the title 'disciple'? Did Palmerston consciously model his foreign policy on Canning's? Or was he sufficiently independent in thought and action for one to reject this description?

Knowledge required

An understanding of the basic principles of British foreign policy. A detailed knowledge of foreign policies and diplomatic methods of Canning and Palmerston.

Suggested essay plan

1   Palmerston had been a supporter of Catholic emancipation and a Canningite before he left the Tory Party in 1830. He had attacked Wellington's government in 1829 for not following Canning's foreign policy over Portugal and his political style, nationalistic fervour and 'liberal' pronouncements indicate him as a 'disciple' of Canning's. His speeches–'We have no eternal allies…' for instance–have about them the ring of Canning's 'Each nation for itself…' and mark Palmerston as vigorous a nationalist as his predecessor. His despatches condemn European autocrats and, like Canning, he displayed a lively contempt for foreigners. Similarly, he voiced his support for liberal and constitutional states claiming that they were the 'natural allies' of

Great Britain. Decisive in action, eager to protect and extend Britain's foreign trade, capable of appealing directly to the public – often over the head of cabinet colleagues (he also wrote for a daily newspaper), tenaciously pursuing the slave traders or castigating Metternich, he did appear to be more a disciple of Canning than Castlereagh.

2   Yet this may be a superficial judgement. An examination of his methods shows a willingness to negotiate and collaborate with the other Powers in a way more reminiscent of Castlereagh than of Canning. In the Belgian question he realized that a satisfactory solution was only possible in concert with the other Powers. He worked with Russia and Austria – both of which he personally disliked – to solve the Mehemet Ali crisis in the Near East in the 1830s. In both he was not afraid to 'throw down the gauntlet' to bring France to her senses, but the ultimate objective was to preserve peace and the balance of power.

3   In the 1848 struggle in Europe between Liberalism and autocracy, Palmerston remained neutral. Despite the popularity in England of the Hungarian cause, he maintained an astute diplomatic silence. He retained his popularity, however, by privately commending Kossuth's bravery and, to the annoyance of the Queen, by condemning Austrian brutality in suppressing the revolt. In other areas Palmerston showed that he was prepared to be more of an interventionist than Canning–*e.g.* in Belgium, the Near East, China, Poland and Schleswig-Holstein. Palmerston also went beyond Canning in his support for the extension of British trade (two China wars) and in giving protection to British merchants legitimately following their business interests abroad. The same can be said for his wide-ranging and effective use of the Navy.

4   On balance it is difficult to argue convincingly that Palmerston was a 'disciple' of Canning, displaying as he did such an opportunistic approach to foreign affairs. Like Canning, he cultivated through his speeches and the press the image of a powerful, independent, liberal Britain. Some of this (as with Canning) was to encourage popular support for himself and his policies. Furthermore, behind the image and the bombast, Palmerston displayed a shrewd understanding of the realities and complexities of European diplomacy. What made him successful, popular and *uniquely* individual was his opportunism – be it in conference, negotiation, bluff or independent action – which best guaranteed the paramountcy of British interests, the balance of power and European stability.

4   Document question (Canning in the 1820s): Study each of these 10 short quotations from the speeches of Canning and then answer questions (a) **to** (h) **which follow:**

1   'I do not believe that I shall learn what England thinks from the King, from great lords and ladies or from great families.'

2   'The middle way means the middle class, which interposes between two extremes... The middle class is the most valuable part of the community, in which its stable interest and sterling good sense reside.'

3   'We are on the brink of a great struggle between property and the populace which can only be averted by the mildest and most liberal legislation.'

4   'Those who reject improvement because it is innovation will have to accept innovation when it is no longer improvement.'

5   'Our business is to preserve peace and the independence of nations.'

6   'Our influence abroad must be secured in our strength at home in sympathy between people and government, in the union of public sentiment in public councils and in the cooperation of the House of Commons and the Crown.'

7   'After being the saviours we have become the model of Europe. Let us hope the model will be generally followed.'

8   'England cannot avoid seeing ranked under her banner all the restless and discontented of every nation.'

9   'We cannot treat as pirates a population of a million and we must try to bring within the bounds of civilization a contest most barbarously conducted on both sides.'

10   'The issue at Verona has split the one and indivisible alliance into three parts as distinct as the constitutions of England, France and Muscovy. So things are getting back to a wholesome state, each nation for itself and God for us all.'

Maximum marks

(a) In view of their content as well as of the fact that he was a leading figure in a cabinet headed by Lord Liverpool what do you consider the most significant single word used by Canning in Nos. 1, 2, 3 and 4? Explain briefly your reasons for this choice.   (3)

(b) What could the Canningites claim to have done in the 1820s to demonstrate Canning's belief that 'The middle class is the most valuable part of the community' (No. 2)?   (4)

(c) What examples can be found, from the years when Canning was in office, of measures by Liverpool's government designed to avert 'a great struggle between property and the populace' (No. 3)?   (4)

(d) What 'innovation' that was not, in his opinion, an 'improvement' (No. 4) did Canning wish to avoid in England?   (2)

(e) Suggest how far Nos. 5, 6, 7 and 8 indicate that Canning's aims in foreign affairs resembled, and how far they differed from, those of Castlereagh?   (5)

(f) What 'contest' is referred to in No. 9; who were the 'population of a million'; and with whom did Canning cooperate in order to bring the contest 'within the bounds of civilization'?   (5)

(g) What was 'The issue at Verona' (No. 10) which 'split the one and indivisible alliance' and which of the other nine quoted sayings of Canning did his action on the issue most clearly illustrate?   (5)

(h) Apart from any difference of aim and policy, what personal asset that Castlereagh lacked do these sayings suggest that Canning obviously possessed?   (3)

20.6 READING LIST

Standard textbook reading

A. Briggs, *Age of Improvement* (Longmans, 1959), chapter 7.

R. K. Webb, *Modern England* (Allen and Unwin, 1980 edn.), chapters 4, 7.

A. Wood, *Nineteenth-Century Britain* Longmans, 1960), chapters 10, 11, 13.

Further suggested reading

O. Anderson, *A Liberal State at War* (Macmillan, 1967).

C. J. Barlett, *Castlereagh* (Macmillan, 1966).

K. Bourne, *The Foreign Policy of Victorian England* (OUP, 1970).

P. Dixon, *Canning, Politician and Statesman* (Weidenfeld & Nicolson, 1976).

J. Joll (ed.), *Britain and Europe 1793–1940* (OUP, 1967).

J. Ridley, *Lord Palmerston* (Constable, 1970).

D. Southgate, *The Most English Minister . . . the Policies and Politics of Palmerston* (Macmillan, 1966).

C. K. Webster, *The Foreign Policy of Castlereagh 1815–22* (Bell, 1925, 1963).

21 Disraeli and Conservatism

21.1 INTRODUCTION

Benjamin Disraeli is one of the most contentious political leaders of the nineteenth century. Contemporaries and historians have never ceased to argue over his contribution to the Tory Party, and over the merits and demerits of his personality and his political beliefs. Did he have political principles, or was he merely a political opportunist whose one settled aim was to reach the top?

Whatever the answer to these questions there is no doubt that his success as a politician is amazing. Even Gladstone, his main rival and enemy, said after his death: 'The career of Lord Beaconsfield is in many respects the most remarkable in our parliamentary history....'

21.2 STUDY OBJECTIVES

1   Disraeli's background and personality; his novelistic and journalistic approach to politics – a combination of extreme egotism and unscrupulousness with political nonchalance.

2   Background of the Conservative Party to 1846; the reasons for regarding Peel as the 'founder' of Conservatism (his pragmatic, nationalist, bureaucratic approach to government); better reasons for regarding Peel as the 'founder' of the Liberal Party and as the antithesis of Conservatism (his moralistic, individualistic, *laissez-faire*, European approach). The 1846 split and Disraeli's part in it. Point out that Disraeli was the government's warmest back-bench supporter during the years 1841–3, so he cannot have attacked Peel merely out of pique at not being given office in 1841. Note that Disraeli did not defend protection in 1846, merely the duty of Party leaders not to betray their supporters.

3   Disraeli's position in the Conservative Party, 1846–74 – he was indispensable but never quite trusted; there were attempts to oust him (in favour of Stanley, Cranborne, etc).

4   Disraeli's political philosophy: he adapts his 'Young England' paternalism from an agrarian, pastoral, nostalgic idyll, as depicted in his novel, *Sybil*, into an urban phenomenon. He was one of the first politicians to see politics as a *medium* for satisfying public opinion, full of symbol and gesture (*e.g.* the purchase of Suez Canal shares, the Empress of India, etc.).

5   The Second Reform Act: the reasons for regarding this as merely a tactical coup ('stealing the Whigs' clothes', gerrymandering re-distribution of seats) or a pragmatic response to disorder (*e.g.* the Hyde Park riots). But there were also serious political motives, such as to tap working-class resentments against the aspiring, lower-middle-class, moralistic, Lord's-Day observing, and temperance fanatics of the new Gladstonian Liberalism.

6   Disraeli's Crystal Palace and Manchester speeches promoted the Empire and social reform; his conquest of the Queen;

his 1874–80 government fulfils this type of Conservatism, with its paternalist legislation (artisans dwellings, trades unions, etc.) and Palmerston-style nationalism.

7   But the actual development of Conservatism after 1870 owed little to Disraeli's romantic, paternalist, vaguely populist philosophy. The main factor was the movement of the middle classes from Liberalism to Conservatism: the business classes were offended by Gladstone's, and later Chamberlain's radicalism (especially the Workmen's Compensation Act), and looked to the Tories for tariff reform and a strong foreign policy; the lower-middle classes turned Conservative as they grew in respectability (what Salisbury called 'villa Toryism').

8   The 1885 election gave the Conservatives a majority in the English boroughs for the first time ever – and a minority in the counties. This transformation had little to do with Conservatism as Disraeli had idealised it. Yet Disraeli has to be credited with having kept Conservatism going through its years in the wilderness, when it seemed likely to disappear without trace, or become merely the agrarian wing of a Liberal-capitalist party, as politics re-oriented themselves horizontally on a 'capital' versus 'labour' basis. That this did not happen owes as much to Disraeli as to anyone.

9   Consider Disraeli's abiding influence in the Conservative Party and on Conservative 'one nation' thinking.

21.3 COMMENTARY

Disraeli's background

Benjamin Disraeli was born in 1804 in Gray's Inn, London. His background was Jewish; his father was a well-known writer. He attended no famous school or university but was self-taught in his father's immense library. His was not the best background for a career in politics. In the 1830s he unsuccessfully fought three elections as an Independent before finally entering parliament, at the fourth attempt, as a Conservative.

Throughout his life politics and literature were his major interests (his first novel, *Vivian Grey*, was published in 1826) and he enjoyed success in both. He eventually decided to make his career in politics and, armed with a burning ambition, superb courage and a brilliant intellect, he was determined to reach the top.

Within 12 years of entering parliament this eccentric MP, who dressed so outrageously and opposed so determinedly, had become leader of the Tory Party in the House of Commons.

The Corn Laws and after

Unhappy with the direction of Peel's liberal-Conservatism, and feeling that Peel and Derby underrated him, Disraeli grew restive and hostile and finally challenged Peel, first on the Maynooth Grant and then on the issue of the Corn Laws. He followed Bentinck in alleging that Peel had betrayed the Conservative Party and rallied the Tory backbenchers to his side. Peel's ideal of non-party government was condemned by Disraeli, who insisted that the Conservatives should 'above all maintain the line of demarcation between the parties.'

His success in overthrowing Peel (whom he called 'a great parliamentary middleman') propelled him into the Party hierarchy – much to the discomfiture of many of the Ultra-Tories, who were not a little suspicious of this Jewish, novelist-dandy. What the Party needed at that time, however, was a leader who shared its apprehensions about Peel's particular kind of Conservatism. Disraeli clearly fulfilled this role. This was not mere opportunism on his part. He had shown sympathy with the aristocratic 'Young England' party of the 'forties and with its idealistic notion about the alliance between the old aristocracy and new industrial classes in the towns. In his novels *Coningsby* (1844) and *Sybil* (1845) he, too, professed belief in the historic factors that had made England great: the Crown, the Church, the constitution and the aristocracy.

The years 1846–66 were critical in Disraeli's career. He dropped his 'Young England' ideals of agrarianism in the late 'forties, abandoned protection in 1852 and introduced a Reform Bill in 1859. In 1866 he accepted a more radical extension of the franchise than Gladstone ('the People's William') would tolerate.

Questions you must research are: was Disraeli re-educating his Party, reverting to 'Peelism' or merely trying to get the Conservatives into power in order to secure his position as leader?

The Second Reform Act

Disraeli's success in passing a Reform Bill in 1867 was a personal triumph achieved with political skill, superb improvisation and not a little deviousness. Neither he nor Lord Derby had any set ideas on reform, nor did they believe in advancing the cause of democracy. After their initial announcement that some measure of reform was necessary, they allowed the Commons and public disorder (though not Gladstone), to dictate the nature of the Bill. The passage was, as Blake perceptively states, 'a moonlight steeplechase'.

It would appear then that their aim was merely to 'dish the Whigs' and thus stay in power as an effective government with a secure majority. Disraeli's performance was dazzling enough for the Party to accept him as leader on Derby's resignation in 1868. He had survived every attempt to oust him from the leadership of the Party and his remark on becoming Prime Minister was characteristic: 'Yes, I have climbed to the top of the greasy pole.' Thus, Disraeli's first major ambition was fulfilled. His initial calculation that the Conservative Party offered the better prospects for political advancement than the more 'pedigree-conscious' Whigs had proved correct.

The passing of the Second Reform Act was the greatest parliamentary success of Disraeli's career. It was also a major milestone in England's advance to democracy. You must know the most important influences that led up to, and influenced the passing of the Act, the general details of the struggle to get it through parliament – particularly the rivalry between Gladstone and Disraeli – and the results of the Act. And you should also now make a careful comparison of the effects of the first and second Reform Acts.

Disraeli's political philosophy

When the electorate failed to give him a mandate to govern in 1868, Disraeli was forced to think about developing a political programme. In the main, however, he reacted to Gladstone's Liberalism, as exemplified in the ministry of 1868–74, rather than innovated. He attacked Liberalism as a peculiarly European, un-English doctrine typified by Gladstone's concern for the rights of every nation but his own. It was also, he alleged, a middle-class, non-conformist, sectarian ideology which excluded too many social groups and concerned itself with narrow, Benthamite administrative reforms rather than with the general condition of the people.

Encouraged by a real Tory democrat, J. E. Gorst, Disraeli set forth his alternative paternalist philosophy in two speeches at Manchester and at the Crystal Palace in 1872. The Tory Party, he maintained, was the national party of England, uniting all groups, all outlooks, all creeds. He stressed the attachment of the majority of English people to the Crown, the Church, the constitution and the Empire: these institutions had made England great, had stood the test of time and should be preserved.

Gorst impressed on Disraeli, however, that the extension of the franchise in 1867 meant that the Tories had to broaden the

basis of their electoral support – the traditional county vote was no longer enough. The problem was how to implement a social reform programme without alienating the country gentry – who were largely unsympathetic to the new democracy.

Disraeli in power, 1874–80

At Berlin in 1878 Disraeli lamented that power had come too late to please him but it could be argued that he was interested in attaining power, not in using it. As a Prime Minister he appeared to pursue without a qualm the policy of liberal-Conservatism for which he had denounced Peel. He formed a strong cabinet and, thanks to the efforts of R. A. Cross at the Home Office, he redeemed his Crystal Palace election pledge to improve the condition of the people. More important, his new Conservatism, a combination of *sanitas et imperium*, gave to the Party a sense of identity and purpose that it had not enjoyed since the days of the Younger Pitt and gave to the new electorate an attractive alternative to Gladstonian Liberalism.

You must now study the government's domestic and social legislation. You should know the main provisions of the Public Health Act, the Artisans' Dwelling Act, Employer's and Workmen's Act, and the Conspiracy and Protection of Property Act – all passed in 1875 – and the Education Act, the Merchant Shipping Act and the Enclosure of Commons Act – passed in 1876. Do not overlook Disraeli's attitude to the Anglican Church (*e.g.* the Public Worship Act), or the agricultural interests, who formed the backbone of the Party and who were now facing a severe economic depression.

Note that many people still felt that state intervention on the scale of the these Acts endangered the individual liberty of the subject. Most of this paternalistic legislation was aimed at the new urban workers and helped to make Disraeli popular in the country. It led Alexander MacDonald, of the Amalgamated Society of Engineers, to declare 'The Conservative Party has done more for the working class in five years than the Liberals have done in fifty.'

Tory democracy?

Were these reforms 'Tory democracy' in action? This is a question which must be carefully considered. First of all, remember that this was a phrase used some years later by Randolph Churchill when describing the work of the ministry. Disraeli himself never used the term. Nor was radical-Toryism a new phenomenon; individual Tory humanitarianism and concern for social reform already had a notable nineteenth-century history. Furthermore, one can easily exaggerate the importance of the social reforms of the ministry and Disraeli's personal contribution. In fact, he displayed little capacity for administration or executive leadership.

It was Gorst's idea that the Party should organize the Conservative working man and develop a programme of social reforms aimed at attracting him. Furthermore, the detailed reform work of the ministry was carried out by the Home Secretary, Richard Cross. Disraeli failed to mention the Conservatives' social reform contribution in the election of 1880 or make it policy thereafter.

In some ways Disraelian Conservatism was a practical response to the needs of the community left untouched by Gladstone's more administrative reforms. Most of the reforms were, however, moderate, limited or permissive. They expressed, as Paul Smith emphasizes, a policy of 'empiricism tempered by prejudice'.

The onset of economic depression, plus difficulties in Ireland (which Disraeli was prepared to ignore) and abroad, also meant that reforms quickly dried up after 1876. Disraeli's social reform proposals were, therefore, primarily electioneering gestures but they did achieve their major objective; they got the Tory Party into power and showed that it was capable of governing.

Despite the usefulness of the Conservatives' reforms it would be wrong to regard their reform outlook as in principle different from that of the Liberals or to see it as an attempt to implement a reform programme based on a belief in 'Tory democracy'. Disraeli knew that such an outlook did not command wide support in the Party but recognized its usefulness in maximizing the tide of Conservative sentiment now flowing away from Gladstone in the English boroughs.

Foreign and imperial policy

In 1874 Disraeli caught the mood of the nation when he said of Gladstone: 'It would have been better for us all if more energy had been expended in foreign affairs and less in our domestic legislation'. He insisted that the Conservative Party was the true representative of Britain's national interests abroad and his own actively interventionist outlook meant that he took over the mantle of Palmerston rather than of Castlereagh.

From the moment he took office his gaze turned to the East. At Crystal Palace in 1872 he had reminded the English people that they were at the centre of a great maritime Empire 'whose flag floated over many oceans.' He wished to encourage (at least verbally) an active policy of imperial growth. Suez canal shares were purchased; slavery was abolished in the Gold Coast in 1874, Fiji was annexed, the Queen was made Empress of India, and a 'forward policy' was pursued in Afghanistan and the Transvaal.

You must know the general details of these events, be able to assess the importance of each and decide whether you think they added up to a policy of 'new imperialism.' You should also consider Gladstone's reactions to this active imperial policy (*e.g.* in the first Midlothian campaign) and why he opposed it. Remember also that he had to deal with the legacy of these events in his second ministry.

Next, you must study Disraeli's handling of the Bulgarian crisis and the Eastern Question. Why did Disraeli get involved in the crisis and why did he support the Ottoman Empire rather than the other European powers? Why did he reject the Treaty of San Stefano (1878) and what did he gain at the Congress of Berlin?

The 1880 election

Gladstone had come out of retirement to attack Disraeli's 'inhuman' Eastern policy and denounced his imperialism in a second Midlothian campaign as aggressive, immoral and expensive. Disraeli failed to refute Gladstone's accusations and clearly outraged Christian and humanitarian sentiment in Britain. Disraeli was also under attack from within his Party because, as the economic depression deepened, this earlier (apparent) champion of protection now resisted every demand for aid. Consequently the more militant agricultural interests eventually abandoned the government for the Liberals.

The impact of the depression was more severe in Ireland, where the collapse in production and rents led to the formation of the Land League and a more aggressively obstructionist Irish Party in the Commons. Tired and in ill-health, Disraeli failed to campaign effectively; an elated Gladstone won the election with an overall Liberal majority of 46. Disraeli did not long survive the defeat and died in April, 1881.

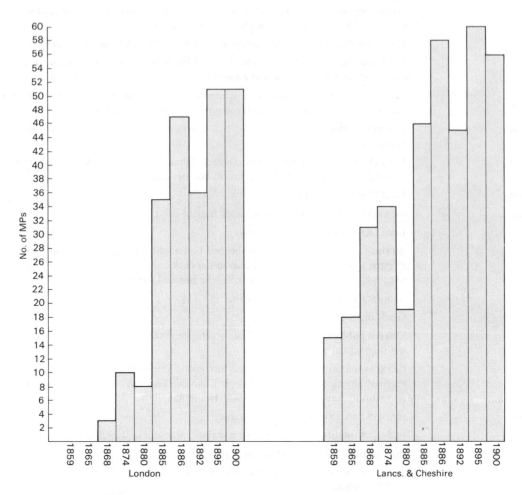

Fig. 14 The 'New Conservative Party' in London, Lancashire and Cheshire

Comparative analysis

Once you have studied the different personalities and political philosophies of these two great rivals – Gladstone and Disraeli, or as Bismarck called them, 'the Professor' and 'the old Jew', you must be able to compare their policies in domestic reform, foreign affairs and imperialism (see unit 22).

There were fundamental differences. Yet, although the Liberals favoured individualism and the Conservatives paternalism in their approach to domestic affairs, there remained some common ground. After all did not Gladstone, as well as Disraeli, firmly believe in the Crown, Church, constitution, aristocracy and reform? Besides, most of his radical outbursts against imperialism occurred while he was in opposition – when in government he was forced to invade Egypt in 1882.

Both have many excellent acts to their credit but neither came near to founding a welfare state and, though both were concerned for democracy, they both kept the people at arm's length. Disraeli certainly had a successful rapport with the Queen, which Gladstone could never match, but perhaps Gladstone's more formal approach was the more prophetic for the future role of a constitutional monarch?

Disraeli's Conservatism was the product of a literary and imaginative mind – a mind that seemed to lack the ability to cope with the detailed legislative and administrative work needed for a reforming ministry. He liked to deal in romantic generalities, but as an intuitive politician he saw the weaknesses of moralistic Liberalism.

Gladstone was a born administrator who possessed the ability to master detail in a manner that astonished contemporaries, but his moral crusades often blinded him to practical realities. Morally and intellectually Gladstone was Disraeli's superior. Disraeli's greatness was as a politician, but overall, his character inspired more distrust than respect. Recent research has shown that most of the worst things said about Disraeli appear to be true, while most of the things said in his favour seem to have little foundation. His one settled aim was to reach the top.

This success in the Victorian era was indeed remarkable. Disraeli had taken a Party which was antiquated and unpopular, held it together in opposition for 20 years, gave it a modern outlook and a popular philosophy which in the end saw it decisively outlast Liberalism. It is not surprising, therefore, that modern Conservatism regards him, and not Peel, as its principal architect.

21.4 CHRONOLOGY

1804	**December**	Disraeli was born, the son of Isaac D'Israeli, a well-known author.
1826	**April**	His book *Vivian Grey* was published; this largely autobiographical work brought him into London society, where his ringlets and extravagant dress made him notorious.
1837	**July**	Disraeli became MP for Maidstone.
1842		JOINED THE 'YOUNG ENGLAND' PARTY, an aristocratic, cavalier movement founded by George Smythe (eldest son of Lord Strangford) and Lord John Manners (second son of the Duke of Rutland). It was against the liberal-utilitarian spirit of the day and offered an alliance of the old aristocracy and the working classes against the new middle-class manufacturers and radicals; it betrayed a nostalgic wish for the pre-industrial age.
1844	**May**	Disraeli published his book *Coningsby*, which was an acid analysis of Peel's Conservatism.
1845	**May**	*Sybil* published. In both novels he attacked the Whigs and advocated monarchy, social democracy and Anglicanism; showed awareness of, and sympathy with, the plight of the poor.
1846	**June**	CORN LAW REPEAL: Disraeli co-operated with Lord George Bentinck and the Protectionists in the attack on Peel; this was the turning point in Disraeli's career,
1849	**February**	DISRAELI APPOINTED LEADER OF THE PARTY after the death of Lord George Bentinck in September, 1848.
1852	**February**	Lord Derby's first administration: Disraeli, Chancellor of the Exchequer, abandoned protection as a policy; his first budget was demolished by Gladstone.
1858	**February**	Lord Derby's second administration.
1859	**March**	DISRAELI'S REFORM BILL, with 'fancy franchises', rejected.
1866	**July**	Lord Derby's third administration.
1867	**August**	SECOND REFORM ACT, which established household suffrage, was passed by Disraeli; this was referred to by Lord Derby as 'a leap in the dark'. REDISTRIBUTION ACT passed to exclude urban and suburban voters from the counties.
1868	**February**	DISRAELI'S FIRST ADMINISTRATION.
	December	Gladstone's first administration.
1870	**May**	Disraeli published his book *Lothair*, which portrayed the aristocracy as attractive and dignified but unpractical and aimless.
1872	**June**	CRYSTAL PALACE SPEECH: Disraeli advocated 'imperialism and social reform'; those policies received strong support from Joseph Chamberlain (and later Lord Rosebery) and proved a popular alternative to Liberalism and Home Rule.
1874	**February**	DISRAELI'S SECOND ADMINISTRATION.
	August	PUBLIC WORSHIP ACT: suppressed ritualistic practices in the Church of England.
1875	**June**	PUBLIC HEALTH ACT: consolidated previous statutes.
	August	ARTISANS' DWELLING ACT: encouraged slum clearance and rebuilding in towns of more than 25 000 inhabitants. CONSPIRACY AND PROTECTION OF PROPERTY ACT: allowed peaceful picketing.

EMPLOYERS AND WORKMEN ACT: made breaches of contract civil and not criminal offences.

SALE OF FOOD AND DRUGS ACT: prevented food adulteration.

MERCHANT SHIPPING ACT: introduced 'Plimsoll line'.

	November	PURCHASE OF SUEZ CANAL SHARES; this secured control of the route to India.
1876	April	QUEEN MADE EMPRESS OF INDIA.
	August	SANDON'S EDUCATION ACT: a step towards compulsory education. Disraeli made Earl of Beaconsfield.
1877	April	ANNEXATION OF TRANSVAAL. Russo-Turkish War.
1878	March	Treaty of San Stefano.
	June	CONGRESS OF BERLIN: a triumph for Disraeli; he returned from Berlin bringing 'peace with honour'.
	November	Afghan War.
1879	January	Zulu War. British defeated at Isandhlwana.
1881	April	Death of Disraeli in London.

21.5 QUESTION PRACTICE

1   Would you agree that Disraeli 'educated his Party' or had a coherent policy of Tory democracy?

Understanding the question

This question asks you to assess Disraeli's contribution to the Conservative Party before and after 1868. Did he take over an antiquated Party and modernize it in preparation for government in 1874? While in office did the Party follow a coherent, *i.e.* consistent, (Disraeli-inspired) programme of Tory democracy? Be able to define and analyse the term 'Tory democracy'.

Knowledge required

The development of Disraeli's political philosophy. His main contribution to Party thinking and action prior to 1868. The major legislative enactments of his ministry. The practical or philosophical reasons for introducing this legislation.

Suggested essay plan

1   Criticism of Disraeli's political philosophy is both easy and justified. He was not a serious politician and, despite occasional enthusiasms, he rarely followed up his ideas. Often he seemed to latch on to slogans ('empire', 'social reform') for their vote-catching potential only. His behaviour was certainly incoherent; he seemed prepared to say anything or adopt any policy to secure parliamentary allies, even such unlikely allies as the Irish or radical parties.

2   But it does not follow that his ideas were necessarily incoherent. In fact they were probably more consistent (unlike his behaviour) than that of most politicians, if only because they were more romantic, less pragmatic. They were worked out at leisure in his books *Sybil* and *Coningsby* and in the rhetoric of 'Young England'. And despite certain adjustments – during the later 1840s Disraeli came to realize that his 'back-to-the-land' agrarianism was impossible – these early theories were held (albeit cynically and sporadically) throughout his career.

3   The word 'democratic' is misleading, however; although Disraeli introduced household suffrage in the boroughs, this was only because he thought the working classes would be more socially conservative, and more easy to manipulate, than the largely dissenting middle and artisan classes supporting Gladstone. He opposed individualism and the ballot, and believed that popular participation should be limited to receiving 'charity' (better housing, protection against adulterated food, etc.) from upper-class paternalists and paternal local cooperations.

4   In fact, Disraeli's 'democracy' consisted in his direct appeal (that is, when he could be bothered) to popular sentiment; he did this occasionally by outdoor speeches, but more by slogans and gestures (his nationalistic foreign policy, the Suez Canal purchase, the Empress of India). Gladstone did the same, but spoke to a different – more moralistic, self-righteous, often dissenting – audience.

5   Of course, he didn't educate his Party. He was deeply mistrusted until the election victory of 1874. The country squires shared his dislikes (*i.e.* Gladstone, political economists, bureaucrats and centralizers) but had no time for his romantic, paternalist idealism.

6   Moreover, the Party was being transformed under Disraeli after 1865, but in quite the opposite direction to aristocratic paternalism. It was becoming a middle-class Party, whose goals included individualism and orthodox political economy. Only on the question of nationalism was Disraeli in line with Conservative thought.

7   And yet, posthumously, Disraeli has proved a potent force in Conservative thinking; the Party has consistently oscillated between 'one-nation' paternalism (as with, in more recent times, Edward Heath) and *laissez-faire*, middle-class competition (as with, even more recently, Margaret Thatcher). The former always appeals to Disraeli's example, although it has abandoned his localism (government by landlords, churches, corporations, trades unions) for government by bureaucracy and the corporate state. But at least Disraeli's influence survives, notwithstanding its lack of widespread appeal in his own lifetime.

2   Compare and contrast the domestic legislation (excluding Irish) of Gladstone and Disraeli's ministries between 1868 and 1880.

Understanding the question

This calls for a detailed knowledge of the main legislative work of Gladstone's ministry of 1868–74 and of Disraeli's ministry of 1874–80. It asks you to emphasize similarities and differences in the legislative programme enacted by the two governments. Note that Irish legislation is excluded.

Knowledge required

The major legislative enactments of Gladstone's first and Disraeli's second administrations. The practical reasons each government had for introducing this legislation. An understanding of the political philosophy which motivated each government and the wishes and demands of its supporters.

Suggested essay plan

1   Because of the personal hatreds of Disraeli and Gladstone, and the highly developed two-party political system, the first example perhaps of 'adversary politics' in Britain, it is easy to imagine that the policies of the Liberal and Tory governments must have been vastly different, but there was much common ground.

2   Both parties accepted the essential features of Victorian society. Few Liberals wished to move society in a more socialist or interventionist direction; few Conservatives still thought of trying to reverse the industrial progress of the previous 70 years. Thus it was Disraeli who refused to protect agriculture against cheap American corn.

3   Moreover, by 1870 there was (thanks to Palmerston's negative influence) a backlog of administrative and legislative reform waiting to be enacted. Much of this was promulgated behind the scenes by civil servants and other public officials. The main thrust of reform was departmental rather than political (for example, Simon at the Board of Health, Sclater-Booth at the Poor Law and Local Government Boards). The main inspiration behind the Conservative reforms was the Home Secretary (Cross), who worked well with official advisers but had few party-political instincts. Disraeli, who had many such instincts, did not even try to claim credit for his domestic reforms in the 1880 election.

4   When Parliament did become involved in reform legislation, it usually operated on a cross-bench rather than a party line – e.g. Pakington and Forster cooperated on educational reform, Mundella and Cross on trade union legislation. Both parties were against centralization and believed that legislation should be permissive rather than mandatory. The fact that the Liberal Education Act of 1870 was passed with a majority of Conservative votes and against a majority of Liberal votes illustrates the cross-bench nature of the proposal.

5   However, there were differences. Liberals, concerned mainly with individual freedom, concentrated on political, institutional, and legal reform (e.g. the ballot, university tests, and the legal rights of trades unions) which would create equality of opportunity, so that the clever, ambitious, and hard-working might thrive. Conservatives had a more paternalist approach and were prepared to see practical improvements imposed on the working classes, even on the less respectable and ambitious of them (*e.g.*, the Artisans' Dwellings Act and the Sale of Food and Drugs Act).

In theory, Sandon's Conservative Education Act was prepared to force parents to send their children to school, which the Liberal Act refrained from doing. More important still, Disraeli was prepared to allow peaceful picketing, whereas Gladstone believed that blacklegs and strikers had equal rights.

6   In conclusion, while there was much common ground between Liberal and Conservative domestic legislation – they were after all fighting for the support of the same new electorate – the main contrast lay in the greater attachment of the Liberals to individualism and of the Conservatives to paternalism.

3   What did the Disraelian Conservative Party stand for?

Understanding the question

This is a straightforward question which calls for an understanding of the political beliefs – in domestic, imperial and foreign policy – of Conservatives under Disraeli's leadership. The question is, did the beliefs of Disraeli and the Party coincide? Did he educate the Party or simply respond to changing political and electoral circumstances?

Knowledge required

An understanding of what policies the main body of Conservatives believed in after 1846. The development of Disraeli's own political beliefs. Disraeli's position in the Party. To what extent the work of the 1874–80 ministry reflected Disraelian and Conservative ideals at home and abroad.

Suggested essay plan

1   Start by pointing out that the Conservative Party only just managed to stand Disraeli. Despite the posthumous adoration, he was never really trusted and only tolerated because he had Lord Derby's support. As late as 1871 there were moves to oust him in favour of Derby's son. So Disraeli and Conservatism should not be regarded as synonymous.

2   Therefore the Party did not *necessarily* stand for what he stood for. Some historians believe it stood for Peelite pragmatism – sound administration, a 'governmental ethic', cautious, un-ideological, essentially middle-class reform. They believe that Disraeli destroyed the Party in 1846, spent 20 years in the wilderness as a punishment, and laboriously put together a Peel-type Party after 1867.

3   In fact, the Tory Party was never Peelite – it was a collection of squires who disliked the pretensions of central government (which is why they ejected Peel) and had no burning desire for office; they were perfectly content with government by the 'conservative' Palmerston; Derby was not particularly ambitious for office, being too fond of his horses, and Disraeli – despite his ambition – was too much a political playboy for single-minded place-hunting.

4   Palmerston's death (1865) and Gladstone's accession (1868) changed all this. Thenceforward Conservatives had to seek office for fear that the 'mad' Gladstone would dismantle all that was dear in society. Office for its own sake, but backed up with a claim that they were the Party of sound government and national security, became the Conservatives' main *raison d'etre*.

5   So Conservatism was essentially a negative reaction to Liberalism. Disraeli himself had a political philosophy, though he was hardly serious or consistent in applying it. It appears in his books *Sybil* and *Coningsby* and in the slogans of 'Young England'. It too was stronger on negatives than positives; it reacted against the utilitarian, centralizing, bureaucratic and individualistic tendencies of the age, and also against its widespread evangelical morality (as expressed in temperance reform and Lord's Day observance); to begin with it was also anti-industrial and agrarian, but Disraeli surrendered these features as they became impractical (declaring the Corn Laws 'dead and damned' in 1852).

The positive ideas were vague, but centred round an ideal of organic communities, in which aristocratic paternalists would protect the poor against the exploiting viciousness of middle-class entrepreneurs. Hence his interest in permissive social reform legislation, like artisans' dwellings and public health. Above all, he preferred group action, even peaceful picketing by organized workers, to individualistic action, such as the rights of blacklegs, which Gladstone defended.

6   Prior to 1868 Disraeli had described the colonies as 'millstones round our necks' but after Palmerston's death he intuitively played on the nation's innate jingoism. His Crystal Palace speech advocated imperialism (at least verbally) and he liked to behave as though Britain was still Europe's strongest power. He had no coherent strategy, and his Palmerstonian-style nationalism was never completely popular in the Party (for example, on the Eastern Question).

7   To conclude, Disraeli was not the Conservative Party and it

is unlikely that many in it shared his vague ideals. All it stood for was power, and its main aim was to limit the damage done by the dreadful 'old man in a hurry'.

4   Document question (Disraeli and the Suez Canal shares): study extracts I and II below and then answer questions (a) to (g) which follow:

Extract I

'As you complain sometimes... that I tell you nothing, I will now tell you a great State secret, though it may not be one in 4 and 20 hours... a State secret, certainly the <u>most important of this year, and not one of the least events of our generation</u>. After a fortnight of the most unceasing labour and anxiety, I... have purchased for England <u>the Khedive of Egypt's interest in the Suez Canal</u>.

'... The day before yesterday, Lesseps... backed by the French government... <u>made a great offer</u>. Had it succeeded, the whole of the Suez Canal would have belonged to France, and they might have shut it up!

'We have given the Khedive 4 millions sterling for his interest, and <u>run the chance of Parliament supporting us</u>....

'<u>The Faery</u> is in ecstasies about 'this great and important event' (and) wants 'to know all about it when Mr D. comes down to-day'.

'I have rarely been through a week like the last – and am today in a state of prostration... sorry I have to go down to Windsor....'
(From a letter by Disraeli to Lady Bradford, dated 25 November, 1875)

Extract II

'Lord Derby... the Foreign Secretary is a responsible statesman ... with that straightforwardness and common sense for which he is eminent, he told us at Edinburgh that the affair which had created so much sensation at home and abroad was not at all the sort of thing it had been represented to be.... <u>He repudiated with scorn the idea that England aspired to an Egyptian protectorate</u>.... <u>What had really been accomplished was a very ordinary affair</u>.... The Khedive had certain shares.. the English government... have acquired them, not to give England any special... <u>influence</u>, nor to secure any <u>exclusive advantage</u>, but to keep open a communication for the benefit of all, which to England is of supreme importance... I gladly take Lord Derby at his word. But now this grand affair is reduced to... moderate dimensions... we may criticize it.... Upon the main ground by which this purchase is justified – namely, the determination to secure a <u>free passage between the Mediterranean and the Indian Ocean</u>, there will be no conflict of opinion. That is a policy in which England is <u>profoundly interested</u>.... But that which has not hitherto been explained, and what remains to be shown, is in what manner and to what extent <u>this investment really does conduce to that desirable object</u>.'
(From a speech by Sir William Harcourt at Oxford, 30 December, 1875)

Maximum marks

(a) What was 'the Khedive of Egypt's interest in the Suez Canal', why was that 'interest' for sale at this time, and why had the French 'made a great offer'?   (5)

(b) Explain the circumstances of Disraeli's government having to 'run the chance of Parliament supporting us'   (5)

(c) Comment on Disraeli's reference to 'the Faery'   (3)

(d) Why did Disraeli claim that the affair was the 'most important of this year, and not one of the least events of our generation' whereas Lord Derby was at pains to say that 'He repudiated with scorn the idea that England

aspired to an Egyptian protectorate' and that 'What had really been accomplished was a very ordinary affair'?   (5)

(e) Why was England 'profoundly interested' in securing 'a free passage between the Mediterranean and the Indian Ocean'?   (3)

(f) What were the grounds for Harcourt questioning whether 'this investment really does conduce to that desirable object?'   (4)

(g) Why and how did England come, during the next ten years, to have the 'special influence' (line 21) and to secure the 'exclusive advantage' which Lord Derby had repudiated?   (7)

21.6 READING LIST

Standard textbook reading

R. K. Webb, *Modern England* (Allen and Unwin, 1980 edn.), chapter 8.

A. Wood, *Nineteenth-Century Britain* (Longmans, 1960), chapters 14, 15.

Further suggested reading

P. Adelman, *Gladstone, Disraeli and Later Victorian Politics* (Longmans, 1970).

R. Blake, *Disraeli* (Methuen, 1966).

M. Cowling, *Disraeli, Gladstone and Revolution* (CUP, 1967).

C. C. Eldridge, *England's Mission: The Imperial Idea in the Age of Gladstone and Disraeli* (Macmillan, 1973).

R. R. James, *Lord Randolph Churchill* (Weidenfeld and Nicolson, 1959).

R. B. MacDowell, *British Conservatism 1832–1914* (Faber, 1959).

F. B. Smith, *The Making of the Second Reform Bill* (CUP, 1966).

P. Smith, *Disraelian Conservatism and Social Reform* (Routledge, 1967).

22   Gladstone and Liberalism

22.1 INTRODUCTION

William Ewart Gladstone, one of England's greatest statesmen, was the archetype of Victorian Liberalism. He regarded politics as a Christian duty in the pursuit of which, according to his biographer Philip Magnus, he 'incurred martyrdom for himself as well as for the Liberal Party, which became his instrument, and which, before it broke in his strong grip, owed its unity and enthusiasm almost entirely to him'.

Yet he began his political career as a most reactionary Tory and High Churchman displaying little sympathy with either Liberalism or democracy. He ended his days as the People's

William, the prophet of Liberalism, and supporter of all nations 'rightly struggling to be free'.

22.2 STUDY OBJECTIVES

1   Gladstone's background and early political beliefs; *The State and its Relations with the Church*, makes him the rising hope of stern and unbending Tories. His abandonment of high Tory principles while at the Board of Trade under Peel; the effect of Maynooth, Corn Law repeal, and the collapse of the Oxford Movement on his ideas; his work as Chancellor under Aberdeen, followed by his resignation in 1855 and four-and-a-half years in the political wilderness.

2   The background to English liberalism; provincial, middle-class, predominantly non-conformist culture; the contributions of the *Edinburgh Review,* philosophic radicalism, and the Whig Party; the contribution of the Peelites; free trade, meritocracy, equal opportunity, 'respectability', self-help, 'peace, retrenchment, and reform' the main Liberal tenets.

3   Political aspirations of middle-class liberals in the 1850s and 1860s; the rise of the *Daily Telegraph* and the provincial presses; the impact of the Crimean War in galvanizing such elements against the old aristocracy; Manchester liberalism (Cobden and free-trade pacifism) against Birmingham liberalism (mainly radical and dissenting); the Education League, Liberation Society, and the United Kingdom Alliance all focus extra-parliamentary liberal enthusiasms; grassroots party organization and the formation of the National Liberal Federation.

4   Formation of the Liberal party in parliament after the Willis's Tea Rooms meeting; the impact of the American Civil War, Italy, Schleswig-Holstein, and Governor Eyre; Gladstone's budgets and the battle against Palmerston on defence spending; his demagogic extra-parliamentary appeals make him the 'People's William'; his alliance with Bright.

5   The death of Palmerston; Gladstone's defeat at Oxford and 'unmuzzling'; put himself at the head of the Radical section of the Party; parliamentary reform, church rates, and Irish disestablishment consolidated this position – though putting him in a false position since he did not wish to travel much further than this; the landslide electoral victory of 1868 and the formation of a balanced 'Whig-Radical' ministry.

6   The first ministry led to the collapse of the alliance between parliamentary and popular liberalism; Irish land, the Education Act, and the Irish Universities were the main areas of controversy; the positive achievements of the ministry (Army purchase, the ballot, Civil Service, and judicature reforms, licensing) could not prevent the defeat of 1874.

7   Gladstone's retirement; his return on the back of the Bulgarian agitation; the Midlothian campaign and second ministry; the Union of Hearts with the Irish Party, and land legislation; problems with Chamberlain and the Radicals; the occupation of Egypt and other colonial entanglements; the third Reform Act and its effects.

8   The 1885 election caused the first-ever Liberal defeat in the English boroughs; Gladstone's adoption of Home Rule; moral conviction, political manoeuvre (especially the wish to subordinate Chamberlain), and administrative convenience (the direct rule of Ireland was too costly for a cheese-paring government) all contributed to Gladstone's decision; the secession of the Liberal Unionists and fall of the third government.

9   Gladstone's final years and brief fourth government; the Newcastle Programme and the possibility that Gladstone became more interventionist in his political philosophy at this time; his powerful posthumous legacy on the Liberal Party was not altogether beneficial.

22.3 COMMENTARY

The rivalry

British politics from 1868 to 1881 were dominated by the rivalry between two of England's most gifted political personalities – William Ewart Gladstone (1809–98) and Benjamin Disraeli (1804–81). This rivalry – arising from differences in race, birth, education and temperament – was both personal and political.

Gladstone was the incarnation of Victorian Liberalism; serious, determined, progressive; powerful in oratory and intellectual energy. Disraeli also had great gifts. His superb courage, driving ambition, exceptional political intelligence and a mastery of clever speech and witty repartee made him a disconcerting political opponent.

The early years

Gladstone, the fourth son of a wealthy Liverpool merchant, had a brilliant career at Eton and Oxford. A deeply religious man (he was originally destined to enter the Church), he became increasingly interested in politics, which he came to regard as a moral crusade based on the best instincts of the people. To Gladstone every important political question was in part a moral one: whatever was morally right could never be politically wrong.

In Lord Macaulay's words, Gladstone, who had begun his political career as 'the rising hope of those stern and unbending Tories,' and had denounced the Reform Act of 1832 as 'anti-Christ', yet ended his days as the prophet of enlightened, Victorian liberalism. He translated into fact the ideas of Smith, Bentham and Mill. More than any other man he was responsible for laying the foundations of modern government and making possible the transfer of power from the aristocracy to the new democracy.

During the 1830s he came under the influence of Peel and, though he abandoned his high-Tory principles, he remained a conservative for the remainder of his life – wanting to preserve a hierarchical society, ruled by an efficient, benevolent and landed élite. As Michael Barker says, Gladstone was 'a curious combination of conservative instincts and radical opinions.' From Peel he learned that politics was not about the pursuit of power but the use of power to achieve noble objectives that were national or, in his own case, Christian. An appreciation of the influence of Peel is crucial to an understanding of the evolution of Gladstone's political thinking.

As Chancellor of the Exchequer under Palmerston in the 'fifties and 'sixties, Gladstone demonstrated his abilities as a parliamentarian and financier and learned about political expediency and restraint. Like Peel, he sought to improve the condition of the people through fiscal and financial measures.

You must know the details of his famous budgets that made England a completely free-trade nation. Note that, although the economic achievement was admirable, the social cost was high – for his policy of retrenchment meant a halt to government expenditure on much-needed social reforms. His clashes with Palmerston over the latter's expensive foreign policy also drove him along the road of political radicalism.

Because of Disraeli's attack on Peel in 1846 he emphatically refused to rejoin the Conservative Party. Instead, after some typical Gladstonian soul-searching, he joined the Whig-Peelite-Radical alliance which formed around the issue of Italian unification in 1859.

Emergent Liberalism

Liberalism was also emerging in the constituencies tutored by liberal editorials in the new, daily provincial press. These consistently advocated civil and religious liberty and the belief in the

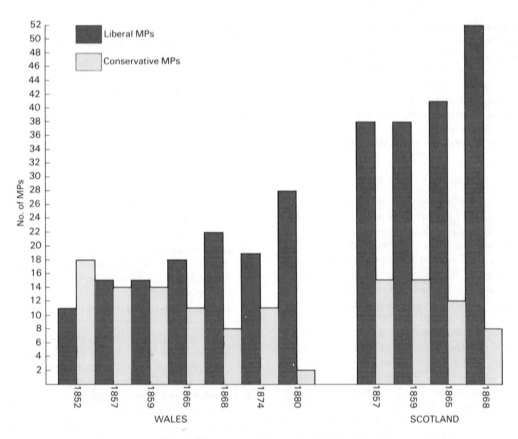

Fig. 15 The growth of 'Celtic fringe' liberalism

idea of progress. Non-conformists also hoped that the new Liberal Party would adopt their progressive policies in education, politics and religion. From their militant religious outlook came their adoption of 'good causes' that gave them their dynamic quality in politics. The new 'labour aristocracy' were also potential supporters. They wanted more legal freedom for trade unions and recognition from the government of their new-found respectability and independence.

The men who noticed the emergence of these groups and the need to bind them to the newly-founded Liberal Party in the Commons were Gladstone and Bright. After Gladstone's 1864 speech ('every man ... is morally entitled to come within the pale of the constitution') he was hailed as the 'People's William.' Bright cultivated this image during the Reform Bill crisis of 1866 – and indicated Gladstone as the natural leader of the next Liberal government.

Gladstone's part in the reform struggle was a confused one. In truth, he advocated only a modest extension of the suffrage but managed to sound more democratic than he was. The people, however, did not forget his 1864 speech and, despite Disraeli's success in passing a Second Reform Act in 1867, voted Gladstone into power in the election that followed in the autumn of 1868.

Gladstone's first ministry, 1868–74

When Gladstone was informed of the Liberal victory at the polls he said: 'Praise be to God. My mission is to pacify Ireland.' Accordingly, his early efforts were devoted to Irish land and religious problems. After these, his carefully-balanced Whig-Radical ministry embarked on a programme of constructive legislation. The new Party stood for progress and improvement.

Uninterested in social reform, Gladstone lacked Disraeli's understanding of the material needs of the working class. To Gladstone this would be taking into the hands of the state the business of the individual man. Like the Whigs of the 'thirties, the Liberals were much better at institutional rather than social

reforms. Nevertheless, the new Cabinet initiated a series of fundamental administrative reforms – attacking entrenched privileges, encouraging individualism and merit and thereby laying the foundations of the modern state.

The Liberal reforms

You must now turn to the reforms of Gladstone's first ministry and familiarize yourself with the main details of Forster's Education Act, Cardwell's Army Act; and the Civil Service, Trades Union, Law, Secret Ballot and Licensing Acts. Try to find out why each Act was introduced – especially the pressures or supposed needs that prompted this major programme of legislative activity. Ask yourself also which social groups benefitted most from the reforms and in what ways they were especially 'liberal' in nature.

Ireland

Note Gladstone's attempts to pacify Ireland by disestablishing the Anglican Church of Ireland (1869), introducing the first Irish Land Act (1870) and his proposed Irish University Reform Bill (1873) which helped to bring about the government's downfall the following year. (See Unit 24).

Foreign policy, 1868–74

Gladstone's main preoccupation was with domestic and administrative matters. He had little interest in foreign affairs, which he approached with some caution. Having like Castlereagh, a European outlook, he was willing to work in concert with the great Powers to maintain peace.

Rejecting Palmerston's jingoism, in matters of Empire he tended to favour non-intervention. His moral instincts, however, were inclined to make him interventionist – for example, in the Bulgarian massacres. Unlike Disraeli, he believed in the right of small nations to govern themselves. Italian unification made him a Liberal – a commitment emphasized by his attitude to Ireland and the Bulgarian Christians.

He had to deal with three major problems during his first ministry: the Franco-Prussian war (1870), the Black Sea clauses and the Alabama affair. After reading about Gladstone's handling of these issues you may feel that his responses were wise and statesmanlike in the circumstances but that the electorate felt that he had failed. Try to discover why this was so. Finally, you must investigate the reasons for the fall of the ministry in 1874.

Assessment of the first ministry

In general the record of the first administration is impressive, but it disappointed the radical members of the new Liberal Associations and as a result broke the alliance between the parliamentary Liberal Party and popular liberalism. Nevertheless, the traditional perks of a caste society had been thrown open to merit. Having by his policy of free trade prepared the way for the emergence of the middle classes economically, Gladstone had now opened the doors of administrative opportunity to the most ambitious.

It is not surprising that the landed classes believed that he was undermining the old order but it was precisely this attack on prejudice and privilege in England and Ireland which was the essence of Gladstonian liberalism. His government had satisfied many middle-class demands and was a triumph of utilitarianism and efficiency – the victory of the Manchester school of values. Little wonder, then, that the parliamentary Front Bench in 1874 looked, in Disraeli's words, like 'a range of exhausted volcanoes'.

Gladstone's second ministry, 1880–85

By 1880 the political situation had changed. Gladstone felt that the essential tasks of liberalism had been performed; but he had been inspired by the Bulgarian massacres in 1876 to return to politics.

In general, his second ministry proved to be a somewhat barren session in domestic affairs, made bitter by the Bradlaugh incident and the Dilke divorce scandal. Gladstone was overwhelmed by difficulties in Ireland and abroad and proceeded cautiously. He relied almost entirely on the old Whigs – much to the annoyance of emerging young radicals like Chamberlain.

Gladstone disliked Chamberlain's brand of 'social-liberalism', with its emphasis on state interference. Such ideas, he believed, might put Party unity at risk. Nevertheless, note that three important democratic reforms were passed – the Corrupt Practices Act (1883), which set a limit on electioneering expenses; the Third Reform Act (1884), which gave the vote to every adult male and thereby enfranchised two million agricultural voters; and the Redistribution Act (1885), which set up single-member constituencies. Salisbury got his reward for insisting on the latter at the 1885 election, when the Conservatives, for the first time ever, won a majority of the English urban seats.

Foreign policy, 1880–85

While in opposition, Gladstone had vigorously opposed Disraeli's jingoistic imperialism and, in 1880, he saw his task as one of returning England's foreign policy to normalcy. However, the legacy of 'Beaconsfieldism' proved intractable; Gladstone's idealistic attempts to create a new, liberal, European foreign policy failed. Gordon's death at Khartoum even managed to dissipate popularity earned by domestic reforms.

Find out why Gladstone's foreign policy in Egypt, the Sudan, Afghanistan and the Transvaal was unconvincing and unpopular.

Ireland

Gladstone's second Irish Land Act 1881 failed to solve the Irish

problem and by 1886 his total commitment to Home Rule had split the Party and had dominated the work of his last two ministries 1886 and 1892–93. (See Unit 24).

Gladstonian Liberalism

What, then, were the main planks of Gladstonian Liberalism? In general, Victorian Liberalism was a mixture of ideology, morality and self-interest. It was a doctrine as much as a political party. It advocated civil and religious liberty.

This religious element often led to the adoption of good causes and occasionally to political militancy. It favoured peace and free trade, a non-aggressive foreign policy, equality of opportunity and material progress at home. It believed in political democracy but not social equality. It accepted selective state intervention but only to organize conditions in which individualism might flourish. Every movement away from laissez-faire was thought to be a movement away from the ideal.

The parliamentary Liberal Party, on the other hand, remained Peelite in outlook and adopted the more practical objectives of sound administration, limited reform, efficiency and careful, but not generous government expenditure. It was still dominated by the great Whig, landed magnates. While these were prepared to accept measures to encourage efficiency and improvement, they found many of the Radical demands unpalatable and some of Gladstone's schemes unacceptable.

Gladstone approved these basic tenets of peace, retrenchment and reform but added his own fundamental belief in morality and dedication to Christian purpose. Although his moral outbursts gained him popular support, he was not a democrat. Even in the 1870s he still believed that the landed classes were the only class trained for government: 'I am a firm believer in the aristocratic principle – the rule of the best.'

Many of his progressive reforms – his attacks on privilege – were made in order to make aristocratic government more acceptable in the eyes of the people. Hence his lack of interest in, or sympathy with, Chamberlain's collectivist policies and his failure to foresee the coming age of mass, class-based politics. In later years he became disenchanted with the selfishness of the ruling classes and, after the success of his crusade against the 'Bulgarian horrors', he directed his appeal more and more to the moral and religious instincts of the masses.

Gladstone's true interests were outside politics – to him politics was simply a means to an end, a method of striving for great and noble achievements such as Home Rule for Ireland or aid for Bulgarian Christians oppressed by the infidel Turk. By the 'eighties he came to appreciate that in these policies he could usually rely on the support of the masses – rarely the classes. It seemed that it was always the politicians at Westminster who let him down and so he grew more and more disenchanted with the narrowness and selfishness exhibited by most of his parliamentary colleagues.

Not surprisingly, Gladstone became increasingly out of place in an era of cabinet committees and disciplined party politics. His sense of mission hindered him as a Prime Minister. Once he convinced himself of the justice of a cause he found it impossible to compromise and often fought a lone battle in the cabinet in an effort to preserve a favourite political policy.

Consequently, although four times in office, as a Prime Minister he was not a success. On such issues as Home Rule he could not compromise or be expedient, nor deal with Radicals like Chamberlain, nor capitulate to the baser needs of party. Gladstone wanted to conceive and execute policy on a grand scale and refused to be tied down by the more mundane matters of party political business.

In John Vincent's words, 'Gladstone ruled the country, not

the party'. He understood man but not men. Gladstone's serious view of life, his passionate devotion to progress, his insistence on justice and right – this made hime a leader who was idolized or hated. In the final analysis, Gladstone, the embodiment of Victorian liberalism, was more important for what he was than for what he did.

22.4 CHRONOLOGY

1809	December	Gladstone born, son of John Gladstone, a Liverpool merchant.
1832	December	Gladstone elected as a (protectionist) Tory MP for Newark.
1838	December	Gladstone's first book published, THE STATE IN ITS RELATIONS WITH THE CHURCH. In this he showed his support for the Oxford Movement and identified himself with the attitudes of the Tractarians on important public issues. Lord Macaulay, in his review of the book, referred to Gladstone as 'the rising hope of those stern and unbending Tories.'
1841	September	Became Vice-President (1843, President) of the Board of Trade in Peel's ministry, 1841–46.
1845	February	Resigned over the Maynooth Grant (see under Unit 19).
1846	June	REPEAL OF THE CORN LAWS (see under Unit 19).
1847	August	Elected 'Peelite' MP for Oxford University.
1852	December	Became Chancellor of the Exchequer under Aberdeen. Whig–Peelite coalition.
1853	April	FIRST FREE TRADE BUDGET: tariffs on 123 articles removed.
1855	February	Palmerston became Prime Minister. Gladstone resigned as Chancellor.
1859	June	Meeting at Willis's Rooms (formerly Almacks) where Whigs, Peelites and Radicals agreed to work together for Italian unification. Foundation of the parliamentary Liberal Party.
1860	April	GLADSTONE'S budget left only 43 articles on the tariff. COBDEN'S FREE TRADE TREATY with France.
1861	April	REPEAL OF THE PAPER DUTIES: made him popular with provincial press. Resisted Palmerston's demands for increased military expenditure.
1863	May	POST OFFICE SAVINGS BANK established.
1864	May	SPEECH IN FAVOUR OF FRANCHISE REFORM.
1865	July	Defeated at Oxford University; elected MP for South Lancashire.
1866	March	REFORM BILL INTRODUCED; defection of Lowe and 'the Adullamites'. Bill defeated.
1867	August	Disraeli's Second Reform Act.
1868	December	Gladstone became leader of the Liberal Party; succeeded Lord John Russell. Elected MP for Greenwich. BECOMES PRIME MINISTER OF THE FIRST LIBERAL GOVERNMENT; 112 majority.
1869	July	IRISH ANGLICAN CHURCH DISESTABLISHED (see Unit 24).
1870	March	IRISH LAND ACT. (See Unit 24).
	June	CIVIL SERVICE REFORM: all public offices (except Foreign Office) open to public competition.
	July	Franco-Prussian War.
	August	ELEMENTARY EDUCATION ACT introduced by Forster. Elected School Boards were to 'fill in the gaps' left in the system. Non-conformists objected to increased rate aid for voluntary Anglican schools. The Act aggravated religious strife.
1871	June	ARMY REGULATION ACT: completed Cardwell's measures to modernize the British Army.
		UNIVERSITY TEST ACT: teaching posts at Oxford and Cambridge open to non-Anglicans.
		TRADE UNION ACT introduced by H.A. Bruce. Made unions legal, protected their funds and allowed the right to strike.
		CRIMINAL LAW AMENDMENT ACT: forbade all picketing.
1872	July	BALLOT ACT: introduced the secret Ballot.
	August	LICENSING ACT: tightened control of licences and fixed the hours of closing for public houses at 11 p.m. for the country and 12 p.m. for London.
1873	April	JUDICATURE ACT: consolidated and rationalized the administration of justice.
1874	February	Gladstone defeated in general election. He said of this: 'We have been borne down in a torrent of gin and beer.
1875	January	Gladstone retired from leadership of the Liberal Party.
	September	Gladstone's book, 'THE BULGARIAN HORRORS AND THE QUESTION OF THE EAST' published.
1878	July	TREATY OF BERLIN.
1879	November	THE MIDLOTHIAN CAMPAIGN: electoral campaign against Disraeli's Eastern and imperial policies; Gladstone came out of retirement for this 'pilgrimage of passion'.

Chornology cont.

1880	**April**	Gladstone again Prime Minister: his second ministry; majority of 46. MUNDELLA ACT: compulsory elementary education.
	May	Charles Bradlaugh newly elected MP (and atheist) refused to take the oath.
1881	**August**	IRISH LAND ACT: conceded 'the 3Fs' – fair rents, fixity of tenure, and free sale – to Irish tenants.
1882	**May**	MARRIED WOMEN'S PROPERTY ACT: granted married women the same rights over their property as single women.
1883	**August**	CORRUPT PRACTICES ACT: fixed a limit for election expenses in proportion to the number of voters.
1884	**December**	THIRD REFORM ACT: adult manhood suffrage; added two million voters to the electorate.
1885	**June**	REDISTRIBUTION ACT: single-member constituencies established. Chamberlain's 'Unauthorized Programme' published.
		Salisbury's first administration.
1886	**February**	Gladstone again Prime Minister; his third ministry.
	April	IRISH HOME RULE BILL defeated; defection of Chamberlain and 93 Liberal-Unionists (see under Unit 24).
	July	Salisbury's second ministry.
1891	**October**	THE NEWCASTLE PROGRAMME: Liberal manifesto, radical in tone; advocated Home Rule, triennial parliaments, disestablishment of Welsh and Scottish Church, 'one man, one vote', House of Lords reform.
1892	**August**	Gladstone yet again becomes Prime Minister; his fourth ministry.
1893	**September**	SECOND IRISH HOME RULE BILL was defeated. Lords reject it 419–41.
1894	**March**	Gladstone retired; succeeded by Lord Rosebery.
	April	Harcourt's budget introduced death duties.
1895	**June**	Salisbury's third ministry.
1898	**May**	Death of Gladstone.

22.5 QUESTION PRACTICE

1   Why was Gladstone able to emerge as unchallenged leader of the Liberal Party by 1868?

Understanding the question

The question is slightly misleading in so far as it implies that the Liberal Party existed and that Gladstone rose to power within it, prior to 1868. In fact, though certain elements of Liberalism existed before Gladstone, the Party and the leader rose to prominence together.

Knowledge required

An understanding of Gladstone's early career and the evolution of his political beliefs. The influence of Peel, and his downfall, on Gladstone's career after 1846. The emergence of provincial, middle-class, non-conformist liberalism in the 1850s and 1860s. Gladstone's fusing of the newly-founded parliamentary Liberal Party and grass-roots liberalism prior to 1868.

Suggested essay plan

1   After Peel's downfall Gladstone became the main keeper of the Peelite conscience (sound administration, free trade, moral rectitude, internationalist approach to foreign affairs) in the 1850s; this 'conscience' became the core of the new Liberal Party. Gladstone held aloof until 1859 because Palmerston went against all these tenets (except free trade). But Gladstone was already the dominant political figure by the middle-1850s, and his inability to choose between Tory, Whig and Peelite factions mainly accounts for the political confusion of those years, as no one else cared to commit himself until Gladstone had.

2   Gladstone finally joined the Whigs (despite Palmerston) because there was really no alternative; also he could not bear Disraeli; even so, his inclinations were mainly Tory in 1859; once in office he dominated Palmerston's cabinet, and forged his own Liberal Party out of the Whig-Liberal, radical-Peelite coalition of the 1850s; the 1861 budget crisis finally established his dominance.

3   Although socially conservative, Gladstone managed to put himself at the head of the radical-liberal grassroots crusade, by keeping politics focused on political and not social issues, and on rather symbolic political issues at that (Church rates, Italy, America, Eyre); friendship with Bright and Newman Hall helped him, though a high churchman, to cement an alliance with dissenting radicals; the paper duties repeal led to his adulation by the popular press, and especially by the *Daily Telegraph*; he became the first politician to speak extensively outside parliament; by attacking privileges he managed to become the 'People's William'; by taking up temperance, non-conformists' rights, anti-jingoism and franchise reform, he became the idol of respectable, moralistic, middle-class Liberals.

4   By the time of Palmerston's death in 1865, Gladstone had defined the issues in such a way that only he could lead the Party. Besides, with Newcastle's death in 1864, the last of Gladstone's contemporaries had disappeared; apart from Russell, Gladstone was now by far the senior politician on the Liberal side, and quite without any rival.

5   Gladstone's adoption of Irish Church disestablishment in 1868 finally made him indispensable; he was motivated by the need to pacify Ireland and by genuine conviction (see his book, *Chapter of Autobiography*), but also by a desire to thwart his rival, Disraeli. He won the premiership, but only by putting himself at the head of forces (radical and dissenting) which would want to go much further than him on education, the Church, the land and on other reform issues. Nemesis came in 1874.

6   As Vincent says, the Liberal Party was mainly formed out-of-doors, and Gladstone came to lead it because he was the only

politician who condescended to appeal directly to extra-parliamentary forces.

2 'The differences between Gladstone and Disraeli in colonial and foreign policy were real and important: their differences on domestic issues were largely a matter of emphasis.' Discuss.

Understanding the question

'Discuss' means that you may consider this (very broad) question from any angle you please. You can agree with the quote and substantiate it with evidence; or disagree with either of the two points made, or even with the whole quote. In fact, one could argue that both were pursuing similar policies abroad (though by different methods) and that it was their domestic policies that were fundamentally different.

Knowledge required

A general understanding of the foreign policies of the governments of Gladstone and Disraeli, 1868–86. The domestic policies and legislative enactments of both governments, 1868–86.

Suggested essay plan

1   It is hard to believe that either Disraeli or Gladstone took external policy seriously. Pandering to the country gentlemen, Disraeli called colonies 'millstones round our necks', an expensive irrelevance; but when Palmerston died he seized the opportunity to take over the 'patriotic card' and to play on the expanded electorate's innate jingoism. Gladstone spoke about the rights of small nations and the duties of large nations to sacrifice their own interests, but in office he found himself bombarding Alexandria and annexing territory nonetheless. Both found themselves helpless in the face of economic forces and of the 'men on the spot'. (Sir Bartle Frere, General Gordon).

2   The point is that the Schleswig-Holstein affair, the Indian Mutiny and the rise of Prussia shattered the 'Pax Britannica'. Britain felt vulnerable. For 20 years she pursued a diplomacy of fantasy. Disraeli's fantasy was to go on behaving (over Afghanistan, Treaty of Berlin, etc.) as though Britain were still the strongest power; Gladstone's fantasy was to make a virtue of weakness and pretend it was self-abnegation (neutrality between France and Prussia, the Alabama claims). Not until Salisbury did anyone work out a coherent diplomacy or strategy. So, yes, Gladstone and Disraeli differed sharply on external policies, but both were play-acting.

3   On domestic policy it is true that Disraeli and Gladstone shared a similar vision of society. Both believed in a hierarchical and organic society, in which property both acknowledged its duties and asserted its rights. Both preferred the agricultural to the industrial mode of life. Despite their bitter opposition, both were arguing within fairly prescribed limits; especially both opposed constructive radicalism, and disliked bureaucratic centralization. Many of their reforms were being pressed by civil servants (licensing, judicature reform, artisan dwellings) and it would have made no difference which government was in power.

4   And yet the domestic policy differences were fundamental. Gladstone was an individualist, believing in equality of opportunity, meritocracy, and self-help; society would 'progress' as the aspiring, respectable, industrious members of society worked hard to improve their situation. Disraeli genuinely disliked respectability, hard work, social aspiration and the sort of improvement which made barriers between people. He preferred people to be lazy, gregarious, community-oriented, unambitious, hard drinking, immoral and happy. If not a philosophy, it was a basic intuition of his, and distinguishes him from Gladstone.

5   So, while the two men complemented each other – Gladstone reformed the law and the institutions, Disraeli took more constructive action in social policy – their domestic differences were more than just a matter of emphasis.

3   What were the main features of Gladstone's Liberalism?

Understanding the question

Once you have defined Gladstone's Liberalism this becomes a straightforward question; the problem is to define it.

Knowledge required

An understanding of Gladstone's political philosophy. The reasons why this philosophy appealed to the Victorian middle-classes. The liberal philosophy in practice in domestic, Irish and foreign affairs.

Suggested essay plan

1   First it is necessary to define Gladstone's Liberalism. The definition should centre on Gladstone's political philosophy from the time that he became a Liberal politician, say 1860. He did not articulate a political philosophy himself, but you might refer to Mill's *On Liberty* and Morley's *On Compromise* as coming closest to Gladstone's own beliefs.

2   Obviously some classes were more likely to support Gladstonian Liberalism than others (the archetypal lower-middle classes, superior and respectable artisans, dissenting shopkeepers, self-educated craftsmen, the more earnest, self-righteous, and 'philanthropic' of the upper classes); also Gladstonian Liberalism can be defined in terms of policies.

3   But in fact Gladstonian Liberalism was not really based on a class appeal or on a programme. It appealed rather to a particular mentality which might be found among all classes; in this sense it was above class, and spawned vertical rather than horizontal loyalties (hence the importance of non-conformity). Also, it was more a political style – a particular way of looking at problems – than a set-piece political programme.

4   The four main elements in the Liberal canon were: that it was possible to achieve appropriate institutional arrangements for the exercise of individual initiative within a minimalist state; that free trade would reconcile classes and nations to each other; that men must stand on their own feet and earn their own comforts by labour and prudence; that progress (mainly of the moral sort, but also material progress) was both possible and indeed inevitable, as individuals learned to behave *conscientiously* in public and private matters.

5   In practice this meant, so far as domestic policies were concerned, securing an equality of opportunity in which merit could be rewarded. Reform of the Civil Service in 1870 thus appealed more for its moral and social aspects than for its effects on bureaucratic efficiency. All special privileges (such as the Anglican monopoly of Oxford and Cambridge, the purchase of Army Commissions) should be removed. And everyone, potentially, had the right to a vote, though in practice he believed in some qualification to ensure respectability (say a £5 franchise).

6   In foreign affairs Gladstone believed in Britain's duty to set an example of self-denial (*e.g.* the Alabama claims), unlike Palmerston and Disraeli. He did not always achieve this in practice (*e.g.* the occupation of Egypt). But he also believed in nations as organic entities; hence he opposed ultramontanism, supported Italy, attacked the Irish church, and eventually adopted Home Rule for Ireland.

7   As Vincent has shown, Gladstone created a revolution in rhetoric and public expectation, but did little in policy terms to change society; believing as he did in strict retrenchment, his governments did not have enough money to make major practical reforms. But he thought men should improve their own

lots and helped to fire his followers with a sense of their own moral worth and importance. He did not, however, aim to change a society which – Anglican and landed as it was in essentials – pleased him very much.

4 Document question (Gladstone Looks Back): study the extract below and then answer questions (a) to (g) which follow:

... Within the half-century a new chapter has opened ... I will refer as briefly as maybe to the sphere of legislation. Slavery has been abolished. A criminal code, which disgraced the Statute Book, has been effectually reformed. Laws of combination and contract which prevented the working population from obtaining the best price for their labour have been repealed ... The scandals of labour in mines, factories and elsewhere ... have either been removed, or greatly qualified and reduced. The population on the sea coast is no longer forced wholesale into contraband trade by fiscal follies ... The entire people have good schools placed within the reach of their children, and are put under legal obligation to use the privileges and contribute to the charge. They have also at their doors the means of husbanding their savings ... under the guarantee of the State to the uttermost farthing of the amount. Information through a free press, formerly cut off from them by stringent taxation, is now at their easy command. Their interests at large are protected by their votes, and their votes are protected by the secrecy which screens them from intimidation ... Nor are the beneficial changes of the last half-century confined to the masses ... Swearing and duelling have nearly disappeared from the face of society ... The sum of the matter seems to be that ... we who lived 50, 60, or 70 years back, and are living now, have lived into a gentler time.

(From an article by Gladstone published in *The Nineteenth Century* in 1887)

Maximum marks

(a) In what ways had the 'criminal code which disgraced the Statute Book' been reformed in the 70 years before 1887? (5)

(b) What had Gladstone's own government contributed towards the repeal of the 'Laws of combination and contract which prevented the working population from obtaining the best price for their labour' and why had he not gone as far as Disraeli on this issue? (5)

(c) What had Gladstone personally contributed to ending the state of affairs in which 'The population on the sea coast' was 'forced wholesale into contraband trade by fiscal follies'? (4)

(d) What steps had been taken in the twenty years before 1887 to ensure that 'the entire people' had 'schools placed within the reach of their children'? (4)

(e) Explain the phrase 'legal obligation to use the privileges and contribute to the charge'. (2)

(f) What part did Gladstone play in giving the people 'the means of husbanding their savings' and in bringing about a 'free press'? (6)

(g) What had Gladstone done towards enabling people to protect 'their interests at large by their votes' and to protect their votes 'by secrecy'? (5)

22.6 READING LIST

Standard textbook reading

R. K. Webb, *Modern England* (Allen and Unwin, 1980 edn.), chapters 8, 10.

A. Wood, *Nineteenth-Century Britain* (Longmans, 1960), chapters 11, 15.

Further suggested reading

P. Adelman, *Gladstone, Disraeli and Later Victorian Politics* (Longmans, 1970).

M. Barker, *Gladstone and Radicalism* (Harvester Press, 1975).

E. J. Feuchtwanger, *Gladstone* (Allen Lane, 1975).

D. A. Hamer, *Liberal Politics in the Age of Gladstone and Rosebery* (Clarendon Press, 1972).

H. J. Hanham, *Elections and Party Management. Politics in the time of Disraeli and Gladstone* (Longmans, 1959, 1978).

P. Magnus, *Gladstone* (Murray, 1954).

R. T. Shannon, *Gladstone and the Bulgarian Agitation* (Nelson, 1964, 1975).

J. Vincent, *The Formation of the Liberal Party 1857–68* (Constable, 1966).

23 Joseph Chamberlain – radical and imperialist

23.1 INTRODUCTION

'Radical Joe' Chamberlain was one of the ablest and most colourful politicians of the late-Victorian period. His family had built an industrial empire on an American screw patent allowing Chamberlain, at the age of 37, to begin his career in politics as the Lord Mayor of Birmingham, the city he loved. Throughout his career his strength was his Birmingham base and he built a following there that never deserted him.

A man of vigour, originality and imagination, he proved a dedicated social reformer and a great imperialist. Like C. J. Fox, he had talents and ability that exceeded those of many a Prime Minister but, though he never attained the highest office himself, he did achieve the unhappy distinction of splitting the two major political parties of the day.

In many ways, however, the man proved greater than his work and, despite his failure to reach the top, non-conformists of every class and workingmen of every creed found him a champion peculiarly their own.

23.2 STUDY OBJECTIVES

1   Chamberlain was the first of the new provincial, non-conformist, professional politicians; he was a self-made millionaire, and his non-conformist commitment and passion soon transferred into politics. He joined the National Education League to campaign for universal, compulsory, free education and led the non-conformist campaign against Forster's 1870 Education Act; the campaign's failure led him to advocate Church disestablishment.

2   Radical reputation was made in the Education revolt and the Birmingham Liberal Association which won the local council, school board and mayoralty for Chamberlain in 1873; ostensibly democratic organization of local Liberal politics – called the caucus – soon extended into the National Liberal Federation of local associations (1877), which threatened to run the Liberal Party. Chamberlain's municipal crusade as mayor; civic gospel at work backed by non-conformist determination and conscience. Birmingham was made into a 'model municipality'.

3   Chamberlain stood as a Radical MP for Sheffield in 1874; he offered as a platform the 'Four Freedoms': free church, free schools, free land and free labour – but was defeated. He threw his energy into Birmingham Liberalism and the NLF; he was aided by Schnadhorst. The Whigs feared the NLF as a movement to radicalize the Party; Chamberlain argued that its aim was to enable the Liberals to discuss policy and make policy. Gladstone and the caucus triumphed in the 1880 election: Chamberlain entered the cabinet as president of the Board of Trade.

4   Chamberlain campaigned vigorously for the Franchise Bill of 1884 (which sought justice for the agricultural labourer). This was held up by the Lords; Chamberlain condemned this hold-up as an 'insolent pretension of an hereditary caste'. Salisbury negotiated a compromise – plus the Redistribution Act. Chamberlain expected the Acts to revolutionize English urban politics and boost radicalism in the Commons (in fact the towns voted Tory in 1885).

The main problem was Ireland, to which Chamberlain was prepared to offer generous powers of local government but would not countenance severance of the Union. He offered his Central Board scheme, which Parnell dismissed as 'a harmless reform'. Neither Parnell nor Gladstone sought to win him over to Home Rule. Backed by the NLF he campaigned for his 'unauthorized programme' in the autumn of 1885. In the election he predicted that 'the Tories will be smashed and the Whigs extinguished'.

5   After Parnell's abortive flirtation with Conservatives in 1885, Chamberlain joined the new cabinet but resigned when Gladstone presented his Home Rule Bill. With him went Gladstone's chance of carrying the Liberal Party on the issue. The Liberals split; Chamberlain and 92 Liberals voted against the Bill. The reasons for his stand must be conjectural – a combination of injured pride, principle, ambition and resentment. Gladstone hoped to reunite the Liberals, Chamberlain hoped to found a new, more national, Party. In fact Chamberlain's future role was to be in the Conservative Party.

6   Chamberlain had sensed the possibilities of Empire in the 1880s; he chose the post of Colonial Secretary in 1895. He saw the colonies, like Birmingham, as areas to be improved and developed by careful concern and investment; his was a positive, constructive approach. Social reform at home was linked to imperial development; he hoped for imperial federation through an imperial customs union (or Zollverein) and imperial defence; he advocated these policies in colonial conferences. On the negative side, Chamberlain was involved with Rhodes in the Jameson Raid, which led to the Boer War and the denigration of Chamberlain's ideal.

7   After the Boer War Chamberlain announced his conversion to imperial preference – income from tariffs would pay for much-needed social reforms. On this the Conservatives split. Chamberlain founded the Tariff Reform League and resigned from the government. The Tariff reform campaign and the landslide defeat of the Unionists at the 1906 election. A stroke ended his political career.

8   Chamberlain's achievements were major at a local and personal level – he became an influential civic leader and an astute political organizer – but much more slight at a national and imperial level, where he split both political parties in turn – a victim perhaps of his own progressive ideas and impatient ambition.

23.3 COMMENTARY

The social and political background

The 'Great Depression' (1873–96) cast its shadow over late Victorian Britain. It hit agriculture particularly hard, stimulated fears about Britain's industrial performance, produced areas of high unemployment and strained relations between management and men.

Many leaders of the 'new unionism' were avowedly socialist, militant and active in politics. Meanwhile the old aristocracy and the new industrial wealth began to merge while the lower middle classes often moved closer to labour.

Politicians during this period were forced to respond to the advent of more national, class-based, politics. They did so either by seeking to control, pacify or exploit this new democracy. The second and third Reform Acts destroyed the old structure of politics – a new structure had to be made.

Early radicalism

Joseph Chamberlain was one of the new politicians who attempted to adapt politics to these changed circumstances. His career presents one of the most remarkable transformations in England's political history. He began as a self-made republican radical, became an aggressive imperialist and ended as the idol of protectionist Tories.

A Birmingham manufacturer and prominent Dissenter, he gained his first experience of public affairs as Lord Mayor of Birmingham from 1873–76. Using Disraeli's legislation, he made the city 'a model municipality' and had it 'parked, paved, assized, marketed, gas, watered and improved' – all as a result of three years dedicated work. The performance of this city-bred, non-conformist, provincial businessman gained him in 1876, a seat in parliament, where he soon became the champion of the new radicalism.

Earlier, his interest had been stimulated by the lack of a national education system. He founded the National Education League in 1869, led the Non-conformist revolt against Forster's Education Act in 1870 and campaigned for a system of 'free, national, secular education.'

He also took over the Birmingham Liberal **Party** organization in 1872, tightened discipline, organized speakers, fund-raising and canvassing on behalf of candidates and threw open membership of the caucus (as it was called) to the newly enfranchised artisans. To co-ordinate the work, a National Liberal Federation was founded in 1877 and was soon propagating its famous radical slogan: 'Free Church, Free Schools, FreeLand.'

Find out what you can about these early political interests and policies of Chamberlain and estimate their political importance.

Ireland

The NLF played an important part in the Liberal victory of 1880 and, despite Gladstone's misgivings about 'Radical Joe', he was obliged to admit him into his cabinet in 1880. His apprehensions proved well-founded. Chamberlain continued to champion the cause of new radicalism, resented the preponderance of landed Whigs in the Party hierarchy and attacked the Tory Lord Salisbury and his class as those who 'toil not, neither do they spin.' He played a major part in pushing through the Third Reform Act of 1884 and, when the Lords resisted, impatiently echoed Morley's phrase 'mend them or end them'.

The big question of the day, however, was Ireland and here Chamberlain (who helped Gladstone oppose coercion in cabinet) offered his limited 'Central Board' scheme. In his 'unauthorized programme' of 1885, however, he insisted that the interests of the British electorate be given priority over Irish affairs. When Gladstone unveiled his Home Rule scheme Chamberlain and 92 others – mostly Whigs – defected and split the Party.

Gladstone was not upset. He believed that Home Rule offered a better platform for uniting the Party and that his continued leadership would prevent it from losing its soul to

Chamberlain's collectivist outlook. This was Gladstone's bid to bind the future.

Find out what you can about Chamberlain's 'unauthorized programme' of 1885. Note how he wanted more state power and was willing to use it to effect social reforms. It is also important that you know the reasons why he opposed the Home Rule Bill of 1886.

Liberal-Unionism

Chamberlain's defection appears to have been a mixture of tactics and principle. His long-term hopes still centred on Liberalism. Gladstone, he believed, would soon retire and then he – Chamberlain – would return to lead the Party. The Conservatives, meanwhile, were happy to work with the Liberal-Unionists, who had the advantage of possessing a popular reform programme.

The alliance was fruitful and produced the Local Government Act of 1888, free education in 1891 and the Workmen's Compensation Act of 1897.

Although Gladstone feared the new radicalism – and its creature the NLF – it was not a revolutionary or a socialist creed; nor was it as dangerous as he supposed: Chamberlain was simply attempting to provide a programme which, by answering some of the legitimate demands of the lower classes, would retain their support for upper-class Parties and politicians. He had a social conscience but also realized that the appeal of Socialism as a mass movement was in the current facts of economic deprivation, not political ideology.

He believed that the working classes could be bought, or at least persuaded, by the offer of much needed social legislation to accept the continued rule of the established orders. His defection in 1886 ended the first stage of his political career but it did not mark the end of his radicalism.

The new imperialism

Chamberlain waited in vain for Gladstone to retire but – when he finally did so in 1893, after the defeat of his second Home Rule Bill – the leadership passed to Lord Rosebery. Meanwhile, Chamberlain had discovered imperialism. In 1895, when the leading Liberal-Unionists took office with the Conservatives, he rejected the offer of the Exchequer and to the surprise of many contemporaries chose instead the Colonial Office. The Queen's Jubilee and Colonial Conference of 1887 had captured his imagination and, after the second Colonial Conference at Ottawa in 1894, he was determined to exploit the possibilities of the new imperialism.

Between 1884 and 1900 Britain acquired a new empire, and as it grew so did the sense of 'imperial mission' among politicians. So also did the belief that somehow the empire was necessary as a means to social reform at home. It was argued that it could provide an outlet for surplus population, for new markets and for new jobs. There was a noticeable and growing tendency for reformers to be imperialists.

The question of why Britain went into empire-building in the late nineteenth century is an important one. Research the main

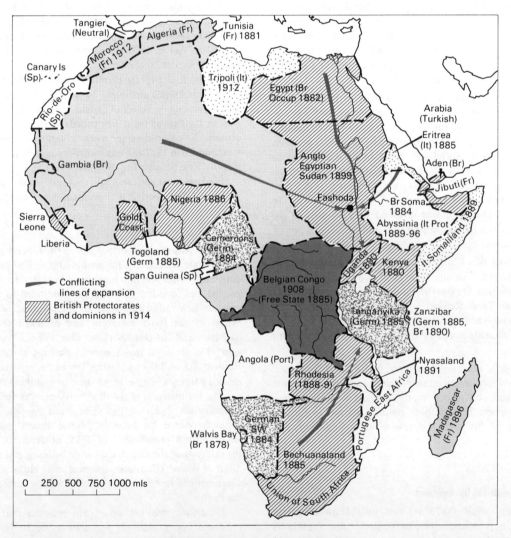

Fig. 16 The partition of Africa 1880–1914

reasons. Finally try to discover what was 'new' about the new imperialism?

Chamberlain's imperial policy

Chamberlain's view of empire marked the end of the old *laissez-faire* attitude to the colonies. His was to be a new, positive policy of intervention and reconstruction, combined with an acceptance of the economic importance of empire.

His first few years at the Colonial Office were devoted to the tropical colonies, which he regarded as estates to be developed by careful planning and expenditure. He was ready to invest capital in any worthwhile scheme – railways in Nigeria, imperial steamship lines, the modernization of sugar production in the West Indies. He also encouraged the founding of Schools of Tropical Agriculture and Medicine at London University.

He saw the British Empire as a force for peace and for the expansion of trade and the development of civilization but there was also an aggressively defensive side to it. Faced by British economic decline and the rise of dangerous European rivals, Chamberlain was determined to hold on to England's imperial dominions, including Ireland, and – something which marked him as a dedicated imperialist – extend them where possible.

In Africa he gave Cecil Rhodes the land on which to build a Cape-to-Cairo railway, he checked French expansion in Nigeria and got involved with the Rhodes-organized Jameson Raid. The latter poisoned Anglo-Boer relations for some considerable time and was one of the reasons for the outbreak of the Boer War in 1899. This event (and especially the British use of 'concentration camps') undermined the heady confidence and sense of duty involved in early attitudes to imperialism and eventually made it a dirty word. Find out as much as you can about Chamberlain's contribution to the Empire during this period. How different was it from Disraeli's imperialism?

Imperial preference

Although the Boer War showed the strength of the Empire, Chamberlain wanted more. The fourth Imperial Conference, called in 1905, convinced him of the impossibility of achieving imperial unity on the basis of Free Trade. His imperial vision was to be realized by the continuation of imperial conferences, by imperial preference in trading, by a new imperial Navy and by new, imperial political institutions to bind the mother country and the Empire together.

After a visit to South Africa in 1903 Chamberlain shocked his cabinet colleagues by suggesting a policy of protection with imperial preference. Tariff reform would also provide the revenue for much-needed social reforms – in particular old-age pensions. The cabinet split and Chamberlain resigned, determined to fight the next election on the issue. He founded the Tariff Reform League and vigorously publicized the policy he hoped would win the working classes to Conservatism.

To the Liberals, squabbling and divided in the wilderness of opposition, Chamberlain's decision was a godsend. It not only bitterly divided the Conservative Party; it provided the Liberals with an issue that they could unite on. H. H. Asquith (a champion of Gladstonian economics) was appointed to dog Chamberlain's heels, challenge his protection arguments and remind the electors of the benefits of free trade – a task he successfully accomplished.

You must find out the main points in these two opposing arguments. Note how Chamberlain's is the more idealistic; Asquith makes it an issue of practical politics – the 'big loaf' of Liberalism versus the 'little loaf' of Conservatism. Asquith's arguments proved the more convincing and the divided Conservative Party was overwhelmingly defeated by the Liberals in the 1906 election.

Chamberlain held Birmingham and some argue that he may even have succeeded Balfour as leader of the Conservatives. On his 70th birthday, however, he had a stroke and, though he lived until 1914, his political career was over.

Radical and imperialist

In parliamentary and platform strength Chamberlain was the equal of Gladstone and Disraeli. He was also closer in touch than either of them with the spirit of the age. Against the old aristocratic order he pitted the growing confidence and egalitarian spirit of the new provincial towns. Capable of imagination and vision, he found, like Gladstone, that it was dangerous for politicians to move faster than public opinion; and like Gladstone he was often seized by a sense of mission in politics, with a nonconformist rectitude underlying much of what he did.

He was not lacking in ideals. He believed in the extension of democracy, in the sweeping away of inherited privilege in the Church of England and in the countryside, in the extension of the franchise, in the destruction of the old Whigs in the Liberal Party and their replacement in the towns by a more democratic party machine – the caucus and the National Liberal Federation.

A second ideal was his belief in the 'civic gospel' – progress and improvement for the citizens of England – better housing, sanitation, education and welfare benefits: 'I would reform and remove ignorance, poverty, intemperance and crime at the very roots.'

His imperial vision was that of a great statesman. It was paradoxical that his Boer policy marked both the apex and the beginnings of decline of the new imperialism. He failed to understand the self-government ambitions of the Boers, of the Irish or of the other dominions. Perhaps, like the Elder Pitt, he came too late in the history of the Empire to hold together those parts which preferred to break away. He did, however, herald the new post-1919 British Commonwealth with its emphasis on unity through common interest.

Assessment

Chamberlain is the greatest 'might-have-been' in modern political history. His character and career will always be something of an enigma. Despite his professional dedication, his vigour, originality and imagination, he failed to get to 'the top of the greasy pole.' He was tough and ruthless, and indeed made some mortal enemies but then, so did Gladstone, Disraeli and Lloyd George; they, however, were not stopped from achieving their ambition.

The main reason for Chamberlain's lack of success is clear. He was the victim of his convictions about Ireland. If he had been prepared to accept Gladstone's Home Rule Bill he could not have failed to inherit the leadership from Gladstone. However, as a staunch believer in the United Kingdom he could not stomach the thought of Irish separation. Whether his break with the Party in 1886 was idealistic or tactical remains an open question but the split proved permanent.

As Colonial Secretary after 1895 he again displayed remarkable ability and, when Salisbury retired in 1902, many people expected him to succeed as Prime Minister. Salisbury saw to it that he did not. His policy of tariff reform must have seriously damaged his chances with the Conservative Party and certainly split them as disastrously as Home Rule had split the Liberals in 1886. Again, was it a question of conviction or a calculated effort to oust Balfour and become Prime Minister?

It is unlikely that it will ever be possible to fully explain Chamberlain's change of allegiance, outlook and policy. It is quite probable that he could not explain it himself. Fundamentally, as Robert Blake points out, there was something about him that made others suspect and dislike him. Many found him too aggressive, too ambitious, too restless, too racist, too much 'on the make'. Lord Salisbury is supposed to have said: 'Mr Gladstone was greatly hated, but he was also greatly loved. Who loves Mr Chamberlain?' Who, indeed? Perhaps in the end that is why he failed to get to the top.

23.4 CHRONOLOGY

1836	July	Chamberlain was born in London.
1868		CHAMBERLAIN ORGANIZED THE BIRMINGHAM LIBERAL ASSOCIATION; from it grew the Liberal 'caucus', which developed and extended in the years that followed.
1869	October	HE HELPED FOUND THE NATIONAL EDUCATION LEAGUE to campaign for free, national, secular education.
1870		He led the 'non-conformist revolt' against Forster's Education Act.
1873	November	Chamberlain became LORD MAYOR OF BIRMINGHAM: he introduced a policy of 'gas-and-water Socialism' and preached the 'civic gospel' of improvement.
1876	June	He became MP for Birmingham.
1877	May	National Liberal Federation founded by Chamberlain to control the work of local Liberal caucuses. The NLF was dedicated to radicalism and to overthrowing the power of the landed Whigs in the Liberal Party.
1880	April	Chamberlain joined Gladstone's second ministry as President of the Board of Trade.
1883	March	Attack by Chamberlain on Lord Salisbury 'and his class'. The CORRUPT PRACTICES ACT rigorously curtailed election expenses.
1884	December	THIRD REFORM ACT: Chamberlain was the driving force behind this law, which conceded adult manhood suffrage.
1885	June	REDISTRIBUTION ACT established single-member constituencies; when the Lords temporized – Chamberlain argued that they should be "mended or ended".
	September	CHAMBERLAIN LAUNCHED HIS 'UNAUTHORIZED PROGRAMME': this included Church disestablishment, land and local government reform, better housing, improved social services, etc.
1886	June	CHAMBERLAIN OPPOSED GLADSTONE'S HOME RULE BILL; he resigned in March and defected with 92 other Liberals – including Lord Hartington and John Bright. This split the Liberal Party. Parnell said of him, 'there goes a man that killed Home Rule'. Liberal-Unionists then became a separate group in the Commons, giving their support to the Conservatives.
1887	April	FIRST COLONIAL CONFERENCE: economic co-operation and imperial defence were discussed.
	June	Queen's Jubilee celebrations.
1888	August	Local Government Act established 62 County Councils to discharge all local administrative duties; councillors to be elected by household suffrage.
1891	April	Education Act made elementary education free.
1894	June	SECOND COLONIAL CONFERENCE AT OTTAWA: imperial trade discussed.
1895	August	CHAMBERLAIN MADE COLONIAL SECRETARY in Salisbury's third ministry.
1896	December	JAMESON RAID: Chamberlain was accused of being involved but exonerated by a parliamentary Select Committee.
1897	June	THIRD COLONIAL CONFERENCE in London under Chamberlain, who put forward a policy of imperial federation.
	August	WORKMEN'S COMPENSATION ACT extended compensation to workmen but excluded agricultural labourers.
1899	October	The Boer War began.
1902	May	Peace of Vereeniging ended the Boer War.
	June	FOURTH COLONIAL CONFERENCE: imperial trade and federation were discussed.
1903	May	CHAMBERLAIN DECLARED FOR PROTECTION AND IMPERIAL PREFERENCE.
	September	Chamberlain resigned from the cabinet, founded the TARIFF REFORM LEAGUE and vigorously campaigned in favour of imperial preference. The CONSERVATIVE PARTY SPLIT on the issue.
1906	January	Conservatives routed at the general election. Chamberlain held Birmingham.
	July	Stroke ended his political career.
1914	July	Chamberlain died.

23.5 QUESTION PRACTICE

1   Discuss Joseph Chamberlain's policies and achievements.

Understanding the question

An uncomplicated question. The only difficulty lies in the use of the word 'achievement' and the need to decide what aspects of his work merits that description.

Knowledge required

It is essential to know the details of his work as Mayor in Birmingham and as the founder of a new form of political organization (the National Liberal Federation). His achievement in giving Liberalism a new social content. The redefining of the importance of the Empire and also of the Colonial Office.

Suggested essay plan

1   Chamberlain was a new type of professional politician; an industrialist who challenged the leaders of the two great Parties and left his mark upon them both. It is questionable whether he had in the end the circumspection and political tact necessary for a Prime Minister or a leader of one of the great Parties.

2   He emerged as a Birmingham city councillor in 1869 and became famous outside the city council as an Education League spokesman and founder of the Birmingham Liberal Association. He became Mayor in 1873 and began by putting the 'civic gospel' of Birmingham non-conformists into practice; the era of 'gas-and-water Socialism' arrived in his two measures to take 'regulated monopolies' out of the hands of private enterprise; he had a great dream to build a rival to the Champs Elysee in Birmingham; Corporation Street, built by clearing the slum dwellings of the city centre, was the result. Birmingham city politics moved (literally) from the age of local public house meetings (the Woodman) to the city hall.

3   Chamberlain took over from the local radical, George Dixon, the leadership of the Education League, which became an engine for challenging the Gladstonian party leadership on education. The Birmingham Liberal Association took advantage of the 1867 Reform Act, which enfranchized the working-class voter and ostensibly established a democratic organization; in reality this was run by Chamberlain and his friends; it was the forerunner of the National Liberal Federation, set up in Birmingham as a means of selecting candidates and controlling MPs: this was to become the model for other Parties. Chamberlain failed to defeat Gladstone (the Party leader) in the Home Rule crisis of 1886 but the NLF remained a powerful institution.

4   Chamberlain advocated a social content to Liberalism – in particular as voice of Birmingham non-conformity and the 'civic gospel', His 'unauthorized programme' argued for better housing for the working-classes; the doctrine of 'ransom' (society owed compensation to the poorer classes); free schools; 'three acres and a cow'; and a graduated property tax. His abandonment of the Liberal Party left this work undone (although this was in part taken up by the Liberal's Newcastle Programme of 1891) but gave to Victorian liberalism a new direction. Chamberlain's own achievement in this direction is (as a member of the Salisbury Conservative government) the Workmen's Compensation Act, 1897.

5   Chamberlain's imperial policies led directly not only to the South African war – hardly an achievement – but also to the Colonial Conferences (forerunners of the Commonwealth conferences) and to the doctrine of imperial preference established in 1932. His other ideas included an imperial parliament and an imperial Navy (both rejected by the colonies, which were anxious to preserve their new-won independence) but his concept of aid to undeveloped colonies (realized with subsidies to shipping lines and for new crops) became a common coin of colonial policy.

His main achievement was to make the Colonial Office of fundamental importance in government. Chamberlain was effectively co-Prime Minister. His Tariff Reform programme split the Conservative and Liberal-Unionists (as his Irish policy had split Liberals) and his negative achievement was to make possible the Liberal landslide in 1906.

6   Chamberlain could never be regarded as negligible. He was often wrong-headed (*e.g.* over Ireland), but his policies were based upon considerable analysis and were expounded in great and noble speeches. His achievements as a politician, as a Midlands mayor, as a social reformer and as an imperialist are all quite remarkable set against the normal inertia of English politics.

2   Compare the imperial policies of Chamberlain and Disraeli.

Understanding the question

This calls for a straightforward comparison which requires that you also identify the major differences. Note that, because of the rise of new Powers in Europe, both men felt that Britain was vulnerable and their policies were a reaction to this fact.

Knowledge required

An understanding of Disraeli's policies: his Crystal Palace speech, the Suez Canal shares, the defence of the Mediterranean (acquisition of Cyprus), his forward policy in Afghanistan and in West and South Africa; and of Chamberlain's policies: the Colonial Conferences and the imperialist concept, his 'forward policy' in Southern Africa, his concept of undevelopment, the Imperial tariff, his 'new imperialism.'

Suggested essay plan

1   Both politicians had much in common – they were 'outsiders' who dominated the Conservative Party; both were capable of advocating original ideas on policy. The imperialism of both was not restricted to Conservative Party or Liberal Imperialists of the period. Both gave a new direction to British imperialism.

2   The concept of Empire was expounded by Disraeli in his Crystal Palace speech in which he put forward a most important idea – an imperial parliament; this idea was taken up and developed by Chamberlain at the Colonial Conferences; it met resistance from dominions such as Canada, who were anxious to maintain their own independence and had no wish to become again a part of the British Empire. Chamberlain developed this concept of Empire to propose an imperial Navy and a *Zollverein*, an imperial free trade plan, neither of which aroused much enthusiasm among colonial statesmen. His Colonial Conferences did in themselves create a new informal imperial institution, which was to develop in our own day into the Commonwealth Prime Ministers' conferences.

3   Disraeli has traditionally been described as the architect of a new expansionist policy in empire-building, and as one who made a final break with the old Liberal policy of inertia. Such a view cannot be maintained. Not only did the Empire keep developing in the mid-Victorian years in the Far East but even in Africa Disraeli's transformation of the Fanti protectorate into a Crown Colony in 1874 had already been decided on by the previous Gladstone government. In Southern Africa Disraeli supported a federation of all four states (Chamberlain's war in South Africa produced this result which Chamberlain approved of and Chamberlain's own Commonwealth of Australia Act of 1900 set up a federation of the Australian colonies).

Both Disraeli and Chamberlain fought wars in defence of African policy – Disraeli against the Zulus (and therefore to protect Boers), and Chamberlain against the Boers. Both were aware of the extent of British economic interests in the area but Chamberlain, as an ally of Rhodes, was committed to a 'forward' policy which led to the Jameson Raid and the South African war.

4   Chamberlain and Disraeli were each involved in different areas. Disraeli's spectacular purchase of the Suez Canal shares in 1875 (his comment to Victoria: 'you have it Madam', which was misleading as his purchase was only 45% with no voting rights) involved Britain in Egypt – leading to occupation – and to Fashoda in 1898, when Chamberlain and his French counterpart Hanotaux hovered near conflict. It was, however, India which was *the* Empire under Disraeli, (this was symbolized by Victoria becoming the Empress of India in 1876). Chamberlain was associated with last stages of the 'scramble for Africa', the great partition of the African continent by the European Powers; these involved ambitious schemes such as the Cape-to-Cairo railway.

5   In economic matters Chamberlain went beyond Disraeli in seeing the mutual advantages of the Empire for trade and development: these included subsidies to shipping lines and for new crops, and to schools of tropical medicine; the importance of the gold and diamond reserves in South Africa. He was very conscious of the decline in Midlands industries and, when *Zollverein* became a non-starter, he modified the policy into one for an imperial tariff as a means of furthering British economic interests. Chamberlain was working within a new imperialist mood and his thinking was symbolized by Birmingham's recommendations to the Royal Commission on the Depression of Trade and Industry, 1885–6, *i.e.* tariffs and the development of colonial markets.

6   Both Chamberlain and Disraeli show similarities in institutional thinking but Chamberlain, coming at a later stage in the development of the imperial mood and the decline of British industries, pushed imperial policies toward the revival of an eighteenth-century type of Empire. He urged, however, that they should run it more in the common interest of all the partners than was the case with the pre–1776 Empire.

Disraeli sketched out the ideas for change; Chamberlain tried to put them into practice. Disraeli developed imperial forward policies within the newly-evolving orthodoxy. Chamberlain pushed beyond, in the headiness of his vision and the weight of his armies. In South Africa this aroused for Britain the kind of hostility in Europe that American policies were to do in Vietnam in a later age. In short, an informal empire was replaced by a formal Empire.

3   To what extent was the new imperialism of the later nineteenth century a product of economic and social factors?

Understanding the question

This question calls for an understanding of the various explanations of imperialism and asks the student to relate the theories and Britain's economic situation to the policies followed by Chamberlain (and to some extent by his predecessors). The words 'To what extent' urge some consideration of the political and diplomatic content of the new imperialism.

Knowledge required

An understanding of Hobson's theory of imperialism, first published in 1902 (*Imperialism, a study*) which attempted to relate the new expansionist policies to the social and economic situation. The Great Depression and the rise of the 'Fair Trade' movement; the Royal Commission on Depression and the

Birmingham arguments; the rapid industrialization of Germany and the USA. Chamberlain's imperial policies and their purpose; and of the European political background, *i.e.* the threats posed by Russia to the route to India, and by France and Germany in Africa.

Suggested essay plan

1   According to Robinson and Gallagher (*Africa and the Victorians*) the scramble for Africa was initiated by Britain's invasion of Egypt in 1885; and her withdrawal from a steadily weakening Turkey to an Eastern Mediterranean base in Cairo, from which she could protect the route to India. From thence she was drawn reluctantly into central and southern Africa and, feeling isolated and vulnerable in Europe, bargained away potential interest in many parts of the new continent to placate her European adversaries. Chamberlain, however, reversed this defensive approach and, in the 'nineties, was determined to consolidate British gains.

2   A change of mood from the mid-1880s onwards. There was the growth of a sense of imperial mission to carry civilization to under-developed parts of the world, and to end slavery, the slave trade and other barbarous practices in Africa. This was paralleled by a sense that the existing Empire was not fully deployed to maintain living standards in England; that the problems of competition could perhaps be solved by protectionism (practised by most other countries). Do these factors explain the new imperialism?

Contemporaries talked of a need to make full use of the Empire to offset adverse trade conditions. Hobson, writing in 1902, advanced the view (a) that imperialism was partly a response to the existence of surplus capital – *i.e.* when capital cannot profitably be invested in England, it will be invested abroad, (b) that the existence of this surplus capital due to underconsumption, (c) that this state was produced by low wages, and (d) that imperialism was therefore of benefit only to capitalists and capitalism, and not the nation as a whole. These conditions in later nineteenth century England – declining industry, the reappearance of the cycle of depressions and underconsumption – were a cause for concern. Rowntree's survey revealed that an estimated 25% of the population were below the level of subsistence. Chamberlain, looking at the economic and social situation, argued, however, that imperialism could benefit the working classes by providing new markets for British products.

3   Even before Chamberlain, the informal empire was giving way to the formal Empire; there was more talk of 'trade following the flag' as mapped out by explorers and missionaries in central, East and West Africa – *e.g.* the arguments for making Zanzibar a protectorate turned upon an analysis of trade. Chamberlain was to give the whole process a new drive and form: in particular, Rhodes' development of Rhodesia and Chamberlain's involvement in forward policies in Southern Africa, and the invasion and conquest of the Ashanti kingdom on the Gold Coast in 1895. But was Africa, potentially, a good market for British goods?

4   The new imperialism: Chamberlain's concept of a *Zollverein* – a free-trade Empire, protected against foreign competition – makes clear the general economic context of his imperialism. Its development-tariff reform, the argument about employment produced by tariff reform, a higher standard of living and an end to declining trades such as sugar-refining, iron and cotton. As early as 1895 Chamberlain had called on all colonial governors to buy British. Equally his approach to the South African problem was largely concerned to make the Transvaal safe for British investors. However, if there had been no Rand gold could Britain have afforded to ignore the scramble?

5   Britain was feeling isolated and vulnerable in the last quarter

of the nineteenth century. Her economy was declining from the heights of mid-Victorian prosperity; her politicians (and businessmen in government) were forced to develop new policies to meet new times. Her imperialism was a complex affair which included (as A. P. Thornton shows) strands of Socialism, humanitarianism and Christian evangelicalism but was also rooted deep in the new economic situation of the United Kingdom. Hobson and Chamberlain represent two different economic approaches to the same phenomenon: one a theory to explain it; the other, practical measures, to a large extent determined by economic forces, to root out the problem.

4   Document question (Chamberlain's Imperial Dream): study the extract below and then answer questions (a) **to** (g) **which follow:**

My... proposition is that a true Zollverein for the Empire, that a free trade established throughout the Empire, although it would involve the imposition of duties against foreign countries, and would be in that respect a derogation from the high principles of free trade, and from the practice of the United Kingdom up to the present time, would still be a proper subject for discussion and might possibly lead to a satisfactory arrangement if the colonies on their part were willing to consider it... it would undoubtedly lead to the earliest possible development of their great natural resources, would bring them population, would open to them the enormous markets of the United Kingdom...

Maximum marks

(a) What did Chamberlain mean by a 'true Zollverein'?   (3)
(b) What practical steps had Chamberlain already taken to develop the 'great natural resources' of some of the colonies?   (3)
(c) Did Chamberlain's imperial dream as reflected in this passage rest squarely on the economic needs of the Midlands?   (4)
(d) What measures did Chamberlain put to the Colonial Conference of 1902 to further his imperial policy?   (4)
(e) Were the colonies, for their part, 'willing to consider it'?   (4)
(f) What effect did Chamberlain's imperialism have on his subsequent career?   (6)
(g) ...and upon the Conservative and Unionist Party?   (6)

23.6 READING LIST

Standard textbook reading

R. K. Webb, *Modern England* (Allen and Unwin, 1980 edn.), chapters 8, 10.

A. Wood, *Nineteenth-Century Britain* (Longmans, 1960), chapters 15, 17, 19, 20.

Further suggested reading

H. Browne, *Joseph Chamberlain: Radical and Imperialist* (Longmans, 1974).

M. E. Chamberlain, *The New Imperialism* (Historical Association, 1970).

D. A. Hamer, *Liberal Politics in the Age of Gladstone and Rosebery* (Clarendon Press, 1972).

A. Jones, *The Politics of Reform, 1884* (CUP, 1972).

D. Judd, *Radical Joe* (Hamish Hamilton, 1977).

W. R. Louis (ed.), *Imperialism: The Robinson and Gallagher Controversy* (New Viewpoints, 1976).

C. J. Lowe, *The Reluctant Imperialists* (RKP, 1967).

R. Robinson and J. Gallagher, *Africa and the Victorians* (Macmillan, 1961).

24 Gladstone and Ireland

24.1 INTRODUCTION

As Professor Mansergh rightly observes, 'the condition of Ireland in the middle years of the nineteenth century, or at least all of it outside Ulster, dismayed and distressed those who saw it – from Italian nationalists to Communist internationalists, to Frenchmen, Germans, Americans and by no means least, to Englishmen, whether economists, officials or more ordinary voyagers'. English rule in Ireland, whether of coercion or conciliation, had been a failure – illustrated above all in Irish eyes by its unsympathetic response to the tragedy of the famine.

Some Englishmen, like John Bright, sought to right Ireland's wrongs but most politicians, like Disraeli, saw Ireland as a problem best avoided or ignored. Gladstone, imbued with a strong moral as well as European sense, proved the exception. He sought to remove Irish grievances primarily to pacify the Irish but also so that Englishmen might be able to look their fellow Europeans in the face.

According to Dr Hammond, Gladstone alone had the breadth of vision and ability 'to change the English temper towards Ireland, to shake fundamental views of property and economics, to overcome all the prejudices that estrange men divided by race, religion and history...' Despite Gladstone's efforts, Home Rule proved unacceptable to the majority of the English people. English politicians failed to offer an alternative solution and the fall of Parnell convinced many Irishmen that the opportunity for a constitutional solution had gone for ever.

24.2 STUDY OBJECTIVES

1   The Irish problem: the importance of the Act of Union; the abject standard of existence for the mass of the people; underdevelopment, complications of tenure and landholding; high rents, evictions and outrages; rule by a small, largely absentee, Anglo-Irish Protestant landlord class; English ignorance of conditions in Ireland combined with religious and national prejudice; the fear that 'Home Rule would mean Rome Rule'.

2   Rise of nationalism in Ireland; the impact of Young Ireland; the Famine; the Fenians; Fenian links with America; the rising in 1867; the outrages in England, in Manchester and Clerkenwell, emphasized the immediacy of the problem to English politicians and people. The Irish problem was a combination of economic, religious and nationalist factors.

3   Gladstone's liberalism and sense of justice encourage him to take up the Irish problem: the pacification measures of his first ministry 1868–74; the effect of the Ballot Act on the Irish Parliamentary Party, the emergence of Parnell, and the policy of obstruction – 'stepping on English toes'.

4   The agricultural depression of the late 'seventies; evictions, outrages, coercion; Davitt and the Land League, Devoy and the Clan na Gael in America; policy of the New Departure. Gladstone concedes the 1881 Land Act; the problem of rent arrears; the Kilmainham Treaty emphasized Parnell's acceptance of constitutionalism; Parnell's tactics; the Phoenix Park murders destroy the possibility of further progress; Parnell's liaison with Kitty O'Shea; pacification removed grievances but the Irish then demanded Home Rule. The Phoenix Park murders, however, convinced the English that the Irish were not ready for self-government.

5 Parnell and the Irish National League; Davitt abandoned; new goal to be Home Rule won from Liberals or Conservatives; the effect of the third Reform Act on the IPP; Parnell's negotiations with the Conservatives in 1885, denounced by Chamberlain as 'flagrant political dishonesty'; the Conservatives drop Parnell after an inconclusive election result and the 'Hawarden Kite'.

6 Gladstone's conversion to Home Rule; the consequences of his tactical secrecy; resignation of Whigs and Radicals – Hartington, Bright, Chamberlain. The Home Rule Bill introduced – limited measure with no Irish MPs at Westminster – provoked violent reaction; Churchill's 'concern' for Ulster ('Ulster will fight and Ulster will be right'), Chamberlain's for Empire; Liberal Party split, Bill rejected. Home Rule henceforth a party-political question. The Irish Party was now dependent on the Liberals.

7 Parnell and *The Times;* the divorce scandal breaks but Parnell was re-elected as leader by the IPP; Gladstone's letter made it clear that it had to be Parnell or Home Rule; IPP rejected Parnell but, aided by some followers, he campaigned against the decision; he married Kitty O'Shea but died shortly afterwards (October, 1891). Although he had set the Home Rule argument on its legs and made it the major issue in British politics, yet effectively it seemed to die with him.

8 Gladstone's final effort; the second Home Rule Bill, 1893, (providing for Irish MPs to remain at Westminster) was passed by House of Commons but rejected by the House of Lords by the largest majority recorded. Home Rule was Gladstone's unhelpful legacy to a divided Liberal Party.

9 Gladstone had progressed from coercion and conciliation to Home Rule; he was convinced that Irish nationalism was not a transitory objective but an inextinguishable passion; Britain's safety and natural justice demanded that the problem be solved. According to Hammond 'he had spent on Ireland strength, health, power and popularity, as no politician had ever done'; his struggle for Home Rule was a gallant attempt but a noble failure.

24.3 COMMENTARY

Irish Nationalism

The belief in Irish separatism and Irish nationalism developed throughout the nineteenth century. O'Connell began the process in 1829 with his success in organizing peasant Ireland to achieve Catholic emancipation. The literary men of Young Ireland continued it in the 1840s. Yet, although the leaders of this movement made a cultural impact, politically they proved ineffective.

Despite their propagation of the romantic glories of ancient, Gaelic Ireland, for most Irish the famine of 1845–49 was the reality. One million died; another million emigrated. The famine became a dividing line in Irish history – above all in the attitude to England that it engendered in the mass of the people of Ireland and their kin in America. Irish policy was henceforth to be one of attack; and in the 'sixties the Fenians, spurred on by US money, opened a new chapter of violence in Ireland and England.

Gladstone's concern

Gladstone fought the general election of 1868 on the Irish issue. Although it was his interest in Italy that aroused his sympathy for Ireland, Gladstone first became aware of the gravity of Irish affairs in 1845, when, in a letter to his wife, he described Ireland as 'that cloud in the West, that coming storm.'

The famine, Fenianism and continued disorder convinced him that a solution to the Irish question was the greatest problem facing the next administration. When he was told in 1868 that he was to become Prime Minister for the first time, he declared, 'My mission is to pacify Ireland.'

Clearly, Ireland was in a distracted state. Reforms were long overdue; but most Englishmen contended that the problem was incapable of solution. We can now see that the root of the problem was nationality.

Ireland wished to be free but England could not allow freedom to a country strategically vital to her own security. Furthermore, by the Act of Union, England was committed to safeguarding the privileges of the Protestant ascendancy in Ireland. Thus, a passionate and growing Irish nationalism was baulked by an equally determined English imperialism.

Major issues

The Irish question in the nineteenth century revolved around three issues:

1   The political and constitutional relations between the two countries.

2   The existence in Ireland of an established Anglican (Protestant) Church. This represented only a tenth of the population but sought to claim goodwill and tithes from the overwhelming Catholic majority.

3   Almost all the land of Ireland (gradually confiscated since the sixteenth century) was owned by Anglo-Irish Protestant landlords. They introduced a system of landholding alien to the Celtic Irish. Furthermore, they were often absentees whose sole interest in their estates appeared to be to maximize their income from rents. Peasant resistance led to eviction, violence, disorder and the introduction of innumerable Coercion Acts by an English government determined to maintain law and order. The resultant conflict between landlord and peasant was eventually translated by nationalists into a conflict between England and Ireland. Nationalist policy oscillated between constitutional demands for some form of Home Rule and revolutionary activities for separation.

Attitudes

The origin of the conflict was English rule in Ireland in accordance with English ideas. Her attitude to Ireland was blinded by her own agricultural and industrial progress. Ireland was blamed for her economic backwardness whereas in fact, it was the economic backwardness of English government in Ireland that was at fault. Irish nationalists felt that the English government was ignorant of Irish conditions and contemptuous of her legitimate aspirations. They did not accept that England had an Irish problem – rather that Ireland had an English problem. In vain therefore did the English government seek to conciliate, civilize or coerce the Irish.

By the 1870s all had been tried and found wanting. Initially it was Gladstone's commitment to Italian freedom from Austria that made him consider the Irish question but it was also an essential part of his belief in liberalism, which, as it matured, convinced him that Home Rule was the only just and honourable solution.

Gladstone's measures

At first Gladstone felt that the desire for separation was merely symptomatic. He recognized that Ireland had two outstanding and genuine grievances – religious and agrarian – and believed that if he could solve these the Irish would stop agitating for repeal of the Union. Fenian outrages in Ireland and England prepared parliament, during his first ministry, for his policy of pacification.

In 1869, despite much opposition, he disestablished the Anglican Church of Ireland and put it on the same footing as the dissenting and Catholic churches in Ireland. Disraeli argued that this weakened the Protestant supremacy, threatened the union of church and state and undermined the rights of property. Although disestablishment pleased the Irish Catholic bishops it did little to pacify the rest of the population in Ireland.

In 1870 Gladstone passed his first Irish Land Act. This gave some protection to tenants and began (with the Bright clause) the process of land transference which, by 1914, had solved the land question. You should know the main provisions of the Land Act and be able to explain why it proved ineffective. It did, however, manage to frighten the English landowning class who, in an age of *laissez-faire*, regarded property rights as sacrosanct and feared that Gladstone's Irish reforms might 'cross the water.'

In 1870 also Isaac Butt founded the Home Government Association which proposed a federal solution to the Irish problem. At the time this seemed a fantastic proposal – Gladstone himself denouncing the idea as absurd. Instead Gladstone proposed his third measure to pacify Ireland – the Irish University Reform Bill – a complicated attempt to set up a non-denominational state university system in Ireland. The Bill was defeated. Gladstone dissolved Parliament but lost the general election that followed.

The agricultural depression

Three events of major importance occurred in Ireland while Gladstone was out of office. The first was the onset of economic depression (primarily owing to American competition) which had disastrous results on Irish agriculture. For the first time since the famine there was starvation in some Western counties. Thousands were evicted for non-payment of rent; inevitably there were outrages – 2,600 in 1880 alone.

Meanwhile, Michael Davitt had founded the Land League to prevent evictions and boycott those who dared take over vacant tenancies. This was a truly original development in that it organized for the first time a secular mass movement to challenge landlordism in Ireland. Fuelled by newspaper propaganda and radical ideas and aided by American dollars, it proved a remarkable successful pressure group. By 1881 it had landlordism in retreat.

The emergence of Parnell

The second event was the emergence of Charles Stewart Parnell a Protestant landowner and MP for Wexford – under whose leadership Irish power was deployed in the House of Commons as never before. Parnell was an unlikely leader of Catholic nationalist opinion in Ireland. He knew little of Irish history, lacked a political philosophy and his political instincts were conservative. Yet his ability for political organization and shrewd use of parliamentary tactics quickly enabled him to wrest power from the kindly but inefficient Butt.

Parnell realized that the key to Ireland's future lay at Westminster. A disciplined and united Irish Party in the House of Commons might come to hold the balance of power and thereby wring concessions from one of the two English Parties as the price of its support. By using obstructionist tactics he had, by 1880, made his presence felt in the House of Commons. He convinced Gladstone that Ireland needed reforming and eventually that she needed liberty. It was one of the great political performances of the nineteenth century.

The New Departure

The third event was the 'New Departure'. This was fashioned by Michael Davitt, who persuaded John Devoy, the leader of the Irish-American Fenians, to throw the full support of Irish Fenianism behind the Land League. The next step was to convince Parnell that the Land League could be 'the engine that would drag the Home Rule movement in its train.'

Though Parnell rejected the violence of the Fenians (even though he exploited their emotional appeal), and also Davitt's policy of social revolution through land nationalization, he realized the enormous popular support that the land question was capable of generating. He therefore became President of the League in 1879. His involvement in the movement was, however, purely tactical, but it made him more powerful in the House of Commons, lending him some of the magic of extremism.

According to MacDonagh, the New Departure allowed the three great forces of Irish nationalism to be brought together for the first time – the revolutionary, the civil and the parliamentary. Their masterful deployment by Parnell in parliament presented the strongest threat to date to English government in Ireland.

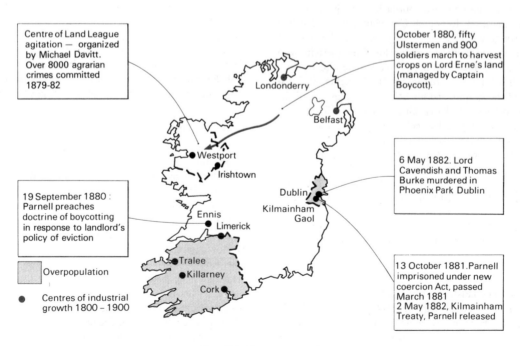

Fig. 17 Irish unrest 1879-82

The Irish Land Act, 1881

The New Departure was a clever and highly efficient instrument of agrarian agitation. Circumstances, however, militated against its success. The severity of the economic depression prevented any quick solution to the land question; and the racist opposition of Chamberlain and the Unionists, coupled with the sectarian bitterness of Ulster Protestants, made a reasoned political solution unlikely. Against this unpromising background Gladstone strove to find a constitutional solution short of separation.

In 1881 he passed a second Land Act which conceded the '3Fs' – fair rents, fixity of tenure and free sale. It was a generous Act which, in the long run, made stable tenant farming possible in Ireland. But its timing was wrong. It did nothing for the enormous and increasing burden of rent arrears and it was accompanied by a Coercion Act. Parnell, therefore, publicly denounced it, while privately he encouraged Irish tenants to test it in the courts. The English ruling class viewed the Act with apprehension. It marked a sharp break with *laissez-faire* and threatened to undermine one of the most basic of Victorian beliefs – the sanctity of private property.

In the short run the repressive Coercion Act created a vicious circle of popular unrest, evictions, boycotts and violence, which in turn led to more stringent measures. Check carefully the details of the Kilmainham Treaty of 1882, which resolved this phase of the struggle. You should also assess the significance of the Phoenix Park murders that followed.

Note that Parnell considered that the land question had been settled. He severed his connections with Fenianism and replaced the banned Land League with the National League, now avowedly dedicated to securing Home Rule.

The Home Rule Bill, 1886

By 1885 Gladstone had come to accept that the Irish problem was basically a question of nationality. The election of the same year gave Parnell 86 seats, the overwhelming majority of Irish seats in the Commons. Gladstone was convinced that Ireland had spoken. He determined, therefore, on a Home Rule Bill, but because Parnell was currently negotiating with the Conservatives, he kept his decision secret. When his son Herbert divulged it most Conservatives and some Liberals – especially Chamberlain – were outraged.

A dispassionate and objective approach to the problem was now out of the question. Randolph Churchill decided that 'the Orange card was the card to play.' Against a background of religious prejudice and political partisanship, Chamberlain, who was prepared to offer local, but not self-government to Ireland, attacked the Bill on three grounds: Ulster's rights, the dismemberment of the Empire and his belief in the inability of the Irish to rule themselves. The Party split, the Bill was defeated and, in an era of growing imperialist feeling, Gladstone's appeal to the electorate was rejected. Henceforth, Irish MPs always cast their vote for the Liberals, and Home Rule now became a Party issue. This was a tragic development.

Thus, in 1886 a great opportunity for solving the Irish question was lost. A modified Home Rule Bill passed the Commons in 1893 only to be rejected by the House of Lords. The matter was then shelved by the Liberals until the constitutional crisis of 1909–11 brought it to the forefront of politics once more.

Motives

It would appear that Gladstone knew precisely what he was doing when he declared for Home Rule in 1886. It was a carefully calculated decision. Having tried coercion and conciliation without success, he felt that *limited* separation was now the only alternative. He was also aware of the feeling that England had much to gain and nothing to lose by devolution. He knew that some Whigs and Radicals in the Party would reject this solution and there is little doubt that his tactical handling of these groups can be faulted. However, like his political master Peel, he was determined to follow a policy which he regarded as just and righteous regardless of Party feelings.

Aware that, in the 'seventies and 'eighties, sectionalism had been a major weakness of the Party, Gladstone hoped that Home Rule might provide an issue around which the majority of Liberals could now unite. Chamberlain, though, knowing that the election of 1885 had shown that the Liberals could no longer rely on the urban working-class vote, sought to rally the Liberals around the more relevant issue of social reform.

Gladstone's Peelite outlook would not allow him to countenance such a programme. His belief that the Party could be reconstructed around the single issue of Home Rule was finally dashed in 1893. His legacy to the Party on his retirement, therefore, was not unity but personal rivalry, divisions over policy and a further decade in opposition. Gladstone may have miscalculated about the future of Liberalism. In terms of electoral appeal, the Party probably needed Home Rule *and* social reform.

Parnell's achievements

Meanwhile, for all his apparent success as leader of the Irish Party in the Commons, Parnell was eventually caught in a web of his own making. By uniting the main strands of Irish agitation he had fashioned a powerful weapon for political action. But in the end extremism, which he rejected, triumphed. He was neither a separatist nor a revolutionary, as he demonstrated in his discarding of both the Land League and the Fenians as soon as he was safely able to do so.

By establishing a disciplined Party tightly bound to the policy of Home Rule, however, he created a series of situations in the Commons in which he appeared to be indispensable. The policy of the New Departure, coupled with his shrewd tactical use of power, allowed him to treat with English politicians on an equal footing and brought him tantalizingly close to his ultimate goal of Home Rule.

Despite the drama and intrigue of his final years, and the suddenness of his fall, his career was not without achievements. He changed the Anglo-Irish balance of power and substantially modified English imperial attitudes to Ireland. Gladstone's Land Acts brought tangible gains in terms of tenant right and land purchase schemes, policies which the Conservative Party under Salisbury continued. After 1886 the Liberals pledged themselves to Home Rule. The discipline and effectiveness of the Irish Parliamentary Party showed that, despite Chamberlain's misgivings, the Irish were capable of self-government.

Parnell's failure

According to Professor Lyons, Parnell was, in the end, largely responsible for his own downfall. His methods and speeches made him many enemies, not least among the ordinary English electors whose feelings he disregarded. He knew what the consequences of his clandestine relationship with Katherine O'Shea would be and, when the divorce scandal made it public, he refused to withdraw from the Party under sympathetic terms. He now appeared to be putting his own career before Home Rule (the Liberal Party – largely Non-conformist – having made it clear that they would not work with him). Despite his achievements it was too high a price to pay for his leadership. His insistence on total Party subservience finally split the Party itself. The Catholic Church then decided against him and his influence steadily slipped away. Determined, and with a certain tragic desperation, he fought against the decision of the Party. Though unwell, he contested a series of by-elections in Ireland, contracted pneumonia and died in October, 1891.

Assessment

Unquestionably, Parnell was the greatest Irish leader since O'Connell and he exercised in the House of Commons a far greater authority than O'Connell ever had. Yet in the end he failed. To some extent the failure lay in himself. With typical Parnellite obstinacy, he had insisted throughout those final months that he alone was right and incurred martyrdom to prove it.

24.4 CHRONOLOGY

1867	**March**	Fenian Rising in Ireland. Fenian outrages in England – Manchester and at Clerkenwell Prison, London.
1868	**December**	Gladstone's first administration.
1869	**July**	DISESTABLISHMENT OF THE IRISH CHURCH. Broke the connection between Church and State. Irish Church also partially disendowed, nearly half its income being allotted to relief. The Bill met strong opposition in the Lords – partly overcome through the intervention of the Queen, who counselled moderation.
1870	**March**	FIRST IRISH LAND ACT. Gave the force of law to tenant-right, provided compensation for unjust disturbance and improvements, and under 'the Bright clause' sought to facilitate land purchase by the tenant. Coercion Act to control crime and agrarian disturbances.
	May	HOME GOVERNMENT ASSOCIATION founded by Isaac Butt in Dublin. The Irish University Reform Bill was defeated by three votes. Parliament was dissolved.
1874	**February**	Home Rule League won 59 seats in the Commons.
1877	**July**	Policy of obstruction began. IRISH AGRICULTURAL DEPRESSION deepened as prices for farm produce continued to fall.
1879	**October**	IRISH LAND LEAGUE founded by Michael Davitt. Parnell became President. THE NEW DEPARTURE; joint programme of action agreed by Davitt, Devoy and Parnell.
1880	**April**	Gladstone's second administration.
	September	Parnell preached the doctrine of 'boycotting' at Ennis, Co. Clare.
	October	Fifty Ulstermen, protected by 900 soldiers, march to Lord Erne's lands in Co. Mayo (managed by Captain Boycott) to harvest his crops. Parnell met and fell in love with Mrs. Katherine O'Shea, sister of General Wood, VC, and niece of Lord Chancellor Hatherley.
1881	**February**	Irish obstruction lasted 41 hours. The 'closure' motion introduced.
	March	Coercion Act. Imprisonment without trial for those suspected of agrarian outrages.
	August	SECOND IRISH LAND ACT conceded the 'three Fs': fair rents, fixity of tenure and free sale. Ignored the problem of rent arrears.
	October	Parnell and other Nationalist leaders imprisoned in Kilmainham Gaol. Land League proclaimed illegal.
1882	**April**	KILMAINHAM TREATY. Government agreed to pay rent arrears – Parnell to dampen agrarian violence.
	May	PHOENIX PARK MURDERS. Lord Cavendish, younger brother of Lord Hartington and Burke, Irish Under-Secretary, were stabbed to death in Phoenix Park, Dublin, by a gang of assassins called 'the Invincibles'. All were eventually captured and five were hanged.. Crimes Act introduced.
	October	National League founded by Parnell to replace the banned Land League.
1885	**June**	Salisbury's first administration.
	August	Ashbourne Act extended the state-assisted scheme of land purchase. The Coercion Acts withdrawn by the government. Parnell entered negotiations with the Conservatives – hoped for some form of devolved government for Ireland. 'CENTRAL BOARD SCHEME': Local, but not self, government advocated by Chamberlain for Ireland. Gladstone was converted to Home Rule but he kept the decision secret.
	November	Encouraged by the government's attitude, Parnell told the Irish in England to vote Conservative. Conceded 30 seats to the Conservatives at the election. Result of election: Liberals 335, Conservatives 249, Irish Party 86.
	December	Herbert Gladstone revealed his father's conversion to Home Rule before parliament met. Conservatives now decided to abandon Parnell.
1886	**January**	Conservatives were defeated on the Royal Address – the famous 'three acres and a cow' amendment moved by Jesse Collings.

Chronology cont.

	February	Gladstone's third administration. Hartington and others refused to serve; Chamberlain and Trevelyan resigned from the cabinet on the Home Rule proposal.
	April	FIRST HOME RULE BILL introduced by Gladstone. Established an Irish parliament in Dublin to deal with all Irish affairs except reserved subjects such as peace and war, defence, foreign affairs, customs and excise. Irish MPs no longer to be present at Westminster.
	June	HOME RULE BILL DEFEATED: 343 votes to 313. Ninety-three Liberals vote against.
	August	Salisbury's second administration.
	October	'PLAN OF CAMPAIGN'. Irish tenants to offer what they considered to be fair rents which, if refused, were to be put in a 'war chest' to fight evictions and aid its victims.
	December	'Plan of Campaign' declared illegal. Balfour takes stringent measures to restore law and order.
1887	**April**	'PARNELLISM AND CRIME': a series of articles in *The Times* alleging that Parnell approved of the Phoenix Park murders. A Special Commission of three judges was appointed to examine the charges.
1889	**February**	Richard Pigott admitted that letters alleging Parnell's involvement were forged; he fled the country.
1890	**November**	Captain O'Shea granted a *decree nisi*, with Parnell cited as the co-respondent. Gladstone, owing to strong pressure from Liberal non-conformists, decided that he could no longer co-operate with Parnell.
	December	SPLIT IN THE IRISH PARLIAMENTARY PARTY; the majority accept the leadership of Justin McCarthy.
1891	**June**	Parnell married Katherine O'Shea.
	October	Parnell died, aged 45. Land Purchase Act introduced by Balfour: widened the scope of the Ashbourne Act. The state advanced to purchasers of land the total price – to be repaid in 49 years.
	December	Congested Districts Board set up in the western counties to encourage investment and employment in these over-populated areas.
1892	**August**	Gladstone's fourth administration.
1893	**February**	SECOND HOME RULE BILL: made provision for the presence of Irish MPs at Westminster. Involved 85 sittings of the House. Passed the Commons with a majority of 34.
	September	Lords reject the Home Rule Bill by 419 votes to 41.
1894	**March**	Gladstone resigned from the Liberal Party.

24.5 QUESTION PRACTICE

1 Why did Ireland need pacifying in 1868?

Understanding the question

The question is a reference to Gladstone's famous comment in 1868 'My mission is to pacify Ireland' and is asking what grievances Ireland had which required redress and also if there was any additional bitterness and urgency given to those grievances by recent events.

Knowledge required

An understanding of the Act of Union, 1800. The position of the Catholic Church in Ireland. The effect of the Famine and emigration on Irish landholding and social structure. The nature of English government in Ireland. Fenianism in 1867.

Suggested essay plan

1   British statesmen were well aware of Irish problems: these were summed up by Disraeli as 'a starving population, an absentee aristocracy and an alien church, and in addition the weakest executive in the world'. The extent of Irish bitterness was revealed by the Fenian outbreaks of 1867. In the election campaign it was clear to Gladstone, and to England, the minimum needed to pacify Ireland. Note, however, that Gladstone ignored Fenian demands for total separation.

2   The alien church: the Church of Ireland was guaranteed by the Act of Union, representing pockets of Protestantism around the great houses, around Dublin and in Ulster; the mass of the population was Catholic, owing allegiance to the Catholic hierarchy but with education, particularly higher education, firmly under Protestant control. Initiative begun by the 1829 Catholic Relief Bill not followed up; the Catholic Church was a bastion of Irish nationalism, giving special urgency to the need for disestablishment.

3   Absentee aristocracy: the land settlement dates back to 17th century, particularly to post-1688; fundamentally it was a transfer of most of Ireland's land to the English and Dutch conquerors; it created the Protestant ascendancy, akin to the Norman aristocracy in England after 1066. Social centre and political centre: London and the shires; usually land holdings in both countries. Pockets of aristocratic society were Protestant, with a life-style similar to that of the English aristocracy, surrounded by oppressed and resentful Irish Catholic tenantry; the situation was worsened by famine, emigration and the consolidation of landholdings, with depopulation of parts of the western counties and a standard of living very near absolute destitution.

4   Weak executive: the union with England had not been a true union between two countries; Dublin Castle basically was still an outpost of colonial government, with an inefficient aristocratic court. Bills passed through parliament at Westminster were not automatically law in Ireland (e.g. the Poor Law Amendment Act, 1834). This could occasionally be beneficial, *e.g.* the setting up of the National Board of Education in 1831 in Ireland, not in England. Fundamentally, however, instrument of anglicization; it was double-edged in that Celtic nationalism was later to reject the whole process. The analogy was with English practice in

India: the argument advanced by Irish constitutional nationalists like O'Connell was, that Ireland was a colonial appendage justified by the separation of control.

5　Fenianism: Fenians were revolutionary nationalists strongly supported by their kin in America. The rising in March, 1867 was followed by widespread arrests (but no death sentences were carried out); the rescue attempt in Manchester (September): the three Manchester Martyrs; the explosion at Clerkenwell prison in December, 1867; this brought the Irish question into the heart of England; the problem now had to be faced. This could lead either to tougher measures or, as in the case of Gladstone, to a search for improvements.

6　Conclusion: New dimension to Irish problem in 1868; American support from Irish-Americans; revival of 1798 tradition of revolutionary nationalism. What was needed were policies which would reduce the legacy of bitterness and the sense of calculated neglect (particularly strong as a result of England's attitude to Irish poverty and her policy towards Ireland in the famine years). The grievances were widely known and acknowledged: Gladstone believed that Ireland had just grievances and that prompt action on these would undermine Fenianism and preserve the Union. Hence his attempt to bring in the healing balm of his Irish policies.

2　In Irish affairs, Gladstone always did too little, too late'. Discuss.

Understanding the question

A straightforward question asking for an assessment of Gladstone's policy in Ireland, relating his policy to Ireland's needs. Was this policy ungenerous, grudgingly conceded and tardily implemented?

Knowledge required

The Act of Union, 1800. Gladstone's disestablishment of the Irish Church, 1869; the Land Act, 1870; the Land Act, 1881; the Kilmainham Treaty, 1882. The treatment of Parnell, 1885; the Home Rule Bill, 1886; the Home Rule Bill, 1893.

Suggested essay plan

1　'Too little, too late' is a comment often made on England's Irish policy as a whole. In general, this is apt and true but when applied to Gladstone's Irish policy it breaks down as an overall generalization. Gladstone came late to Irish affairs but in the next 25 years he addressed himself to all the main aspects of the Irish problem, with effect and despatch in some areas and disaster and delay in others.

2　Disestablishment: the Act of Union had guaranteed the Protestant ascendancy in the Church, in property and in the state. Gladstone set about each of these problems in turn. The position of the Church was eroded by the Catholic Emancipation Act; English non-conformists through the British Liberation Society pushing for general disestablishment, as was the Irish Catholic hierarchy. Gladstone made this his first task; he vigorously piloted the Bill through parliament; severed the link between the Irish Church and the state; sought justice for Irish Catholics and a fair settlement for Irish Protestants. This measure was in no sense too little and could be regarded as too late only if the whole span of nineteenth-century policy in Ireland is included. Certainly not late in Gladstone's achievements.

3　Agrarian reform: England's great achievement in Ireland was to carry out a profound revision of the structure of landed property without revolution. Completed by the Conservatives in Wyndham's Act (1903), it was Gladstone's pioneering work which set the policy in motion. The basis of the Protestant

ascendancy in land goes back to the seventeenth century and essentially created an occupying power throughout the land acting through the normal means of civil government. To tamper with this was to undermine the English position in Ireland. Certainly the 1870 Act was too little – clauses were evaded and tenants were still evicted for the non-payment of rent – but it was a further example of the sense of urgency which Gladstone brought to Irish affairs; it was revolutionary in its implications. The 1881 Bill established the novel principle of co-partnership in the soil between landlord and tenant.

4　Gladstone and Parnell. The wisdom of the Kilmainham Treaty; which gained Irish, Liberal support; the Irish worked within the constitution until 1916; Gladstone failed to give Parnell a firm commitment on Home Rule in 1885 (potentially difficult with the Liberal Party) but decision in 1886 brought Gladstone into power to try to rectify the last of Ireland's grievances. The Home Rule Bill was too little in two respects: the lack of Irish MPs in the Commons and the limitation on the Dublin parliament's powers, but it would have been a precedent. Its failure may have been due, in part, to Gladstone's tactless handling of Chamberlain, the new power in the Liberal Party. Liberals were now committed to Home Rule. In 1893, and again in 1912, Home Rule was carried in the Commons. In 1886, or even 1893, Home Rule would have been neither too late nor, on the whole, too little.

5　Conclusion. Gladstone's major achievement in Irish policy was to set course in the right direction: towards a free Church, towards agrarian reform, towards political autonomy. The judgement 'too little, too late' might be the comment expected from extreme Irish nationalists such as the Fenians, and may to some extent be justified in the details of his policy, but not in the grand sweep of his inspiration.

3　Assess the strengths and weaknesses of Parnell as a leader of the Irish cause.

Understanding the question

The question calls for an understanding of Parnell's personality, character and private life – in so far as they affected his leadership of the Irish Party. It also requires an understanding of what the Irish cause was, how Parnell led it and finally an assessment of the tactics he used to achieve his objectives.

Knowledge required

A knowledge of Parnell's character and personality. His involvement in the Land League; the 'Boycott' initiative; Parnell's insistence on the priority of political, *i.e.* constitutional, methods – as in the Kilmainham Treaty (1882). The negotiations with the Conservatives and Liberals (1885–86) over Home Rule. The year 1890 and the crisis, personal and political, which led to Parnell's loss of the leadership.

Suggested essay plan

1　Parnell emerged in the 1870s as a great Irish leader to stand with O'Connell in the Irish imagination; he advocated Irish self-government, and impressed English politicians. By his style and appearance, he was cast in the mould of a leader but basically he was a diffident, nervous man, driven by passion and a commitment to master all the public arts of politics and to accept all the necessary tedium of political life. What were his strengths and weaknesses as an Irish leader?

2　His major strength was his commitment to a single goal: self-government. He took over from Isaac Butt, replacing the latter's unsatisfactory federalism with a policy of Home Rule for Ireland. This commitment was of great importance in the agrarian disturbances of the period (1879–82). Parnell became President of the Land League (seeing it as useful in 'laying the foundation in the movement for the regeneration of legislative independence') this allowed him to share in some of the excitement of extremism and violence but did not, despite the Coercion Bill and Davitt's arrest, tempt him to secede from parliament to lead a potentially violent agrarian campaign in Ireland.

Parnell accepted the 1880 Land Act and the Kilmainham Treaty, promising an end to 'outrages' in return for a settlement of the rent arrears question. Parnell made an important promise to co-operate with the Liberals; from then on, the Irish Party worked strictly within the constitutional context to achieve the political ideal of Home Rule. His unwillingness to support the Plan of Campaign (after 1886) demonstrated that his association with the Land League was tactical.

3 Parnell's strength was also seen in his astuteness, flexibility and political insight. Sometimes he was seen as obdurate, by reason of his obsessive commitment to his political goal; the other side of the coin was his shrewd political insight coupled with flexibility used for tactical purposes. His call to send to a 'moral Coventry like a leper of old' anyone who moved onto a farm from which a tenant has been evicted highlighted the campaign against the growing evictions of the late 'seventies. The same political insight and flexibility showed in his willingness, in 1885, to negotiate with the Conservatives (through Carnarvon) with an offer to support them if they would introduce Home Rule. Parnell knew that the Conservatives could more readily get a bill through the House of Lords – as did Gladstone, who refused therefore to commit the Liberals. Parnell's instructions to the Irish in England to vote Conservative lost perhaps 30 seats to the Liberals. Gladstone's conversion in 1886 swung Irish MPs behind the Liberals.

4 Parnell's weakness: the impact of his private life on his political life. Neither Catholic Ireland nor non-conformist England could overlook Parnell's co-habitation with a married woman; this misalliance (despite the personal comfort it brought to him) proved fundamentally injurious to his leadership when revealed publicly in 1890. Although his health was failing, yet he refused to see that his only course was resignation – for (as Morley and Gladstone saw) a Liberal Party based on English non-conformity would not be able to accept an Irish leader who was involved in a divorce case.

Parnell's intransigence put at risk the Liberal Irish alliance and the Home Rule policy which Liberals had adopted. His unwillingness to accept the logical consequences of his action split the Irish Parliamentary Party and reduced the strength of the powerful machine he had created.

5 Irishmen, like the editor D. P. Moran, were to criticize Parnell for ignoring Irish civilization in the pursuit of the principle of Home Rule. However, not all goals may be pursued at the same time with success. Parnell made all England conscious of the Irish problem, created a great parliamentary Party and set Irish politics on the road of constitutionalism where they remained until 1916.

A combination of charm, passionate conviction, political skill and astute parliamentary tactics took Parnell (and with him the Irish cause) to the forefront of British politics. Protected by his bitter anti-Englishness from the danger of being won over by the English establishment, he was defeated in the end by his pride and his failure to see where his duty to Ireland lay.

4   Document question (Gladstone and Ireland): study the following extract and then answer questions (a) **to** (g) **which follow:**

'Today's *Times* throws back Mr Gladstone's recent conduct into the perspective of history, saying: "The first Minister of a great nation, having just admitted to political rights a great mass of voters entirely unpractised in political affairs, appealed to the country as a whole for power to deal independently with a conspiracy which he had himself denounced in the strongest language, the leaders of which he had put in prison, and against which he had repeatedly employed exceptional powers. No sooner were the elections over than that Minister allied himself closely with the leaders of that conspiracy, accepted their traitorous plans for the disruption of the empire, and fought 'shoulder to shoulder' with them in Parliament and afterwards in the country, a desperate battle for the attainment of their ends. In the interests of his new allies he turned ruthlessly against all the men who had done most to build up the party from which he derived his power, and whose single offence was that they guarded the great interests which he betrayed. The most elementary right of the nation was to have the whole question laid before it with the most scrupulous fairness and the most absolute clearness; but this Minister, on the contrary, resorted to every trick, device and stratagem to conceal from the people the nature, cost and consequences of submission to the conspiracy he had joined. The touching confidence reposed in him by the masses of his countrymen was by this Minister deliberately used without stint or scruple to blindfold, mislead and betray the nation which had made him the chief guardian of its interests. Great as is Mr Gladstone's fall measured by numerical standards, it is altogether eclipsed by the tremendous moral descent, which we venture to think cannot be paralleled in English history."'

(From 'The St James's Gazette')

Maximum marks

(a) From the information and accusations contained in this extract, in what year do you consider that it first appeared in print?   (2)

(b) Explain to what 'having just admitted to political rights a great mass of voters' refers.   (3)

(c) Show clearly to what *each* of the following refers:
   (i) 'conspiracy';
   (ii) 'leaders of which he had put in prison'
   (iii) 'repeatedly employed exceptional powers'.   (6)

(d) What 'traitorous plans for the disruption of the empire' did Gladstone accept from 'the leaders of that conspiracy'?   (2)

(e) Of which individuals and groups is the writer thinking when he refers to 'all the men who had done most to build up the party from which he derived his power'?   (6)

(f) What is meant by saying Gladstone 'fought "shoulder to shoulder" with them in Parliament and afterwards in the country'   (4)

(g) The writer accuses Gladstone of acting in a manner calculated to 'blindfold, mislead and betray the nation'. Discuss the bias shown in the report and the real reasons why Gladstone acted as he did.   (8)

24.6 READING LIST

Standard textbook reading

R. K. Webb, *Modern England* (Allen and Unwin, 1980 edn.), chapter 8.

A. Wood, *Nineteenth-Century Britain* (Longmans, 1960), chapters 15, 17.

Further suggested reading

J. C. Beckett, *The Making of Modern Ireland 1603–1923* (Faber, 1966).

D. A. Hamer, *Liberal Politics in the Age of Gladstone and Rosebery* (Clarendon Press, 1972).

J. L. Hammond, *Gladstone and the Irish Nation* (Cass, 1964).

J. Lee, *The Modernization of Irish Society 1848–1918* (Gill and Macmillan, 1973).

F. S. L. Lyons, *Ireland Since the Famine* (Collins, 1971, 1973).

F. S. L. Lyons, *Charles Stewart Parnell* (Collins, 1977).

B. L. Solow, *The Land Question and the Irish Economy 1870–1903* (Harvard UP, 1971).

25 The Liberal governments, 1906–14

25.1 INTRODUCTION

The Edwardian age often conjures a colourful, nostalgic image of an apparently stable and happy society soon to be shaken by the horrors of the Great War. Yet although the Liberals laid the foundations of the welfare state and righted many political wrongs, it was conflict which characterized almost every aspect of Edwardian life. Feminist frustration, labour unrest, a rebellious House of Lords and Ulstermen in revolt – all this made the Edwardian period an epilogue of political violence, confusion and pessimism.

Fervently-held beliefs – in *laissez-faire* and free trade, in capitalism and in Empire – were being challenged as never before. Everywhere unwelcome change was in the air; Victorian certainty had slipped away. To contemporaries it became clear that the Queen's death had, indeed, marked the end of an era.

25.2 STUDY OBJECTIVES

1   Problems of the Liberal Party to 1914: the Party's split on Home Rule in 1886, and Boer War divisions; the loss of middle-class business support weakened Party finances; it could not, therefore, afford to adopt working-class candidates as MPs; the Party's severe organizational weaknesses, despite the efforts of Herbert Gladstone.

2   Progress of the Labour Party to 1914: 'a coagulation of incompatible elements' such as the Social Democratic Federation, the Fabian Society, Labour churches, and the trade unions; the rise of the 'new unionism' in the 1880s, the employers' counter-attack (the Free-Labour Associations) in the 1890s, and the hostility of judges to the unions, culminating in the Taff Vale case; the formation of the Independent Labour Party (1893) and of the Labour Representation Committee (1900); the Gladstone-MacDonald Pact (1903) ensured Labour a parliamentary foothold in 1906; Hardie's and MacDonald's leadership; the rise of unionism during the 1900s; the continuing uncertainty, even by 1914, of whether Labour would try to permeate or replace the Liberals.

3   The Liberal landslide victory of 1906 occurred on issues like free trade, Chinese labour, non-conformity and education; was this a last flicker of the 'old' Gladstonian Liberalism or an indication of future political strength? Efforts to revitalize the Liberal Party: the 'New Liberalism', evolving from the 'Liberal Imperialism' of Asquith, Rosebery and Grey, was then taken further by Churchill, Lloyd George and Scott's *'Manchester Guardian'*; demands for purposive intervention in the economy to achieve greater 'national efficiency'.

4   Campbell-Bannerman and Asquith governments ushered in a period of Liberal reforms – old age pensions, national insurance, etc.; there was also the bitter constitutional dispute over the 'People's Budget' and the Parliament Act – this was instigated by Unionist determination to retain their monopoly of power despite the 1906 defeat. Why did Balfour adopt these tactics? To appease his critics? Or was he a victim of the prevailing mood of political extremism?

5   The Parliament Act marked the zenith of Liberal success; the Party had already exhausted its programme, and now its radical energy in the struggle; the Unionists under Bonar Law were more disaffected than ever. The Irish question was re-opened due to the Parliament Act; the Liberal Party in the 1910 general election advocated Home Rule – in return for Irish Party help in the constitutional crisis. The prospects seemed good but Liberal and Nationalist hopes were dashed by Ulster's intransigence under Carson; openly encouraged by Bonar Law's Conservative and Unionist Party ('there are things stronger than parliamentary majorities', Blenheim, July, 1912).

6   The Ulster rebellion undermined government confidence during the years 1911–14. There were other problems too – *e.g.* the tense situation in foreign affairs and, at home, industrial unrest, syndicalism, the 'Triple Alliance'. Askwith and Lloyd George occasionally rescued the government here but militant labour (*i.e.* the Shop Stewards' Movement) were clearly disenchanted with Liberal efforts to tackle the economic and social problems of the day.

7   Positive in their handling of labour unrest, the government was weak, irresponsible and divided in their treatment of women's suffrage. Asquith's procrastination discredited the constitutional approach and encouraged WSPU militancy. On the other hand, the suffragettes' unwillingness to compromise, and the violence of their campaign, was ultimately self-defeating.

8   Assessment of the Dangerfield thesis. Were the Liberals in the throes of a 'a strange death' during the years 1906–14? Were they ceasing to govern? Was their replacement by Socialism and the Labour Party inevitable? The answer must be negative but the Liberals were clearly on the defensive during the period 1911–14 and the omens were not good.

25.3 COMMENTARY

The end of an era

The death of the Queen on 22 January, 1901, marked the end of an era in British politics. *Laissez-faire*, British economic supremacy, imperialism – all were under attack; and the conviction was growing that to solve the industrial and social problems of England the state must accept more responsibility for the welfare of the people.

During their period of dominance (1886–1905) the Conservatives had failed to come to terms with this development. In 1906, for the first time in English history, there was a government which not only represented the majority of the people but which consisted largely of the men of the people. The pre-eminence of the landed classes was successfully overthrown. It was now the turn of a Liberal government and its Labour allies to confront these problems.

The 1906 election

The election of 1906 was one of the most dramatic in British history. It gave the Liberals a landslide victory and an absolute majority in the Commons. But it would be a mistake to interpret this as a massive vote of confidence in the Liberal Party and its electoral programme.

A vote against Conservatism, reflected, in particular, nonconformist fury against the Education Act of 1902. Labour supporters were angry at the Taff Vale judgement, 'Chinese slavery' and growing unemployment. But above all it was a vote against protection. The electors were punishing the Conservatives, who had neglected the middle ground, rather than positively supporting the Liberals. It is important to remember this when one examines the difficulties which the Liberal Party encountered during the next eight years.

Old Liberalism or new?

The Party which took over government in 1906 loosely represented this mixed force of Liberal, Labour and non-conformist dissidents. Would the Lib.-Lab. electoral pact endure? Could the government satisfy the demands and hopes of its many supporters? Would the Labour Party replace a Liberal Party unable to cope with the divisive politics of class?

The new administration was certainly talented but was by no means united – nor could it clearly see the way ahead. Churchill and Lloyd George wanted the Party to adopt the cause of the working classes and, by introducing valuable social reforms, show them that the Liberals and not the Labour Party could best be relied upon to advance their interests.

Despite the misgivings of some Right-wing members, they seized the initiative in the field of social policy and introduced government measures on trade unions, wages and conditions of work, health insurance and old age pensions – the latter to be paid for by increasing income tax, death duties and land taxes. Clearly these were policies designed to adapt to the increasing importance of class, steal the thunder of the new Labour Party and re-emphasize the progressive and reforming nature of Liberalism. 'Socialism,' Churchill declared, 'seeks to pull down wealth; Liberalism seeks to raise up poverty.' Many in the Party were reluctant, however, to sanction more state interference or agree to any further taxation for social reforms. Instead they reiterated their dedication to the fundamental principles of *laissez-faire* and the capitalist economy. Despite Lloyd George's efforts, therefore, many Liberals were still committed to the policies of 'Old Liberalism' – Cobdenite free trade, Gladstonian economy and non-conformist conscience.

The origins of the reforms

The social policies of Churchill and Lloyd George were not original. They were able to draw on much creative social thought generated since the 1870s by Chamberlain, the Fabians, philanthropists, administrators and civil servants. Reforms were long overdue. The social surveys of Booth and Rowntree showed that in the world's richest empire a third of the population was living below the breadline.

Recruiting statistics in the Boer War appeared to confirm the findings of these surveys. Social reform – in the interests of 'national efficiency' and British military power – was suddenly more respectable. German strength also seemed to emphasize the benefits of Bismarckian social welfare which, incidentally, provided Lloyd George with a model for his National Health Insurance scheme. There were also other continental social reform systems to emulate.

Nevertheless, it still needed authority and stamina to get things done and Lloyd George and Churchill, well endowed in both, were determined that the Liberal Party, and not Labour, should make full use of the 'new mandate' to improve the conditions of the working class.

Lloyd George believed that the main social purpose of the New Liberalism was 'to draw a line below which we will not allow people to live and labour'. In the attainment of this humane purpose, Gladstonian ideas of self-help had to be modified or discarded: Booth and Rowntree had documented circumstances – of childhood, sweated trades and old age, for example – in which it could not conceivably operate.

You will need to know the details of the social reform legislation of the Liberal governments from 1906 to 1912. Note that this legislation, though at first narrow in its application, laid the foundations of the post-1945 welfare state. And do not overlook the fact that these reforms were carried out only because progressive and enlightened administrators, such as Morant and Beveridge, were prepared to challenge Civil Service traditionalism.

Labour support

If the policy of the Liberals was meant as an antidote to Socialism it would appear to have been effective. The Labour Party in the Commons during these years was content to support the programme while concentrating its efforts on attaining measures for the trade unions. The Liberal victories in the elections of 1910 also demonstrated the ability of the New Liberalism to win working-class votes while converting the new Labour Party in the Commons to a piecemeal approach to poverty, sickness and unemployment.

By 1914, the Labour Party – out-trumped by the social reforms of Lloyd George and Churchill – was, according to Brand, 'dependent on the Liberals, dissatisfied with its achievements, unsure of its aims and apparently in decline'. Other historians, however, emphasize the growth of trade-union membership and increasing Labour representation in local government during these years and insist that these developments indicate a growing future threat to Liberalism. Up to 1910, therefore, the New Liberalism appeared to provide a satisfactory ideological alternative to Socialism.

The House of Lords

Labour supporters appeared to have been satisfied with this programme of social reforms. But what happened to traditional Liberal demands? The Unionists, dismayed at their overwhelming defeat at the polls in 1906, decided to use the House of Lords as a second Conservative opposition, selectively savaging proposed Liberal bills, while at the same time carefully refraining from antagonizing Labour.

Impotent in the Commons, Balfour was using the House of Lords as a means of perpetuating Conservative rule and appeasing his own critics. The Liberal cabinet was alarmed by this policy. Accordingly, before Asquith became Prime Minister, they agreed that if the legitimate aspirations of Liberal supporters were to be satisfied the power of the Lords would have to be reduced to a suspensory veto.

By 1908 the Liberals were frustrated and in disarray. They needed a popular issue on which to challenge the Lords. Home Rule and Welsh Church disestablishment clearly were not popular; defence and social reform were proving more expensive than many Liberals wanted but all were agreed that these policies at least were vital.

In the event it was Lloyd George's search for the £15 million to cover the cost of old age pensions and the new naval programme which provided the issue – the 'People's Budget' – on which to challenge the 'unconstitutional' actions of the Lords. Thus it was that the policy of defence and social reform precipitated a fundamental constitutional conflict which dominated the years 1909–11.

The People's Budget

Lloyd George designed a budget calculated to infuriate the Unionist peers. It did. Increased income tax, death duties and the introduction of a new Land Tax, coupled with the Chancellor's biting invective against the peers (*e.g.* at Limehouse in July), goaded them into rejecting the budget. The trap was sprung on 'Mr Balfour's poodle'.

In the ensuing events the worst fears of the more moderate Unionists in the Commons were confirmed. There followed two electoral defeats in 1910, the passing of the budget, the curtailment of the Lords' power, the prospect of more social reform and, even worse, Home Rule for Ireland.

According to Dangerfield, the struggle with the Lords was a triumph for Asquith, both as a Party leader and constitutional lawyer, who had treated the King with consideration. Lloyd George also had reason to be pleased.

Study carefully the details of the contest and the main provisions of the Parliament Act of 1911. Note that although the Liberals won a great victory over the aristocracy, the loss of their overall majority in the elections of 1910 marked the end of their Commons' supremacy. From then on they were dependent on the Irish Party – and the price of that support was Home Rule. It was now no longer possible to keep the Irish Question out of English party politics.

The elections of 1910

Why was the Liberal majority reduced in 1910? Despite their legislative successes they antagonized many sections of the community. First of all, wages were falling behind rising prices. Dissatisfied workers were beginning to lose faith in the Liberals' ability to look after their interests. And they insisted that the Liberal social reforms did not go far enough and were simply palliatives designed to ease, but not cure, poverty and unemployment. They correctly gauged that, because of opposition in the Party, this was as far as the Liberals were prepared to go.

On the other hand, industrialists, manufacturers and farmers felt that these social reforms had gone too far and began to fear 'Socialism by the back door'. They had also lost faith in free trade and, encouraged by the prospective benefits for all classes, were beginning to look more favourably on the Conservative and Unionist policy of protection. Right-wing Liberals were alarmed at the size of government expenditure on the Army, Navy and social reform – policies hardly in keeping with the Liberal tradition.

Note that although Haldane's spending was stringently limited, he brought the Cardwell system up to date so that, when war broke out in 1914, Britain was militarily better prepared than she had ever been in her long history.

The Ulster Question

The year 1910, then, was a watershed and after this date the Liberal momentum appeared to slow down. A growing estrangement between the Liberals and Labour became noticeable as the government failed to contain, or to satisfy, a series of challenges made to parliamentary democracy itself. The Parliament Act of 1911 had solved the constitutional crisis, but the Irish Party in the Commons now demanded that Home Rule be implemented.

The Conservatives, having discarded Balfour, took a militant stand under Bonar Law and, in resisting the Bill, identified themselves with Ulster loyalist preparations for civil war. The actions and speeches of Carson and Bonar Law began increasingly to seem treasonable – 'there are things stronger than parliamentary majorities'.

Asquith seemed unable to face the Ulster situation squarely. He saw the impossibility of coercing Ulster but did not know how to deal with the implications. In the end, the Liberals' nerve failed but the outbreak of the First World War rescued them from the immediate consequences of their failure.

Votes for women

The Irish problem also complicated the issue of women's suffrage; for the Irish Party refused to consider any measure of electoral reform (presuming that the Liberals were willing to introduce such a measure) before Home Rule was conceded.

Meanwhile, the suffragettes, having failed to gain their objectives by persuasion in parliament, began to appeal, like the Unionists, to the argument of force outside it. They, therefore, waged a campaign of violence from 1912 to 1914. The Liberal leaders opposed the extension of the suffrage for party political reasons, believing that women, being naturally 'conservative', would vote for the opposition. Frustrated by a combination of Liberal male chauvinism and selfish political calculation, suffragette militancy added to the atmosphere of social crisis in the years before the war.

Labour unrest

The most serious challenge to Liberal power, however, came from the working class. In one industry after another strikes dominated the scene from 1911 to 1914. Unofficial strikes also became more common. Union membership doubled during these years and with it there developed a new awareness of the power of trade unionism. Though the government made considerable progress in introducing techniques of industrial conciliation under George Askwith, syndicalist ideas continued to spread. These encouraged the belief that national strike action, especially when strengthened by industrial rather than craft organization, would force the Liberals to intervene more positively on behalf of the workers.

The unrest also threatened to undermine the stability of the Lib.-Lab. pact (on which the future of Liberalism now depended) and precipitate a split between Liberals and Labour which would open the way for a Conservative victory at the polls. The depth of labour unrest, however, convinced many Liberals that a more independent, socialist Labour party would soon emerge, whatever measures the Liberals adopted.

Strange death?

George Dangerfield imputes that these events turned a gradual Liberal decline into a rout from 1911–14. He alleges that the Liberal leaders, handicapped by an outdated philosophy, were 'airy, remote and irresponsible' and had lost the ability and the will to govern. This inability, he argues, provoked much of the violence of Unionists, workers and women.

Trevor Wilson disputes this interpretation. He acknowledges that the Liberals were faced with formidable and dangerous problems but asserts that their policies and actions demonstrate that they were far from being exhausted in 1914. Furthermore, it was the First World War, with its massive programme of state intervention, and the ending of peace, retrenchment, reform and free trade, that destroyed liberalism as a philosophy – quite apart from Lloyd George's ambitious activities in parliament after 1914.

Though the period ended in social conflict and the Liberal future looked uncertain, much had been achieved. The Liberals had laid the foundations of the welfare state, modernized the Army and Navy, tackled the wrongs of labour – though not of women – curtailed the power of the House of Lords and introduced an Irish Home Rule Bill. For some it was not enough; for others, too much. Dissatisfied Ulstermen and suffragettes, peers and trade unionists continued to challenge the established order. Thereby they ensured that the years of liberalism's decline would be years of turbulence.

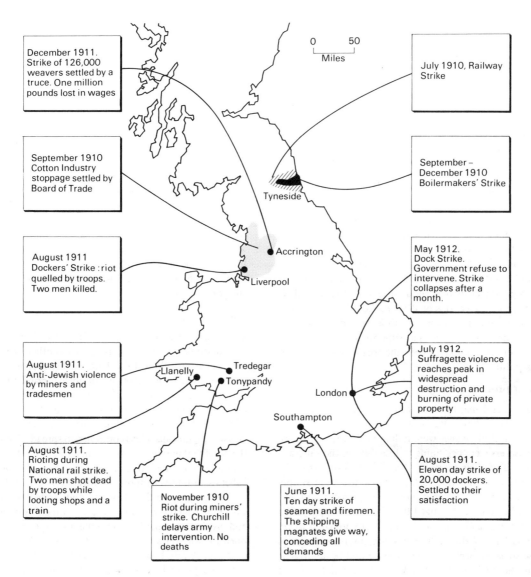

December 1911.
Strike of 126,000
weavers settled by a
truce. One million
pounds lost in wages

July 1910, Railway
Strike

September 1910
Cotton Industry
stoppage settled by
Board of Trade

September –
December 1910
Boilermakers' Strike

Tyneside

May 1912.
Dock Strike.
Government refuse to
intervene. Strike
collapses after a
month.

August 1911
Dockers' Strike : riot
quelled by troops.
Two men killed.

Accrington

Liverpool

August 1911.
Anti-Jewish violence
by miners and
tradesmen

Llanelly   Tredegar
          Tonypandy

London

July 1912.
Suffragette violence
reaches peak in
widespread
destruction and
burning of private
property

Southampton

August 1911.
Rioting during
National rail strike.
Two men shot dead
by troops while
looting shops and a
train

November 1910
Riot during miners'
strike. Churchill
delays army
intervention. No
deaths

June 1911.
Ten day strike of
seamen and firemen.
The shipping
magnates give way,
conceding all
demands

August 1911.
Eleven day strike of
20,000 dockers.
Settled to their
satisfaction

0   50
Miles

Fig. 18 Labour unrest in Britain 1910–12

25.4 CHRONOLOGY

1906   **January**   Campbell-Bannerman's first administration.

February   First Dreadnought built.

December   TRADE DISPUTES ACT reversed the Taff Vale decision of 1901. Trade Union funds not liable to actions for damages.
WORKMEN'S COMPENSATION ACT: made compensation payable for accidents arising out of, and in the course of, the workman's employment.
SCHOOL MEALS ACT: introduced in a Labour initiative, allowed local authorities to arrange for school meals.
Education Bill rejected by the Lords.
Plural Voting Bill rejected by the Lords.
Land Bill rejected by the Lords.

1907   **January**   ARMY REFORMS begun by Haldane. Size of the Regular Army reduced but an Expeditionary Force set up under a new General Staff and backed up by a Territorial Army at home. Officer's Training Corps also founded. In August, 1914, some 26 divisions were quickly mobilized.
SCHOOL MEDICAL INSPECTION introduced.

December   Port of London Authority established by Churchill.

1908   **April**   Asquith's first administration.

November   Licensing Bill rejected by the Lords.

1909   **January**   OLD AGE PENSIONS ACT implemented: pensions of 5s. per week to persons over 70 years of age, 7s. 6d. to married couples.
Royal Commission on the Poor Law: report published. The majority wanted to retain a re-organized Poor Law; the minority, headed by Beatrice Webb, wanted to abolish it. Both accepted the need to extend the state machinery to help the poor and the unemployed.

	March	COAL MINES ACT: limited the working day of coalminers to eight hours.
		TRADE BOARDS ACT: fixed minimum wages in the sweated trades, *e.g.* tailoring and lace-making.
	April	THE PEOPLE'S BUDGET: a bold measure introduced by Lloyd George to raise £15 million for defence and old-age pensions. He also wanted fresh sources of revenue to finance social insurance. The Budget also increased income tax and death duties, as well as licensing, tobacco and spirit duties; there were new taxes on cars, petrol and land – the latter being the most contentious because it necessitated an inquisitorial valuation of land.
		CHILDREN'S ACT: forbade child begging, children in public houses or the sale of alcoholic liquor to children under 16.
	August	Labour Exchange Act passed.
	November	PEOPLE'S BUDGET REJECTED by the House of Lords by 350 votes to 75.
	December	OSBORNE JUDGEMENT: forbade the trade unions to raise a political levy.
		Asquith dissolved parliament. General election scheduled for January.
1910	**January**	Asquith's second administration. Liberals lost their overall majority in the Commons.
	April	PEOPLE'S BUDGET passed by the House of Lords.
	May	PARLIAMENT BILL introduced. The Lords to lose control over money Bills but allowed the power to delay legislation for two years.
		Death of King Edward VII. Accession of George V.
	June	Constitutional Conference – to resolve the deadlock between the Lords and the Commons (over the Parliament Bill) – began but failed, after five months' negotiations, in October, 1910.
	December	General election.
1911	**February**	Asquith's third administration.
	August	PAYMENT OF MPS introduced: £400 a year.
		Railway strike.
	November	PARLIAMENT ACT passed by the Lords, 131 votes to 113.
		Cambrian miners' strike against wage cuts. Troops used at Tonypandy; two fatal casualties; miners blamed the Home Secretary, Winston Churchill.
	December	NATIONAL INSURANCE ACT (introduced by Lloyd George) provided insurance for the poor against sickenss for a contribution of 4d.; employers paid 3d and the state 2d. Unemployed workmen (in certain trades) were to receive 7s. per week for 15 weeks. By 1914 thirteen million people were insured.
1912	**February**	Coal strike: 2 million men now unemployed.
	March	MINIMUM WAGES FOR MINERS.
		Dockers' strike in London.
	April	IRISH HOME RULE BILL introduced: similar to the 1893 Bill; attacked for its injustice to Protestant Ulster.
	September	Ulster Covenant pledged resistance to Home Rule.
1913	**January**	Home Rule Bill passed by the Commons, rejected by the Lords.
		Plural Voting Act prohibited an elector from voting in more than one constituency.
		TRADE UNION ACT: nullified the Osborne Judgment of 1909.
		'Cat and Mouse Act': passed to deal with the suffragette hunger-strikers.
	July	Ulster Protestants founded the Ulster Volunteers to resist Home Rule by force of arms if necessary.
		Home Rule Bill again rejected by the Lords.
1914	**May**	Home Rule Bill passed the Commons.
	August	BRITAIN DECLARED WAR ON GERMANY.
	September	HOME RULE ACT passed but suspended until after the war, when the question of Ulster's exclusion from the Act could be further discussed. The Act never came into operation – it was replaced by the Home Rule Bill of December, 1920.
		Welsh Church Disestablishment Act passed but suspended for the duration of the war.

25.5 QUESTION PRACTICE

1 Explain the origins and development of the Labour Party down to 1914.

Understanding the question

This is a straightforward narrative question but you must remember to include all aspects of the Labour movement and support – not just the narrowly organizational development of the parliamentary Party; remember to point out the weaknesses of the rival parties.

Knowledge required

Social and economic developments c. 1870–1914 and their effect on politics at grass-roots level. The history of trade unionism and the development of socialist organizations. Political history c. 1890–1914.

Suggested essay plan

1   The 'great depression' of the 1880s ends the long mid-Victorian boom; the effect on living standards and employment

levels; middle-class observers and social theorists 'discover' unemployment as a structural problem; the poverty surveys (Booth, Rowntree) and propaganda awakens sensitive middle-class consciences; anger at the absorption of politicians with 'human rights' issues (Home Rule especially) instead of social reform and reconstruction.

2   The influence of socialist writers like Morris; the rise of middle-class socialist societies; the Social Democratic Federation (Hyndman) develops the Marxist critique; the Fabian Society (the Webbs) advocates 'permeation' (Lib-Labism) and piecemeal reforms; the Labour churches; the Co-operative movement; Socialism on the LCC and in the provinces (Blatchford in Manchester, Jowett in Bradford, etc); the foundation of the Independent Labour Party (1893) under Keir Hardie; the Colne Valley by-election.

3   The rise of trade unionism 1825–1880, the effects of Disraeli's 1875 legislation; the growth of the 'new' unionism among unskilled labourers, starting with the gas workers (Will Thorne) and the dockers (Ben Tillett); the impact of the 1889 dock strike. Many unionists, however, still hoped that the traditional Parties (especially the Liberals) would attend to their interests.

4   The 1890s undermined the newly-won trade-union confidence; few gains were made *despite* full employment; the employers fought back with the Free Labour Associations; judge-made law supported the employers (Lyons *v* Wilkins, Allen *v* Flood) culminating in the trauma of the Taff Vale decision; the decline of real wages after 1900, and the development of syndicalism.

5   The trade union crisis led to the foundation of the Labour Representation Committee (1900); the role of MacDonald in its policies; the Gladstone/MacDonald pact (1903) gave Labour 53 MPs in 1906 (including 24 Lib-Labs); the 1906 Trades Disputes Act enabled the rapid rise of trade unionism during the period 1906–14; the failure to make parliamentary headway, in 1910 (40 MPs) and the Osborne Judgement (1908–9) encouraged more direct action (*e.g.* sydicalism; the Triple Alliance); the Liberals contained Labour in 1910–14 by-elections; even the Irish Home Rulers had more MPs than Labour.

6   The long-term forces favouring Labour against the Liberals; the rise of unionism, the greater industrial concentration, the new management structures, the increased geographical mobility of labour – all this led to the gradual break-down of the old 'vertical' deference politics, based on community relationships, places of worship, and paternalist employers, and to the arrival of 'horizontal' class-based politics looking for an extension of workers' power and comforts. The affiliation to Labour of the Miners' Federation in 1909 was a milestone; in this light, the rise of Labour was inevitable and awaited only the increased electorate (achieved in 1918) for its consummation.

7   And yet, Labour continued much of the ethos of the old Liberalism: free trade, a League-of-Nations type of pacifism, a secularized revivalism, teetotallism, the dignity of man and of labour, the old non-conformist conscience; Hardie and Snowden both preached a 'come-to-Jesus' type of evangelical Socialism, which had little in common with the collectivism of the unions or the full-blooded Socialism of the ILP; and since, down to 1914, the Liberals proved to be capable of responding to specific working-class needs, it is possible that, without the tumult caused by the war, the Liberals would have held off the Labour challenge.

2   Account for the crushing defeat of the Conservatives in the general election of 1906.

Understanding the question

The general election of 1906 saw one of the greatest electoral transformations in modern British political history, The question asks for a discussion of the reasons which lay behind the Conservative defeat and the remarkable recovery of the Liberal Party.

Knowledge required

The election results in the 1895–1906 period. The changes in the position of the Conservative, Liberal and Labour political parties in the years before 1906. An understanding of the main issues in early twentieth-century politics.

Suggested essay plan

1   The 1906 general election brought to an end to a long period of Conservative dominance in British politics. The Liberal split over Home Rule, together with the rise in imperialist sentiment, ensured a string of Conservative electoral victories, punctuated only by the weak Liberal government of 1892–5. The 1900 general election, held at the height of the Boer War and with British armies sweeping to victory in South Africa, brought only a slight diminution of the government's majority.

2   A major cause of Conservative success was the continued disarray in the ranks of their opponents. The Liberal Party had formed strong, successful governments in the heyday of Gladstone's political career; it was not clear, however, where the Party was to go thereafter. Gladstone had nailed Irish Home Rule (never an electoral winner in Great Britain) to the Party's banner; there were also serious disagreements between Liberals who favoured, and those who opposed, further government intervention in society. From the late 'eighties the Conservatives, with their Liberal Unionist allies, seemed a more united bloc, being anti-Irish Home Rule and pro-maintenance of the Empire.

3   In the early twentieth century this situation was transformed by the re-emergence as a major political issue of the question of tariff reform. Free trade, since the struggle over the Corn Laws, was almost an ingrained dogma of national belief; efforts to move away from free trade (the Fair Trade movement of the 1880s) made little impact. In 1903 Joseph Chamberlain, the main spokesman for the Liberal Unionists, came out openly in support of a policy of selective tariffs with these twin objectives: (1) the cementing of imperial bonds by strong economic links, and (2) helping British industries hard-hit by foreign competition in an unprotected home market. Chamberlain's tariff reform crusade broke the unity of the governing Party. Some Conservatives eagerly embraced this new doctrine, others refused to drop free trade, while Balfour and a middle group strove unconvincingly to paper over the cracks and preserve unity within the Party. On the other hand, almost all Liberals – however much they might differ on imperialism and on the role of government – happily came together in defence of free trade, to which Britain's increase in wealth and power was attributed. This issue was the biggest single factor in the Conservative catastrophe in 1906.

4   This was not, of course, the only factor tending towards a Conservative defeat; it was always easy for a government, especially in a very prolonged office, to accumulate enemies. The long-drawn-out last stages of the Boer War – the lack of spectacular victories impaired the patriotic fervour; the 'Chinese slavery' arising from the post-war settlement in South Africa – all this played a part in the 1906 electoral propaganda. Militant non-conformists expressed a strong dislike of the 1902 Education Act, with its increased aid to Anglican schools. The government could not point to a wide range of popular and exciting reforms for which it could claim credit.

5   Developments in the Labour area of politics also played a part, though it is important not to exaggerate the position of the infant Labour Party in these years. The creation of the Labour

Representation Committee in 1900 provided a new basis on which to build, while the Taff Vale judgement recruited important new trade union support. The electoral pact between the Liberal and Labour parties gave the new Party a chance to increase the number of MPs, while doing much to prevent competition for votes among opponents of Conservatives.

6 Although many subsidiary factors played a part in determining the result of the 1906 election, the principal cause of the Conservatives' disaster was that the free trade *v.* tariff reform debate arrived on the scene as the major political issue of the day, effectively smashing Conservative unity and credibility while allowing the Liberals to unite in defence of a popular traditional policy.

3 Why, and with what consequences, did the Conservatives oppose the People's Budget?

Understanding the question

Why did the Conservatives dislike the provisions of the budget? What did they think the Liberals were aiming to do in the budget, and why did they disapprove? What were the political repercussions of the Tories' stance?

Knowledge required

The provisions of the 1909 budget and narrative of the events following the Conservative rejection. The Parliament Act crisis (1909–11).

Suggested essay plan

1 The best way to explain this Conservative opposition is to consider Liberal motivation: perhaps Lloyd George was deliberately setting a trap for the Lords, hoping for their rejection of the budget as an excuse for him to attack them on a 'Peers-*v.*-People' basis and so, by emasculating the Lords, pave the way for Home Rule, which the upper House had blocked. If so, Conservative opposition is easily understandable.

2 Home Rule had created a sense of apprehension and bitterness in the Conservative *and Unionist* Party, and the Lords seemed to be the only weapon with which to preserve the union; there was a feeling that the whole Empire depended on 'confidence' in British strength and that, if Ireland could 'opt out' against imperial wishes, India and the rest would soon follow; Balfour's 1906 boast was: 'The Conservative Party still controls the destinies of the nation through the Lords.'

3 But there is little evidence that Lloyd George *was* setting a trap; Asquith, the previous Chancellor and the then current Prime Minister, also wanted a strong budget to bolster liberal morale as well as intimidate the Lords.

4 The long battle between the Commons and Lords for constitutional supremacy: Lloyd George exploited the Commons' rights to exclusive cognizance of finance bills by tacking on blatantly *social measures* (especially the taxation of land values) to an ostensibly financial bill. Seen in these terms the Lord's opposition can be explained as a defence of their own powers.

5 Lloyd George was anxious to show that it would be possible to pay for social reforms like OAPs (and dreadnoughts) out of *direct* taxation without recourse to tariffs – a vital tenet of the 'new liberalism'. If so, many Conservatives may have opposed the budget in order to demonstrate that tariff reform really was essential to national efficiency and social reconstruction.

6 Perhaps Lloyd George genuinely saw the budget as socially reformist, especially the principle of graduation and the taxation of land values. If so, Conservative opposition may have been equally genuine. However, the radical element in the budget was limited, since it was to be balanced by increased taxation of such working-class 'luxuries' as tobacco and alcohol.

7 Conservative opposition made no difference to the outcome of the budget, which passed in 1910 and set a trend towards increased direct taxation – and also (eventually) towards greater fiscal graduation.

8 The political and constitutional consequences of rejection: the Parliament Act, the elections of 1910, the action of George V, the threat to create peers, the 'last-ditchers' stand and the Conservative surrender. The Parliament Act reduced the power of the Lords to a suspensory veto; this was a triumph for the democratically elected Liberal government in the Commons. Asquith and Lloyd George had revitalized Liberal morale; the Party could now embark on further reforms – including Home Rule.

9 The Liberal victory was deceptive – the government's reforming energy was exhausted; the Lords could still delay legislation for two years. Defeat made the Unionist Party even more irreconcilable; Balfour was forced to resign, and Bonar Law initiated a new generation of professional businessmen-politicians, like Baldwin and Neville Chamberlain. The Party of law and order now also supported extra-parliamentary activities like gun-running in Ulster and encouraged plans by Ulster extremists for civil war in Ireland if Home Rule were allowed through the Commons.

10 The crisis still left unsolved the exact constitutional role of the reformed Lords. Reforms since 1911 have reduced the suspensory veto to one year and have widened the upper chamber's membership by the nomination of life peers. The Lords still await their true constitutional reformation.

4 Document question (The Powers of the House of Lords): Study the extract below and then answer questions (a) to (g) which follow:

'... What meaning does the supremacy of the House of Commons convey to the minds of the House of Lords? In the first place, it is a matter of common knowledge that its working varies according to circumstances. When their own Party are in power ... there is never a suggestion that the checks and balances of the Constitution are to be brought into play; there is never a hint that this House is anything but a clear and faithful mirror of the settled opinions and desires of the country, or that the arm of the executive falls short of being the instrument of the national will ....

'... Witness the transition that takes place the moment a Liberal House of Commons comes into being. A complete change comes over this constitutional doctrine of the supremacy of this Chamber .... Now they challenge it; and it becomes a deferred supremacy – a supremacy which is to arrive it may be, at the next election, or the election after that, or may be never at all .... I have never been able to discover by what process the House of Lords professes to ascertain whether or not our decisions correspond with the sentiments of the electors; but what I do know is that this House has to ... carry on its existence in a state of suspense, knowing that our measures are liable to be amended, altered, rejected and delayed. It is a singular thing, when you come to reflect upon it, that the respresentative system should only hold good when one Party is in office, and should break down to such an extent that the non-elective House must be called in to express the mind of the country whenever the country lapses into Liberalism ....

'... Let the country have the fullest use in all matters of the experience, wisdom, and patriotic industry of the House of Lords in revising and amending and securing full consideration

for legislative measures; but, and these words sum up our whole policy, the Commons shall prevail.'
(Extract from a speech by Sir Henry Campbell-Bannerman on 24 June, 1907)

Maximum marks

(a) Why did the House of Lords habitually behave in the manner referred to 'when their own Party' were 'in power'? (2)

(b) What is the meaning of 'the checks and balances of the Constitution'? (3)

(c) Explain the passage from 'a deferred supremacy' to 'or may be never at all'. (5)

(d) What evidence from the previous half-century could Campbell-Bannerman draw on to support his view of how the Lords behaved towards Liberal administrations? (5)

(e) Explain how conflict between the Lords and the Liberal government developed into a major political conflict during 1909–10. (6)

(f) What legislation did the Liberals pass to ensure that 'the Commons shall prevail'? (5)

(g) What is the connection between the passage of that legislation and the conduct of the Conservatives in the years 1912–14? (5)

25.6 READING LIST

Standard textbook reading

T. O. Lloyd, *Empire to Welfare State* (OUP, 1970), chapters 1, 2.

R. K. Webb. *Modern England* (Allen and Unwin, 1980 edn.), chapters 10, 11.

A. Wood, *Nineteenth-Century Britain* (Longmans, 1960), chapter 21.

Further suggested reading

P. Adelman, *The Rise of the Labour Party 1880–1945* (Longmans, 1972).

C. Cook, *A Short History of the Liberal Party 1900–1976* (Macmillan, 1976).

G. Dangerfield, *The Strange Death of Liberal England* (Paladin, 1936, 1970).

R. Jenkins, *Mr. Balfour's Poodle* (Heinemann, 1954).

K. O. Morgan, *Lloyd George* (Weidenfeld and Nicolson, 1974).

K. O. Morgan, *The Age of Lloyd George* (Allen and Unwin, 1971).

E. H. Phelps Brown, *The Growth of British Industrial Relations* (Macmillan, 1965).

D. Read, *Edwardian England* (Harrap, 1972).

C. Rover, *Women's Suffrage and Party Politics in Britain 1866–1914* (RKP, 1967).

P. Thompson, *The Edwardians* (Paladin, 1975).

26 Britain's interwar economy

26.1 INTRODUCTION

The First World War made obvious Britain's relative decline as an industrial nation. The war put an enormous strain on her re-

sources. She emerged with much of her export trade lost, her capital equipment depleted and an industrial structure clearly out-dated. The structural roots of the interwar crisis lay in the pre-war inheritance of a narrowly based and increasingly archaic export trade. This was exacerbated by the general decline in world markets between the wars.

Britain now paid the price for her businessmen's preference for under-developed and imperial markets since 1870 and her failure to encourage the more technologically advanced industries such as motor cars, machine tools, electrical goods and so on. Unemployment was the result – over 9% throughout the period and reaching 23% in August, 1932. The politicians had no effective remedies.

Economic orthodoxy triumphed over Keynesian and socialist alternative policies. Large-scale government expenditure and public works were rejected as possible solutions. Meanwhile politicians argued, irrelevantly, about the benefits to be gained from protection as opposed to those which might accrue from the continuation of the pre-war policy of free trade.

26.2 STUDY OBJECTIVES

1   The effects of the 1914–18 inflation and monetary depreciation on the economy and on the different social groups; the loss of markets and increased foreign competition; the reasons for the difficulties of the staple industries, especially cotton, coal, shipbuilding, and iron and steel; the wage cuts and their effect on industrial relations.

2   The 'new' industries of the 1930s – especially vehicles, chemicals, electrics, synthetic fibres, housebuilding and the service industries – were made possible by cheaper money after 1931, by greatly increased industrial concentration (mergers, trade associations, etc.) in the 'old' and 'new' industries alike, and by modest improvements in technology, productivity, and scientific research during the 1930s.

3   It is difficult to discern whether these 'new' industries amounted to an industrial 'recovery' or merely to a cyclical upswing; the problem of distinguishing short-term fluctuations (the 1918–20 boom and inflation, the early 1930s consumer boom, the late 1930s rearmaments-led boom) from underlying structural trends; the good industrial growth rate of 1924–39 and the improved per capita income; but reasons for seeing the recovery as a mainly defensive retreat from the nineteenth-century international economy to the protected home and commonwealth markets, which often fostered inefficient production.

4   Britain's position in the world economy: the depression was less sudden and severe than elsewhere, which perhaps inhibited a thorough revitalization in the 1930s; the dependence on America and the effects of the Wall Street Crash; the export trades were hit by the reduced purchasing power of the world's primary producers, in whose economic system the Victorian economy had become entwined.

5   The depressed areas and patterns of unemployment: the distribution of wealth and condition of the unemployed; the policies on unemployment insurance, pensions and other welfare provisions; one third of British families were near the official poverty line in 1939, while the remainder enjoyed an unprecedentedly high standard of living; the contrasts in the 1930s between the depressed areas (Scotland, Wales, Northern England) and the prosperous South and Midlands, which benefited from the domestic consumer boom.

6   The pressure for retrenchment and for pre-war normalcy or 'decontrol', especially from Liberals and Conservatives but also from Snowden, who repealed the McKenna Duties in 1924; this undermined 'War Socialism' or government intervention in industry; the establishment in 1926 of the BBC and the CEB merely highlighted a lack of centralized planning elsewhere; the Geddes' Axe undermined Lloyd George's and Addison's plans for social reconstruction (Homes Fit for Heroes, the Fisher Education Act); Party disputes over fiscal and financial policies, 1926–30.

7 Civil Service (Niemeyer) and banking (Norman) pressure for a return to gold at the old parity raised interest rates and aggravated industrial difficulties; the decision to abandon gold and reintroduce protection (especially the Ottawa agreements) contributed to the economic upturn in 1930s; invisible earnings remained high but were not now (unlike before the war) sufficient to compensate for visible trade deficit.

8 An analysis of the 'balance of power' within the economy – was, for example, the Macmillan Committee right in saying that industry was being sacrificed to the financial sector? The influence of 'The City' and also of the Chambers of Commerce in policy-making circles, compared with National Employers' Confeder-ation (but note the increased role of NEC and the TUC in the 1930s); the 1920s revival of agriculture was not maintained in the 1930s.

9 The 'Keynesian' policy alternative called for public works, deficit financing and counter-cyclical controls; the work of the 1930 Economic Advisory Council began the process of the grad-ual conversion of official thinking to such economic planning; there is reason for believing that senior Treasury and Bank of England officials (*e.g.* Phillips and Hopkins) were converted from classical economics to 'Keynesianism' before the publication of the *General Theory* in 1936 – but the government machinery and the political will to implement the new policies were both lacking before the 1939–45 war; anyway, the size of the budget was still too small in the 1930s for deficit financing to have had much im-mediate impact. In short, 'recovery' (if that is what it was) took place without a 'New Deal', but benefitted from cheap money, protection and increasingly favourable terms of trade.

26.3 COMMENTARY

The First World War stimulated war-time industries and agri-culture and ensured full employment. It also caused inflation with a consequent rise in prices and profits. The war cost the Exche-quer £9000 million, and dictated economy for the remainder of the interwar years. After the brief post-war boom, which lasted until 1920, came the inevitable slump, as wartime demands and contracts for industrial goods and agricultural produce fell.

Optimism, however, prevailed and most people still hoped for a return to pre-1914 'normalcy'. Paradoxically, the difficulties of the 1920s, including the General Strike, simply strengthened this feeling. Britain returned to the Gold Standard in 1925 at the pre-war parity of pound to dollar. Even in the early 'thirties, after the onset of the world economic depression occasioned by the Wall Street Crash in the US, most voices still maintained that Britain's difficulties were mainly temporary, caused by external events and forces outside her control.

Slow decline

The uncomfortable truth was that Britain had been in decline for some considerable time – the First World War had simply em-phasized this fact. This decline had begun in the 1870s when Britain was first challenged by Germany and the US in foreign markets. The latter had considerable advantages over Britain – bigger home markets, modern technology and up-to-date equip-ment. Their entrepreneurs could produce on a large scale and reap the benefits of economies of scale primarily through lower costs.

Britain's early start left her with outdated plant and technology, and dependent on industries (coal, iron and steel, textiles and shipbuilding) which reflected the past rather than the present. The future lay with the newer industries (chemicals, electrical, petro-technology and motor cars) which formed an insignificant part of our Gross National Product.

Effects of World War I

Britain's decline in the world economic league was seriously hastened by the Great War. During the period 1914–18 our over-seas customers had to buy from other nations (Japan and the USA), develop the commodity themselves, or learn to use a substitute. Wartime allies also borrowed heavily from Britain (£1825 million) and from the USA, and in order to repay these debts after the war they reduced their imports – imports which Britain relied on selling to them. They also protected their new industries with tariffs and subsidies, thereby further undermining Britain's vitally important exports – a trade on which rested the viability and prosperity of her basic industries.

The staple industries

Other factors accentuated these changes. The nature of demand changed. Britain had won her industrial supremacy in the nine-teenth century by selling, primarily, producers' goods. After the Great War, demand for these staple goods was not so great and consequently our export trades of coal, iron, shipping and textiles suffered. This decline could, of course, have been offset by the rise of new industries (and to a limited extent it was) but we were slow to develop these and had not made as much progress as had our nearest competitors. The result was a decline in our export trade accompanied by a fall in the National Income, a rise in unemployment and a growth of depressed areas where the old staple industries were concentrated.

The chief of these depressed areas (or, as they were euphem-istically called, 'special' areas) were Lancashire, the North-East, South Wales and South-West Scotland. They contained only one quarter of the population but nearly one half of the unemployed. The emergence of those depressed areas emphasized the con-tinued existence of the division of the people of the United Kingdom into 'two nations' – the rich and the poor – soon to be joined (in Ernest Bevin's words) by the 'third nation', the un-employed. One of the main early reasons for this, as Mowat explains, was that throughout the 'twenties, 'Britain's industrial machine was throttled down; the world's workshop worked only on short time'.

The return to gold

This situation was compounded by the effects of the revaluation of the dollar in 1925 at $4.86 to the pound. This was largely a political decision – people wanted to see the pound 'look the dollar in the face' but the effect was to *increase* the price of Britain's exports and further weaken her position in markets in which she was already struggling to retain a much diminished share. Equally, this decision *lowered* the price of one of our major competitor's imports into the United Kingdom with disastrous effects on certain home industries now unable to compete.

It is clear that circumstances were against Britain's export trade anyhow – the world needed less of what she had to sell – but over-valuation, by keeping export prices artificially high, only made matters worse. It called forth an enormous volume of criticism. In a pamphlet, *The Economic Consequences of Mr. Churchill*, Keynes had clearly warned what the result would be, but economic orthodoxy died hard. In 1930 Montagu Norman continued to insist that the subsequent depression in industry was simply due to bad luck! Coal miners in particular, the first to suffer the inevitable wage cuts, would have had little sympathy for such a convenient analysis.

The coal industry

The coal industry is a good example of an industry which was suffering acutely from the onset of depression. Because Britain had been the first nation to industrialize, her pits were older and deeper, and geological formations often meant that in many areas mechanization was impossible. British miners were also paid higher wages than their German and Polish counterparts and, consequently, our exports dropped.

Furthermore, in the home market substitutes for coal, such as electricity and oil, were being increasingly used. Improved industrial techniques also provided output from a smaller input of coal, as did newer and more economical grates in the homes of domestic consumers. Finally, the switch from coal to oil fuel in the Navy and merchant marine further exacerbated an already difficult situation. Consequently, after the war the British coal industry found itself with a large surplus of coal for which no market existed.

These circumstances, made even more intractable by poor management, to a great extent explain the problematical political and social nature of the industry in the 'twenties. Remedies were, of course, proposed – nationalization, the amalgamation of pits, the closing of uneconomic mines, reorganization and so on – but the owners usually resorted to the more traditional methods of wage cuts and longer hours, methods which defeated the miners in 1926.

Apart from the coal industry it is essential that you have a good understanding of the comparative strength and weaknesses of the staple industries between the wars.

Economic problems

The word *depression* needs defining. There are many criteria which economists may use to interpret it. Some will emphasize long-term figures, others concentrate on whether the depression is continuous or cyclical. For example, examination of growth rates during the period reveals a much healthier picture than concentrating on cyclical down-swings or on unemployment figures.

Some sectors of the economy were growing – chiefly the consumption goods industries such as motor cars, bicycles, aircraft, man-made fibres (rayon), plastics, electrical goods and so on. The tertiary (or service sector) was also expanding but, as yet, these newer industries only accounted for around 16% of Britain's exports in 1930, while the staple industries still accounted for over 60%.

This development was a welcome one but employing, as it did, only 8% of the workforce, it was hardly a substitute for the declining staple giants. (Note that there remains the occasional extra problem of separating old and new industries – for example, are chemicals and textiles old or new? Both had experienced important changes over a long period of time.)

Another difficulty for economic historians is to define precisely what is meant by the interwar years. Because of the postwar boom they rarely wish to begin with 1918 or end, because of rearmament, with 1939. The choice of the base year is vital. If, for example, we were to choose the period 1920 to 1937 one would get a different picture than by taking the years 1921 to 1937 because whereas 1920 was a boom year, 1921 was a year of deep slump. Growth statistics, in particular, would inevitably be affected by the choice of the base year.

There is also the problem of what statistical criteria to use to measure the depression – industrial production, per capita productivity, Gross Domestic Product, or unemployment? Finally, when we use the word *depression* are we simply referring to the interwar period in comparison to Britain's previous economic performance, or are we comparing Britain with other countries during the same or earlier periods?

Modern research shows that Britain's industrial growth and productivity performance (1924–39) *vis-a-vis* other countries was about average. Even old industries, such as shipbuilding and textiles, had a high output per capita in the 'thirties – but, of course, at the cost of shedding labour and thereby creating pools of unemployment. Ironically, many of the older industries performed well in productivity terms and could not be defined as depressed but, in terms of overall growth and employment, they were obviously in 'depression'.

Unemployment

It is not surprising that unemployment is regarded as the critical problem of the interwar period. The average unemployment rate over the whole period was above 14% of the insured population, varying from 395,000 in 1920 to 2.8 million in 1932. Note that these are figures for those insured – 12 million out of a working population of 19 million. These tend to exaggerate the percentage of those unemployed because there was a lower rate of unemployment amongst the uninsured. It has been calculated that the insured unemployment rate in 1931 was 21%, but only about 15% if one includes the uninsured. Whatever figures one uses, the two main types of unemployment between the wars were structural and cyclical in nature.

Structural unemployment occurred where entire industries contracted due to economic or market change. This was the case with the old staple industries which were export-orientated and labour-intensive. Technological and productivity improvements often merely aggravated these structural problems by causing more unemployment. Cyclical unemployment, on the other hand, was caused by major downturns (or slumps) in the economy, the two most notable being in 1921 and (much the more severe one) in 1932. The latter was largely the result of the Wall Street Crash of 1929 and led to unemployment figures of the order of 46% in coal, pig-iron and shipbuilding in 1932. Unemployment was not equally distributed throughout the country. The rate varies from 67% in an old staple town like Jarrow to a mere 3% in a new industry town such as High Wycombe. Those industries, not tied to export markets, therefore fared much better. In the consumption goods industries (irons, vacuum cleaners, refrigerators, wireless sets, washing machines etc.) and in the distributive trades, unemployment averaged 12–14%.

Suggested remedies

Certain historians have argued that the migration of labour, underemployment and an acceptance of lower wages might have reduced the numbers of unemployed still further. There was considerable voluntary internal migration from the depressed areas towards the South Midlands and outer London, but hardly on the scale required to solve such a massive problem. Furthermore, in the face of such profound structural and cyclical problems it is doubtful if this would have had much effect – especially during a period of enforced technological and industrial change, during which time the population between the ages of 15 and 65 increased by over 4 million.

To have solved the unemployment problem it would first have been necessary to generate sufficient industrial growth to absorb the increasing number of new people coming into the labour market. Interwar governments failed, however, to realize the need for a positive labour policy, embracing overall unemployment, labour migration and planned location of industry, and appeared content to extend unemployment benefits and accept a situation of maintaining in enforced idleness a large section of the country's most highly skilled workers.

Benefits dulled the pain of economic distress but did little to cure the widespread despair caused by long-term unemployment. Demonstrations by the unemployed, hunger marches, deputations to Downing Street, etc. all showed that what the unemployed wanted was not state maintenance but work.

Government response

As stated earlier, the governments' economic policy reflected Britain's position in the world economy and was biased, quite understandably, towards the international context. Despite the more direct involvement by these governments in the domestic economy during the Great War, they carried this outlook into the 'twenties and 'thirties, often with disastrous results – as in the case of the revaluation in 1925. In budgetary and fiscal policy their stance was basically orthodox and they refused to consider Keynesian (or 'Mosleyian') reflationary policies.

The Labour Government of 1929–31 accepted Snowden's view that public works were mere 'palliatives' and that deflation was the correct reponse to the depression. There was to be no 'New Deal' in Britain. However, even if they had implemented these reflationary policies, structural economic problems of the magnitude Britain was experiencing (compounded by cyclical downturns) would have been almost impossible to solve in such a relatively short period of time.

What did governments do? They imposed tariffs, made bilateral trade agreements (*e.g.* with Scandinavian countries), lowered the Bank rate, encouraged schemes for rationalization and made some attempt to create employment in depressed areas. At best these measures enjoyed only limited success. Find out what you can about the drawbacks which resulted from the implementation of each of these policies.

Recovery

Britain's recovery in the 1930s has been the subject of much research in recent years and certain points have been stressed. Some recovery was inevitable as the world moved out of the 1929–32 slump. (Nevertheless this recovery was by no means a complete one). The abandonment of the gold standard and the imposition of tariffs in 1932 by the National government certainly helped. The former encouraged a cheap-money policy while the latter aided some of the older industries such as iron and steel.

The main stimulus, however, was given not to the protected industries but to the building, distribution and new industries, such as motor manufacturing and electrical – all based on a home demand activated by the introduction of 'hire-purchase' buying. This increase in demand in turn stimulated the older, staple industries. Furthermore, the housing boom was critical as this was a labour-intensive industry (note in particular the extraordinary growth of the middle-class suburbs and their stimulus to the development of both public and private transport).

The governments' role in the upswing, therefore, was permissive rather than positive. The partial economic recovery that occurred may have occurred in any case, and owed little to the deliberate actions of the state. It would appear that the main political debate on economic policy between the wars was irrelevant, concentrating as it did on the issue of free trade versus protectionism.

The alternative of progressive capitalism or Socialism (Keynes and Mosley) were never clearly posed at the centre of the political debate. It took the Second World War to put these alternatives seriously on the political agenda.

In conclusion, it is important to note that the interwar years, for all their tragic unemployment and distress, were not years of unrelieved economic stagnation and decline. In spite of mass unemployment, hunger marches, etc., there was a substantial rise in the volume of industrial production, (both from old and new industries), a fall in the cost of living (cheaper imports due to more favourable terms of trade), a decline in the number of families below the poverty line and a welcome expansion of the social services by governments who were not allowed to discontinue the work begun by the Liberals from 1906–14.

26.4 CHRONOLOGY

1917	**July**	MINISTRY OF RECONSTRUCTION, headed by Dr C. Addison, set up to consider post-war schemes for demobilization, health, housing, education and employment insurance.
1918		Wartime price controls abandoned.
1919	**February**	TRIPLE ALLIANCE (miners, railwaymen, transport) resurrected. The miners demanded a 30% pay increase and a seven-hour day; the government offered a Royal Commission under Mr. Justice Sankey. National Industrial Conference to promote better relations between employers and employees.
	March	SANKEY COMMISSION recommended partial nationalization of the mines and a 20% increase in pay; this was rejected in September by Lloyd George. Government control continued.
	July	Army demobilization was completed; most of the 4 million men released were absorbed into industry. HOUSING ACT (Addison Act) gave powers and subsidies to local government for the building of council houses. 170 000 houses were built by 1923.
	September	RAILWAY STRIKE against a threatened wage reduction; settled by Lloyd George in favour of the railwaymen. Gold Standard was suspended. Ministry of Health created; included the Poor Law, Housing and Local Government Boards.
1920	**February**	Government decided to ignore the recommendations of the Sankey Commission; the coal mines were to be returned to the owners, who immediately reduced wages.
	October	Miners Strike. The 'Slump' began. UNEMPLOYMENT INSURANCE ACT: extended to cover all insured men.
	November	Agriculture Act, to deal with post-war problems in agriculture, included a guaranteed price for wheat and oats, but was later to be repealed, due to the slump, in 1921.

Chronology cont.

1921	**April**	MINERS STRIKE/lockout: the railway and transport unions refused to strike in sympathy (Black Friday, 15 April).
	July	Miners were defeated and returned to work.
	November	Unemployed Workers Dependents Act: gave relief for wives (5s per week) and children (1s per week).
		Safeguarding of Industries Act: selected import tax to prevent dumping.
1922	**February**	GEDDES AXE: aimed to save £75 million by economies. Cuts in the Army and Navy, education, housing, public health and teachers' salaries were recommended by the Geddes Committee on Government Expenditure.
	November	UNEMPLOYMENT INSURANCE ACT: this consolidated the Acts of 1920 and March 1921. It made the dole general for an unlimited period for all unemployed, and recognized the responsibility of central Government to help the unemployed.
1923	**July**	CHAMBERLAIN'S HOUSING ACT: £6 million subsidy, primarily for private housebuilding; produced 362 000 houses.
1924	**September**	WHEATLEY HOUSING ACT, designed to meet the needs of working-class tenants, increased the government subsidy to house builders (who built to rent only) to £9 million per annum; 504 000 houses were built by 1933.
		Dawes Plan, for the payment of German reparations, encouraged period of economic recovery.
1925	**April**	RETURN TO THE GOLD STANDARD by Churchill, ending the wartime and post-war prohibition on the export of gold; J. M. Keynes opposed the proposals in his pamphlet *The Economic Consequences of Mr Churchill.* He argued that the return would lead to further deflation and unemployment.
	May	Widows, Orphans and Old Age Pensions Act passed by Chamberlain; provided (contributory) widows' pensions of 10s per week.
	September	Samuel Commission set up to investigate coal industry problems; government subsidy of £23 million.
1926	**March**	SAMUEL COMMISSION reported: advocated the nationalization of royalties, the amalgamation of smaller pits, etc. Also called for wage reduction. The Commission's recommendations were rejected by the miners.
	April	MINERS LOCKED OUT (30 April). The government declared a general emergency.
	May	GENERAL STRIKE began (3 May) in support of the miners. The strike ended (12 May). The government brought back the eight-hour day for miners.
	November	MINERS RETURNED TO WORK on the owners' conditions, *i e.* longer hours and lower wages.
1927	**May**	TRADE DISPUTES AND TRADE UNION ACT: made *general* strikes illegal; forbade civil servants to form trade unions; made it necessary for a trade union member to 'contract-in' in order to contribute through his union to Labour Party funds.
1928	**January**	Mond Committee met to investigate ways of improving the efficiency of British industry.
1929	**April**	CHAMBERLAIN'S LOCAL GOVERNMENT ACT: its main measures were the abolition of the 1834 Poor Law Act; the transfer of poor relief powers to county and borough councils; and derating to help industry and agriculture.
	October	WALL STREET CRASH: markets collapsed, unemployment grew to almost 3 million by 1932.
1930	**August**	Coal Mines Act: cut half-hour off the miners' working day; also set up a scheme to close inefficient pits.
		GREENWOOD HOUSING ACT continued subsidized housebuilding.
1931	**July**	May Committee Report recommended public expenditure cuts, including unemployment relief.
	August	Cabinet failed to agree on the May Committee cuts. The Labour government resigned. MacDonald formed the National Government.
	September	BRITAIN ABANDONED THE GOLD STANDARD: the pound fell from $4.86 to $3.40.
		Means test introduced to determine eligibility for unemployment benefit.
1932	**February**	IMPORT DUTIES ACT: imposed a general tariff of 10% on all except imperial imports.
	June	Bank rate reduced to 2%. The recovery began as world trade improved, but its effects were uneven.
	August	Ottawa Agreements: established imperial trade preference for the Dominions.
1933	**June**	Children and Young persons Act passed: 'a landmark in the history of child protection'.
1934	**May**	UNEMPLOYMENT ACT: restored the 10% benefits cuts but retained the means test.
	December	UNEMPLOYMENT ASSISTANCE BOARD set up with a staff of 6000 to help relieve the pressure on Labour Exchanges.
		Special Areas Act: aimed to stimulate investment in depressed areas.

1935		HILTON YOUNG'S HOUSING ACT: obliged local authorities to clear slums and end overcrowding.
1936		Public Health Act: a major consolidating measure.
	March	Rearmament began, especially aircraft production.
1937	July	Factory Act: a major consolidating measure.
1938	July	COAL ACT: nationalized coalmine royalties, with compensation given to the pit owners. The Act brought full nationalization nearer.

26.5 QUESTION PRACTICE

1   To what extent did government policies help Britain's economic recovery in the 1930s.

Understanding the question

This question demands a knowledge of the various economic policies which the government used to influence the economy between 1929 and 1939. It requires an assessment of whether the policies were specifically aimed at recovery and also whether such policies were capable of success, given the structure of the British economy in the 1930s.

Knowledge required

An understanding of the National government's role in restoring confidence by abandoning gold (1931) and introducing tariff protection (1932). A brief analysis of the government's fiscal and monetary policies in the 1930s. Other factors influencing this limited economic recovery – often irrespective of government policy, e.g. Britain's improved terms of trade position.

Suggested essay plan

1   The problem of the 1929–32 world depression led Britain to abandon the gold standard and introduce tariff protection. These two factors may have restored confidence in the British economy, as it was now possible to let rates of interest fall from the higher levels of the 1920s; the pound fell to \$3.40, thus reducing the price of British goods and boosting our export trade. The introduction of the tariff helped some industries, *e.g.* steel, but some retaliatory action and diversion of trade occurred. Also, the economic recovery of the 1930s was not based solely on these newly-protected industries.

2   The introduction of 'cheap money' meant that investors could now borrow money at lower rates of interest (2%). The figures of bank advances in the 1930s, however, do not show a substantial increase as much business investment was still derived from reinvested profits. The policy of cheap money was, in general, a passive factor in the recovery of the 1930s. The government's fiscal policy was still orthodox in the sense that attempts at balancing the budget were paramount in the minds of successive Chancellors. They favoured public expenditure cuts and new taxes (*viz.* the May Committee). There was no real counter-cyclical government fiscal policy during the depression.

3   Britain's recovery in the 1930s was due to more specific circumstances. This country had not suffered such a severe economic downturn as had the US and Germany; confidence was not so low. Even at the trough of the cycle real incomes and consumption were relatively high; this gave a greater chance for a reasonably quick recovery. In 1932 there was a distinct shortage of housing: also scope for the growth of the 'new' industries. The basis of recovery was to be seen more in the growth of these two areas than in specifically-designed government policies. Nevertheless, cheap money and low interest rates (especially through building society loans) encouraged the private building boom which stimulated other sectors of the economy through the multiplier process. Tariff reform signalled a retreat from competitive international markets to protected imperial and home markets; it encouraged recovery but also inefficient production.

4   Some government assistance was given to depressed industries; attempts were also made to rationalize industry e.g. textiles, and alleviate unemployment e.g. the dole. Unlike the US, Britain did not attempt to reflate the economy by budget deficits and preferred deflation; therefore there was no increase in state interference in, or control of, the economic system. This was left to private enterprise, aided only by protection and cheap money. The government failed to take advantage of the cheap money to introduce a national public works programme of the kind that occurred in the US, or to raise the school-leaving age; what was attempted (roads, public buildings, etc.) was totally inadequate. The government was aided, however, by a temporary revival of world trade in the mid-thirties and by rearmament which proceeded rapidly after 1936.

5   To conclude: recovery from the world depression was expected as the economy naturally moved out of the recession phase. Government policy may have been a contributory factor but, in some cases, was not particularly beneficial (*e.g.* the fiscal stance). The recovery was also incomplete in the sense that the growth of the new industries could not compensate for the decline of the old industries. Structural problems reflected in unemployment levels of 13.8% in 1938, could not be alleviated by the government's policy. Most of the time the government reacted to international factors rather than to internal needs, to orthodoxy rather than to active economic management.

2   Account for the high rate of unemployment in Britain between the world wars.

Understanding the question

This is a reasonably straightforward question which requires an analysis of the various types of unemployment which Britain experienced between the wars and an assessment of the main reasons for this depressing phenomenon.

Knowledge required

The decline in Britain's relative share of world trade and production in the interwar period and the particular role of World War I in disguising pre-1914 trends. An analysis of structural unemployment and its causes. The emergence of regional, technological, cyclical unemployment and their effects. Government attitude to these varying levels of unemployment.

Suggested essay plan

1   Britain's share of world trade, particularly in manufactured goods, declined during the interwar period; this set an overall constraint on the growth of production and on the maintainance of employment. The experience of World War I at first hid the unemployment problem, but after 1918, to some extent, exaggerated its effects. Even in the best interwar years the volume of exports never again exceeded 80% of the 1913 level. The trend after 1918 should have been away from the international market, with more concentration on potential domestic demand, because the international market was in depression. The return to gold (1925) further undermined Britain's competitiveness in the staple industries.

2   The major reason for Britain's unemployment problem was the slow pace at which the structure of the economy adjusted to

meet the changing demand pattern of British and world economies. Most 'new' industries (using new fuels like oil and electricity) were demand-orientated (*i.e.* geared to consumer wants) and were less labour-intensive, but they could not compensate adequately for the decline of the old labour-intensive staples which produced capital goods for export. Even by 1937, the old industries still accounted for most of Britain's industrial capital. Industrial adaptation to new conditions remained incomplete and the result was severe structural unemployment on Tyneside, Merseyside and in South Wales, etc.

3   Many of the traditional staples – coal, shipbuilding, textiles – were concentrated in specific areas of Britain, basically near sources of raw materials and power. Hence, the effect of structural unemployment was exaggerated. The fact that unemployment in the coalfields was twice the average for other areas illustrated the problem of regional unemployment. The presence of regional unemployment meant that frictional unemployment frequently resulted. Often, short-term unemployment can be caused by workers moving from job to job. But the difficulties created by strong ties to a specific area greatly inhibited mobility and thereby increased longer-term frictional unemployment during the interwar period. Government measures sought to deal with symptoms rather than the cause of unemployment and although they adopted a cheap money policy the government refused to adopt Keynesian reflationary schemes, having an unrealistic desire to maintain pre-war policies.

4   The emergence of the 'new' industries, which made good use of labour-saving techniques, sometimes meant further redundancies as capital was increasingly substituted for labour. Ironically, old industries often increased their productivity and streamlined their production by adopting new techniques, but only at the cost of shedding labour. Thus, structural change during the inter-war years involved technological changes which often further increased unemployment.

5   In conclusion, unemployment arose because of decreased demand, accentuated by government unwillingness to adopt certain policies that might have alleviated the situation. Not all was depression, however. Recovery was uneven (until rearmament, post-1937) and technological advances were patchy but the overall growth record of the economy was better than before 1914; there was a marked increase in per capita industrial production. By the mid-thirties real income per head had increased by about 30%. The middle classes, and those in employment, made the greater gains but the trend towards smaller families and a shorter working week, plus the growth of social services, ultimately benefited all. It was only with the coming of the Second World War that the problem of high unemployment was solved.

3   **'A period of economic gloom and increasing prosperity?' Is this a good description of Britain in the 1930s?**

Understanding the question

This question requires an analysis of the paradoxical nature of the interwar economic experience characterized by depression and gloom but, after our abandonment of the gold standard, by symptoms of dynamism and increased prosperity.

Knowledge required

An understanding of why the period has been designated as one of gloom – *i.e.* of high unemployment, slow industrial transformation, inadequate change in the distribution of income, a high incidence of poverty, and problems of international trade. An analysis of the positive features of the period – faster growth in Britain than the average of the other major countries, rising

real incomes and consumer expenditure, an increased emphasis on housebuilding, the rapid introduction of electricity, an extension of statutory holidays, the gradual involvement of the state in providing help for the unemployed, increased expenditure on entertainment and holidays, and the introduction of the motor car.

Suggested essay plan

1   The 1930s are often described as 'gloomy' because of the waste of resources involved in unemployment – which ranged from a high of 22% in 1932 to just under 11% in 1937. This high unemployment rate was intimately bound up with the structural problems of the staple industries during inter-war period, which the upswing in the economy in the 'thirties could not totally alleviate. This created a mood of despair and disillusionment. Successive governments appeared unable to cure the unemployment problem.

2   The gloomy description of the period is also linked to the comparatively slow change in the distribution of wealth towards the lower-income groups and to the persistence of poverty in the depressed areas. The overall nature of the depression was exaggerated by Britain's relatively poor international trade performance and her inability either to reduce imports or to increase substantially her invisible earnings.

3   However, it is important to note that the symptoms listed above underestimate Britain's real progress during the 1930s. Britain's overall growth rate was higher than the average of other industrialized countries. There was an increase in GDP; also, rising real incomes and consumer expenditure created a demand for the products of the new consumer-orientated industries. The increase in housebuilding and the spread of electricity contributed to improving the general standard and quality of life. Unemployment insurance was widened and the government became more generous in the provision of welfare facilities. There was also a gradual increase in the length of holidays, and greater expenditure on leisure (*e.g.* Butlin's Skegness holiday camp opened in 1937). At the outbreak of World War I the average working-class family spent 60% of its income on food and 16% on rent and rates. By 1935 only 36% was spent on food and 9% on rent and rates – leaving more to spend on other goods. This resulted in the growth of consumer stores such as Woolworths, Marks and Spencers, Sainsbury's, etc. Furthermore, new technologies, plus economies of scale, meant cheaper prices for consumer goods in the shops.

4   Finally, both assessments of the 'thirties are valid but the experience of that part of the population living on unemployment benefit/dole was not similar to that part of the workforce in constant employment – *e.g.* in the 'new' industries, the rapidly growing consumer-goods industries and in the distributive trades. It has been estimated that three-quarters of Britain's population benefited significantly from a rising standard of living in the 1930s, but a quarter still lived on the bread line – some experiencing abject poverty.

4   **Document question (The General Strike): study the extract below and then answer questions** (a) to (h) **which follow:**

'The miners were greatly heartened by the support of the TUC. On 4 May jubilation in the mining village was at its height, though all knew the going would be tough. The Government had already indicated that it had plans in preparation to oppose the TUC's efforts in their support of the miners. Glowing speeches of support for the miners from TUC leaders, from Bevin, Ramsay MacDonald (the traitor), Thomas (another one) were good for sustaining morale... since the Government had now declared a State of Emergency. My family deduced correctly that troops would be used to break the strike.

'Councils of Action were set up in our area . . . to set up at strategic points barriers to be manned by pickets to control the movement of any vehicle in and out of the village.

'The news media were already blazing out unprincipled lies about breakaways from the strike and denunciations against the strikers. . . . There was total paralysis of the mines, railways and factories. But at this vital stage, grave doubts given great publicity by the TUC leaders, began to appear. Meetings behind the scenes of the union leaders supporting the miners became common. It was no secret that the way was being prepared for retreat. Warrants issued for the arrest of several trade union leaders in the North aroused fierce opposition. The Council of Action in Newcastle were dealing quite effectively with the Government's Commissioner . . . It was the same elsewhere. Boards of Guardians, long hated for their means-test manner of doling out food to the sick, injured and miners' families, were the target of many demonstrations. . . . We had our share too of the Specials and Volunteers . . . the press played its usual role. Through it I learned first about . . . 'Moscow reds' and 'alien workshy agitators' who were supposed to have infiltrated our British way of life . . . .

'The collapse of the General Strike, while disheartening to the miners did not lead to any weakening of resolve . . . As strike funds dried up, a growing number of men did return to work . . . . By November . . . the Government offered what they thought were reasonable terms as conquerors: (*a*) immediate resumption of work (*b*) longer hours to be discussed at district level, (*c*) rates of pay temporarily at pre-strike level (but conditional on acceptance of longer hours), (*d*) No guarantee against victimization . . . On November 29 it was all over.'

(From an account of the General Strike of 1926 by Bill Carr in 'The General Strike' ed. J. Skelly, 1976)

Maximum marks

(a) Which Prime Minister of which party was in power this time? (2)

(b) What light does line 1 throw upon the origins of the General Strike? (2)

(c) In what ways does this extract illustrate differences in attitude between the ordinary mineworkers and the official leaderships of the TUC and of the Labour Party? (5)

(d) What distinctive view of the General Strike does the extract offer? (3)

(e) Who were 'Boards of Guardians', and what was meant by 'their means-test manner'? (4)

(f) What light does the extract throw on the reasons for the failure of the General Strike and for the prolongation of the coal strike? (4)

(g) Comment on the terms which were accepted by the miners in November, 1926. (5)

(h) What steps were taken by the Government in 1927 to prevent a repetition of the General Strike? Comment on the political significance of the fact that such steps were taken. (6)

26.6 READING LIST

Standard textbook reading

T. O. Lloyd, *Empire to Welfare State* (OUP, 1970), chapters 4, 5, 6, 7.

L. C. B. Seaman, *Post-Victorian Britain* (Methuen, 1966), chapter V.

R. K. Webb, *Modern England* (Allen and Unwin, 1980 edn.), chapter 12.

Further suggested reading

D. H. Aldcroft, *The Interwar Economy: Britain 1919–39* (Batsford, 1973).

S. Constantine, *Unemployment in Britain Between the Wars* (Longmans, 1980).

S. Glyn and J. Oxborrow, *Interwar Britain: A Social and Economic History* (Allen and Unwin, 1976).

J. Lovell, *British Trade Unions 1875–1933* (Macmillan, 1977).

C. L. Mowat, *Britain Between the Wars* (Methuen, 1955).

S. Pollard, *The Development of the British Economy 1914–67* (Arnold, 1962, 1968).

R. Skidelsky, *Politicians and the Slump* (Macmillan, 1967).

J. Stevenson and C. Cook, *The Slump* (Cape, 1977).

L. J. Williams, *Britain and the World Economy 1919–1970* (Fontana-Collins, 1971).

27 British politics in the 1920s

27.1 INTRODUCTION

The first World War had a profound effect on British politics and society. It was a severe test of Britain's social stability and effected major political changes – the most important being the eclipse of the Liberals as a major party of government and its replacement by a reunited and reconstructed Labour Party committed, at least in theory, to implementing a comprehensive programme of Socialism.

During this period the three-cornered nature of the political struggle (plus the number of elections) kept party politics alive, exciting and unpredictable. For some time it appeared that Lloyd George would resist the challenge of Labour and establish a centre party with progressive goals. His downfall in 1922, however, encouraged Baldwin, in the electoral stalemate of 1923, to give Labour its chance.

The failure of direct action appeared to suggest that Labour could be trusted; MacDonald could be relied on to control the militant class-conscious forces of labour and, in the process, perhaps destroy Liberalism. While the collapse of the general strike in 1926 confirmed MacDonald's position on the former, the rout of the Liberals in 1924 seemed to indicate the success of the latter.

Though beset by challenges from Right and Left, Baldwin and MacDonald restored the two-party pattern of pre-war government and, in doing so, contained the working-class challenge to the established order.

27.2 STUDY OBJECTIVES

1 The consequences of the first World War; the intrigues leading to the formation of the coalition and to Lloyd George's replacement of Asquith; Labour Party divisions (over pacifism) healed in time for 1918 election, but divisions within the Liberal Party were emphasized by the use of the coupon; the Conservatives, in a demoralized state before the war, were revitalized by participation in government and by the patriotic mood in the country.

2   The structure of politics: the 1918 Reform Act nearly tripled the electorate; the likely social composition of new voters; the effects of the female franchise; the frequency of elections (1918, 1922, 1923, 1924, 1929) heightened tension: in voting terms England now had a genuine three-party system which, without proportional representation, led to the gradual elimination of the weakest party, the Liberals.

3   The heightened tension and the polarization of politics: the Spen Valley by-election (1919) and post-war industrial unrest (the Triple Alliance, the 1921–3 strikes, the general strike) made 'the impact of Labour' or 'Labour versus Capital' or 'the Bolshevist menace' the major factors in politics and contributed to the elimination of the middle party, the Liberals.

4   Weak political leadership, composed mainly of pre-war survivals: the 'sale of honours' by Lloyd George typified the smell of corruption in post-war politics.

5   The Conservative Party: back-bench opposition to and overthrow of Lloyd George in 1922; the struggles for leadership, 1922–3, and the return of the Coalition Unionists, 1924; the accession of Churchill; disputes on fiscal and imperial (especially Indian) policies; the 'Empire Free Trade' group attacks Baldwin; the nature of the latter's leadership in the twenties – especially his 'astute' or 'feeble' role in the 1931 crisis.

6   The eclipse of the Liberal Party: the 'coupon election' confirmed the split; MacDonald saw the Liberals as a more immediate threat than the Tories; the 1923 reunification and Lloyd George's accession to the leadership in 1926 both came too late; the Liberals' poor organization; the Summer School initiatives and 'We Can Conquer Unemployment' create a modest but temporary Liberal revival in the 1929 election; reasons for thinking that Liberal economic policy would have been inadequate to deal with unemployment.

7   The Labour Party became the alternative 'party of government', which had been MacDonald's main objective; the major electoral advances in 1922; the 1918 Constitution confirmed trade union dominance (the block vote, etc.) in return for a vague ideological commitment to Socialism; the establishment of a General Council and of relations between PLP and TUC (weakened by the general strike); the declining position in the Labour movement of the ILP and the socialist societies; the influence of the *Daily Herald* and improved organization; the Party was outflanked on the Left by the Communist Party (1921); MacDonald succeeded Clynes, his qualities of leadership and his socialist philosophy – which was of the organic evolutionary, 'one-step-enough-for-me' type of gradualism; the attitudes of Snowden, Thomas, the Webbs, Henderson, etc.; the desire for social spending, to be paid for by a capital levy on rentiers.

8   Party policies and attitudes in the 1923 election; no party majority: Asquith decided to back Labour because of the danger of Liberal absorption by the Conservatives; Labour established its competence to govern – the Party's achievements were modest at home but more impressive in foreign affairs. A *Workers Weekly* article brings the government down, impact of Zinoviev letter.

9   Baldwin returned to office; the eclipse of the Liberals; the return to the gold standard exacerbated the depression; the Liberal public works proposals and 'proto-Keynesianism'; the effect of revaluation on coal-mining. The causes and course of the general strike of 1926; the political consequences for the trade union and Labour movements. Chamberlain extended Welfare provision; the Conservatives' public enterprise measures – establishment of the BBC and the CEB; parliamentary duels between Churchill and Snowden.

10   The 1929 election: the growing problem of unemployment; the 'safety first' policy of Labour and the Conservatives; the election result was inconclusive but the Liberals again backed MacDonald; Labour was unable to deal with unemployment, to balance the budget or to maintain the external value of the pound. The economic crisis, the May Committee recommendations, the divisions in party and in the cabinet; and did MacDonald's behaviour 'betray' the Labour Party?

11   Foreign policy and politics: the Anglo-Irish war and Lloyd George's 'settlement' of the Irish problem by partition; his pacific attitude towards Germany alienated Tory back-benchers, and was anyway undermined by Chanak, which made him appear a warmonger. MacDonald's successful and pacific foreign policy – his handling of Franco-German relations and the reparations problem; the Geneva Protocol and Locarno; the Zinoviev letter and its part in the Labour government's downfall; the Kellogg Pact and the Hague Conference marked an easier foreign situation; imperial politics were more important than foreign in the later 1920s.

27.3 COMMENTARY

A new political stage

British politics in the immediate aftermath of the war were characterized by a heady optimism about the future. As early as 1919, however, the Lloyd George coalition's taste for social reform was disappearing, and working-class unrest only convinced the Conservatives that firmer government was needed. By 1921, furthermore, the post-war boom had ended, and politicians, though coming to terms with a three party system, were baffled by the onset of an economic depression whose effects were to dominate interwar politics.

The traditional parties had also to recognize the threat posed by the rapidly growing Labour Party, a threat given particular edge by a new political style which drew heavily on class-based rhetoric to foment party unity. In the event, the main consequence of the rise of Labour was political not economic: it killed off Liberalism as a political force, without finding any way out of the economic tangle in which the country found itself, with the problems of ageing industries, mass unemployment and international indebtedness throughout Europe.

The reaction to these events was overwhelmingly political. In five general elections the major Parties disputed who should have power, but had nothing to offer on the economic questions which faced the country except caution and financial orthodoxy. Only the Liberals, by then a fading force, came up with a comprehensive alternative economic policy (inspired by Keynes) in 1929, and they were brushed off by the voters.

Following Lloyd George's departure in 1922, politics were dominated by two men who, in practical terms, were indistinguishable: Ramsay MacDonald and Stanley Baldwin. Both wanted England to recover, and neither had the slightest idea how she might. Baldwin offered political stability and, although MacDonald stood against him in parliament, what he sought for Labour was a share in maintaining that stability, not a mandate for revolution or class war. The limitations of Baldwin and MacDonald was the price England had to pay for maintaining its political freedom and stability at a time when, everywhere else in Europe, these things were fast disappearing.

Reconstruction, 1918–22

The general election of December, 1918 returned to power the Lloyd George coalition, which had made extensive promises of social improvement. These were never fulfilled for both economic and political reasons. The cost of the war had been enormous, far beyond what anyone had imagined and the House of Commons was obsessed with the need to reduce expenditure.

Lloyd George had only a small political following in parliament. He was dependent for survival on his coalition allies, the Tories, who increasingly demanded the decontrol of industries taken over by the government during the war and resisted proposed expenditure on social reform. The Labour Party and the trade unions, on the other hand, wanted to extend the more egalitarian aspects of war collectivism. Lloyd George himself was by now prepared to trade principles for power and he went along with the Tories' arguments. As a consequence, retrenchment, not reconstruction, dictated government thinking, and inevitably it was the newest areas of state activity which suffered most.

The cause of social improvement was not helped by Lloyd George's failure to make it a cabinet priority. The commitment to social improvement through vigorous government action was eventually abandoned in 1921, when the Geddes Committee was appointed to recommend cuts in expenditure. The Tories finally ditched Lloyd George in October, 1922 but this was largely a consequence of his personal unpopularity and of resentment of the Irish treaty. He had cast aside social reform long before then.

After the election that followed (1923), the Conservatives took a conscious decision to destroy the Liberals by putting a minority Labour government in office. Baldwin knew that the Labour Party could be trusted. Better a short period of 'responsible' Labour rule than to drive Labour back to supporting 'direct action' by the trade unions. Labour was, indeed, contained and its challenge to the established order muted. Between them Baldwin and MacDonald engineered a return to the two-party system, offering the electorate a choice between 'decent' Conservatism and 'respectable' Labour.

The first Labour government (1924)

In March, 1923 a Conservative peer told a friend: 'I think that people are rather sick about the King and Queen going to dine ... to meet a lot of Labour bounders'. Within a year, and to the dismay of many, the 'bounders' took office in the first Labour government under Ramsay MacDonald. He chose an ineffective political team; his cabinet was almost totally inexperienced.

Dependent on the Liberals, 'Labour were in office but not in power'; this goes some way to explain their modest performance (and also their growing resentment of the Liberals). MacDonald was anxious, above all, to rebut Churchill's sneer that Labour was not fit to govern. He succeeded and, as Prime Minister and Foreign Secretary, he made a considerable personal impression both in Britain and abroad. But, like Baldwin's Conservative administration, his government was devoid of ideas on how to remedy Britain's deep-rooted economic problems.

In the circumstances of the time, with many Tories still paranoiac about the dangers of Bolshevism, MacDonald's caution was understandable. It did, however, cause some resentment among the TUC and Left-wing Labour MPs. Apart from his own success in foreign affairs, and some improvements in social policy, his government's record was uninspiring; it was notable chiefly for the fact of its existence.

Labour had made remarkable progress in a very short time, from being largely the spokesmen for the trade unions (who favoured direct action) to being a major parliamentary and national force now campaigning to gain political power so that they might change (constitutionally) the political system. MacDonald's major intention was to consolidate these gains by making Labour an acceptable party of government. According to one historian, 'he had tried to prove that Labour was fit to govern; all he did in the end was to prove that MacDonald was fit to be Prime Minister'.

Consider now some fundamental problems of interpretation: did Socialism replace Liberalism, or was it merely that Labour replaced the Liberals and continued much of the old Liberal policy (free trade, little Englandism, etc.)? Was the triumph of Labour the inevitable consequence of long-term trends – the growth of class politics, the rise of organized labour, the decay of old Liberal *laissez-faire* values – or was it the fortuitous and chance result of war and of personality differences among the Liberals?

Baldwin

Baldwin's air of good-humoured laziness was misleading. He appreciated the fragile nature of post-war politics where dynamic forces like Lloyd George and Churchill could prove liabilities. Mediocrity also had its function; hence Baldwin stood for conciliation all round. He had a clear sense of how Britain should develop, and made political stability a key objective at a time when much of Europe was experiencing political and economic upheavals. He appreciated that Labour had become part of the political system and determined to resist their effort to make class the main factor in politics.

He strove to make himself the acceptable leader of all non-Labour opinion. His success is reflected in the fact that his severest critics were in his own Party. On occasion he could act with speed and decisiveness – see for example how effectively he led the revolt against the Lloyd George coalition. As Prime Minister in 1923 and from 1924 to 1929, he aimed to preserve a balance in the country's affairs. Despite economic difficulties and the general strike, in political terms he succeeded. His style of leadership, however, did have repercussions on the Conservative Party – note in particular the disputes on fiscal and imperial policies.

Where the economy was concerned, however, his government had nothing to offer except traditional remedies. Unemployment and depression were regarded as unalterable facts of life; Baldwin lacked political imagination and took the line that the country could only wait until things got better. His government fell, not because it did not come up with the right answers – but because it failed to come up with new ideas to tackle the problems. Its only positive economic action was the return to the gold standard in 1925, a decision which made things even worse for Britain's export industries. Apart from Neville Chamberlain at the Ministry of Health, his cabinet had few achievements to record in their five years in office. But politics remained orderly and broadly consensual, which was Baldwin's particular aim.

The general strike of 1926

The general strike was not the climax of post-war political, ideological or even industrial confrontation and tension. It was something which few people wanted to happen. In the end it was forced on the TUC by Baldwin's decision to suspend negotiations. Basically, he was not prepared to back down from a confrontation with the labour movement either by enforcing a fair settlement on the mine owners or by continuing to subsidize wages while negotiations went on.

The government had learnt the lesson of 'Red Friday', when it had been forced to give way in the face of the threat of widespread industrial action. By May, 1926 adequate preparations to combat a general strike had been made. The trade unions by contrast, had done little to prepare for action. In the event, once the labour movement blundered into open conflict with the government both sides had reason to be pleased: the government's plans to maintain essential services worked almost without a hitch, while nearly all trade unionists unhesitatingly obeyed the strike call.

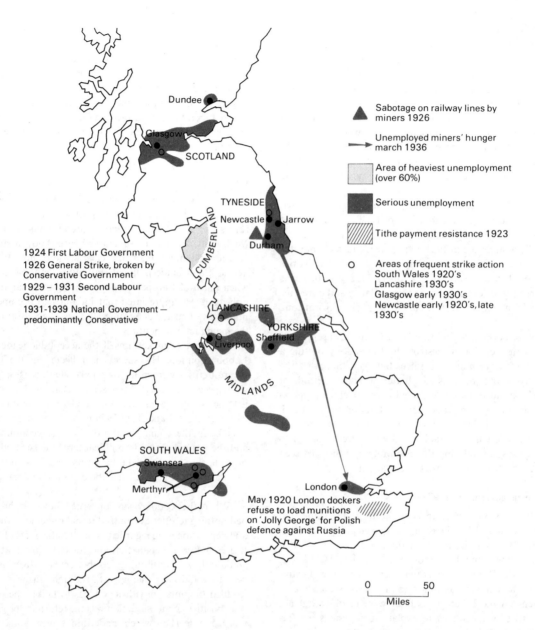

1924 First Labour Government
1926 General Strike, broken by Conservative Government
1929 – 1931 Second Labour Government
1931-1939 National Government – predominantly Conservative

Sabotage on railway lines by miners 1926

Unemployed miners' hunger march 1936

Area of heaviest unemployment (over 60%)

Serious unemployment

Tithe payment resistance 1923

Areas of frequent strike action
South Wales 1920's
Lancashire 1930's
Glasgow early 1930's
Newcastle early 1920's, late 1930's

May 1920 London dockers refuse to load munitions on 'Jolly George' for Polish defence against Russia

0   50
Miles

Fig. 19 Unemployment and unrest during the interwar years

The strike was generally an orderly affair: there was little serious violence and no bloodshed. Nevertheless, for the TUC it was a traumatic episode. They insisted that the reasons for the strike were economic; the object was to get the government back to the negotiating table. Their problem, however, was that in order to do this they had to defeat the government in the strike. The government regarded it as a political action designed to usurp the authority of parliament (the constitutional issue was always uppermost in Baldwin's mind), and took the line that it could not possibly negotiate at the point of a gun.

Neither the TUC nor the vast majority of the strikers had any wish to bring the government down, and the unions had no clear idea of what to do if the dispute dragged on for any length of time. In the end, faced by the government's intransigence, the weakness of their own policy, the fear of permanent Communist domination of local trade councils (which had run the strike) and by the miners' obstinacy, the TUC called off their action in response to a vague promise of intercession from the Liberal MP, Sir Herbert Samuel, who had led an earlier enquiry into the coal industry. The miners stayed out, but eventually were forced to go back to work for more hours and less pay.

The outcome of the strike has often been portrayed as a massive weakening of the trade union movement, which persisted throughout the interwar years. As A. J. P. Taylor points out, however, it was 'a warning to other employers not to push their workers too far'; in fact, wages in Britain held up well in the difficult years that followed. Although the cost for the unions in terms of money, membership and employment was great, it was at least a positive response to the country's economic problems. During the protest they also demonstrated how responsible and law abiding they were – no longer could they be accused of fomenting revolution every time they took industrial action. According to Lloyd 'a whole cycle of working class militancy...had come to an end...For the next five years immobility reigned triumphant and almost unchallenged'.

The second Labour government (1929)

MacDonald's second government brought few changes for Britain. Although alternative economic policies were advocated by the Liberals prior to the election, Labour in office would have nothing to do with them and regarded the depression with the same fatalism which had characterized Baldwin's government. Skidelsky has described Labour's failure to adopt these

Keynesian policies as 'the major missed opportunity of the interwar period'. But, would they have worked?

The Labour government was more innovative in foreign policy, with Henderson, the Foreign Secretary, pushing for disarmament and winning French confidence. The question of European security, however, was tied up with the complex problem of reparations. MacDonald and the Treasury believed that only when the tangle of international indebtedness was finally cleared away should a settlement of the political tension between France and Germany be attempted. Henderson took the contrary line that Germany's economic problems represented an opportunity to force her into political concessions and thereby ease French fears.

The 1931 crisis

This row was complicated in August, 1931 by the dramatic collapse of the pound; Henderson and his followers refused to accept the proposed drastic economies. These included cuts in unemployment relief, and were held to be necessary to balance the budget – thus restoring the confidence of holders of sterling and preventing a financial collapse. As the cost of living had fallen substantially, some argued that the cuts were not entirely unjustified. MacDonald, who undoubtedly liked office and who was tired of his pereptual battle with Henderson, split decisively with his Party by staying on as Prime Minister of a 'national' government formed to see out the immediate crisis. The shocked surprise of his colleagues indicated his isolation in the Party and demonstrated in their eyes his love of power. Equally, however, it could be argued that their actions during the crisis demonstrated their fear of exercising it. In considering this emotive episode, two questions should be borne in mind. First, why did the government of 1929–31 not take up interventionist economic policies such as those put forward by the Liberals? Second, given the conventional wisdom of the day, what other course could the government have taken in August, 1931 if a financial collapse was to be avoided?

27.4 CHRONOLOGY

1918	**August**	FISHER EDUCATION ACT raised the school leaving age to 14.
	November	Armistice ended the fighting in the Great War.
		REPRESENTATION OF THE PEOPLE ACT introduced full manhood suffrage and gave the vote to women aged 30 and over.
	December	Lloyd George Coalition won the 'Coupon Election'. First vote for more than 6 million women. Sinn Fein won 73 seats in Ireland and decided to boycott the Westminster parliament.
1919	**January**	Sinn Fein set up their own assembly, the Dail, in Dublin. A guerrilla war between the IRA and the Royal Irish Constabulary (the British armed police) followed.
	June	Allies concluded the world war with a peace treaty at Versailles. The huge reparations demanded of Germany created immense problems in international finance during the following years.
1920		The breakdown of law and order in Ireland: Lloyd George sent supplementary British forces (the Auxiliaries and the 'Black and Tans') to Ireland to enforce a policy of coercion.
	August	War between Poland and Russia: the British government was deterred from intervention by the threat (at home) of a general strike.
	December	GOVERNMENT OF IRELAND ACT: provided for Ireland to remain in the UK but for Southern Ireland and the Six Counties each to have their own elected assemblies. This was ignored by Sinn Fein.
1921		Post-war boom ended; the onset of economic depression.
	March	UNEMPLOYMENT INSURANCE ACT: introduced the 'dole'.
	April	Black Friday; the railway and transport unions refused to support the miners by coming out on strike.
	July	Truce declared in Ireland.
	December	IRISH TREATY signed between the government and Sinn Fein representatives; gave substantial independence to Southern Ireland, now to be called the Irish Free State, but not the republican status desired by De Valera. The Six Counties were allowed to contract out. De Valera denounced the Treaty and a civil war followed between the pro-treaty (Free Staters) and the anti-treaty (IRA) parties.
1922	**February**	GEDDES COMMITTEE called for major cuts in public expenditure.
	October	Conservative backbenchers rebelled against the coalition at a Party meeting, rejecting the advice of most of their leaders.
		Lloyd George resigned; he was replaced by his former colleague, Andrew Bonar Law.
	November	Conservatives won the general election.
1923	**January**	The war debt with the US settled by the Chancellor, Baldwin, on what were regarded as unfavourable terms.
	May	Baldwin became Prime Minister, succeeding Bonar Law.
	October	TARIFF REFORM: Baldwin announced his conversion to protectionism.
	December	Conservatives lost their overall majority at the general election. Baldwin declined to continue in office.
1924	**January**	FIRST LABOUR GOVERNMENT: a minority administration formed under James Ramsay MacDonald.

Chronology cont.

	June	Dawes Plan: new proposals for the payment of German reparations.
	August	WORKERS' WEEKLY: offices raided. Editor charged with incitement to mutiny.
	October	ZINOVIEV LETTER, apparently from the Comintern and signed by Zinoviev, called on English Communists to overthrow the Labour government; published by the Foreign Office. Conservatives returned to power under Baldwin.
1925	**April**	BRITAIN RETURNED TO THE GOLD STANDARD at pre-war parity of $4.86, to the pound.
	July	RED FRIDAY: the government backed down in the face of the threat of a general strike over problems in the mining industry.
	September	SAMUEL COMMISSION set up to investigate the problems of the coal industry. Baldwin bought time by granting a subsidy of £23 million.
	October	Treaty of Locarno: began an era of tranquility in international affairs.
1926	**March**	Government failed to persuade the miners and the mine owners to accept a compromise on pay and reorganization of the coal industry.
	1 May	Miners locked out.
	3 May	GENERAL STRIKE began.
	12 May	General strike ended; miners decided to stay out – and did so until December.
1927	**May**	TRADE DISPUTES and TRADE UNION ACT: reduced the income of the Labour Party by forcing trade unionists to 'contract in' to the political levy raised by the unions. This was resented by the Labour movement as an attack upon fundamental union rights and as a breach of faith by Baldwin and the Conservatives.
1928	**April**	REPRESENTATION OF THE PEOPLE (EQUAL FRANCHISE) ACT: gave the vote (dubbed the 'flapper vote') to women on the same terms as men.
	August	Kellog-Briand Pact: Britain, as one of the signatories, renounced war as an instrument of national policy.
1929	**April**	LOCAL GOVERNMENT ACT, introduced by Chamberlain, abolished the 1834 Poor Law Act.
	May	Labour won most seats in a three-party general election; Liberals seats were reduced to 59.
	June	SECOND LABOUR GOVERNMENT formed.
	October	WALL STREET CRASH: securities at market value crashed by almost $16,000 million; this immediately affected world trade and investment. British exports began to fall from a current £839 million to a 1931 low of £461 million; growing unemployment was the result.
1930	**May**	Cabinet rejected the Mosley reflation scheme.
	August	COAL-MINES ACT: reduced miners' working day by a half-hour.
1931	**July**	MAY COMMITTEE recommended public expenditure cuts. Foreign investors withdrew funds. Bank of England lost £2½ millions per day.
	August	Cabinet failed to agree on the cuts. Labour government resigned. NATIONAL GOVERNMENT FORMED by MacDonald.
	September	Naval mutiny at Invergordon; caused a run on the pound. BRITAIN ABANDONED THE GOLD STANDARD; the pound fell from $4.86 to $3.80.

27.5 QUESTION PRACTICE

1   Why did the second Labour government fail to solve the problems posed by the depression?

Understanding the question

You must review the ways in which the problems might best have been solved, and also the restraints imposed on the government both from within the Labour movement and from without.

Knowledge required

The causes and course of the economic depression of the 1920s and 1930s. The response of the Labour Government. Alternative policies proposed for dealing with it.

Suggested essay plan

1   Causes of the depression: the loss of traditional markets in wartime; the difficulties of staple industries in the cost price squeeze of the 1920s; dependence on world trade meant Britain

could only recover in the wake of a general world recovery; this restricted the freedom of manoeuvre in dealing with the domestic economy, since Britain dared not take measures (*e.g.* competitive devaluation) which might impede world recovery; the 1925 return to gold exacerbated the depression; the course of the depression, with unemployment figures, etc.

2   The long-term constraints on economic policy: the unrealistic desire to return to the favourable conditions of Edwardian England; hence the gold standard and the 'politics of nostalgia'; free trade as an article of faith with the electorate (cf. the 1923 election), which hampered all governments' attempts to deal with the economic problem.

3   External constraints on the Labour government: pressure from bankers like Norman ('Bankers' ramp') and civil servants like Maybury, who were both in collusion with Tories like Chamberlain to undermine ministers; the government's minority position and dependence on the Liberals.

4   The internal constraints on the government: the indecisive-

ness of MacDonald; Snowden was rigidly locked into the orthodox Treasury view (balanced budgets, free trade, sound money);
Thomas was irresponsible and lazy (and responsible for severe
cabinet leaks); the PLP was indisciplined; the ILP was alienated
from the government; the TUC was suspicious since the general
strike; Henderson was hostile to MacDonald; ultimately the
cabinet divided on the question of unemployment benefits; the
Labour Party had inherited the old Liberal/Cobdenite free-trade/
League-of-Nations tradition, which further hampered its freedom of action.

5   Was there really a better alternative? Sweden, New Zealand
and Australia do not provide satisfactory analogies; the small
size of the English public sector made an interventionist policy
hard to operate; there were legal and administrative difficulties
in the way of public works; 'Keynesianism' became the great
alternative, but it was not codified until 1936; advisers Keynes
and Hubert Henderson were both very confused and self-
contradictory throughout 1930/1 – they did not realize until
1931 that protection and devaluation were required; in 1929 the
Liberal Party, under Lloyd George, proposed public works but
was against protection and against cuts in the dole, so reflecting
only a feeble grasp of 'Keynesianism'; reflation without protection would merely have sucked in imports; TUC policy under
Bevin and Citrine (to suspend the national debt and to tax
rentiers) would have been dangerous; only Mosley had a properly
worked-out alternative strategy, and it was unacceptable to the
Party.

6   So, while the government may be blamed for its policy of
'drift' when *either* full-blooded deflation *or* reflation might have
been preferable, the problems it faced were immense; and it did
do some positive things (*e.g.* the Economic Advisory Council
began industrial reconstruction).

**2   Examine the causes and consequences of the general strike in
1926.**

Understanding the question

This is a straightforward question on a single event in political
and working-class history; remember that, while its causes were
mainly local and industrial, its consequences may be traced over
a wide area of British life.

Knowledge required

Thd economic, social, industrial and political events of the interwar period. A fairly detailed narrative of the general strike itself.

Suggested essay plan

1   The general strike has given rise to a most consolatory
myth – *i.e.* one in which both parties to the dispute (bourgeoisie
and workers) can look back on it as a moment of supreme
triumph (for 'individualism' and for 'collective solidarity'
respectively); as such it had a usefully purgative effect on the
bitter politics of the 1920s; indeed, all except the miners could
derive some comfort from the event.

2   The fundamental causes of industrial unrest in the 1920s:
the rise of the trades unions, 1914–20, and the rise of shopfloor
power; the increase of Syndicalism in the union movement; the
establishment of the TUC General Council (1920) and the
Communist Party (1921); the depression and unemployment,
1922–5, put trades unions on the defensive, starting with
'Black Friday'; the losses of membership 1920–5, and concentration within the movement (Triple Alliance); the increasing
power of national officials *vis-à-vis* the rank and file; wage-cuts
caused a wave of disputes, 1921–3; then in 1925 the return to
gold at old par made further wage cuts seem necessary.

3   The specific problems of the mining industry: the workers

desired nationalization; the Sankey Report and Lloyd George's
decision to decontrol (leading to Black Friday and to great
bitterness); renewed competition from Polish and German coal
hit the market in 1924; A. J. Cook led the resistance to the pro-
jected wage cuts; Bevin and other unionists were anxious to show
that direct action could accomplish more for workers than could
'woolly' Labour politicians (even when in government), so the
General Council took up the miners' cause; Red Friday and the
Samuel Commission; cabinet 'hawks' like Joynson-Hicks and
Churchill finally provoked the general strike by breaking off
negotiations after an 'overt act' on the part of *Daily Mail* com-
positors.

4   A brief narrative of the events of the general strike, of volunteer strike-breaking and the subsequent miners' lock-out.

5   The political consequences: the alienation of the TUC from
Labour Party over the attitude to the strike of MacDonald and
others; Baldwin threw away his reputation for fairness with the
Trades Disputes and Trade Union Act (1927); membership of
the Communist Party collapsed in 1926; the Conservatives
looked for recovery through planning rather than renewed confrontation.

6   The consequences of the strike for the mining industry: the
miners were demoralized by the General Council's decision to
end the strike and did not recover their former power within the
movement (until 1972–4); the large wage cuts and consequent
bitterness led ultimately to nationalization in 1947.

7   The consequences of the strike for the trade union movement: membership was falling anyway (1920–5) and the general
strike merely reinforced the trend (1925–8) – especially on the
railways and in printing, which had done relatively well during
1921–5; except in mining, membership (and funds) were to
recover, 1933–9; there was a decline of Syndicalism and in use of
the strike weapon generally – which may have occurred anyway
during the depression – but it must be significant that the main
strikes in 1929–33 were in an industry largely unaffected in 1926
(textiles); shop stewards lost in influence to the national officials;
the General Council (under Bevin and Citrine) turned against
the Communist Party, and agreed to participate in industrial
rationalization and planning (or state monopoly capitalism)
instead of trying to overthrow capitalism completely; the Mond-
Turner talks paved the way.

8   So the general strike marked a transition in the Labour
movement towards a more willing cooperation with capitalism
– but it was probably happening anyway; more significantly, it
made employers (except the mine owners) more moderate in the
future – there was little wage-cutting (except according to negotiated sliding-scales) in the 1929–33 depression.

**3   Consider the achievements and limitations of Baldwin as
Conservative Party leader.**

Understanding the question

This is a question about the Conservative Party's political success
in the interwar years and Baldwin's contribution to it. Do not
waste time discussing Baldwin's qualities as a national leader.

Knowledge required

The party-political history of the 1920s and 1930s.

Suggested essay plan

1   The considerable success of the interwar Conservative
Party compared with its demoralization and disarray before
1914; the benefits to the party from the 1918 and 1928 franchise
extensions and from the Liberal decline; the divisions within the
Party.

2   Baldwin's industrial background and swift political rise; his career at Board of Trade and Exchequer; the settlement of the war debt to America brings him prominence at a critical time; his part in the Carlton Club's rejection of Lloyd George (1922); he was preferred to Curzon as Party leader in 1923.

3   Baldwin's leadership was constantly challenged by apparently cleverer men, who could never understand his success, and who chafed at his 'masterly inactivity'; his deliberate rejection of Lloyd George's 'dangerous dynamism' – such opposition culminated in the 'Empire Free Trade' group's (Rothermere, Beaverbrook, Churchill) attack on his Indian and fiscal policies in 1930–1; in thwarting this group Baldwin showed himself to be a master of political tactics; his handling of the abdication crisis revealed the same.

4   Yet he himself mendaciously claimed: 'I am not a clever man. I know nothing of political tactics.' His greatest contribution as leader was to project and personify the Conservative image of morally serious, commonsensical, anti-intellectual, unflappable, honest, pipe-smoking, rural-utopian decency; this 'Salisbury-style' of Conservatism, an effective antidote to Lloyd George's flippant insincerity, attracted many middle-class and female voters, and some deferential worker voters; through association with Reith's BBC, Baldwin became the first modern media politician.

5   As leader Baldwin made many apparent blunders: he threw the 1923 election away 'needlessly' by espousing protection, squandered a reputation for social 'fairness' by agreeing to the Trades Disputes and Trade Union Act (1927), misjudged the electorate's mood in 1929 by relying on 'Safety First' and his own persona, and he might have done better, from a party viewpoint, to have refused support to MacDonald's National government, since the Conservatives would certainly have won handsomely in 1931 on their own; his only really successfully-managed election was that of 1935.

6   Yet, looked at in a longer perspective, all these actions revealed his tactical astuteness; protection provided a clear line of demarcation between the Conservatives and the Coalition Liberals, prevented Chamberlain and Birkenhead from joining Lloyd George, and so helped to heal old party wounds; it was probably a wise move to 'lose' in the depression of 1929, and by supporting a National government Baldwin was able to settle the protection issue – the Conservative's biggest albatross – in a non-party political way.

7   But in longer terms still, Baldwin's homely image worked against the Party, making it seem too stupid for the dangerous 1930s, and also dishonest (*e.g.* the 1935 election and rearmament); his failure to encourage younger moderates and reformers (Macmillan, Boothby, Stanley) was also costly; Baldwin's Party collapsed in bitterness in May, 1940; he had managed to keep it together for a decade, but had associated it too closely with his own personal political 'style'.

5   Document question (Labour and Liberalism in 1924): study the extract below and then answer questions (a) **to** (h) **which follow:**

'My difficulty about the Liberal Party lies further back than yours. I doubt if it any longer stands for anything distinctive. My reasons are on the one side that moderate Labour – Labour in office – has on the whole represented essential Liberalism, not without mistakes and defects, but *better* than the organized party since Campbell-Bannerman's death. On the other side the Liberal party, however you divide it up, never seems any better agreed within on essentials. Of the present fragment part leans to the Tories, part to Labour, part has nothing distinctive, but is a kind of Free Trade Unionist group. The deduction I draw is that the distinction between that kind of Labour man who does not go whole hog for nationalization on the one side and the Liberal who wants social progress on the other is obsolete. I myself have always felt it was unreal and that if we divided parties by true principles the division would be like this

Communist	ordinary Labour	Bad Liberal	Diehards
Theoretical	Good Liberal		ordinary Tory
Socialist			

'But tradition and class distinction kept many good Liberals outside Labour. Now Labour has grown so much that it tends to absorb them and to leave only the 'bad' Liberals who incline to the Tories and a mass of traditional Liberals who can't desert a party of that name . . . .

'For a moment fate seemed to be avoided by the decision of the Liberals on the *Manchester Guardian* lead to support Labour. Labour responded badly, and the Liberals then drew away and inevitably gravitated to the other side. They failed to present a third view because outside the extremists there is really no third view. Liberals may be full of fight but as against the main body of Labour what have they to fight for? Internationalism? Free Trade? Ireland, India, any particular kind of Social Reform? No, on all these there is agreement. There is really nothing, till you come up against doctrinaire Socialism, which is really outside 'moderate' Labour.'

(*J. A. Hobhouse to C. P. Scott, 7 November, 1924*)

Maximum marks

(a) Who was Prime Minister in the Labour Government of 1924? Who was 'Campbell-Bannerman'; and who was leader of 'the other side'?   (3)

(b) Explain the comment 'Labour responded badly'.   (2)

(c) To what extent had the 1924 Labour Government's policies 'represented essential Liberalism'?   (7)

(d) Comment on the division of 'parties by true principles', laid out above.   (4)

(e) How apt is the writer's assertion that there was 'no third view', in the light of the party politics of the 1920s?   (3)

(f) Why had Labour 'grown so much' since 1918?   (2)

(g) The writer suggests that the Liberals' main problem was the common ground between Labour and Liberal policies. What other factors are relevant to Liberal decline up to 1924?   (4)

(h) The writer clearly found little 'doctrinaire Socialism' in the first Labour government. How much would he have found in the second?   (6)

27.6 READING LIST

Standard textbook reading

T. O. Lloyd, *Empire to Welfare State* (OUP, 1970), chapters 4, 5, 6.

L. C. B. Seaman, *Post-Victorian Britain* (Methuen, 1966), chapters III, IV.

R. K. Webb, *Modern England* (Allen and Unwin, 1980 edn.), chapter 12.

Further suggested reading

R. Blake, *The Conservative Party from Peel to Churchill* (Fontana, 1972).

C. Cook, *A Short History of the Liberal Party 1900–1976* (Macmillan, 1976).

C. Farman, *The General Strike* (Hart-Davis, 1972).

F. S. L. Lyons, *Ireland since the Famine* (Collins, 1971, 1973).

C. L. Mowat, *Britain between the Wars* (Methuen, 1955).

R. K. Middlemas and A. L. Barnes, *Baldwin, a Biography* (Weidenfeld, 1969).

D. Marquand, *Ramsay MacDonald* (Cape, 1977).

R. Skidelsky, *Politicians and the Slump* (Macmillan, 1967).

A. J. P. Taylor, *English History 1914–45* (OUP, 1965).

28 The National governments, 1931–9

28.1 INTRODUCTION

The starting point for an understanding of British politics in the 1930s is the political crisis of 1931, which led to the formation of the National Government dominated by the Conservatives. The way was now open for them to introduce their favoured policies of tariff reform and imperial preference in the Ottawa agreements of 1932. In practice, the Ottawa agreements fell a long way short of Empire free trade but at least they had implemented pre-war Tory aspirations without splitting the Party.

Both the domestic and foreign policies of the 1930s were influenced by this victory of Empire free trade over pre-war Liberal internationalism. Political opposition to the National Government was weak. The only challenge to the government, such as it was, came from outside parliament – from Oswald Mosley's British Union of Fascists and, to a lesser extent, from the Communist Party.

The main problem faced by the Conservatives in the general election of 1935 was still unemployment and the need to justify their policy of appeasement – which Baldwin cleverly accomplished by stealing Labour's 'collective security' approach for the purposes of electioneering.

Though the Munich agreement in September, 1938 was greeted with much pacifist relief, and some popular approval, it became clear – as Czechoslovakia was dismembered in the months that followed – that appeasement as a policy was doomed. Winston Churchill, who had led the opposition to appeasement within the Party throughout the 'thirties, finally became Prime Minister in May, 1940 on the downfall of the government whose every move 'had seemed stamped with futility and failure'.

28.2 STUDY OBJECTIVES

1 The formation of the National Government: background to the financial crisis of 1931; the collapse of the Vienna Credit Anstalt (May), and the May Committee Report (August) led to a run on sterling and to a cabinet split over the budget; the role of opposition leaders (especially Chamberlain) and of the King in helping to 'persuade' MacDonald.

2 The composition of the National Government: Conservatives predominated at ministerial level and overwhelmingly so at the parliamentary level; the failing powers and political weakness of MacDonald, the nominal premier, and the growing dominance in cabinet of Chamberlain; the National Liberals increasingly submerged within the Conservative coalition; the exclusion of Churchill and Amery; Baldwin's succession, then Chamberlain's.

3 The electoral history of the National Government: the abusive nature of 1931 and the scare campaign against Labour's 'Bolshevism run mad'; the essential dishonesty of MacDonald's demand for a 'doctor's mandate', since a protectionist solution to the economic problem was already intended; the divergent nature of the manifestos of the different parties to the Coalition; the reasons for its enormous victory; a class and regional analysis of the 1935 election; Baldwin's astute handling of it, again on the somewhat dishonest slogan, 'Collective Security through the League'. Labour's recovery in London, Yorkshire, Scotland and Lancashire, and the collapse of the Liberals.

4 Political opposition to the National Government: the lack of government dynamism stemmed in part from the weakness of the opposition; the problems of the Labour Party under Henderson and Lansbury; the latter's pacifism and resignation over Mussolini; the recovery under Attlee and the likelihood by 1939 that Labour would win the next election. Samuelite Liberals (and Snowden) quit the National Government over the Ottawa Agreements in 1932, leaving the Simonite National Liberals inside; bleatings of 'the Goat in the Wilderness' (Lloyd George), mainly for public works and 'Keynesianism', could not prevent a further squeeze of the Liberals in 1935.

5 In the absence of political opposition, most attacks on the government came from outside the Commons; the hunger marches, demonstrations and growth of political violence, especially from Mosley's Blackshirts (BUF, founded in 1932) and the Communist Party; this enabled the government to pose as the bastion of law and order, by restoring general warrants in 1934 and by the 1936 Public Order Act.

6 The domestic policies of the National Government: the abandonment of gold and free trade; the Imperial Economic Conference revealed little accord but did lead to the Ottawa Agreements; the protective tariff and colonial preference played an important part in 1930s economic recovery; note also the importance of rearmament (from 1934, seriously from 1936); Chamberlain as Chancellor, balancing rearmament against social spending, and the influence here of Horace Wilson. There was some very modest deficit financing (1932–3) and cheap money (a low Bank rate), but few public works; there was some economic regulation (marketing boards) and nationalization (London Transport, BOAC); the school-leaving age was raised to 15; the 1934 Unemployment Act regulated relief and created political controversy. There was a steady, unnoticed, growth of Treasury influence in government decision-making during the 1930s, especially under the Treasury-minded Chamberlain. The TUC was also playing a more independent role in the developing 'corporate' state and was less reliant on the Labour Party.

7 For foreign policy and 'appeasement', see Units 12, 13 and 29, but note the domestic political aspects of appeasement: opposition to the National government inside the Conservative Party grouped round Churchill and focused on appeasement; note the roles of Eden, Amery and Boothby; Churchill's shortcomings as leader of this opposition – his rhetoric was splendid but he lacked credibility; he was personally discredited by his stand on Indian nationalism and the abdication; look at the pivotal role of Halifax in supporting this Churchillian opposition in 1939; the events leading to the collapse of the National Government in May, 1940, and its replacement by a more genuine, less Conservatively-dominated, coalition.

28.3 COMMENTARY

The emerging danger

The 1930s were at first characterized by the same economic problems which had dominated politics in the 'twenties. The rise to power in Germany of Hitler, however, was a turning point; from then on, the government was increasingly concerned with questions of foreign policy and defence. As the economic situation gradually improved for Britain, so international relations steadily deteriorated.

Until 1937 the public was largely unaware of the danger to Britain posed by German expansionism. Pacifism and disarmament were popular. These were, nevertheless, unrealistically combined with the belief that Britain should support the League of Nations against those countries which broke their pledges to the international community. Yet no one could explain how the League's supporters could defeat military aggression without themselves rearming.

Defence was, therefore, politically unpopular – although a national necessity – and so the government had to begin rearmament slowly and cautiously. The result was that foreign policy evolved from a position of weakness in relation to the dictator states; the temptation for the government was continually to buy time by making concessions.

This process culminated in the Munich agreement in 1938, widely welcomed in Britain at the time, but soon to be regarded as a national disgrace. It is correctly associated with the Prime Minister of the day, Neville Chamberlain, and it is for this policy that he is now chiefly remembered and villified.

What should be borne in mind, however, is that in its best sense appeasement was a popular preference for peace and conciliation and was a policy for which the nation as a whole bore considerable responsibility. After all, no one of importance had publicly aired a realistic alternative foreign and defence policy in the mid-thirties and it was not until 1939 that Churchill, Eden and the Labour Party began to emerge as a credible opposition to Chamberlain.

National politics, 1931–5

The 'National' government, formed in August, 1931, failed in its immediate object of protecting the pound – after a brief revival of confidence, the Invergordon mutiny panicked the market and forced Britain off the gold standard. (According to Seaman, the Navy saved the country from 'the economic consequences of Mr Churchill'!). Despite this, the government continued in office, and grew in strength as its overwhelming electoral victory of October, 1931 testified.

In considering why it was so successful politically, it should be noted that, in general, people trusted the National Government. It contained well-known political figures from all three Parties. This emphasized the gravity of the crisis; it also meant that alternative policies lacked prominent spokesmen; Henderson, who led the split against MacDonald, lost his seat. It meant, too, that opposition to the government came from Parties themselves split by its formation.

From the start, the government was predominantly Conservative in outlook, and grew more so as the immediate crisis receded. By 1932 its leaders were sufficiently confident to introduce measures for tariff reform and Imperial preference.

Protectionism

Since 1903 the central demand of progressive Conservatives had been for tariff reform. In the 'twenties this proved impossible but, after the collapse of Churchill's attempt to return to a policy of Liberal internationalism, the way was now open for a policy of imperialist protectionism. The collapse of the Labour government in 1931 allowed the Tories to get their protective tariff in 1931, and the Ottawa agreements that followed in 1932 moved Britain nearer the concept of Empire free trade.

This victory for the 'new imperialists' explains much of the domestic and foreign policies of successive governments in the 'thirties but especially their conciliatory responses to the colonial nationalisms in Ireland, India and the Middle East. It was also their starting point for dealing with Hitler, for it emphasized the defence of overseas empire, which Germany did not threaten, but which Japan, for example, did. Fearing Japan more meant that confrontation with Germany *had to be* avoided. Eventually many Conservatives, including Chamberlain, came to believe that such a clash *could be* avoided.

Snowden and two Liberals resigned from the cabinet in protest at this protectionism, but their departure had no effect on the government's standing. The sheer size of its majority made parliamentary opposition appear futile, and on Right and Left reaction against it became focused in political movements which took their inspiration from the supposed achievements of Fascism and Russian communism.

Electorally these movements were insignificant, but they were a source of continual worry to those in power. Mosley's Fascists were hampered by their own absurdity as much as by government action, but the intellectual challenge from the Left was much greater. Nevertheless, its popular appeal, even in areas of mass unemployment, was slight. The Labour Party recovered from the shock of 1931, and in the election of 1935 increased its vote considerably, mainly at the expense of the Liberals.

The politics of unemployment

Since the collapse of the post-war boom in 1921, unemployment had been a permanent feature of British economic life. But the rapidity with which it increased in 1930 was entirely unexpected. Although the country recovered slowly from the shock of the 1931 crisis the numbers of unemployed remained obstinately high. They never fell below 1 400 000 and reached almost 3 million in 1932.

Dates of departure for arrival in London on Oct 27

Glasgow Sept 26

Newcastle Oct 2

Burnley Oct 9

Manchester Oct 11

Sheffield Oct 14

Liverpool Oct 11

Mansfield Oct 15

Stoke Oct 15

Birmingham Oct 18

Hereford Oct 17

Norwich Oct 16

Cardiff Oct 15

London

Bristol Oct 16

Canterbury Oct 21

Plymouth Oct 10

Brighton Oct 23

Southampton Oct 21

Fig. 20 The National Hunger March on London, 1932

In studying this phenomenon and its effect on national politics, it is important to look at the figures region by region; while South Wales and parts of the North had huge numbers of unemployed for most of the decade, much of the country was relatively unscathed, especially London and the South-East. This may explain why, in spite of the scale of the problem and of considerable political agitation, the unemployment situation never threatened to split the government or to produce an electoral revolution against its cautious and rather limited policies.

Government remedies were traditional. By the 1934 Unemployment Insurance Act they accepted responsibility for looking after the unemployed and tried to establish for them a minimum standard of existence. This was their chief policy. Apart from this they reintroduced tariffs and encouraged bilateral trade agreements, left the gold standard and lowered Bank rate, encouraged all reasonable schemes for the rationalization of industry and made some attempts to create employment in the depressed areas. The expenditure allowed, however, was meagre and the results, therefore, were predictably disappointing.

Two other factors kept unemployment figures high. First, the continual influx of school-leavers into industry and second, the rise in the number of immigrants (especially Jews) from Europe during the thirties.

The failure to adopt a more positive attitude to the problem of unemployment – for example, by raising the school-leaving age or attempting a wider scheme of public works, as Roosevelt did in the US – did not affect the government's general popularity very much. The depressed areas were by now probably safe Labour seats and it is a fact that, for those in jobs, real wages increased. For them, the National Government's timid policies brought tangible improvements, all the more valued because of the well-publicized plight of those out of work.

Rearmament

Planning of rearmament began as early as the winter of 1933, in reaction to Hitler's rise to power, but it was not until March, 1935 that a White Paper on defence gave a cautious warning to the public. Pacifist feeling was very strong in the country, and the opposition was critical of all arms expenditure. Surprisingly, in view of his performance as Prime Minister, the minister most enthusiastic for rearmament against Germany was the Chancellor, Neville Chamberlain. He urged Baldwin to fight the 1935 election on this issue; Baldwin refused. The consequence was that the public remained very ignorant of the danger from abroad.

In July, 1936 a Labour MP wrote in his diary: 'The Hitler rearmament races on. Few people in the Labour Party seem to know or care anything about it'. By 1937, when the first major programme was announced, opinion had shifted decisively: now the government was criticized for proceeding too slowly.

Rearmament was, however, a huge undertaking, not a tap which could be turned on at will. Industry had to re-equip itself, and Britain's manufacturing output could not be switched overnight to making arms since it produced the wealth which financed government spending. Ironically at a time of mass unemployment, there was also an acute shortage of skilled workers, and the unions were understandably reluctant to abandon established working practices for what might be only a temporary boom.

Finance was an additional constraint, although this has been exaggerated. Each service department saw their own needs as paramount, whereas only the Treasury had an overview of the whole financial and industrial situation and was in a position to impose priorities on the three services. In fact, it was only due to the Treasury (and Chamberlain) that in 1938 a reluctant RAF was forced to give higher priority to fighter production.

By the time war broke out, Britain had just enough arms to secure her survival. Note that rearmament was essentially defensive. No expeditionary force was planned before 1939. The development of air power was concentrated on fighters, not on bombers. Britain was equipping herself for national defence – not for resistance to German aggression on the Continent.

The National government, 1935–9

In June, 1935 Baldwin took over as Prime Minister from the ailing MacDonald, now merely a figurehead. In November, his Party won a majority in this important general election. Within a few weeks, his new government was attacked from every side because of the Hoare-Laval pact. The storm blew over, however, once the scheme was abandoned. Baldwin's personal stock had never been lower: his public image of decency and honesty was in tatters. Nevertheless he recovered his prestige in 1936 with his masterly handling of the abdication crisis, and early in 1937 announced his intention to retire.

His departure in a blaze of glory brought Chamberlain to No. 10. He was unlike his predecessor in almost every respect: hard-working, efficient, aloof. Contemptuous of Labour, he aimed to succeed by accomplishment not by amiability. He had been waiting a long time for his chance, but was already an elderly man. Furthermore, the talents which had made him a good Minister of Health and Chancellor of the Exchequer were not those required of a party leader. Chamberlain failed to realise that in the volatile sphere of foreign policy different rules applied. Ironically, his greatest mistake, the Munich agreement, at first brought him mass adulation; ultimately it totally discredited him, obscuring his earlier support for rearmament and the general rise in the standard of living for which he deserved some credit.

Fascists and Communists

The depression of the 1930s saw the growth of an extra-parliamentary group of some importance, the British Union of Fascists. Its leader was Sir Oswald Mosley, aristocrat and ex-Conservative; he had also been a cabinet minister in the Labour government, but had resigned in protest at his colleagues' passivity in the face of the economic crisis. He eventually came to argue that only drastic political action, on the lines followed by Mussolini in Italy, could save the country and give employment to all. Thus he drifted into Fascism, and moulded the hitherto insignificant Fascist groups in Britain into a relatively organized movement.

It soon became clear, however, that the electorate had no taste for Italian-style corporatism, and Mosley's party resorted more and more to anti-Jewish propaganda, to marches and to public meetings intended to provoke a response from the Left and thus to force the government to act against what he saw as the dangers of Communism. The climax was a mass meeting held at Olympia in June, 1934, when Fascist stewards treated hecklers with un-English brutality. Deliberately provocative Fascist marches in the East End of London later resulted in some violent rioting but the only important consequence was a further diminution of the movement's credibility.

In 1936 the government passed an act forbidding the wearing of political uniforms and granting the authorities power to ban public marches and displays which were judged a threat to public order. This straightforward measure was enough to kill off the BUF as a political force. Denied their black shirts and parades, Mosley's supporters melted away.

Communism was intellectually much more respectable than Fascism, and in the 1930s evoked admiration in some circles – as demonstrated by the success of the Russians (through the Cambridge connection) in penetrating British intelligence. Its working-class support, however, was never strong, despite the influence of some front organizations such as the National Unemployed Workers Movement.

The problem both for Communism and for Fascism was that the weakness or unpopularity of the government was never serious enough to threaten the downfall of the existing political order.

Another difficulty for both was that they were identified with foreign powers – the Fascists with Mussolini's Italy and the Communists with Soviet Russia. The Communists in particular took their line from Moscow, with the result that their policy on international events changed (often dramatically) whenever Stalin's did. This was particularly marked when the Germans and the Russians signed their non-aggression pact in August, 1939, after which the British Communists opposed the war until 1941, when Hitler suddenly invaded Russia.

In the long term, it was Socialism, not Communism, which was the main beneficiary of the intellectual ferment of the 1930s in Britain. This bore fruit in the radical programme of the Attlee government elected in 1945.

Conclusion

Until recently the National governments have usually been subjected to adverse criticism by historians. Firstly, for their failure to solve Britain's economic problems and their willingness to accept unemployment and its social consequences. Furthermore it is alleged that they were Tory dominated, complacently middle class in outlook and feeble in action both at home and abroad. Recent research has tended to modify this picture. Unemployment and industrial decline was never widespread in the 'thirties while economic recovery and a return to full employment had been achieved by 1937. Nor was the government ungenerous in its response to the plight of the unemployed – there was a marked increase in expenditure on the social services. These governments may have appeared feeble in action but they preserved political stability and if they excluded talent (e.g. Churchill and Lloyd George) they also excluded extremists (e.g. Mosley and Maxton). European Fascist and Communist governments could hardly be regarded as attractive alternative propositions, while the benefits ascribed to Keynesian policies (like the New Deal in the United States) were hardly impressive. Clearly, it would be unjust to see the ministers of the National governments as the only failures during the 'thirties while their impressive election victories in 1931 and 1935 demonstrated that they enjoyed enormous popular support.

28.4 CHRONOLOGY

1931	**August**	NATIONAL GOVERNMENT formed under MacDonald; the cabinet included Baldwin, Neville Chamberlain, Samuel, Snowden, Thomas and Lord Sankey.
	September	Naval Mutiny at Invergordon. GOLD STANDARD abandoned; the pound fell from $4.86 to $3.80.
	October	The government won a huge victory in the general election.
1932	**February**	IMPORT DUTIES ACT: the government abandoned free trade; a general tariff of 10% on all except imperial imports. BRITISH UNION OF FASCISTS founded by Sir Oswald Mosley. Four hundred branches by 1934: 20 000 estimated membership.
	July	ILP disaffiliated from the Labour Party.
	August	Ottawa Agreements: imperial trade preference meant that Britain obtained more of her food imports from the Dominions.
1933	**January**	Adolf Hitler became Chancellor of Germany. Children and Young Persons Act.
1934	**May**	UNEMPLOYMENT ACT took the issue of unemployment out of the political arena by removing it from local control and putting it under an Unemployment Assistance Board from December, 1934.
	June	BUF meeting at Earls Court.
	August	Government of India Act: gave dominion status to India.
	October	Peace Pledge Union formed by Canon Sheppard.
	December	SPECIAL AREAS ACT: to stimulate investment in depressed areas; a grant of £2 million allowed. UNEMPLOYMENT ASSISTANCE BOARD set up to administer unemployment relief on a national level.
1935	**March**	Government White Paper, *Statement Relating to Defence*, was published; this emphasized the nation's inadequate defences.
	May	MacDonald resigned as Prime Minister; succeeded by Baldwin (June).
	June	Peace Ballot result published; high-water mark of post-war pacifism.
	October	Italy invaded Abyssinia.
	November	Conservatives won the general election with 432 seats; Labour recovered with 154 seats; the Liberals won 21 seats.
	December	Hoare-Laval Pact provoked an outcry against the government by a public newly-devoted to the aims of the League of Nations.
1936		Beginning of serious rearmament. Left Book Club founded. (Between 1935 and 1937 nearly one million Communist pamphlets were sold in Britain)
	July	Spanish civil war: the British government adopted the controversial policy of non-intervention.
	October	Jarrow hunger march.

	December	ABDICATION OF EDWARD VIII because of his love for (twice-divorced) Mrs Wallis Simpson. The crisis was handled brilliantly by Baldwin; Edward's only prominent supporter was Churchill. Edward was succeeded by his brother, George VI. Public Order Act, prohibiting the wearing of political uniforms, was aimed at the BUF. Irish Dail removed the King from the Constitution.
1937	May	Neville Chamberlain became Prime Minister on Baldwin's retirement. He adopted a policy of rearmament and conciliation (appeasement) toward the dictator states.
	December	Irish Free State adopted its Gaelic name, Eire, but did not yet formally contract out of the Commonwealth.
1938		International affairs dominated domestic politics.
	February	Anthony Eden resigned over the policy of appeasement.
	July	COAL ACT: brought nationalization of the coal industry nearer.
	September	MUNICH AGREEMENT: Chamberlain twice visited Hitler; conceded most of his demands; Czechoslovakia was dismembered; the agreement at first was warmly welcomed by the British public; but it was soon seen to have bought time but not peace, and at a very high price.
1939	March	Poland's territorial integrity was guaranteed by Britain after Germany had seized the remainder of Czechoslovakia.
	September	Germany invaded Poland. WAR DECLARED by Britain on Germany. 'Phoney War' began (September–April 1940).
1940	May	Chamberlain lost the support of the Commons. Churchill became Prime Minister.

28.5 QUESTION PRACTICE

1 Why was a National Government formed in 1931?

Understanding the question

You have mainly to consider the motives of those individuals most responsible for setting up the National Government; look also at the long-term problems which precipitated a crisis in the old Party system.

Knowledge required

A detailed knowledge of the Second Labour Government and of the political crisis of the summer of 1931. A general knowledge of political events between the wars.

Suggested essay plan

1   A brief narrative of the Labour Government up to the divisions of August, 1931: its inadequate attempts to solve the unemployment problem; the collapse of Credit Anstalt and the run on sterling; the financial crisis and the May Committee Report; the demand by Opposition leaders for cuts in the unemployment fund as the price of their support for the budget, and Snowden's willing compliance; Bevin and Citrine, on the TUC General Council, demanded the suspension of the national debt and the sinking fund, and also called for a tax on rentiers, instead; this prompted Henderson and eight other ministers to withdraw their agreement to Snowden's cuts, leading to a cabinet deadlock (22–23 August).

2   A brief narrative of the events leading from here to the formation of the National Government: the meetings between MacDonald, Snowden, Baldwin, Chamberlain, Samuel, McLean and George V; the decisive role of the latter in persuading MacDonald.

3   The government's immediate rationale was the overriding national need for orthodox financial policies to restore foreign confidence in sterling; in this respect Snowden was central – he was an ardent free trader (the only man in the Labour cabinet who had opposed a revenue tariff) and was a believer in balanced budgets; he was not disloyal to his class and sincerely believed in inaugurating Socialism by confiscating the profits of successful capitalism; but, first, capitalism had to be made to succeed, which required orthodox policies leading to a restoration of confidence among capitalists.

4   Snowden, and probably MacDonald, genuinely believed this rationale for the National Government, but as a defence it was undermined by the fact that the new government quickly abandoned orthodox policies (the gold standard and free trade).

5   The central and most enigmatic figure was MacDonald: perhaps he *did* betray the movement he had so much shaped, as opponents alleged; his vanity made him susceptible to royal flattery and the 'aristocratic embrace'; but there is no evidence that he deliberately planned a National Government before the cabinet deadlock; yet he undoubtedly relished the role of national saviour, and by now regarded nearly all his Labour colleagues with contempt.

6   The possibility that MacDonald was deliberately playing the role of martyr, knowing his career was nearly over anyway, in order to save the Labour Party: there is some (albeit flimsy) evidence that he discouraged young Labour politicians like Morrison from joining him in the National government; certainly MacDonald's act enabled Labour to maintain its essential unity and to be spared further unpopularity for the Slump, and after the 1931 electoral disaster it quickly began to recoup; but that this was MacDonald's motive seems unlikely in view of his vitriolic attacks on his former colleagues in the October, 1931 election.

7   There is little evidence of a Bankers' ramp (either by Norman and the City bankers, or by the New Yorkers who imposed conditions on further credits to Britain); the City was genuinely alarmed by the possibility of devaluation – needlessly, as it turned out.

8   Baldwin, perhaps, was happy to form a National Government (though he would have won the next election anyway) because of his recent difficulties with Empire Free Trade Conservatives; his affection for MacDonald; Snowden also had been cooperating with Chamberlain for some time – the latter, especially, helped to sow the seeds of the 'coalition' idea, perhaps even in George V's mind; did the Conservatives see a National government as the only way of pushing fiscal reform (protection) past the electorate?

9   The Liberals saw it as a way of getting a toe-hold back on power; in fact it marked their virtual absorption by the Conservatives; they were crushed in 1935.

10   To conclude: the National Government was really a Conservative government in disguise; the national divisions and economic crisis, however, demanded a veneer of non-partisan respectability, and this MacDonald provided by consenting to lead it.

2   How 'national' was the National Government, 1931–39?

Understanding the question

You must first consider how much general public support there was for the National Government, both in geographical and class terms; secondly, how far its policies were in the national interest.

Knowledge required

The political history of the 1930s and the policies of the National Government.

Suggested essay plan

1   Two opposing views have dominated the historiography of the National Government: one – that it was a thinly-disguised cover for Conservative rule which, by pursuing deflationary policies (reduction of public salaries and unemployment benefits) and an ungenerous social welfare policy (especially the hated 'means test') at a time of between two and three million unemployed – was clearly acting in a partisan and sectional (*i.e.* promiddle class) way; the other – the 'official' view – is that its policies were in the national interest and were broadly correct, and that, by attracting widespread support, it managed to heal the bitter socio-political divisions of the 1920s.

2   A brief account of the circumstances leading to the formation of National Government; of the composition of the government and of its parliamentary support in terms of previous Party allegiance.

3   The October 1931 election showed overwhelming national support for the government (67% of votes on a 76.3% turnout); but note that Conservative voters made up 55.2% out of this 67%, Liberals 10%, and National Labour only 1.6%, while there were no *non*-national Conservatives and (at this stage) very few *non*-national Liberals; Conservative voting strength was reflected in the number of MPs (473 out of 554 supporting the government, out of 615 altogether); 1935 gave the government 55% of the vote, an impressive victory but hardly one of 'national' dimensions.

4   Regional analysis of the elections: Durham and South Wales were persistently hostile to the government, but otherwise in 1931 it had some claim to be called 'national'; Clydeside, Tyneside, West Riding, Lancashire and London showed a considerable Labour revival in 1935, however; there was now support for the government concentrated in prosperous Southern England and the Midlands, reflecting the economic division of Britain into 'two nations'; the professional middle classes and many white collar workers probably had 'never had it so good', and provided the backbone of the government's support.

5   Also, the two individuals most conspicuously excluded from the government (Lloyd George and Churchill) were, of all politicians, the ones who had most hankered after coalition or consensus politics, and the only two with the stature to embody (the one previously, the other in the future) the national will; therefore in retrospect the National government, in its rejection of Churchill and his timely warnings against Hitler, seemed to be anything but national.

6   In standing for compromise the government undoubtedly reflected the rather lack-lustre national mood; Baldwin skilfully exploited George V's Silver Jubilee to capitalize on the patriotic sentiment and devotion to the Monarchy; the BBC and Dawson's *Times* contributed to the government's image here.

7   Beneath the Party battle, somewhat attenuated anyway because of the one-sided 1931 election, the 1930s saw the silent advance of the 'corporate state'; the Civil Service, the TUC and the FBI probably counted for more in decision-making than did the cabinet and politicians; this was especially noticeable under Chamberlain, who strictly controlled the flow of information to the cabinet; in so far as this burgeoning bureaucracy reflected a national, rather than Party, viewpoint, the 1930s probably marked an advance in 'national' solutions.

8   MacDonald, on forming the National Government, denied an intention of holding an election under 'national' auspices, but Conservative backbenchers forced him to hold an early election; the different elements in the Coalition issued different manifestos, so that the government could claim no mandate to do anything; what it did was, of course, Conservative policy (especially protection); Snowden's Land Valuation policy, for example, was discarded. One may defend the National Government's policies but there can be no doubt that they were more Conservative than 'national'.

3   'The National Government solved the immediate problems but left the fundamental ones unsolved.' Discuss.

Understanding the question

One is tempted to point out that most governments leave most fundamental problems unsolved, since such problems are often insoluble – this is really a straightforward question on the achievements of the National Government.

Knowledge required

The social, economic and political difficulties facing Britain in the 1920s and 1930s. The policies of the National government.

Suggested essay plan

1   Perhaps the main immediate problem of the 1920s was political: how to fit three Parties into a two-party, 'first-past-the-post', political system. This had led to weak governments and frequent elections, which meant a lack of continuity in policy at a time when the economic slump made such a thing highly desirable; the Second Labour Government merely took a decade of 'drift' and 'wait-and-see' policies to extremes.

2   At least the National Government solved this problem; it was formed as a short-term emergency administration which would not even fight an election – but in the event it stayed in power for a decade, and at least provided more consistent, if hardly more decisive, policies.

3   A more fundamental political problem in the 1920s was the bitter polarization of politics; to some extent the National Government, by reflecting a national mood (perhaps exaggerated by Dawson's *Times* and Reith's BBC), contributed to national reconciliation.

4   In terms of policy, it is perhaps not true to say that the Government even solved the immediate crisis, which was financial. Ostensibly it was formed to save the gold standard, yet it went off it almost immediately after the Invergordon mutiny (September, 1931); moreover, this short-term 'failure' – *i.e.* devaluation and cheap money – probably contributed as much as anything that the Government did, or could do, to 'fundamental' economic recovery in the 1930s; the lesson was, of course, that only a government which has the confidence of investors can, in a free enterprise capitalist system, *dare* to follow unorthodox policies.

5   Again, the National Government was formed to balance the budget; yet the 1932/3 Chamberlain budgets were mildly inflationary with modest deficits; this was hardly full counter-cyclical economic control, but it was pragmatic and a sensible response to a fundamental problem of public finance.

6   By 1931 it was clear to many, including Keynes, MacDonald and Hubert Henderson, that the fundamental problem of economic policy lay in the electorate's refusal (see 1923) to countenance an abandonment of free trade; by laying this particular nineteenth century ghost, the National Government made it possible for the British economy to enter the twentieth century.

7   It is true that the absence of a strong parliamentary opposition made the National Government more supine in dealing with new dangers (like Nazism) as well as with old sores like unemployment; but it seems unlikely that a more energetic government with the provision of public works could have reduced the dole queues, given the small size of the British public sector – and it is unlikely that public opinion would have allowed a much less craven attitude to Hitler.

8   The history of the National Government was in many ways one of decline after a relatively successful beginning. Its economic policies – given ministerial beliefs in palliatives rather than government cure (*i.e.* that there was little that government, as opposed to natural economic forces, could do) – were modestly successful. The worst of the unemployment was over by 1933. Economic recovery, however, had little to do with government action. The ultimate failure in foreign affairs lost the government whatever credit it deserved for its earlier modest successes – as Mowat remarks, Neville Chamberlain 'succeeded to a barren inheritance' when he became Prime Minister in 1937.

4   Document question (Crisis in 1931): study the extract below and then answer questions (a) to (f) which follow:

line

'At 10 a.m. the King held a Conference at Buckingham   1
Palace at which the Prime Minister, Baldwin and Samuel
were present. At the beginning, the King impressed upon
them that before they left the Palace some communiqué must
be issued which would no longer keep the country and the   5
world in suspense. The Prime Minister said that he had the
resignation of his Cabinet in his pocket, but the King replied
that he trusted there was no question of the Prime Minister's
resignation: the leaders of the three Parties must get together
and come to some arrangement. His Majesty hoped that the   10
Prime Minister, with the colleagues who remained faithful
to him, would help in the formation of a National Government,
which the King was sure would be supported by the
Conservatives and the Liberals. The King assured the Prime
Minister that, remaining at his post, his position and repu-   15
tation would be much more enhanced than if he surrendered
the government of the country at such a crisis. Baldwin and
Samuel said that they were willing to serve under the Prime
Minister, and render all help possible to carry on the
Government as a National Emergency Government until an   20
emergency bill or bills had been passed by Parliament, which
would restore once more British credit and the confidence of
foreigners. After that they would expect His Majesty to grant
a dissolution. To this course the King agreed. During the
Election the National Government would remain in being,   25
though of course each Party would fight the election on its
own lines.
'At 10.35 a.m. the King left the three Party leaders to settle
the details of the communiqué to be issued. About 11.35 the
King was glad to hear that they had been able to some extent   30

to come to some arrangement. His Majesty congratulated them on the solution of this difficult problem and pointed out that in this country our constitution is so generous that leaders of Parties, after fighting one another for months in the House of Commons, were ready to meet together under   35
the roof of the Sovereign and sink their own differences for a common good and arrange as they had done this morning for a National Government to meet one of the gravest crises that the British Empire had yet been asked to face.'

(Memorandum by Sir Clive Wigram, Private Secretary to King George V, 24 August 1931)

Maximum marks

(a) Who was the Prime Minister at the time of this meeting and why were 'the country and the world in suspense'? (6)

(b) Why did the Prime Minister have the 'resignation of his Cabinet in his pocket' and why, apparently, would only some of his colleagues remain 'faithful to him'? (5)

(c) What possibility was so greatly feared at the time that it should cause the king to speak of 'one of the gravest crises that the British Empire had yet been asked to face'? Why was fear of this possibility so great? (4)

(d) What was the National Government (i) able to do and (ii) unable to do, when taking steps to 'restore once more British credit and the confidence of foreigners'? (6)

(e) What aspects of their characters and policies would make the three Party leaders at this meeting unusually ready to be influenced by the King's assurance in lines 14–17 and his congratulations in lines 29–37? (5)

(f) What were the effects of the decision arrived at by this meeting on the position of the three main Parties during the rest of the 1930s? (5)

28.6 READING LIST

Standard textbook reading

T. O. Lloyd, *Empire to Welfare State* (OUP, 1970), chapters 6, 7.

L. C. B. Seaman, *Post-Victorian Britain* (Methuen, 1966), chapter V.

R. K. Webb, *Modern England* (Allen and Unwin, 1980 edn.), chapter 12.

Further suggested reading

R. Benewick, *The Fascist Movement in Britain* (Penguin Press, 1972).

S. Constantine, *Unemployment in Britain between the Wars* (Longmans, 1980).

K. Feiling, *The Life of Neville Chamberlain* (Macmillan, 1946, 1970).

D. Howell, *British Social Democracy* (Croom Helm, 1976).

R. Miliband, *Parliamentary Socialism* (Merlin Press, 1964, 1973).

C. L. Mowat, *Britain between the Wars* (Methuen, 1955).

H. W. Richardson, *Economic Recovery in Britain 1932–39* (Weidenfeld and Nicolson, 1967).

J. Stevenson and C. Cook, *The Slump* (Cape, 1977).

A. J. P. Taylor, *English History 1914–45* (OUP, 1965).

29 British foreign policy in the 1930s

29.1 INTRODUCTION

British foreign policy in the 1930s must be understood in the context of public opinion, which was intensely opposed to any repetition of the horrors of the 1914–18 war and which was demoralized by the severity of the world economic depression. At government level there was an acute awareness of the limitations of British power – of the fact that the burdens of imperial and home defence were increasingly more than resources could meet.

Practical considerations of strategy and finance, and of the likely impact of a future major war, combined with a public mood resistant to the idea of war to produce the policy of appeasement. The three 'revisionist' Powers – Germany, Italy and Japan – presented the greatest threat to British interests during this period, which saw the progressive destruction of the Versailles Settlement and the discrediting of the notion of collective security under the League of Nations. Though the Munich Conference was popularly applauded as a vindication of appeasement, Chamberlain's hopes of long-term peaceful coexistence with Germany collapsed with the dismemberment of Czechoslovakia in March, 1939 and the German invasion of Poland in September of that year.

29.2 STUDY OBJECTIVES

1   As background: the Versailles Settlement, particularly the territorial, disarmament and reparations clauses; criticisms of these terms, especially those of J. M. Keynes; and British relations with France, Germany and the Soviet Union, 1918–30.

2   Public opinion and pacifism: the mood of 'never again' after the First World War; the League of Nations Union; the East Fulham by-election; the Peace Ballot; the Oxford Union 'King and Country' debate; the hostility of the Dominions to the prospect of involvement in another European war.

3   Britain and the problems of defence: the 'twin pillars' of home and imperial defence; the Ten-Year Rule; the difficulty of preserving both the Empire and the home islands from attack – the respective threats of Germany, Italy and Japan to British interests and the insistence of British military planners that Britain could not defend herself against these powers simultaneously – the concept of 'limited liability'.

4   Rearmament: priority given to the Royal Air Force; the Defence White Paper of March, 1936; the two-power standard for the Navy; the rejection of the idea of a field force as involving a continental commitment that was too expensive, given Britain's limited resources, and unacceptable to British public opinion.

5   British conduct of foreign policy during (a) the Manchurian crisis, (b) the Abyssinian War, and (c) the Spanish civil war. Note particularly the Hoare-Laval Pact and the role of public opinion during the Abyssinian war; on the Spanish civil war special attention should be given to the reasons for the British policy of non-intervention *and* the unofficial involvement of British volunteers in the war; note also the attitudes of the political Right and Left in Britain to the Spanish conflict.

6   The significance of (a) the German reoccupation of the Rhineland; (b) the Anschluss; (c) the crisis over Czechoslovakia, 1938, and the German occupation of Bohemia and Moravia,

1939; and (d) the guarantees of the British government to Poland, Greece and Rumania.

7   The reasons for the mutual distrust between the British and Soviet governments and the failure to reach an understanding to resist Hitler in 1939; the Molotov-Ribbentrop Pact; the immediate pre-war crisis over Danzig and the course of events leading to the outbreak of war.

8   An overall assessment of the foreign policy of Neville Chamberlain: a comparison of his approach with that of Stanley Baldwin; Chamberlain's view of the Powers and his meetings with the dictators.

9   The justifications offered for the appeasement of Italy and Germany, particularly at the Munich Conference; the role of the anti-appeasers, especially Churchill and Eden; the revulsion against appeasement after the final collapse of Czechoslovakia in March, 1939.

10   The circumstances of Chamberlain's resignation and of Churchill's assumption of power.

29.3 COMMENTARY

Background

The British policy of appeasement in the 1930s has often been portrayed in simple terms, with the cowardice and wrongheadedness of the 'guilty men', Chamberlain, Halifax and Hoare – who failed to stand up to the dictators until it was too late – contrasted with the courage and vision of Churchill. This is, in fact, a very simplistic view and fails to take into account the extremely tight constraints on British policy-makers in these years.

Britain during the 1930s was strongly influenced by the fear of war and looked to governments to ensure that the horrors of 1914–18 would never recur. The League of Nations Union, which advocated the renunciation of war and campaigned in favour of collective security, helped to organize a 'peace ballot' in June, 1935: over $11\frac{1}{2}$ million people demonstrated their support for the League and about 10 million voted to reduce armaments and abolish military aircraft. This last vote reflected the universal terror of bombing, expressed in the statement by the Prime Minister, Stanley Baldwin, that the 'bomber would always get through'.

The East Fulham by-election convinced politicians that the public would not tolerate a massive rearmaments programme. The general mood, however, was not one of out-and-out pacifism, which was a minority cause, but a rejection of the idea of *national* war for King and Country in favour of collective security through the League of Nations.

Another sentiment, which was quite widespread, was regret over the treatment of Germany in the Treaty of Versailles – something bitterly attacked by J. M. Keynes in his book *The Economic Consequences of the Peace*. The German people themselves never accepted the post-war Eastern boundaries, nor the scale of disarmament imposed upon them. And though the deeply-resented reparations payments were effectively abolished at the Lausanne Conference, Hitler was able skilfully to exploit the British sense of injustice at the settlement. He also set out to convince the West that a strong Nazi Germany was the only bulwark against the spread of Soviet Communism.

The question of defence

Britain's defence policy was based upon the 'twin pillars' of home and imperial defence. This was a period in which the power of the Empire as a military bloc was being undermined by the development of air and submarine warfare and by the relative decline of the British Navy itself. Although Britain had always concerned herself with the balance of power in Europe, she had traditionally avoided involvement unless that balance was endangered by the dominance of a single major Power.

In the 1920s Britain had reverted to non-involvement as far as was practicable. This was encouraged in the interwar years by the growing independence of the Dominions, which was recognized in the 1931 Statute of Westminster. They made very clear their resistance to being drawn into a future European war, thereby powerfully reinforcing British hostility to European entanglements.

Urged on by Churchill, (then Chancellor of the Exchequer), Britain drastically reduced her armaments under the Ten-Year Rule. Later, the slump made increased arms expenditure even less attractive. The Chiefs of Staff, as well as the Treasury, were strongly opposed to burdening Britain with commitments greater than her resources could bear. Foreign and defence policies were dominated by the concept of 'limited liability'.

It was regarded as self-evident that Britain could never successfully combat three major enemies simultaneously – and yet, Germany, Italy and Japan all posed a threat. Some experts, like Hankey and Chatfield, considered Japan to be the main threat to Singapore and other Far East interests. However – to give one instance – the Navy would find itself unable to guarantee the safety of British possessions in the Far East if we had to withdraw ships from Far Eastern waters to combat Mussolini in the Mediterranean. Consequently, reasonable relations with Italy appeared essential.

Appeasement was, among other things, a reflection of Britain's defence weakness. There was a further reason for appeasing Mussolini – the realization that Germany posed the greater long-term threat to British interests. The Stresa Front – between Britain, France and Italy – was brought into existence in 1935 in order to contain growing German power.

The Abyssinian crisis

The unity of Stresa, such as it was, was broken by the Abyssinian crisis. Mussolini's invasion of Africa in quest of Italian empire led the Abyssinians to appeal to the League of Nations – but the League was more preoccupied with Hitler's renunciation of the disarmament clauses of the Versailles Treaty and his reintroduction of conscription.

In June, 1935 Eden visited Rome to explain Britain's signing of the Anglo-German Naval Treaty, which permitted Germany's Navy to reach 35% of Britain's, and Eden added a suggestion that Abyssinia make Italy some small concessions. When war broke out, the League, including Britain, imposed sanctions on Italy – though these excluded the crucial item of oil, which could have halted Mussolini.

The British Foreign Secretary made a pact with the French Prime Minister, Laval, whereby Abyssinia would have to cede much of its territory to Italy. This deal entirely contradicted the principle of collective security which Hoare had recently, and openly, upheld at Geneva. There was uproar, followed by Hoare's resignation.

In March, 1936, partly encouraged by this international distraction, Hitler reoccupied the Rhineland. The British government offered no firm opposition to this and many shared the view of Lord Lothian that it was hardly reprehensible for Germany to re-enter 'her own back garden'. Eden did add that Britain would support France and Belgium if Germany invaded them.

Subsequently, many shared Churchill's view that this was a missed last chance to stop Hitler before he began his major rearmament programme. At this time, however, there was no likelihood that British public opinion would have supported a war against Germany over the Rhineland.

Rearmament

The British government began a major rearmament programme after the 1935 general election. Defence expenditure rose from £103 million in 1932/3 to £186–7 million in 1936/7. Chamberlain tried to raise a national defence contribution from business in 1936, though this scheme failed.

The Defence Requirements Committee had, in November, 1935, recommended a programme that included 1736 front-line aircraft by 1939 and a two-power standard Navy. The suggestion that there should be a field force for the Army was squashed, Chamberlain arguing that the public would not tolerate 'continental adventures' and that national resources could not sustain increased expenditure in all the Services. Britain's opposition to the idea of a 'continental commitment' was very deeply entrenched.

The Labour Party, which – under its pacifist leader Lansbury, who resigned in 1935 – had opposed all military increases, slowly moved towards support for rearmament in the later 1930s. This was strongly encouraged by the Spanish civil war, which broke out in 1936. Many socialists supported, or (in the International Brigades) fought for, the Republic against the Nationalists under Franco, who were supported by Hitler and Mussolini. The British Government's policy was one of non-intervention, one which the French followed reluctantly and the dictators ignored.

Britain's policy of rearmament in the later 1930s was essentially defensive. The RAF expansion, for instance, was concentrated on fighters, not bombers. Britain was equipping herself for national defence, narrowly defined, but not for resistance to German aggression on the Continent.

The Anschluss and Munich

In 1937 Neville Chamberlain succeeded Baldwin as Prime Minister. He set out, in contrast with his predecessor, to pursue an active foreign policy. He believed in the value of personal talks with the dictators. He wished, at all costs, to avoid the division of Europe into ideological blocs. War, in his view, would mean the end of the British Empire. He had little trust in the US, which – with her two Neutrality Acts of 1935 and 1937 – seemed gripped by a public mood of profound isolationism from European concerns.

In March, 1938 Hitler invaded Austria in what, to Sir Winston Churchill, was only part of a 'programme of aggression . . . unfolding stage by stage'. This very seriously weakened the position of Czechoslovakia, a pro-Western democracy with an army of 35 divisions. Hitler's demands from Czechoslovakia centred on the 3½ million ethnic Germans living in the Sudetenland, which he wished to see incorporated in the Reich. The issue endangered the general peace since France had an alliance with Czechoslovakia and, since 1935, had had a pact with the Soviet Union.

On 15 September, 1938 Chamberlain flew to see Hitler at Berchtesgaden, where the dictator made it quite clear that he was prepared to go to war over this issue and demanded the annexation of the Sudentenland. The French ministers, Daladier and Bonnet, were persuaded by Chamberlain to agree to a peaceful transfer of territory – in return for which Britain would guarantee the new Czechoslovak frontiers.

Though this concession to Hitler meant France reneging on her treaty commitment to Czechoslovakia. Chamberlain was confronted by even more extreme demands from Hitler at Bad Godesberg. War seemed imminent and the British Navy was mobilized, but Chamberlain was determined to avoid armed conflict – at least at this stage.

Appeasement in this crisis meant discouraging East Europeans from resisting Hitler's expansion and thus preventing the French from getting themselves embroiled in a war to defend their Eastern allies.

This was the essence of his policy at the subsequent Munich Conference. This, despite the criticisms which can be levelled against it, won Britain a breathing-space in which to rearm, and Chamberlain's declaration of 'peace in our time' was accorded very wide public support at home.

From Munich to the outbreak of war

When Hitler marched into Prague in March, 1939 Chamberlain was not at first alarmed, for he had not believed that the guarantee to the remainder of Czechoslovakia would prove durable – but this event marked a turning-point in public opinion, which now, increasingly, turned against the policy of appeasement.

The Government had now to declare its opposition to any further German expansion and a guarantee was given to Poland, followed by undertakings to Greece and Rumania. Negotiations were also begun with Russia. Chamberlain was highly distrustful of Russia and harboured an intense dislike of Soviet Communism. But what also complicated the discussions with Russia was the deep fear and hostility towards her of Britain's new allies, Poland and Rumania. Furthermore, Chamberlain still hoped for a basis of agreement with Hitler.

These Anglo-Russian negotiations ended when the Nazi-Soviet Non-Aggression Pact was signed. This was the immediate prelude to the German-Polish crisis and the German invasion of Poland on 1 September, 1939. Any hope of further appeasement was ruled out by the robust stand of the House of Commons; Chamberlain admitted that everything he had worked for had 'crashed in ruins'.

He remained Prime Minister, however, until May, 1940. Then – following the Norway disaster – many Conservatives voted with the Opposition, as a result of which he was obliged to resign. As Chamberlain had anticipated, the war exposed the vulnerability of the British Empire and accelerated its end. It marked the end of Britain's role as a really major world Power.

29.4 CHRONOLOGY

1931	**November**	MacDonald's Second National Government.
1932	**February**	Second Disarmament Conference opened in Geneva under Arthur Henderson; Germany made claim to equal status.
	March	The Cabinet dropped the TEN-YEAR RULE.
1933	**January**	Hitler became German Chancellor.
	February	THE LYTTON REPORT ON JAPANESE AGGRESSION AGAINST MANCHURIA: endorsed by the League of Nations; Japan rejected it and left the League. THE 'KING AND COUNTRY' DEBATE at the Oxford Union: interpreted abroad as a sign of British unwillingness to resort to war.
	October	Germany left the League. EAST FULHAM BY-ELECTION: victory of a pacifist candidate convinced many politicians, including Baldwin, that the British people would not support large-scale rearmament.
1934	**July**	Baldwin announced a major increase in expenditure on the Royal Air Force, which from now on took priority in rearmament.
	December	Clash between Italian and Abyssinian troops at Wal-Wal.
1935	**April**	Britain, France and Italy formed the STRESA FRONT against German expansion.
	June	Baldwin became Prime Minister. ANGLO-GERMAN NAVAL AGREEMENT: the Germans agreed to limit their navy to 35% of the British. THE PEACE BALLOT: $11\frac{1}{2}$ million supported the League and 10 million voted for disarmament, but $6\frac{3}{4}$ million supported military action to prevent aggression.
	September	Hoare pledged Britain to collective action against aggression.
	October	Italy invaded Abyssinia. Britain joined in League sanctions against Italy – but no sanction on oil.
	December	THE HOARE-LAVAL PACT: a substantial reduction of Abyssinian territory suggested, leaving her 'a corridor for camels' to the sea. Hoare resigned as Foreign Secretary; succeeded by Eden.
1936	**March**	HITLER REOCCUPIED THE RHINELAND. DEFENCE WHITE PAPER: substantial construction of British front-line aircraft proposed.
	May	Italy formally annexed Abyssinia.
	June	Chamberlain denounced sanctions as the 'very midsummer of madness' and Britain withdrew them.
	July	THE SPANISH CIVIL WAR BEGAN: Britain and France adopted a policy of non-intervention.
1937	**May**	Chamberlain succeeded Baldwin as Prime Minister.
	November	Halifax visited Hitler at Berchtesgaden.
1938	**February**	Eden resigned as Foreign Secretary; succeeded by Halifax.
	May	FIRST CZECH CRISIS: Britain and France stood firm over German threats.
	July	Runciman visited Czechoslovakia as mediator.
	September	Hitler called for self-determination for Sudetenland.
	15	Chamberlain visited Hitler at Berchtesgaden: Hitler stated he would annex the Sudetenland.

	18	Daladier, the French Premier, and Bonnet, his Foreign Minister, visited London: it was agreed to ask the Czechs to accept the terms in return for a guarantee of the new frontiers.
	22	Chamberlain visited Hitler at Bad Godesberg, where Hitler made extreme demands.
	27	Royal Navy mobilized.
	29	Chamberlain flew to the MUNICH CONFERENCE: Britain, France, Italy and Germany signed agreement.
	1 October	Duff Cooper, First Lord of the Admiralty, resigned.
	10	German forces occupied the Sudetenland.
1939	January	Chamberlain and Halifax visited Mussolini.
	February	Britain recognized the Franco regime in Spain.
	March	GERMAN TROOPS OCCUPIED BOHEMIA AND MORAVIA: Czechoslovakia ceased to exist. THE BRITISH GUARANTEE TO POLAND.
	April	THE BRITISH GUARANTEE TO RUMANIA AND GREECE. Britain introduced conscription.
	May	Anglo-Turkish Mutual Assistance Pact signed. Germany and Italy signed the 'Pact of Steel'.
	23 August	Molotov and Ribbentrop signed NAZI-SOVIET NON-AGGRESSION PACT in Moscow.
	27	Chamberlain warned Hitler that Britain would stand by Poland.
	1 September	HITLER INVADED POLAND.
	3	BRITAIN AND FRANCE DECLARED WAR ON GERMANY: SECOND WORLD WAR BEGAN; Churchill was made First Lord of the Admiralty.
1940	10 May	CHAMBERLAIN RESIGNED AS PRIME MINISTER, following the Norway debate: Churchill became Prime Minister and Minister of Defence.

29.5 QUESTION PRACTICE

1   Examine the rapid rise and fall in the popularity of Neville Chamberlain 1938–40.

Understanding the question

You must be willing to enquire into Neville Chamberlain's policy in the light of public opinion and its expectations. Do *not* simply record the course of his foreign policy and its reverses – explain why the consensus of opinion running strongly in his favour at the time of the Munich Conference progressively deserted him over the next 18 months.

Knowledge required

Neville Chamberlain's foreign policy objectives and his style of diplomacy. The course of events from the first Czech crisis to Churchill's assumption of power. The changing attitudes of public opinion and the political parties towards appeasement.

Suggested essay plan

1   *Introduction.* Neville Chamberlain's skills as a politician were essentially those of an administrator who was most at home in matters of finance or local government. When he succeeded Baldwin as Prime Minister, however, he found political debate increasingly dominated by foreign affairs, with which he had far less familiarity. Moreover, he had to deal with demagogic dictators whose style and ambitions and political systems were totally alien from his own.

2   In contrast with Baldwin, he intended to pursue an active foreign policy, with personal discussions with the dictators. He had encouraged British rearmament and he hoped that an armed peaceful coexistence would provide the basis for long-term détente with Germany. He continued to believe this (quite unrealistically, which helps to explain his later unpopularity) even *after* the Second World War had broken out. He was acutely aware of British vulnerability. His desire to preserve peace was fully in harmony, in 1938, with British public opinion. Those who opposed appeasement at this stage neither had the consensus, nor did they appear to be entirely consistent amongst themselves.

3   *The Czech crisis and Munich.* The problem was to prevent war breaking out over German ambitions in Eastern Europe – primarily to prevent the French from blocking Hitler in such a way as to provoke a general war – hence the strategy of appeasement. Munich was the peak of Chamberlain's popularity. The policy of appeasement, with which he was identified, *seemed* to have succeeded. The popular acclaim for his announcement of 'peace in our time' expressed immense public relief. So long as Hitler's policy could be interpreted simply as one of putting right the 'wrongs' of the Versailles Treaty, of which one was the situation of the Sudeten Germans, then Chamberlain's missions to Berchtesgaden, Bad Godesberg and Munich could be seen to have been vindicated. In any case, Chamberlain was seen as having prevented war; 'justice' had been done to Germany, and if, as Hitler maintained, Germany had no more territorial claims, peace would be maintained.

4   After Hitler's occupation of Prague in March, 1939 there was a marked revulsion against appeasement, which was now seen to be a craven retreat in the face of force rather than magnanimous. Hitler's ambitions now seemed to be much more far-ranging than redressing Versailles. The British government was now forced to declare resistance to any further German expansion: Chamberlain, in a Birmingham speech, announced that the 'democracies would resist'. The guarantee to Poland was an unprecedented alliance for Britain to have with an Eastern European state, and was intended to incline Hitler to moderation. The Labour Party opposed appeasement and Halifax informed Chamberlain that no government which continued to pursue appeasement – *in public* – could hope to win the next election.

5   While openly taking a firm stand, Chamberlain still hoped for success from appeasement after the fall of Prague. As public opinion became more and more impatient, his foreign policy in 1939 seemed increasingly characterized by failure – in A. J. P. Taylor's words, the Government 'drifted helplessly'. Note the

unenthusiastic pursuit of an understanding with the Soviet Union – *e.g.* the Prime Minister's unwillingness to send a senior statesman to negotiate; the Molotov-Ribbentrop Pact. The Polish crisis and the unpopularity of Chamberlain's continued attempt to find a compromise after the German invasion of Poland; the revolt of the House of Commons against any further appeasement, and Chamberlain's confession that his policy had 'crashed in ruins'.

6   *His war policy.* The 'phoney war' – the uncertainty of purpose, with no sense of forceful direction. The German attack on Norway and Denmark and the British failure, due to lack of air cover rather than to Chamberlain. However, he is now seen as the discredited figure of a failed strategy of appeasement, while Churchill is regarded as the prophet of the 1930s and the courageous man of the moment. Leo Amery's appeal: 'In the name of God, go.' – though note that the Conservative Party preferred Halifax (another appeaser) to Churchill.

2   Why did British governments of the 1930s follow a policy of appeasement towards the dictators?

Understanding the question

Note that this is about the policy of British governments. Ensure that you give a full and reasoned account of the practical power-political constraints on policy, such as defence strategy, as well as of the influence of public opinion.

Knowledge required

Government foreign policy in the 1930s. The course of international relations over the same period. The arguments offered in favour of appeasement.

Suggested essay plan

1   Britain in the 1930s was slowly climbing out of a very grave economic depression and was faced, at the same time, with world-wide defence commitments. Strategic thinking was based on home and imperial defence and on the need to preserve the balance of power on the Continent – *i.e.* the need to prevent any single power (in the later 1930s, Germany) from dominating Europe. Given the resources available, these objectives imposed extremely burdensome demands; this was particularly so since the Dominions had made clear their reluctance to become drawn into any European entanglements. Note that the estimated cost of a major rearmament programme presented a slowly-recuperating economy with great difficulties.

2   The national mood of 'never again', after the horrors of the 1914–18 war, found expression in the League of Nations Union, the Peace Ballot, etc. and in the literature of the interwar period, *e.g. Journey's End* and *Testament of Youth*. Note the influence of this on governments.

3   Another factor influencing British attitudes was the view (forcefully advanced by J. M. Keynes in *The Economic Consequences of the Peace*) that Germany had been unjustly treated in the Versailles Settlement; hence Hitler's moves to revise the Treaty in the German interest were regarded with some sympathy.

4   Many in governing circles took the view that the greater long-term danger was the spread of Soviet Communism, which, Hitler claimed, only a strong Nazi Germany could resist. Chamberlain was deeply distrustful of Russia – something which was intensified by the purges of the 1930s under Stalin. Mussolini, Hitler and Franco found support for some of their objectives (if not for their methods) among Right-wing parliamentarians.

5   Anglo-French relations: the background of mutual distrust in the 1920s; Britain was unwilling to be dragged into a war as a result of French commitments in Eastern Europe. The Rhineland crisis of 1936: the French wanted sanctions against Germany but the British were unwilling to support them. This was also the case, during the Spanish civil war, with the policy of non-intervention – Britain indicating to France that she would not support her in a war with Germany arising out of French involvement in Spain.

6   Illustrate the strategy of appeasement by examining two crises: (a) Abyssinia; and (b) Czechoslovakia. In (a) the traditional policy of friendship with Italy was endangered. Britain did not want naval confrontation in the Mediterranean – her priority at this stage was defence of her Far Eastern interests against the possibility of Japanese attack. Hence the Hoare-Laval Pact with its major concessions to Mussolini was the logical outcome of this priority of imperial defence.

7   The Czech crisis – Hitler claimed the Sudetenland to be his 'last territorial demand'. German economic interests and territorial expansion in this area could be seen as legitimate and as posing no threat to British interests ('a far away country of which we know nothing.'). But if the East Europeans resisted Hitler's expansion and France became embroiled in a war with Germany, Britain would not be able to stand aside – hence a policy of active appeasement, 'Munich sprang from a mixture of fear and good intentions' (A. J. P. Taylor). Note that the Cabinet's military advisers at this time drew up the balance of advantage of war in 1938 or a year later: a crucial consideration was that British air defences would be radically improved within a year.

8   To conclude: the turning-point, with public opinion abandoning appeasement, came with the German invasion of Prague and the British guarantee to Poland. It was now clear that Hitler's ambitions went far beyond any justifiable revision of Versailles – but note that Chamberlain still pursued a policy of appeasement, believing that a major war would involve the end of the British Empire and, quite probably, bring about the Soviet domination of Eastern Europe.

3   Why did Britain go to war in 1939 over Poland but not over Czechoslovakia?

Understanding the question

You should avoid offering *simply* a narrative account of the Czech and Polish crises and the course of events leading to war. Consider how far the circumstances of September, 1939 differed from those of September, 1938 and be prepared to assess the change in public opinion towards appeasement and the reasons for it.

Knowledge required

The course of British foreign policy since 1933, particularly during the crises over Czechoslovakia and Poland, and the events leading to the outbreak of war. The considerations underlying the policy of appeasement. The attitude of British public opinion towards German demands in this period.

Suggested essay plan

1   View the later 1930s as a period of continuous crisis, with Germany and Italy seeking to expand and Britain and France making concessions to them. If Hitler's demand for the Sudetenland in 1938 had been his 'final territorial demand', then the Munich Conference could have inaugurated a new period of international co-operation. But with his occupation of Prague in March, 1939 his ambitions were proved to be more extensive than a mere revision of the Versailles Treaty. Thus British opinion rapidly hardened against him.

2   Give a brief account of the motives underlying the policy of appeasement: (a) strategic necessity – possible aggression (from Japan, Italy and Germany) dictated conciliation wherever possible; British policy in Europe was concerned to avoid a war for which she was not prepared in terms of home air defence of the homeland; (b) the prevalent mood of 'never again' after the First World War; (c) the 'justice' of the German case against the Versailles Settlement; and (d) the fear of Soviet Communism, which was seen in some quarters as a far greater danger than Nazi aggrandizement, led to the rejection of the idea of collective security with Russia.

3   The Munich Conference and the Czech crisis: Both Britain and France were fundamentally unsympathetic towards the Czechs over the Sudetenland. It was not Hitler's claims, but his brutal, threatening presentation of them which affronted Chamberlain, who supported a *peaceful* revision of Versailles. He was eager at all costs to prevent a general war breaking out over France's commitments in Eastern Europe; a further practical consideration was the British need for more time in which to build up its air defences. Also, there was no enthusiasm among the Dominions for a war on behalf of the Czechs. The public response to Munich one of great relief – but note that Chamberlain nevertheless took the precaution of speeding-up the armaments programme.

4   The 'final settlement' of Munich was broken by the annexation of Czechoslovakia: independence had been guaranteed by the four signatory Powers but there was no question of Britain making this guarantee an occasion for war. The British government took the view that, as the Czech state no longer existed, the guarantee was invalid. However, in the light of Hitler's clear betrayal of the previous understanding, public opinion shifted to one of opposition to further concessions; the 'triumph' of Munich was now regarded by British people with a degree of self-disgust.

5   Opposition to further German expansion now became the first principle of *declared* British policy: Chamberlain's speech and the guarantee to Poland. Britain did not, however, succeed in bringing the Soviet Union into a European system to make the Polish guarantee a military reality (Chamberlain confessed to 'the most profound distrust of Russia'); here note the Molotov-Ribbentrop Pact.

6   Hitler invaded Poland on 1 September, 1939, expecting British acquiescence. As at Munich, Mussolini was ready to act as the middle man, but Britain insisted on German withdrawal from Poland. Despite the cabinet's agreement to the sending of an ultimatum, the government continued to seek a negotiated agreement to a conference. In the end, the House of Commons revolted against any further procrastination and the ultimatum was sent. The British government went unwillingly to war; the ultimatum flowed both from the guarantee to Poland and from the pressure of public opinion, as represented by the House of Commons, and Hitler failed to assess the British reaction accurately.

4   Document question (Britain and Europe in the 1930s): study Extracts I, II and III below and then answer questions (a) **to** (g) **which follow:**

Extract I
'The Members of the League undertake to respect and preserve as against external aggression the territorial integrity and existing political independence of all Members of the League. In case of any such aggression the Council shall advise upon the means by which this obligation shall be fulfilled.'

(The Covenant of the League of Nations, Article 10)

Extract II
'Should any member of the League resort to war in disregard of its covenants under Article 12, 13 or 15 it shall *ipso facto* be deemed to have committed an act of war against all other Members of the League, which hereby undertake immediately to subject it to the severance of all trade or financial relations, the prohibition of all intercourse between their nationals and the nationals of the covenant-breaking State, and the prevention of all financial, commercial or personal intercourse between the nationals of the covenant-breaking State and the nationals of any other State, whether a Member of the League or not.
'It shall be the duty of the Council in such case to recommend to the several Governments concerned what effective military, naval or air force the Members of the League shall severally contribute to the armed forces to be used to protect the covenants of the League.'

(The Covenant of the League of Nations, Article 16, paragraphs 1 and 2)

Extract III
'In October 1934, the so-called 'Peace Ballot' was held, and the results were made known in June 1935. It consisted of six questions. Over 11 million people 'voted'. The questions and the approximate number of 'yes' votes for each were as follows:

1. Should Great Britain remain a member of the League of Nations? (*over 11 million*).
2. Are you in favour of an all-round reduction in armaments by international agreement? (*over 10 million*).
3. Are you in favour of an all-round abolition of national military and naval aircraft by international agreement? (*just under 10 million*).
4. Should the manufacture and sale of armaments for private profit be prohibited by international agreement? (*over 10 million*).
5. Do you consider that, if a nation insists on attacking another, the other nations should combine to compel it to stop by
    (*a*) Economic and non-military measures? (*over 10 million*);
    (*b*) If necessary, military measures? (*under seven million*).

(G. M. Gathorne Hardy: 'A Short History of International Affairs, 1920–39')

Maximum marks

(a) What comment, in the light of the European situation in the autumn of 1934, would you make on the large majority who answered 'yes' to question 2 of the 'Ballot'?   (3)

(b) What may be deduced about general opinion at the time from the inclusion of question 3 in the 'Ballot'?   (3)

(c) From an examination of Extracts I and II, to what extent do you think there is inconsistency in the fact that there was a large majority in favour of question 1 of the 'Ballot' and a considerably smaller one in favour of question 5(*b*)?   (4)

(d) In what way did Stanley Baldwin's pledges during the 1935 election campaign suggest he was influenced by the size of the 'yes' vote to most of the questions in the 'Peace Ballot'?   (2)

(e) What did the policy of the British government in 1935 towards Mussolini's invasion of Abyssinia reveal about the application in practice of the principles of Extracts I and II?   (6)

(f) It is often claimed that the voting in the 'Peace Ballot' of 1934 showed that the British people were 'pacifist' at that time. How far do the figures given in Extract III justify this claim?   (5)

(g) What other evidence from the 1930s is used to support the claim that the British people were 'pacifist', and why do you think they nevertheless supported war against Nazi Germany in 1939? (8)

29.6 READING LIST

Standard textbook reading

T. O. Lloyd, *Empire to Welfare State* (OUP, 1970), chapters 7, 8.

L. C. B. Seaman, *Post-Victorian Britain* (Methuen, 1966), Part VI.

R. K. Webb, *Modern England* (Allen and Unwin, 1980 ed.), chapter 12.

Further suggested reading

Lord Avon, *Facing the Dictators* (Collins, 1962).

W. S. Churchill, *The Gathering Storm* (Cassell, 1948).

P. Hayes, *The Twentieth Century 1880–1939* (A & C Black, 1978), (Modern British Foreign Policy).

W. N. Medlicott, *British Foreign Policy since Versailles* (Methuen, 1968).

W. R. Rock, *British Appeasement in the 1930s* (Edward Arnold, 1977).

A. J. P. Taylor, *The Origins of the Second World War* (Penguin, 1964).

C. Thorne, *The Approach of War 1938–39* (Macmillan, 1967).

30 Labour in power, 1945–51

30.1 INTRODUCTION

The Second World War brought about fundamental changes in the nature and structure of British politics and society. Labour had made a notable contribution to the Churchill coalition government and, in 1945, won its first parliamentary majority in a landslide victory. War had stimulated collectivist ideas and government intervention in every aspect of social and economic life.

A White Paper in 1944 had accepted government responsibility for maintaining full employment in the future, and in 1943 the predominantly Conservative government had agreed to implement the main principles of the Beveridge Report after the war. Nevertheless, reconstruction promises had been made before and one reason for the Labour victory in 1945 was the knowledge that such promises were often forgotten once war was over. But, unlike 1918–22, this time there was no equivalent retreat from wartime collectivism.

The Labour government continued the mixed economy through the policy of nationalization and, through its social reforms in health etc., established the welfare state. These reforms fell short of the socialist state the Labour Party, in theory, was committed to building, but at least the Labour government did accomplish the extensive programme of reform that they had promised in *Let us Face the Future*. This was not an inconsiderable achievement.

30.2 STUDY OBJECTIVES

1 Background to the 1945 election: the Labour 'broad church' of the 1930s, taking in Cripps's ethical Socialism and Morrison's pragmatic 'Fabianism'; the development of agreed policies based on 'planning', nationalization and government intervention; the participation of Attlee, Bevin, Dalton, Morrison and others in the wartime coalition; the 'Left Book Club' and other intellectual socialist organs contributed to the wartime socialist mood.

2 Reasons for Labour's 1945 victory: the evidence of Mass Observation; 'Let us Face the Future' appealed to the progressive, vaguely Left-wing, consensus in the nation; Attlee's quiet competence contrasted favourably with Churchill's scare-mongering; the Service vote (influenced by the Army Bureau of Current Affairs) and a general pro-Russia sentiment probably helped Labour; younger voters also supported Labour, and Labour's emphasis on such issues as housing (while the Tories were trying to emphasize foreign policy) also contributed to victory.

3 Even so, the 1945 election result was not the popular mandate for Socialism that it was often supposed to be; despite a huge popular vote, Labour benefited mainly from Tory defections, a massive protest vote against the economic and foreign policy record of the Tories in the 1930s; it was especially the middle-class white-collar vote in the south and east which deserted the Conservatives.

4 The 'balance of power' within the Attlee government: the efficient but unassertive leadership of 'chairman' Attlee; the general dislike of Morrison; the domination of the Right-wing in the cabinet isolated Bevan; the use of cabinet committees and of standing committees of the House (with powers of delegated legislation) to push through the government's crowded programme.

5 Economic policy and performance: full employment, an 8% rise in industrial production, and a 25% rise in exports during 1945–51 could not compensate for the widespread feelings of deprivation and austerity, especially among the middle classes who had voted Labour in 1945; the suspicion that Cripps even relished and welcomed such austerity did not help; the impact of food rationing; difficulties caused by dependence on American loans; the financial crises, the cheap money policies, and the 1949 devaluation; Wilson's 1948 'bonfire of controls' and moves towards deregulation.

6 Foreign policy under Bevin: his patriotic imperialism and close relationship with Foreign Office officials; he moved towards support of America in the 'Cold War' against Russia, leading to a 'Keep Left' reaction (Foot, Mikardo), and supported the policy of American defence of Western Europe (leading to NATO, 1949); the reluctant commencement of decolonization in the Indian subcontinent (Mountbatten and Partition) and in Palestine; despite Bevin's 'Western Union' speech (1948), Labour used imperial commitments as an excuse not to participate in plans for European unity; Britain opposed the 1950 Schuman Plan.

7 Social policies: Bevan's policies on housing and the struggles with SMA over the National Health Service; Griffith's National Insurance policies; the Trades Disputes and Trade Unions Act (1946) reversed Conservative legislation of 1927; town and country planning.

8 The piecemeal and uncoordinated nature of the nationalization programme; Morrison's influence, and the decision to provide fair compensation; Dalton's nationalization of the Bank of England, Shinwell's of coal; then the nationalization of civil aviation, gas, railways and road haulage; finally, the controversy and political complication roused by the bills to nationalize iron and steel.

9 The developing splits in the government and in the Labour Party: the loss of morale following the 1947 winter, the coal crisis, and the disastrous 'groundnuts' scheme; the Cripps and Dalton intrigue to replace Attlee by Bevan; Dalton was forced to resign over budget leaks.

10 The Conservatives in opposition: organizational improvements under Lord Woolton; also sensible enough not to oppose the more 'defensible' parts of Labour's programme; the emergence of progressive young Conservatives like Butler, Macleod and Macmillan.

30.3 COMMENTARY

Background

The Labour government elected in 1945 was quite unlike those which had taken office in 1924 and in 1929. Attlee's government had a parliamentary majority, an abundance of talent and experience, and a definite and well-thought-out programme. Unlike the Conservatives, the Labour Party in 1945 knew what they wanted to do and, when elected, they did it.

Although faced with economic problems far beyond what had been expected, the government did not sacrifice its policies in the difficult years after 1945. The things for which it is most remembered are the creation of the welfare state and the nationalization of major industries, but other achievements should not be forgotten: the country came through crippling balance-of-payment difficulties, and the government repaid in a few years a huge proportion of Britain's vast overseas debts while maintaining full employment at home and putting through its domestic policy.

Britain's foreign policy under Ernest Bevin was also notably successful in achieving its aims, though there were divisions within the Party about what those aims should be.

The economic problems inherited by the Labour government were formidable and it was inevitable that, as time went on and hardship persisted, the government should lose popularity and, eventually, support. Nevertheless, the final defeat in 1951 came as a great shock to many Labour men, just as Churchill and the Conservatives had been stunned by the result of the 1945 election. In terms of positive action, Labour had promised and achieved far more than any other peacetime government, and it set the pattern for post-war politics in Britain.

The 1945 general election

Labour's victory in 1945 surpassed all expectations. The King commented that Attlee arrived at the Palace 'looking very surprised indeed'. The Conservatives had been confident that the electors would vote for Churchill, the man who had saved the nation. Instead they turned him out and gave Labour a massive mandate for change. There were a number of reasons for this: Labour's programme as expounded in *Let us Face the Future* was more sweeping and better worked out than the Tories', though in fact, with the exception of nationalization, its proposals were quite similar to those which the Conservatives offered.

To the public, however, it appeared that Labour had a positive commitment to social reform, whereas the Conservatives endorsed it, as they had in the 'coupon election' of 1918, only because they thought it politically necessary. In addition, although Churchill was personally popular because of his wartime leadership, his attitude towards the Labour movement

in the decades before the war had not been forgotten or forgiven. Furthermore, the Conservatives, as the principal party of government between the wars, were blamed for the failures of British foreign and domestic policy during the 1930s; as Blake puts it, 'it was the Conservatives who were in' in the thirties, 'and they were bound to take the rap for what went on'.

There were other factors also: the experience of war had altered attitudes greatly. Evacuation in particular, had revealed the deprivations of urban working-class life. People were unwilling to abandon what the war had brought about, such as the commitment to increased social and economic equality and the mobilization of the country's resources by the government in the interests of all.

The egalitarian spirit built up in wartime, and the determination that the sacrifices of war should be worthwhile, naturally inclined the voters towards Labour. As Bevan remarked, 'the British people have voted deliberately and consciously for a new world'. It had been the 'people's war'; now they wanted a people's peace, not a return to the pre-war pattern of government. They could no longer be frightened by the spectre of collectivist, Socialist government: what Labour offered above all was the continuance of wartime egalitarianism and state intervention.

Labour's domestic policy

In considering Labour's domestic record, therefore, the impact of war upon the expectations of the people must be remembered. They had seen the beneficial effects of state intervention in industry, in the labour market and in welfare, and they wished it to continue. The Beveridge Report of 1942 provided a blueprint for politicians to follow, and it was Labour which embraced its proposals most enthusiastically. New Liberals and progressive Conservatives had been associated with welfare provisions as well as Labour before 1945, but this was not so afterwards.

The creation of the welfare state was immensely popular, although the Conservatives objected to the structure of the National Health Service. It was, however, a pragmatic development of the state's responsibilities, not a revolutionary step into the unknown; that Labour's approach was not doctrinaire was shown by its introduction of contributions and charges for those who could afford them in 1950.

Labour also put a great deal of effort into housing, although their record was somewhat disappointing. There were acute shortages of essential building materials, and it was widely believed that progress was slow because of the insistence of Bevan, the Minister of Health, on quality rather than quantity in the housing programme. Although unemployment never became a major problem, unemployment benefit and pensions were greatly improved.

The other main element in Labour's domestic programme was nationalization, and this was more controversial. Schemes for public ownership were proposed, however, in terms more of efficiency than ideology. Its approach was piecemeal, uncoordinated and technocratic rather than socialist. The Bank of England and the airlines were taken over without much trouble: most Western countries already had state-run central banks, while it was generally agreed that a national airline could not operate satisfactorily without some form of government support.

It soon became clear that the two classic issues of nationalization would be coal and the railways. Neither had been profitable before the war, and their workforces were heavily unionized and in favour of the abolition of private ownership. Yet, as Lloyd says, 'it was the performance of these two industries that did more than anything else to reduce enthusiasm for public ownership, and this played its part in weakening the Labour government'.

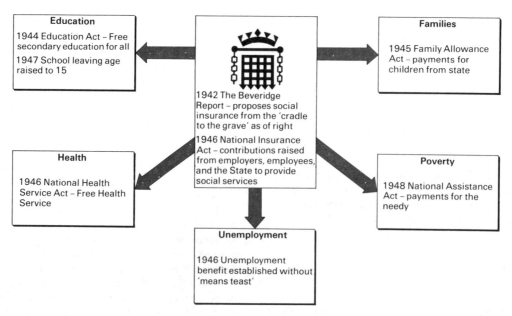

Education

1944 Education Act – Free secondary education for all

1947 School leaving age raised to 15

Families

1945 Family Allowance Act – payments for children from state

1942 The Beveridge Report – proposes social insurance from the 'cradle to the grave' as of right

1946 National Insurance Act – contributions raised from employers, employees, and the State to provide social services

Health

1946 National Health Service Act – Free Health Service

Poverty

1948 National Assistance Act – payments for the needy

Unemployment

1946 Unemployment benefit established without 'means teast'

Fig. 21 Labour in Power 1945–51

The coal industry's problems were deep rooted, and nationalization could not work wonders overnight. The government was unlucky in that the takeover was followed by the severest winter for decades, creating a demand for fuel which could not be met. According to Dalton, the major casualty of the crisis, 1947 was Labour's *'annus horrendus'*. It was doubly unfortunate that the minister responsible had disregarded warnings about production levels and failed to take action to improve them.

The railways also had a bad public image, and in nationalizing them the government took on their unpopularity. Electricity and gas, by contrast, were nationalized without much difficulty in 1948, though the takeover of steel provoked fierce controversy even within the government and was easily undone by the Conservatives when they returned to power. The problem with all these industries was that they had never been very profitable or rationally organized, and the state – in assuming responsibility for them – could do little to change this.

Clement Attlee as Prime Minister

When Attlee became leader of the Labour Party in 1935, in succession to the ineffectual Lansbury, his appointment was generally regarded as a temporary measure. Few people within the Party or elsewhere thought of him as a future Prime Minister. Attlee was seen as inoffensive, dull and uninspiring – yet his leadership was never seriously challenged. He sought to hold the various groups in the Party together, not to secure the supremacy of either Left or Right, and, because he could command trust on all sides, he succeeded. Like Morrison and Bevin he had a preference for pragmatism rather than doctrinal purity.

As leader and Prime Minister he was as unlike Ramsay MacDonald as it was possible to be. His personal modesty was matched by his practical commonsense, something for which most of his challengers were not noted, and this made him an effective wartime minister and a good leader in the difficult postwar years. Despite his lack of personal charisma, he won his internal battles in the Party and in the cabinet against more forceful and colourful men.

He had, however, the defects of his virtues. He was not an exciting political figure, and as Prime Minister he conveyed an air of drabness which was in keeping with the difficult times in which Britain found herself, but which was not very cheering for the people. Even after Labour's extraordinary success in the 1945 election, he was regarded as a political pygmy, and despite the achievements of the following years this image has persisted.

While it is true that he was more the chairman of the Party than its leader, it is hard to see how another man would have done any better in the circumstances. Under Attlee's leadership, the Labour government put through the radical legislation it had promised. Its economic record was good. That the voters (and especially the middle class) did not thank them for it by renewing their mandate was hardly surprising, given the inevitable economic hardship of the immediate post-war period.

Foreign and defence policy

The economic problems inherited by the Labour government were enormous. As Keynes noted, they forced Britain into dependence on the Americans – who drove an excessively hard bargain, hence the sterling crisis of 1947. One price of the economic dependence was a foreign policy influenced by American susceptibilities and pressure. Despite Labour's commitment to extensive social reform at home, Britain under Attlee began spending a larger proportion of her national income on defence than any other Western European country. This was the result, however, not merely of her overseas obligations, but of a belief that she remained a world power.

Clearly the government believed that England could still play a decisive role in any future war. Both Attlee and Bevin (like Churchill) accepted that the wartime alliance with America had to be retained, but they felt that a more satisfactory relationship could be worked out from a position of independence and strength: hence the decision that Britain should have her own atomic weapons.

In the wake of the Berlin blockade and the onset of the Korean War, two important developments occurred: in 1949 the North Atlantic Treaty Organization was founded, an event which Bevin regarded as the crowning point of his career, and in September, 1950 a massive rearmament programme was announced. These left Britain and Western Europe firmly aligned with the United States against the Soviet bloc, a pattern which has persisted ever since.

American pressure, together with a simple inability to hold on, also had more to do with Labour's decolonization programme than any socialist belief in anti-imperialism. In general, Attlee's government handled the withdrawal from

Empire in an entirely pragmatic, almost a ruthless, fashion. It refused to abandon Malaya (because of the Communist threat to the strategically and economically invaluable supplies of rubber and tin) but hastily abandoned India and Palestine to predictable post-colonial violence and unrest.

The defeat of Labour, 1951

The defeat of Labour in the 1951 election came in spite of its achieving the highest number of votes in its electoral history. Yet the defeat was not unexpected. The 1950 election had sapped the Party's strength and morale; the leadership appeared jaded; policy and personality differences in the Party were deep and plain to see. The closeness of the result offered little consolation; defeat was still a profoundly disillusioning experience.

Yet the achievements of the government were not inconsiderable. It had set the scene for a new brand of post-war orthodoxy – policies based on the mixed economy, the creation of a welfare state at home, decolonization and collective security abroad. It was not a revolutionary government. The welfare state was created but the class system continued. Also there were limits to nationalization – despite clause IV. Eatwell points out that the major centres of private power were left untouched, and certainly those who in 1945 thought that they were about to witness the advent of Socialism were to be disappointed. Nevertheless, the government did seem to have an impressive unity of purpose and its record compared very favourably with previous post-war administrations.

The pattern of future British politics was firmly set by the innovations in peace-time state control and provision of welfare services – and, beyond the largely symbolic denationalization of certain industries, there was little that the Conservatives could do to alter it. Labour could take satisfaction from the fact that the events of the 1940s had given rise to a profound and irreversible change for the better in the position of the working class in Britain.

30.4 CHRONOLOGY

1945	**July**	Labour won the general election with 393 seats (145-seat majority).
1946	**February**	BANK OF ENGLAND nationalized. Labour divided about the necessity of this measure. Dalton believed that power had been transferred from 'the City to the Exchequer'. Bevin and the TUC were sceptical. Trade Disputes Act of 1927 was repealed: restored the status quo. The Act's main result was the doubling of Labour Party income from the political levy.
	July	NATIONAL INSURANCE ACT: provided for compulsory payments to finance the new National Health Service; also provided various other benefits – *e.g.* unemployment, sickness, maternity, widows', retirement etc.
	August	COAL INDUSTRY nationalized (with compensation for the owners) and placed under National Coal Board, which was allowed to borrow £150 million during 1946–51 for modernization. The Act was fervently welcomed by the unions. HOUSING ACT: Bevan began the task of rebuilding; but the record in 1946 was unimpressive; Conservatives accused Labour of 'socialist inefficiency'.
	November	NATIONAL HEALTH SERVICE ACT: based on the Beveridge Report; the whole population was to be covered, regardless of income. Nationalization of hospitals; free drugs on prescription; free dental and optical treatment. Doctors' resistance to the scheme was eventually overcome. The Act came into operation in July, 1948. New Towns Act: sought to place house-building powers in public hands. New towns, such as Stevenage and Harlow, were built.
	December	Civil Aviation Act: BEA and BSAA were added to the existing corporation, BOAC.
1947	**June**	Marshall Plan: American financial aid for Europe.
	August	India granted independence. Britain to withdraw in 1948. ROAD, RAIL AND CANAL TRANSPORT nationalized (with compensation allowed) under the British Transport Commission. This was strongly supported by the unions – but fiercely opposed by road hauliers and the Conservatives. ELECTRICITY industry nationalized under the British Electricity Authority, with subsidiary Area Electricity Boards. This was vigorously opposed by the Conservatives. Fuel Crisis and the introduction of Labour's 'austerity programme'. The Conservatives had a slogan, 'Starve with Strachey and shiver with Shinwell'. Town and Country Planning Act: all development land was henceforth to be subject to planning permission. Agriculture Act: retained the wartime government control of prices and output by retaining subsidies and the Marketing Boards.
1948		NATIONAL ASSISTANCE ACT: complemented the National Insurance Act of 1946; the Poor Law was finally abolished. National Assistance Boards henceforth handled all cases of hardship, paid weekly subsistence rates, etc.
	May	Palestine mandate ended: the British withdraw; war between Jew and Arab followed.
	July	GAS INDUSTRY nationalized – an uncontroversial measure; Area Gas Boards were set up under the British Gas Council. Berlin Airlift began (ended May, 1949). British Nationality Act: all citizens of the Commonwealth made British subjects.

Chronology cont.

	August	India and Pakistan granted independence: Britain withdraws; followed by rioting and bloodshed between Hindu and Muslim.
		Local Government Act: introduced by Bevan; it regulated finance between central and local government; allowed the introduction of local Citizens Advice Bureaux.
	December	National Service Act: provided for compulsory military service for men aged 18 to 26.
		REPRESENTATION OF THE PEOPLE ACT: abolished university votes.
		Criminal Justice Act: abolished corporal punishment, and generally allowed for greater leniency towards criminals.
1949	**April**	Eire declared itself an independent Republic.
		Housing Act: sought to improve Labour's housebuilding programme.
	June	Dock Strike: led to a proclamation of a state of emergency.
	September	DEVALUATION of the Pound from $4.03 to $2.80.
		Legal Aid and Advice Act: allowed legal aid for the poor.
	November	IRON AND STEEL industry nationalized and placed under the British Iron and Steel Corporation; this controversial measure was strongly resisted by the Conservatives and the owners. It was Labour's final nationalization measure.
1950	**January**	Britain recognized Communist China.
	February	Labour won the general election but its majority was reduced to seven.
	April	Britain recognized Israel.
	June	Korean War began: Britain supported the UN and the US; she subsequently contributed naval and ground forces.
	December	Bevan and Wilson resigned over the introduction of Health Service charges to meet the rise in rearmament spending proposed by the government (£4700 million from 1951–4)
1951	**April**	Anglo-Iranian oilfields in Abadan taken over by Iran.
	October	Conservatives won the general election.
		Churchill became Prime Minister.

30.5 QUESTION PRACTICE

1   To what extent was the Labour government of 1945–50 an innovator in social policy?

Understanding the question

You must place the social policies of the Labour government in the context of twentieth-century thought and policy on social problems. You must consider especially how far pre-war and wartime 'blueprints' influenced the legislation of 1945–48.

Knowledge required

The domestic legislation of c. 1918–50. The intellectual and political forces which contributed to the development of theories of social welfare.

Suggested essay plan

1   It has often been claimed that Attlee's government established the 'welfare state'; its commitment to full employment, attempts (not wholly successful) to establish a coherent housing policy, its schemes for national assistance independent of a means test, and – above all – its establishment of a free National Health Service, contrast so greatly with the situation in the 1930s that the claim has much apparent validity.

2   An assessment of the Attlee government's commitment to social welfare – especially the attitudes of Bevan, and also of Chancellors Dalton and Cripps, who successively controlled the purse-strings on which social policy depended.

3   The fact that 'welfare state' legislation was put through by the first Labour government to possess a parliamentary majority strengthens the claim that such a policy was innovatory; the fact that the Conservative opposition supported much of the legislation (though not the Health Service), perhaps undermines it; it is also the case that the government was only able to achieve full employment thanks to Marshall Aid.

4   In fact there was considerable continuity; even before the war the Treasury had been slowly adopting Keynesian ideas, though the first full-scale Keynesian budget was Kingsley Wood's of 1943. Despite the demand of armaments spending, Chamberlain had been developing policies on housing and insurance during the years 1937–9.

5   But the main anticipations of Labour's policy are to be found in wartime; though little (except Butler's 1944 Education Act) was actually *done* under Churchill, much was being prepared. In order to encourage support for the war, the government subscribed to the popular belief that it was being fought, not to beat the Nazis, but to 'build a better Britain'; Sir John Anderson directed Civil Service plans for social reconstruction, while Beveridge's 1942 Report popularized the idea of cradle-to-grave social insurance; despite the government's disapproval of its emotive overtones, the media took up its message, and by the time the war ended the build-up of expectations was such that no government could have ignored it.

6   An analysis of the social legislation of 1945–8: Cripps and the Treasury maintained a high commitment to welfare and social spending, so there was no 'Geddes Axe' to undermine policies as after World War One; Bevan's National Health Service went beyond wartime blueprints in securing the public ownership of hospitals, which Bevan fought for in cabinet, but otherwise followed wartime precedents; thus Bevan accepted the policy of contributions which the Labour Party had long opposed; Attlee also accepted Butler's tripartite division of the schools; the 1947 Town and Country Planning Act followed the wartime Uthwatt Report closely; and though there was a great ethos of 'planning' during 1945–8, in reality it amounted to little more than the (necessary) continuation (and extension) of wartime controls.

7   Despite its achievements, the Attlee government lacked any carefully thought-out plans of social reconstruction, and so

followed rather haphazardly along the lines which the wartime coalition had stumbled on during a national crisis.

8 Setbacks for Labour in the 1950 election; the impact of the Korean war on the foreign policy of the second Attlee government; Morrison as Foreign Secretary; inflation and balance of payments difficulties led to the Bevanite/Gaitskellite split over terms of the 1951 budget (especially the Health Service charges); Bevan, Wilson and Freeman resign.

9 An analysis of the election campaign and results of 1951: Labour's popular vote held, but the middle-class 'idealists' of 1945 reverted to Conservative allegiance; the defeat of the Labour government provoked heart-searching in the Labour Party as to whether Attlee's government had failed because it had been too socialist or not socialist enough (Croslandite revisionism as against Bevanism). Despite its achievements, it had failed to capitalize politically on a golden opportunity (a large parliamentary majority, a pliant Right-wing TUC, etc.); probably, as in 1929–31, economic difficulties had proved decisive.

2 Why did the Labour Party win the election of 1945 and lose the election in 1951?

Understanding the question

This is a straightforward question on the political fortunes of both main Parties during the 1940s; remember to provide a background to the Labour victory of 1945.

Knowledge required

The political and electoral history of the 1940s.

Suggested essay plan

1 Both the 1945 and 1951 election results were once thought surprising; that of 1945 seemed to Churchill, the war hero, to be one of outrageous ingratitude; the 1951 result seemed unfair on a government which had broadly carried out its popular mandate of 1945.

2 The 1945 result had, therefore, to be explained by special factors: perhaps there were naive people who voted for Labour not realizing that Churchill would have to resign if Labour came in; perhaps the soldiers wished to register a protest against their commanding officers; perhaps Churchill's 'Gestapo' speech and generally alarmist campaign alienated moderate and middle-class voters; perhaps the Conservative organization had run down while its personnel were at the front, allowing Labour's more home-based activists (trade union officers, etc.) to 'steal a march'.

3 Recent historians have rejected these special factors, and see the 1945 result as unsurprising; many contemporaries thought Labour would have won an election in 1940 if there had been one; Conservative unpopularity seems to have peaked in 1940–41 (according to Mass Observation), and to have recovered thereafter, even during the course of the 1945 campaign.

4 Perhaps the main point is that by refusing to take Labour members (or even Conservative rebels) into his administration, Chamberlain ensured that his fall in May, 1940 and replacement by a Churchill coalition, seemed to contemporaries to be a total replacement of the discredited Conservatives by their bitterest rebel; thus, though Churchill won credit for winning the war, the Conservatives did not; meanwhile Labour leaders (like Attlee, Morrison and Bevin) participated in the government, mainly on the home front, and made themselves 'known' to the public; since Labour also provided the leader of the opposition, it gained the sympathy of both supporters and opponents of the coalition.

5 Reasons for the reversal of electoral fortunes, 1945–51: was it merely a 'swing of the pendulum' effect? Personal squabbles (often involving Morrison), and later the Bevanite/Gaitskellite split, gave the government an air of incompetence; Bevan's 'vermin' speech gave its image a veneer of class hatred hardly warranted by its policies, and frightened moderates; despite the government's accomplished economic policies (in the face of crippling financial difficulties), nevertheless the 1947 winter, the coal and food shortages; general austerity (rationing, etc.) and devaluation made it unpopular; meanwhile, Conservative organization improved under Lord Woolton, and a fairer face of Conservatism (Butler, etc.) began to emerge; the 1948 Redistribution probably gave the Conservatives about 40 more seats.

6 But Labour's *popular* vote remained steady even in 1951. Despite its large popular vote, Labour won in 1945 because normally Conservative, middle-class, professional and white-collar voters in London and the dormitory suburbs of the south and east (Newbury, Dover) deserted the Tories; the Liberal vote (not seats won), and that of the minor parties, also improved, suggesting that it was as much a protest against Conservative rule in the 1930s as a mandate for radical change. However, it may be that a new mood of socialist idealism, fostered by wartime propaganda ('Fair Shares for All', the Left Book Club, J. B. Priestley, etc.), did infuse the middle classes with reformist ideals; if so, they were quickly disillusioned by austerity, and reverted to the Conservatives in 1950/51; by then, a BBC-inspired nostalgia for Churchillian glories (strengthened perhaps by indignation over the abandonment of India) helped to revive their conservatism.

7 Perhaps the most telling explanation is that both Parties went into the 1945 and 1951 elections saying the same things; in 1945 both were paying lip-service (at least) to social reforms, 'fairness' and full employment; by 1951 both parties were recommending deregulation and economic expansion (Wilson's 'bonfire of controls', Morrison's attack on further nationalization); on each occasion (1945 and 1951), the electorate put its trust in the Party which was able to put the common message across most sincerely; in 1945 that was Labour – in 1951 it was not.

3 How 'socialist' were the policies of the Attlee governments?

Understanding the question

You must consider the various types of Socialism to be found in the Labour movement and assess the Attlee governments' record in the context of each.

Knowledge required

The development of socialist ideology in the Labour movement down to 1945. The ideas, policies and subsequent recriminations of the Labour ministers of 1945–51.

Suggested essay plan

1 Morrison later defined Socialism as 'what Labour governments do'; since Labour, in 1945–51, carried out an extremely full manifesto, it could on this definition be regarded as extremely socialist. However, there was no agreed socialist tradition in the Labour movement, or socialist programme; Marxism had played an insignificant role (SDF), and the ILP, and later Cripps's Socialist League, had lost influence to the unprogrammatic and anti-ideological trade unionists; MacDonald's evolutionary gradualism and Lansbury's ethical Socialism (best represented then by Cripps) survived to complicate the position.

2 The composition of Attlee's ministry, and the differences of opinion within the cabinet and the Party; his own social con-

servatism and 'moderate' approach to policy; despite a parliamentary majority and an apparent electoral mandate, constraints on policy were imposed by civil servants, by dependence on American aid, by financial difficulties, etc.

3   Socialism was enshrined in 1918 constitutional commitment to Clause IV: in this context Attlee's nationalization programme seemed effective; 10% of all workers were in nationalized industries by 1951 – *but* they were mainly weak industries, not the 'commanding heights'. A brief account of the nationalization of coal, gas, rail and road haulage, iron and steel, etc.: despite the political controversy, the compensation provided was usually generous; no progress was made towards worker control or 'cooperatives', so that workers' positions in the industries involved did not improve; it was all done *ad hoc*, at the behest of trade union pressure, and without any planning.

4   Socialism can be defined in terms of welfare: Bevan introduced the National Health Service (conceding, however, the unsocialist 'dual system' or right to private practice by NHS consultants) and improved National Assistance. However, similar policies had been developed by the wartime coalition government, and were equally compatible with Liberalism (Beveridge Report) and with the Conservative determination to make the workers more contented with capitalism. Conservative opposition was muted; there was also a disappointing Labour performance (after high promises) on housing policy.

5   Likewise, increased state control and planning, Keynesian budgetary techniques, town and country planning regulations, etc. all followed imperceptibly from wartime precedent, when it had been undertaken out of necessity, not from socialist ideals; it had the full backing of the Treasury, and had more to do with 'efficiency' than with Socialism.

6   Attlee's government accomplished a modest redistribution of wealth through fiscal policy; but full employment and improved living standards for the working classes contrasted with a feeling of 'relative deprivation' among the middle classes, who perhaps suffered more austerity (food shortages, rationing, etc.) than in the flourishing black-market days of wartime. However, the tangible redistribution was limited, and far less than contemporaries supposed was taking place.

7   Under Bevin a 'socialist' foreign policy was lost sight of: there was an increasing identification with America against Soviet Russia. However, decolonization was begun and hostility was shown towards the nascent 'capitalists' club' in Europe.

8   Perhaps Socialism is mainly about power? In 1918 the trades unions had accepted Clause IV in return for consolidating their constitutional position within the Party (the block vote); the Trades Disputes and Trades Unions Act reversed the inter-war Conservative attack on the unions; close relations were maintained with a 'Right-wing' union leadership (Deakin, Lawlor, Williamson). However, both the government and the TUC fiercely opposed unofficial strikes, making great use of strike-breaking emergency powers and sending troops to the docks and to the railways; this led to rank-and-file disaffection among trade unionists, which in turn led to a Bevanite revolt against the government (which was ironic, as Bevan personally was 'hawkish' on the subject of unofficial strikes).

9   Probably, then, Attlee should be seen as continuing the wartime consensus in favour of state-run capitalism and the mixed economy, plus a dash of welfare; it was later taken over by the Conservatives and called 'Butskellism'. Only in the 1950s did any serious rethinking of what Socialism means (Crosland's egalitarian version, Crossman's worker-power version) take place, though it split the Party in the process.

4   **Document question (Aneurin Bevan in 1951): study the extract below and then answer questions** (a) **to** (g) **which follow:**

'Aneurin Bevan was born in 1897 at the mining town of Tredegar. The social setting in which his adolescent character . . . matured was the South Wales coalfield before and during the First World War. It was the grimmest part of the United Kingdom, the part that felt itself most disinherited, least connected with the war against the Kaiser. While the great North Welsh rebel, Lloyd George, was becoming the father of his country in its hour of need – taking confidence from the two thousand years of Celtic peasant history behind him – the young Aneurin, watching him with something of the ideology of an industrial "dead-end kid", was rejecting the ways of his fathers. He felt that he knew better what were the real needs of his like, and that patriotism was not a useful emotion. . . .

'The only part of his father's outlook he adopted was that expressed by the Tredegar Workingmen's Medical Aid Society (a miniature National Health Service). His father was one of its founders, and Aneurin fought his first battle with a local outpost of the British Medical Association when they wished to boycott the miners' society. The only ideas he accepted from Lloyd George were those of his National Insurance Act.

'In the 'thirties he did not visit the countries threatened or seized by Fascism, as Ellen Wilkinson did, but consolidated his position in Monmouthshire and spoke in the House on coal. He will not be remembered for his warning speeches against Hitler, but for his violent war-time onslaughts on Churchill. And, since the war, his intensity has remained concentrated on domestic issues – despite the evident crisis of the world. The great totalitarian issue of our time has always seemed to be outside his ambit.

'There are only fairly exotic members of the Labour movement among those who have advised him to take his recent decision – it is the secession of a group that can be compared to the friends of the brilliant but ill-fated Lord Randolph Churchill (whom Bevan, in some respects, resembles).

'Much the most solid and constructive effort of his political career is, of course, the establishment of the National Health Service. It is easy to see how his early training had equipped him to out-manoeuvre the doctors – he turned their flank and captured them by playing to the calloused appetite for power and money of some great consultant physicians. And his driving motive was plain – his own experiences had given him ample reason to believe sincerely in the need for a free medical service for the poor.

'What is more surprising is his administrative success. He not only established the Service promptly, despite all obstacles, but earned the regard of his own civil servants. This may be the one episode in his career which justifies comparisons in stature between him and Lloyd George.'

('The Observer', 1951)

*Maximum
marks*

(a) Explain the meaning of 'boycott'.   (2)
(b) What was the 'recent decision' which prompted the writing of this article?   (3)
(c) How does the writer of the article show himself hostile to Bevan?   (5)
(d) What 'early training', apart from that cited in the extract, 'equipped him to out-manoeuvre the doctors'?   (4)
(e) How far did Bevan extend the 'ideas he accepted from Lloyd George' in creating the National Health Service? (5)
(f) What light is cast, by the final paragraph, upon the relative roles of parliament and the Civil Service in modern Britain?   (5)

(g) How does this extract exemplify the dilemmas of the Labour Party in the post-war period? (7)

30.6 READING LIST

Standard textbook reading

T. O. Lloyd, *Empire to Welfare State* (OUP, 1970), chapters 10, 11.

L. C. B. Seaman, *Post-Victorian Britain* (Methuen, 1966), chapter IX.

R. K. Webb, *Modern England* (Allen and Unwin, 1980 edn.), chapter 13.

Further suggested reading

E. E. Barry, *Nationalization in British Politics* (Cape, 1965).

S. Beer, *Modern British Politics* (Faber, 1965, 1969).

R. C. Birch, *The Shaping of the Welfare State* (Longmans, 1974).

R. Eatwell, *The 1945–51 Labour Governments* (Batsford, 1979).

D. Fraser, *The Evolution of the British Welfare State* (Macmillan, 1973).

A. Marwick, *Britain in the Century of Total War* (Bodley Head, 1968; Penguin Books, 1970).

R. Miliband, *Parliamentary Socialism* (Merlin Press, 1964, 1973).

F. S. Northedge, *Descent from Power: British Foreign Policy 1945–73* (Allen and Unwin, 1974).

L. J. Williams, *Britain and the World Economy 1919–1970* (Fontana-Collins, 1971).

Index